Climbing Medicine

Volker Schöffl
Isabelle Schöffl • Christoph Lutter
Thomas Hochholzer
Editors

Climbing Medicine

A Practical Guide

Springer

Editors
Volker Schöffl
Center of Sports Medicine, Department of Orthopedic and Trauma Surgery
Sozialstiftung Bamberg
Bamberg, Bayern, Germany

Isabelle Schöffl
Department of Pediatric Cardiology
Friedrich-Alexander Universität Erlangen-Nürnberg
Erlangen, Bayern, Germany

Christoph Lutter
Department of Orthopedic Surgery
University of Rostock
Rostock, Germany

Thomas Hochholzer
Orthopädie und orthopädische Chirurgie
Privatklinik Hochrum
Innsbruck, Tirol, Austria

Translation from the German language edition: Klettermedizin by Isabelle Schöffl, et al., © Springer-Verlag GmbH Deutschland, ein Teil von Springer Nature 2020 2020. Published by Springer Berlin Heidelberg. All Rights Reserved.

ISBN 978-3-030-72183-1 ISBN 978-3-030-72184-8 (eBook)
https://doi.org/10.1007/978-3-030-72184-8

This Springer imprint is published by the registered company Springer Nature Switzerland AG
The registered company address is: Gewerbestrasse 11, 6330 Cham, Switzerland

Preface

Rock and sport climbing is currently evolving from an individual sport into a mainstream sport. The worldwide boom of bouldering has especially contributed to this increase in popularity, and the number of indoor climbing centers is constantly growing. Unfortunately, climbing's 2020 Olympic debut in Tokyo had to be postponed but the whole climbing community is eagerly awaited the event in 2021.

We, as editors and contributors of this book, are all avid climbers ourselves and have been practicing climbing medicine for many years (some of us approximately 30 years or more). In this capacity, we have not only been treating patients with injuries in climbing clinics but have also been monitoring national teams and supervising competitions. The climbing sport provides many professional challenges, as its pathologies are very sport specific and unique to the sport. Since climbing has only been a professional sport for the last 30 years, many questions about long-term developments and consequences remain unanswered. Based on our clinical knowledge as well as our scientific research of climbing pathologies, we are proud to present the current standards of research in this book.

The English edition of "Climbing Medicine" aims to demonstrate anatomical and biomechanical characteristics of the sport and to analyze the specific injuries and overstrains. Furthermore, we will present the medical, gynecological, and pediatric aspects of the sport. While our book "One Move Too Many" (Schöffl V, Hochholzer T, Lightner S Jr) is in its third edition and is addressed to trainers, athletes, and laypeople, this book targets medical professionals and scientists with a focus on evidence-based knowledge in climbing medicine, covering a wide range of topics and current research. We are privileged to have received the support of the famous physiologist Professor Phil Watts for Chap. 4, "Physiology." Phil Watts is a remarkable researcher and his work on climbing physiology has set the standard in climbing. In addition to Prof. Watts, Y. El-Sheikh, the Canadian team doctor, contributed to the chapter on finger injuries, and Thomas Bayer presents diagnostic imaging and all of the newest research from the field of radiology. Our British friends Gareth Jones and Ugo Ehiogu present data and recommendations on rehabilitation and prevention. The very important topic of medical issues in climbing medicine is covered by Thomas Küpper. Michael Simon and Jens Liße contributed valuable insight to various chapters. Our special thanks go to all of the contributors for this great joint effort as well as to Rosamunde Pare for the English language support. Michael Simon, Enrico Haase, and Kilian Reil contributed with numerous photos to illustrate this work, as well as Tiffany Fung did provide illustrations. Finally, we want to thank all of our patients for their trust and support in allowing us to carry out our research with and on them, as well as the teams from our clinics and institutes, research staff, and chiefs of staff for making it possible for us to conduct all of this research.

Photographs by

Enrico Haase: Instagram: rico.haase.photo

Kilian Reil: www.kilianreil.com, Instagram: kilioreilly

Michael Simon: www.michaelsimon.org, Instagram: Michael_simon_photography

Bamberg, Germany — Volker Schöffl
Erlangen, Germany — Isabelle Schöffl
Innsbruck, Austria — Thomas Hochholzer
Rostock, Germany — Christoph Lutter

Disclaimer

WARNING!! THIS BOOK DESCRIBES TRAINING TECHNIQUES FOR DANGEROUS ACTIVITIES. BOTH THE ACTIVITIES AND THE TRAINING TECHNIQUES DESCRIBED CAN CAUSE SERIOUS INJURY AND/OR DEATH.

By your use of this book, you agree to the following:

USE THE INFORMATION CONTAINED IN THIS BOOK AT YOUR OWN RISK.

The techniques, ideas, and suggestions presented in this book should not be considered medical advice, and they are not intended to replace consultation with a qualified medical professional. Any application or use of the techniques, ideas, and suggestions contained herein is at the reader's sole discretion and risk.

THE AUTHORS, PUBLISHER, DISTRIBUTOR, and/or anyone else connected with the creation, marketing, and distribution of this book (collectively referred to as the "Authors") MAKE NO WARRANTIES, EXPRESS OR IMPLIED, OF ANY KIND IN CONNECTION WITH THIS BOOK, INCLUDING WARRANTIES OF SAFETY OR FITNESS FOR ANY PARTICULAR PURPOSE, AND EXPRESSLY DISCLAIM ANY WARRANTY REGARDING THE ACCURACY OR RELIABILITY OF THE INFORMATION CONTAINED HEREIN.

The authors assume no liability or responsibility of any type to any person or entity for errors contained herein, or for any special, incidental, or consequential damage caused or alleged to be caused, directly or indirectly, by the information presented herein.

WARNING! The activities described in this book carry a significant risk of personal injury or death. DO NOT participate in these activities unless you are an expert, have sought and obtained qualified professional instruction or guidance, are knowledgeable about the risks.

involved, or are willing to assume personal responsibility for all risks associated with these activities. The learning and/or practice of suitable techniques and safety measures is your sole responsibility, and you assume any and all risks associated with the practice of techniques discussed herein.

If you are unwilling to assume these risks, do not use this book or undertake the activities or the training techniques described herein.

Contents

Contributors

Thomas Bayer, MD, PhD Department of Radiology and Neuroradiology, Klinikum Fürth, Fürth, Germany

Uzo Dimma Ehiogu, MSc, BSc, BSc, MMACP Research and Training Department, Birmingham Royal Orthopaedic Hospital, Birmingham, UK

Birmingham Medical School, College of Medical and Dental Sciences, University of Birmingham, Birmingham, UK

School of Clinical and Applied Sciences, Leeds Becket University, Leeds, UK

Y. El-Sheikh, BSc (Kin), MD, FRCSC Department of Surgery, North York General Hospital, Toronto, ON, Canada

Division of Plastic & Reconstructive Surgery, Department of Surgery, University of Toronto, Toronto, ON, Canada

Climbing Escalade Canada, Ottawa, ON, Canada

Thomas Hochholzer, MD Private Clinic Hochrum, Innsbruck, Austria

M. I. Johnson, PhD School of Clinical and Applied Sciences, Leeds Becket University, Leeds, UK

G. Jones, PhD, MSc School of Clinical and Applied Sciences, Leeds Becket University, Leeds, UK

T. Küpper, MD, PhD Institute of Occupational, Social and Environmental Medicine, RWTH Aachen Technical University, Aachen, Germany

Jens Liße, MD Department of Orthopedic and Trauma Surgery, Klinikum Bamberg, Bamberg, Germany

Christoph Lutter, MD, PhD, MHBA, MSc Department of Orthopedic Surgery, University Hospital Rostock, Rostock, Germany

A. Morrison, BSc, RNutr Royal Free London NHS Foundation Trust, London, UK

Isabelle Schöffl, MD, PhD, MSc Department of Pediatric Cardiology, Friedrich-Alexander Universität Erlangen-Nürnberg, Erlangen, Bayern, Germany

Volker Schöffl, MD, PhD, MHBA, FAWM Department of Orthopedic and Trauma Surgery, Center of Sportsmedicine, Klinikum Bamberg, Bamberg, Germany

School of Clinical and Applied Sciences, Leeds Becket University, Leeds, UK

Section of Wilderness Medicine, Department of Emergency Medicine, University of Colorado School of Medicine, Denver, CO, USA

Department of Orthopedic and Trauma Surgery, Friedrich-Alexander Universität Erlangen-Nürnberg, Erlangen, Germany

M. Simon, MD Department of Orthopedic and Trauma Surgery, Center of Sportsmedicine, Klinikum Bamberg, Bamberg, Germany

P. B. Watts, PhD School of Health and Human Performance, Northern Michigan University, Marquette, MI, USA

English Language Editing

Rosamunde Pare, MSc, Bern, Switzerland

Gymnasium Lerbermatt, Bern, Switzerland

1 Introduction

Volker Schöffl

Volker Schöffl in „El Dolphin", Rodellar, Spain, Photo Kilian Reil

V. Schöffl (✉)
Department of Orthopedic and Trauma Surgery, Center of Sportsmedicine, Klinikum Bamberg, Bamberg, Germany

School of Clinical and Applied Sciences, Leeds Becket University, Leeds, UK

Section of Wilderness Medicine, Department of Emergency Medicine, University of Colorado School of Medicine, Denver, CO, USA

Department of Orthopedic and Trauma Surgery, Friedrich-Alexander Universität Erlangen-Nürnberg, Erlangen, Germany

V. Schöffl et al. (eds.), *Climbing Medicine*, https://doi.org/10.1007/978-3-030-72184-8_1

Rock and sport climbing is currently evolving from an individual fringe sport into a popular mainstream sport. The worldwide boom of the popular subdiscipline of bouldering (ropeless climbing on low height boulders) is especially contributing to the overall increase in popularity. Indoor climbing centers are becoming more common and are no longer solely training facilities for specialists but have become a meeting place for leisure athletes, "weekend warriors," and afterwork enthusiasts. The climbing gyms offer activities as children's birthday parties, amateur competitions, and beginner classes for all age groups. Due to inclusion of climbing as a competition sport into the Tokyo Olympic program and the regular livestreaming of main climbing events such as World Cups, climbing now has a larger audience than ever. This upward trend has been constant over the last decade; there are more than 500.000 active climbers in Germany and more than two million in Europe [1]. Studies from the American Outdoor Industry Foundation report as many as nine million participants in rock climbing in the USA. The number of people who identify themselves mountaineers raise these numbers even higher, although the exact definition of a "mountaineer" can greatly influence the scope of the group [2].

In addition to the classic, world-renowned outdoor climbing areas such as Yosemite Valley (USA), the Alps (Europe), the Peak District (UK), and the Frankenjura (Germany), many new outdoor climbing areas have been discovered around the globe; rock climbing is now present on all continents and in almost every country. Many remote places such as Laos, the Seychelles, Oman, Reunion, or Antarctica have become well-known climbing destinations. While this globalization of outdoor climbing activities continues, more and more indoor climbing gyms are opening. Indoor climbing walls are present in almost every major city around the world, as well as in schools, rehabilitation centers, hospitals, army camps, and cruise liners. Consequently, indoor competition climbing has become popular not only on an international level but on local and regional levels as well. In the international competition scene, the climbing and training intensity is constantly growing, and almost all top-level climbers are now full-time professional athletes. While this development is understandable in a Western society with a stable financial foundation, it is even more exciting to see the climbing boom spillover to developing countries. Climbing is a new sport in many countries such as India, China, Laos, or the Philippines. It is great to see that despite limited financial means, there is a strong and growing local climbing scene driving the sport forward in these places, especially since the sport was mostly inaugurated through travelling Western pioneering climbers like ourselves in Thailand, Laos, and other destinations in Southeast Asia in the early 1990s. We regularly receive emails and social media messages from injured climbers around the globe who have little access to sports medical facilities in developing countries. All of these climbers are connected by their love for climbing movements, thirst for adventure, and their desire to set and achieve new goals. These common interests provide a common ground for communication within the community between people of all ages from all over the world.

For sports medical doctors, climbing provides substantial professional challenges as its pathologies are very sport-specific and many of them are not present in other athletes. Although a common ankle sprain, a frequent injury in many other sports as well as climbing (especially bouldering), is treated by general sports medical and orthopedic trauma guidelines, other injuries such as pulley tears are climbing-specific and almost only seen in these athletes. Additionally, climbing has only been established as a professional sport for approximately 30 years, and there are still many unanswered questions about long-term developments and consequences. The increasing professionalism of Olympic competition climbing has led to a drastic increase of the overall training load and intensity [3, 4]. This will certainly lead to an increase in injury and overstrain incidence and could also affect long-term consequences [4, 5]. The change of the competition format to the "Olympic combined" mode for Tokyo 2020 (all athletes take part in all three competition disciplines: lead, boulder, and speed) also influences the impact on the athletes; we found a certain change in injury epidemiology, e.g., an increase in hamstring or knee injuries and more frequent overstrain edema of the midcarpal bones [4–7]. On the other hand, the increased popularity of climbing as a leisure and recreational sport also leads to an increased occurrence of more severely graded trauma in beginners [8]. Based on the fact that many of these beginners are nonathletes, minor falls which would probably not injure a well-trained athlete may have serious consequences for these beginners. This may be due to a lack of core-stabilizing muscles, a lack of coordination, or other motoric deficits. The term "newbie syndrome" has been used to describe high-grade injuries in beginner climbers, for example, vertebral or tibial plateau fracture as a consequence of minor trauma (such as a fall from only 0.5 m height onto good protective mattresses) [8]. This also demonstrates that the demands on the sports medical teams increase and adapt with the changing direction of the climbing sport itself.

The aim of this book is to demonstrate the climbing-specific anatomical and biomechanical specifics and to analyze the specific injuries and overstrains as well as the medical, gynecological, and pediatrical aspects of the sport. While our book "One Move Too Many" (Schöffl V., Hochholzer T., Lightner S. Jr.) is aimed at trainers, athletes, and laypeople, this book is based on the German "Klettermedizin" (Springer) and targets medical professionals and scientists with its main focus on evidence-based knowledge in climbing medicine.

1.1 History

The various versions of modern sport climbing have all developed from mountaineering, a sport which was established in the European Alps. By the mid-1980s, the popularity of climbing had spread globally and had diversified to include new categories, such as ice climbing, bouldering, speed, and aid climbing. In many instances, the style in which a route is climbed and the difficulty of the route itself became more important than reaching the summit itself [9]. Style refers to either free climbing, which is defined as a climb completed using only the rock face formations, or aid climbing, in which various devices such as a sling ladder can be anchored into the wall in order to assist the ascent [10]. Simultaneously to the developments in sport climbing, the mountaineering routes became increasingly challenging, both mentally and physically, and started to fall into the definition of extreme climbing.

The basic idea of "free climbing" probably originated from the Elbsandstein region in the Eastern part of Germany, where the modern style-oriented free climbing approach has been used for over 100 years. In the Elbsandstein region, the use of pitons and bolts for upward ascent was considered cheating, so only the athlete's hands and feet could be used to scale a face. In 1918, Emanuel Strubich free climbed the "West Arete" at the Wilde Kopf, establishing the first free climb to be graded UIAA 7–, or 5.10 on the Yosemite decimal scale (YDS). It was a monumental achievement upon which other free climbers like Berndt Arnold would build in the future. However, with the beginning of the Cold War, the East German climbing scene was lost to the rest of the world, and the concept of trying to "free" a climb disappeared with it [2].

In other climbing areas of the world, the sports of rock climbing and mountaineering went through a radical transformation in the late 1960s and early 1970s. The stone master generation of climbers in Yosemite Valley began looking at the route rather than the summit as the most important part of the climb. With that, the style of climbing was brought into question, and the quest for ascending a route specifically by "free climbing" was born. The rope, pitons, and new pieces of clean-climbing protection called nuts (and later friends) were soon being used to arrest a fall rather than to aid the ascent. The grades of difficulty in free climbing consequently began to jump off the scale [2].

Two Americans, both climbing in relative obscurity to the well-known scene in Yosemite, made some of the biggest steps in free climbing. John Gill brought gymnastic skills and many gymnastics training techniques to the sport in the early 1960s. Using the previously unseen gymnasts' chalk to keep his hands free of sweat, Gill established boulder problems that would not be repeated for a decade or more. One of these problems, "The Thimble," would be a four or five bolt sport route today and was undoubtedly the first UIAA 8, or 5.11+ YDS [2]. Tony Yaniro followed in his footsteps, approaching the sport with training techniques he had learned from the US Olympic Team. Yaniro approached training systematically and used specially designed equipment, including an early version of the campus board, to improve his skills as a free climber. All of this training paid off in 1979 with the first ascent of "The Grand Illusion," the world's first UIAA 10- (5.13c YDS) [2]. In Central Europe, Reinhard Karl and Helmut Kiene opened the previously limited UIAA scale with their legendary first ascent of the "Pumprisse" (Fleischbankpfeiler, Wilder Kaiser). The seventh UIAA grade had finally also been climbed in an alpine environment. The British climber Jerry Moffat added the route "The Face" (Germany) to the tenth grade in 1983 before the German climber Wolfgang Güllich began to push the grades even higher. He established routes such as "Wallstreet" (UIAA 11-, 5.14b YDS) in 1987 and "Action Direct" (UIAA 11, 5.14d YDS) in 1993, a route that didn't see a second ascent for years [2]. Most recently, Chris Sharma, Adam Ondra, and Alex Megos have further raised the bar with routes that are solidly graded in the UIAA 12 (YDS 5.15) range. Adam Ondra's route "Silence," Flatanger, Norway, established in 2017, is currently considered the hardest rock climb in the world (UIAA 12, YDS 5.15d).

1.2 Terminology

The vocabulary of rock and sport climbing uses many different terms both for its subdisciplines and for the various ways a in which a climb is ascended. Rock climbing is a multidiscipline sport and as such, different levels of risk must be considered. The risks depend on the subdiscipline examined, the climber's experience and skills, the grade of a route's difficulty, equipment, climbing surface (type of rock or ice, artificial indoor wall, scree), remoteness of location, altitude, and weather [11]. It is common for climbers to regularly participate in more than one climbing subdiscipline, raising the overall participation time in the sport and subsequently increasing the risk of injury. In the following paragraphs, the various subdisciplines will be discussed in further detail.

Sport climbing or **free climbing** (Fig. 1.1) is similar to gymnastics in many ways. It requires a combination of flexibility, finger strength, overall upper body strength, and endurance in a manner dictated by the route climbed. The climbing is slightly prescriptive as the climber ascends toward mostly permanently fixed anchors, such as predrilled bolts, to clip their rope into for protection. The route length can range from 10 m over 100 m with fixed anchors usually placed 2–5 m apart. Falls are frequent, are practiced, and are mostly harmless and considered a common part of the discipline (Fig. 1.2). Physical hazards (rock fall, weather changes, etc.) are minimal, and neglecting to wear a climbing helmet is widely accepted. The term free climbing is often misused and misunderstood; it reflects the fact that the climber does not use any artificial aid but climbs a route from ground to top without loading the belay equipment (rope, quickdraws, etc.). The climber does not pull or rest on gear and bolts, not climbing unprotected but free from technical aid. Climbing a route without any protection is referred to as "free solo climbing," whereas "solo climbing" is defined as a single

Fig. 1.1 Sport climbing. Photo Kilian Reil

Fig. 1.2 Normal falls into the rope are frequent and trained for in modern sport climbing. Photo Kilian Reil

Fig. 1.3 Bouldering. Photo Enrico Haase

person completing the climb, protected by rope and gear in a very specialized technique. The opposite of free climbing is aid climbing, in which hooks, pitons, bolts, and other kinds of gear are used as active aid to ascend a route. A sport climb (free climb) is only counted as completed if it is performed as a so-called redpoint, defined as having climbed the route from the bottom to the top and without loading the belay devices. If a climber climbs a route on the first try without any prior knowledge of the route, it is called "on sight"; if the route is climbed on the first try with information on the route (e.g., seen someone climb it, analyzed video footage, etc.), it is called a "flash" ascent. Climbing a route with rope protection from above is called "top rope."

Bouldering is defined as ropeless climbing involving a usually short and low sequence of powerful and technical moves to complete the graded route on large boulders at a height from which the climber can safely jump down. However, these boulders can be up to 10 m high, classifying these as a "**highball boulder.**" Bouldering (Fig. 1.3) can be performed without a partner and with minimal equipment—climbing shoes and a crash pad (a portable mattress for fall protection). Falling onto one's feet or body is common in bouldering, whether a route is completed (defined as "topped") or not.

Traditional (alpine) climbing emphasizes the skills necessary for establishing routes in an exploratory fashion outdoors, performed normally in a group of two climbers. The lead climber typically ascends a section of rock while

Fig. 1.4 Traditional climbing. Alex Luger in Moab, USA. Photo Archive Adidas

placing removable protective gear where possible along the climb (Fig. 1.4). Therefore, falls can be much longer than those experienced when sport climbing. Unreliable fixed pitons may occasionally be found on older established routes. As physical hazards are likely, the use of a helmet is considered mandatory (Fig. 1.5). Above altitudes of around 2500 m, physiological altitude-induced adaptations must also be factored into the climbs [11].

Indoor climbing is performed on artificial structures that try to mimic outdoor climbing in a more controlled environment. As physical hazards are almost totally eliminated, indoor climbing has become an extracurricular sport in many countries. National and international competitions are held on indoor walls and involve three major disciplines—lead climbing (i.e., sport climbing), speed, and bouldering (Fig. 1.6).

Fig. 1.5 Alpine climbing. Mayan Smith-Gobat, El Capitan, Yosemite Valley, USA. Photo Archive Adidas

Fig. 1.6 Indoor bouldering. Photo Kilian Reil

Bouldering is performed above thick foam mat flooring, speed climbing in top rope, and lead climbing in an on-sight style [11].

1.3 Grading

The most common of the many various grading scales are the UIAA, the French, and the Yosemite decimal scale. In bouldering, different grading scales exist; the French Fontainebleau scale (Fb scale), and the US Vermin scale (Hueco scale or V-scale) are the most important. For scientific analysis, the decimal UIAA scale [12] or the International Rock Climbing Research Associations scale [13] is recommended.

1.4 Equipment

In the early times of climbing, classic heavy mountaineering boots were worn for climbing in alpine regions as well as on rock faces. The first real climbing shoe (the EB) with a friction sole only entered the market at the beginning of the 1980s. A common characteristic of all climbing shoes is that they have to fit very tightly in order to obtain optimal contact to the rock, which often leads to health problems, such as callosity, toenail infections, or even long-term conditions such as hallux valgus deformities [14]. The introduction of bolts was an important factor for the explosive development of the sport climbing grades; falls into the rope are now common for sport climbers [15]. The structure of climbing harnesses also has changed dramatically, from the combination of a chest and seat harness traditionally used in mountaineering to the pure seat harness currently used sport climbing [15, 16]. This allows injury-free falls with the maximum freedom of movement while climbing. Modern belay devices are semiautomatic, targeting the reduction of human error while belaying. Nevertheless, the correct techniques of lead and boulder falling, spotting, and belaying are essential. The bolts, ropes, harnesses, and other equipment used for climbing should be approved by the UIAA safeties commission [17]. In bouldering, crash pads, portable foldable mattresses with various foam layers, are placed underneath the climber to protect the falls (Fig. 1.7). "Spotting" is also an important component injury prevention, in which other climbers protect the falling climber by directing the climber's fall toward the mats in order to minimize the severity of the impact. There have also been reports of injuries to the "spotter" caused by the falling climber (Fig. 1.8).

Fig. 1.7 Crash pads for injury prevention in bouldering. Photo Enrico Haase

Fig. 1.8 Spotting in bouldering. Photo Kilian Reil

1.5 Competition Climbing

Since the 1970s, speed climbing competitions have been held regularly in East European regions. In the mid-1980s, climbing competitions became more frequent, with the first ever German Climbing Championship held in 1985 in a beer hall in Munich ("Salvator Keller"). The first, still unofficial World Championship took place in 1991 in Frankfurt. Since then, World Cups, national and international, championships have taken place regularly. The World Championships in Munich in 2005 already hosted more than 500 participants from 55 countries. The international body of competition climbing is the International Federation of Sport Climbing, founded in 2009 in Paris. Parallel to these official IFSC competitions, various large events, so-called masters are also held, receiving substantial media coverage, such as the "Adidas Rockstars" event in Stuttgart, Germany, or the "Rockmasters" in Arco, Italy. All competitive events are performed on artificial walls, both indoors and outdoors. The increased presence of climbing as a competition sport peaks with the inclusion into the Olympic program of Tokyo 2020 and Paris 2024.

1.6 Disciplines

Three disciplines are currently being held in national and international competition formats: **lead** (climbing with rope protection), **speed** (maximum speed climbing on a standardized route in one-on-one mode), and **bouldering** (climbing at lower heights with matrass floor protection). In addition, a triathlon of all three disciplines (**Olympic Combined**) has been launched recently, representing the Olympic competition format. In addition to the official disciplines, various other events such as marathon climbing or boulder nights have become increasingly popular, hosted mostly in local and regional competitions with a focus on entertainment and recreational sports. Some international master competitions are also offering new disciplines such as "duel climbing" or "after work" formats. In "duel climbing," both athletes compete on identical routes, either in boulder or lead, similar to the classic speed climbing competitions. In the "after work" format, climbers have the chance to practice the route for a certain time frame prior to the competition and have one "redpoint" attempt for the final competition. The three standard competition disciplines will be further explained below.

1.7 Lead Climbing (On-Sight)

In lead climbing, the athletes rope climb a route on-sight as far as they can, ideally all the way to the top of the artificial competition wall. International competition routes are a minimum of 15 m in length and are set specifically for the event by certified route setters. Prior to climbing, all athletes participate in a collective route inspection lasting 6 min. The athletes are then placed in an isolation zone until their turn to climb, guaranteeing that they cannot see how the other

Fig. 1.9 German Championship Lead Climbing (DAV, German Alpine Club). Photo Marco Kost

Fig. 1.10 Competition bouldering at "Adidas Rockstars." Photo C. Waldegger

climbers attempt the route and removing an unfair advantage for the later climbers. All climbers have the same time limit of 6–8 min on the route. The first criterion for the result is the overall climbing length, measured to the last handhold the climber held. If there is a tie based on progress up the route, the time taken to reach the hold determines the ranking (Fig. 1.9).

1.8 Bouldering

Bouldering is performed without a rope above thick foam mats, usually at a height from which the climber can safely jump down. The boulder problems are also set specifically for the competition and are between 3 and 4 m high. The results are calculated based on the number of completed boulders, number of bonus holds reached (earlier called "zone"), and the number of attempts to complete the boulders as well as the number of attempts to reach the bonus holds. A bonus hold is usually a hold halfway through the boulder problem and is specially marked [18] (Fig. 1.10).

1.9 Speed

The overall goal of speed climbing is to top-rope climb to the top of the route and to hit the buzzer before the opponent. Per round, two climbers compete against each another on identical adjacent routes in a knockout system. Both climbers climb both sides once with a short break in between. The times from both ascents are summed and the faster climber reaches the next round. In contradiction to lead and boulder, the speed routes have remained exactly the same

Fig. 1.11 German Championship Speed Climbing (DAV, German Alpine Club). Photo T. Schermer

since the 2005 World Championships, with standardized wall angulation, normed handholds, and either 10- or 15-m length. This standardization allows for the times to be compared over all competitions and for world records to be set [18] (Fig. 1.11).

In addition to these standard disciplines, climbing is constantly developing and new disciplines are evolving. "Deep-water soloing" or "psicobloc" is a new subdiscipline in which routes are free solo climbed above water, usually oceans. These attempts often end in spectacular falls into the water below. There are also psicobloc competitions held on artificial walls above public pools. Additionally, solo and speed big-wall ascents are evolving and drawing attention from the media. As climbing doctors, we need to adapt to the dynamic evolution within our sport, which consequently presents us with new pathologies and injuries. This is the fascinating part!

References

1. DAV. Spitzenbergsport: Deutscher Alpenverein; 2011. www.alpenverein.de.
2. Schöffl V, Hochholzer T, Lightner S. One move too many. Boulder, CO: Sharp End Publishing; 2016.
3. Lutter C, El-Sheikh Y, Schöffl I, Schöffl V. Sport climbing: medical considerations for this new Olympic discipline. Br J Sports Med. 2017;51(1):2–3.
4. Lutter C, Tischer T, Schöffl V. Olympic competition climbing: the beginning of a new era. Br J Sports Med. 2020;55(15):857–64.
5. Lutter C, Tischer T, Hotfiel T, Frank L, Enz A, Simon M, et al. Current trends in sport climbing injuries after the inclusion into the Olympic program. Analysis of 633 injuries within the years 2017/18. 2020. (In press).
6. Schöffl V, Lutter C, Popp D. The "heel hook"—a climbing-specific technique to injure the leg. Wilderness Environ Med. 2016;27(2):294–301.
7. Lutter C, Schweizer A, Hochholzer T, Bayer T, Schöffl V. Pulling harder than the hamate tolerates: evaluation of hamate injuries in rock climbing and bouldering. Wilderness Environ Med. 2016;27(4):492–9.
8. Schöffl V, Lutter C. The "Newbie" syndrome. Wilderness Environ Med. 2017;28(4):377–80.
9. Schöffl V, Morrison A, Schöffl I, Küpper T. Epidemiology of injury in mountainering, rock and iceclimbing. In: Caine D, Heggie T, editors. Medicine and sport science—epidemiology of injury in adventure and extreme sports, vol. 58. Karger; 2012. p. 17–43.
10. Schöffl V. Rock and Ice Climbing In: adventure and extreme sports injuries, Ed: Omar Mei-Dan, Michael R. Carmont, Spinger, London, Heidelberg, New York, Dordrecht. 2013; S.7–36
11. Schöffl V, Morrison AB, Schwarz U, Schöffl I, Küpper T. Evaluation of injury and fatality risk in rock and ice climbing. Sport Med. 2010;40(8):657–79.
12. The UIAA Medical Commission Injury Classification for Mountaineering and Climbing Sports [Internet]. Official Standards of the Union Internationale des Associations d'Alpinisme (UIAA). 2010 [cited 15.05.2010]. www.theuiaa.org/medical_advice.html.
13. Draper N, Giles D, Schöffl V, Fuss K, Watts P, Wolf P, et al. Comparative grading scales, statistical analyses, climbers descriptors and ability grouping: International Rock Climbing Research Association position statement. Sports Technol. 2016;8(3–4):88–94.
14. Schöffl V, Küpper T. Feet injuries in rock climbers. World J Orthop. 2013;4(4):218–28.
15. Schöffl V, Küpper T. Rope tangling injuries—how should a climber fall? Wilderness Environ Med. 2008;19(2):146–9.
16. Hohlrieder M, Lutz M, Schubert H, Eschertzhuber S, Mair P. Pattern of injury after rock-climbing falls is not determined by harness type. Wilderness Environ Med. 2007;18(1):30–5.
17. Schöffl V, Lutter C, Woollings K, Schöffl I. Pediatric and adolescent injury in rock climbing. Res Sports Med. 2018;26(sup1):91–113.
18. Schöffl V, Schöffl I. Competition climbing. Sportortho Sporttrauma. 2012;28:22–8.

Part I

Basics

Injury Statistics

2

Volker Schöffl, Christoph Lutter, and G. Jones

Yasser El-Sheikh at Rifle canyon, USA, Photo Michael Simon

V. Schöffl (✉)
Department of Orthopedic and Trauma Surgery, Center of Sportsmedicine, Klinikum Bamberg, Bamberg, Germany

School of Clinical and Applied Sciences, Leeds Becket University, Leeds, UK

Section of Wilderness Medicine, Department of Emergency Medicine, University of Colorado School of Medicine, Denver, CO, USA

Department of Orthopedic and Trauma Surgery, Friedrich-Alexander Universität Erlangen-Nürnberg, Erlangen, Germany

C. Lutter
Department of Orthopedic Surgery, University Hospital Rostock, Rostock, Germany

G. Jones
School of Clinical and Applied Sciences, Leeds Becket University, Leeds, UK
e-mail: g.j.jones@leedsbeckett.ac.uk

V. Schöffl et al. (eds.), *Climbing Medicine*, https://doi.org/10.1007/978-3-030-72184-8_2

2.1 Introduction

Accurate estimates of injury burden are required to develop appropriate preventative and rehabilitative strategies. Many studies have already presented injury analysis for sport climbing, bouldering, and alpine climbing [1–22]. Unfortunately, many studies fail to differentiate between the various types of climbing disciplines. Some studies focus specifically on indoor climbing [16, 17, 23–25] or competition climbing [17, 22, 26, 27]. Concerning the specific injury incidence rate, sports medicine performs analyses of the injury risk per 1000 h of sports performance, a procedure which can be difficult in climbing, as the climbing time of a single climbing day is more difficult to analyze than the 90 min of play during a soccer game [28–30]. In other mountain sport activities such as high-altitude mountaineering, injury risks differ largely from those present in climbing [28–30]. To objectively analyze and compare injuries from different sports, a common scoring system for the grading of the injuries is essential. In general, there should be a distinction between overstrain (overuse) injuries and acute injuries or accidents when assessing whether a sport is "high-risk" or not.

Overstrain injuries are usually less severe and can generally be avoided with informed training. An examination of the injury rate for acute sport-specific injuries, especially their severity, is crucial. When considering the question of whether a sport is high-risk, overstrain injuries are to be neglected. A standardized scoring scale is essential to compare injury risk and severity.

2.2 Injury Grading and Scoring

In classic outdoor rock climbing, the historic studies of Schussmann et al. [3] and Bowie et al. [2] were the first to grade registered injuries with a scoring model (ISS-score) to investigate the connection between specific injury risk and climbing days [2] or climbing time [3]. Unfortunately, the ISS-score showed weak validity for injuries in a self-recall [31]. Other studies used the NACA score (National Advisory Committee for Aeronautics) [29, 32–34] or other general trauma scores. The NACA score, which originated in the field of aviation rescue, is the most popular pre-hospital emergency medical score in Germany and is an essential part of the pre-hospital emergency rescue documentation [35]. In Europe, this scale is also recommended internationally for alpine trauma and its evaluation [36]. The disadvantage of the NACA score is that it is a pre-hospital score and does not consider the final patient outcome. A NACA score of 3–6 can either have a good or a fatal outcome and is therefore not appropriate [30]. An inter-study comparison both within the sport and with other sports is difficult to execute since many studies in the past used different scoring models. All of these scoring systems have proven flawed for the evaluation of mountaineering and climbing injuries [28, 34], as they do not present treatment outcome, give sport-specific references such as climbing grades, or assess the climbing risk. Therefore, the Medical Commission of the UIAA (Union Internationale des Associations d'Alpinisme) developed its own scoring model which allows a sport-specific injury grading [30] (Table 2.1). This score was recommended for climbing and mountaineering injury and accident analysis to guarantee an inter-study comparability. In an extension of this classification, the IFSC (International Federation of Sport Climbing) presented their own scoring model analyzing the World Cup injuries in 2012 [17]. For a fatality analysis in climbing and mountaineering accidents, the UIAA Medical Commission also presented a risk classification [30] (Table 2.2). This is comparable to the British "E" scale of climbing risk, which considers not only the physical demand of the climb but also its danger, exposure, and rock quality.

In order to compare the risk of injury between climbing sports and other sports, studies should calculate an injury risk per 1000 h of sport-specific performance (time-related injury risk, morbidity rate) as this measure is increasingly used in scientific analysis. As sport-performance time analysis in climbing presents some difficulty, the UIAA Medical Commission (UIAA MedCom) proposed the following:

Table 2.1 UIAA MedCom Score 2011 (Schöffl et al. [30])

Injury and Illness Severity Classification (IIC) – UIAA MedCom Score
0. No injury or illness
1. Mild injury or illness, no medical intervention necessary, self-therapy (e.g., bruises, contusions, strains)
2. Moderate severe injury or illness, not life threatening, prolonged conservative or minor surgery, outpatient therapy, doctor attendance within a short time frame (days), injury-related work absence, heals without permanent damage (e.g., undisplaced fractures, tendon ruptures, pulley ruptures, dislocations, meniscal tear, minor frostbite)
3. Major injury or illness, not life threatening, hospitalization, surgical intervention necessary, immediate doctor attendance necessary, injury-related work absence, heals with or without permanent damage (e.g., dislocated joint, fractures, vertebral fractures, cerebral injuries, frostbite with amputations)
4. Acute mortal danger, polytrauma, immediate prehospital doctor or experienced trauma paramedic attendance if possible, acute surgical intervention, outcome: alive with permanent damage
5. Acute mortal danger, polytrauma, immediate prehospital doctor or experienced trauma paramedic attendance if possible, acute surgical intervention, outcome: death
6. Immediate death

Table 2.2 Fatality Risk Classification (FRC) (UIAA MedCom (Schöffl et al. [30]))

I	Fatalities are technically possible but very rare. No objective danger, e.g., indoor climbing
II	Few objective dangers, fatalities rare, falls are not very dangerous, risk is mostly calculable, e.g., sport climbing, low elevation, and technically easy peaks
III	High objective danger, risk is difficult to calculate, falls lead frequently to injuries, fatalities more frequent, e.g., traditional climbing, high Himalayan (7000–8000 m), or difficult peaks
IV	Extremely dangerous, falls have a high fatality rate, totally unjustified to normal mortals

> "If the hours are not collected for individual rock climbing days, each day can be calculated as four hours for sport and traditional climbing, eight hours for alpine climbing, two hours for indoor climbing, six hours for ice climbing and sixteen hours for an expedition day. The sixteen-hour expedition day estimate is based on twelve hours of climbing time, four hours of camp activity and eight of hours sleep and reflects the very small risk of an injury being sustained during camp time and sleep (e.g. avalanche). All of these hours (especially for the expedition day) are estimations and have been agreed upon by the UIAA Medical Commission to enable calculations of injury risk per 1000 hours of sport-specific performance" [30].

Table 2.3 Functional outcome after finger injuries in climbers (Lutter et al. [22])

Excellent: free range of motion, no objective strength deficit, normal motion pattern
Good: extension or flexion deficit of the PIP joint up to 10°, minor strength deficit, normal motion pattern
Satisfactory: extension or flexion deficit of the PIP joint up to 20°, clinical strength deficit, minor irritation of motion pattern
Fair: extension or flexion deficit of the PIP joint of more than 20°, major strength deficit, major irritation of motion pattern

PIP proximal interphalangeal joint
This score is based on the original functional outcome score after pulley injuries (Schöffl et al. [39])

Other concepts use a climbing intensity score (CIS) [37] or calculate athletic load. The latter combines weighted operation measures of performance including frequency of ascent, specific climbing behavior, grade of difficulty, and rate of perceived exertion (RPE) to produce an arbitrary unit [38].

Specific outcome scores have been proposed and implemented for the climbing-specific outcome in analytic studies of injuries and therapeutic concepts. Schöffl et al. [39] showed a functional as well as a sport-specific outcome score in the analysis of finger pulley injuries in climbers. This scoring system was later extended to include all finger injuries [40] (Tables 2.3 and 2.4).

2.3 Analysis of the Climbing Subdisciplines

For further analysis, an exact differentiation in between the various climbing subdisciplines is necessary. The geographic origin of the data also must be considered in order to account for the different situations present in various geographic regions and climbing areas. Serious climbing injuries may have a worse outcome in

Table 2.4 Sport-specific outcome score after finger injury in climber (Lutter et al. [22])

Excellent: full load capacity of the former injured finger after 12 months (with or without tape), no subjective strength deficit of the former injured finger, regain of full climbing ability/pre-injury climbing level, no pain
Good: full load capacity of the former injured finger after 12 months (with tape), subjectively minor strength deficit of the former injured finger, regain of full climbing ability/pre-injury climbing level, minor pain
Satisfactory: minor restricted load capacity of the former injured finger after 12 months (with tape), subjectively strength deficit of the former injured finger, regain of full climbing ability/pre-injury UIAA climbing level minus one UIAA grade, minor pain
Fair: major restricted load capacity of the former injured finger after 12 months (with tape), strength deficit and restricted ability to use the former injured finger while climbing, major decrease in climbing ability and grade, frequent pain
Poor: climbing is not possible anymore

This score is based on the original sport-specific outcome score after pulley injuries in climbers (Schöffl et al. [39])

Fig. 2.1 Alpine climbing with objective danger (rock fall, weather conditions) and potential long falls due to mostly clean protection with nuts and friends. (Credit: Archive Adidas Terrex)

a developing country or a remote climbing location when compared to an injury in the Alps or other areas with a well-developed mountain rescue infrastructure and optimal medical care [28, 29, 41]. We are going to distinguish in the following as much as possible between alpine climbing (Fig. 2.1), sport climbing (Fig. 2.2), bouldering, indoor climbing, and competition

Fig. 2.2 Sport climbing with low risk of objective danger. (Image credit: Michael Simon)

climbing. For completeness, water ice climbing will be also presented.

2.3.1 Alpine Climbing, Sport Climbing, and Bouldering

Unfortunately, analysis of the literature shows that few studies differentiate precisely between these disciplines [42]. Therefore, these disciplines are investigated collectively. Many studies focus on hand and finger injuries and only give little data concerning injuries [28, 29]. Looking at analyses of the most frequent injuries in climbing, the injury distribution is very much depending on the participant selection. Buzzacott et al. [8] reported on the NEISS database (US) for the years from 2008 to 2016. The NEISS database reports all emergency medical patients of the US NEISS hospitals, which then can be analyzed through keyword tracing (e.g., "climbing") [43]. Buzzacott et al. [8] found that 47% of all injuries are to the lower extremities and are mostly fractures and sprains. This finding is similar to the prior years of 1990–2007 in the NEISS database as shown by Nelson et al. [43]. These data collections do not account for minor and very sport-specific injuries (e.g., finger pulley injuries or finger joint capsulitis), as affected patients normally do not present to an emergency department, but will be seen in specialized sport-orthopedic centers. Looking at our database of climbing-specific patients over various time frames, we mostly see overstrain injuries to the upper extremities, mostly to the hand, fingers, and shoulders [15, 22, 44]. This corresponds with the findings of many other studies which state that 58–67% of the injuries are to the upper extremities, examining climbers in cross-sectional and longitudinal studies [4–6, 9, 10, 12, 37, 44, 45]. Nevertheless, our data also have a selection bias, as we see many climbers to provide a second opinion on very sport-specific injuries (e.g., pulley injuries), whereas acute traumas (e.g., ankle fracture) receive primary treatment close to the geographic location of the accident or the patient's residence. In conclusion, it can be said that most acute traumatic injuries (fractures, sprains) occur to the lower extremities, while most sport-specific injuries occur to the upper extremities and are often chronic and related to overstrain. Sport climbing injuries occur most frequently during lead climbing [1, 2, 4, 7, 9, 22, 28, 29]. The most frequent cause of injury is a fall [1, 2, 4, 7, 9, 28, 29]. Looking at the subdisciplines, alpine climbing injuries are more likely to be to the lower extremities, usually caused by falls. In sport climbing, most injuries occur to the upper extremities, usually while performing a strenuous move [4, 6, 9, 10, 12, 46]. Bouldering shows an almost even injury distribution between fall injuries (sprains and strains) and overstrain. Josephsen et al. [47] found no difference between indoor and outdoor bouldering. Recently, an the incidence of bouldering injuries has increased [14]. The overall injury incidence is minor and was recently reported by Jones et al. [9] through a meta-analysis of eight studies with 2.7 ± 4.5 per 1000 h of sport. These injury rates are lower compared to those of motorcycling with 13.5 per 1000 h of sport [48], soccer with 31 per 1000 h [49], handball with 50 injuries per 1000 h [50], ice hockey with 83 injuries per 1000 h [51], or rugby with 286 injuries per 1000 h of sport [52].

Concerning the injury severity, rock climbing studies show mostly minor injury severities (NACA-score 1 and 2) with a fatality rate of 0–28% [1–3, 5, 6, 9, 44, 47, 53]. The vast span between these numbers must be further evaluated through ongoing studies and may reflect the bias of injuries recorded in the study. For alpine climbing, a 1988 study from Bowie et al. [2] reported a rate of 13 fatalities out of 220 climbing accidents. Eleven climbers died on site and the overall "case-fatality-rate" was 6%. It needs to be stated that these data are over 30 years old and the safety equipment has largely improved since then, decreasing the injury risk [41]. Gatterer et al. [54] reported that the mortality during hiking, trekking, and biking in the mountains was lower compared to that during paragliding or during rock, ice, or high-altitude climbing. Traumatic deaths were more common in activities primarily performed by young adults.

Most studies are performed retrospectively, which frequently causes subsequent systematic errors in the fatality analyses [41]. Historically, older studies show higher grades of injury severity and higher fatality rates [1–3]; the only prospective study on bouldering showed no fatalities at all [47]. These comparisons are hard to draw and have a high bias as the older studies examine high-altitude alpine terrain with high external and objective danger (e.g., weather, temperature, rock fall), whereas Josephsen [47] investigated bouldering, with negligible external dangers. He also found no difference in between indoor and outdoor bouldering, a finding which is also reported by Gerdes et al. [6]. Climbing frequency and climbing level correlates positively with the incidence of overstrain injuries in many studies [9, 15, 19, 23, 53, 55]. Only a few studies did not show this correlation [56]. In summary, Schussmann [3] had already concluded in 1990 that rock climbing has a lower injury risk than football and horse riding, with the obvious difference that the latter sports rarely result in fatalities, although this is arguable for horse riding [57].

2.3.2 Indoor and Competition Climbing

Several studies explored injuries and injury rates in both indoor climbing and indoor competition climbing. Wright et al. [23] evaluated the frequency of overuse injuries during the 1999 indoor

World Cup Championships ($n = 295$), in which 44% of the respondents had sustained an overuse injury and 19% had sustained an overuse injury at more than one site. Wright et al. [23] found an independent correlation to increased injuries ($p < 0.01$) when climbing harder routes, bouldering or leading versus top rope climbing, and climbing for over 10 years. Multivariate analysis removed the effect of sex as an independent predictor.

Jones et al. [55] similarly found increased numbers of overuse injuries or injuries caused by strenuous moves and fewer from fall-related injuries than in traditional and outdoor sport climbing [1, 2, 4, 7]. Two large-scale studies [24, 25] analyzed indoor climbing injuries. Limb's survey reported 55 accidents and no fatalities from 1.021 million climbing wall visits [25]. Schöffl and Winkelmann prospectively surveyed 25,163 registrants at ten climbing walls [24]. Only four significant injuries (NACA 3) were found, and no fatalities occurred; the injury risk per visit was 0.016% or 0.079 injuries per 1000 h of performance [24]. A higher injury risk rate of 3.1 per 1000 h was found at the 2005 World Championships [26], during which eighteen acute medical problems were treated, including 13 cases of skin bruises. The IFSC Medical Commission evaluates all of its competitions and has published the results of the 2012 World Cups [17]. Overall, they counted 1362 climbing days (3405 h) of lead climbing, 1083 days (27,075 h) of bouldering, and 255 days (6375 h) of speed climbing. Only five notable injuries occurred; all were grade 2 injuries. The overall injury risk was 0.74 per 1000 h. The injury risk in lead climbing was 0.29 per 1000 h, 1.47 per 1000 h in bouldering, and zero for speed climbing (Fig. 2.3).

In summary, these indoor climbing studies demonstrated a minor injury risk and severity in comparison to traditional climbing and various other sports [24–26]. Overuse injuries were commonly reported in the upper limbs, with the fingers being mostly affected.

No study reported a fatality rate, even though fatalities do occur when climbing indoors. The cause of these rare fatalities need to be investigated in future studies to determine whether climbing misadventure or pre-existing comorbidities contribute significantly to any death [58].

Fig. 2.3 Elbow dislocation during a boulder competition. (Credit: DAV, German Alpine Club)

Nevertheless, some of these findings may need to be re-evaluated or are in progress of change. Just recently, we reported an increase of injury severity in indoor bouldering, mostly in beginners ("newbie syndrome") (Fig. 2.4) [14]. This increase comes parallel to the advent of many new modern bouldering gyms. Up to about 5–7 years ago, most indoor gyms were classical rope climbing gyms or bouldering gyms in a traditional way. As these traditional bouldering walls are already referred to as "old style," we will continue to use this term. "Old style" bouldering gyms have a large number of holds in a grid style on wooden panels (Fig. 2.5). While this high density of holds allows a high variety for training in limited space, it also demands that the climber independently finds and defines the routes for training. Even though this is still considered to be an effective training method, it holds little interest for recreational climbers and beginners.

Within the last 5–7 years, a vast number of modern "new style" bouldering gyms have evolved with a completely different approach and set-up. Boulder problems are preset in various colors which reflect the grade (Fig. 2.6). These grading colors vary between the gyms. This style is rewarding even to the newest boulderers ("newbie") because it provides an immediate sense of accomplishment. When these first "new style" gyms evolved in Europe, they not only quickly motivated boulderers from all over the area but

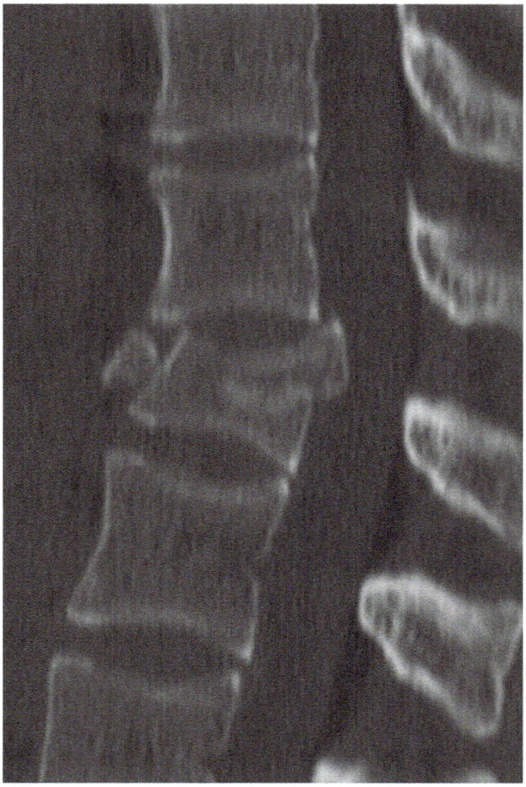

Fig. 2.4 High-grade spine fracture due to boulder fall from minor height (1.20 m) in a beginner climber

Fig. 2.5 "Old school" style indoor bouldering wall

also attracted former non-climbers. Currently, most of the bigger European cities have modern "new style" boulder gyms. These gyms are meeting venues for competitive athletes to train, and plenty of fun competitions take place (Fig. 2.7). Furthermore, these venues also attract leisure

Fig. 2.6 New, modern bouldering wall with color-coded preset boulder problems. (Image credit: Enrico Haase)

Fig. 2.7 Competition at an indoor bouldering wall. (Image credit: Enrico Haase)

climbers and beginners due to the aforementioned immediate success experienced when reaching the top. Thus, it is common that staff outings, manager meetings, and children's birthday parties are being held in boulder gyms. This causes a new injury mechanism, which we called the "newbie" syndrome.

Many beginners are absolutely untrained when they start climbing. They lack the basic trained muscular system needed to stabilize the body, especially the skeletomuscular system. Due to this, a comparably minor fall can already cause major injuries. Additionally, the beginners' equilibrium, body perception, and control are less developed, resulting in a lack of control of the body position during falls and while landing. We are seeing an increased number of patients with higher-graded injuries from falls in indoor bouldering gyms. While falls in classic old-school boulder gyms are mostly onto the back and can be practiced, beginners are now falling from greater heights in an uncontrolled way, causing serious damage. Many of the athletes presenting with these injuries were new to the sport and had been climbing for less than 1 year [14].

In addition to the advent of boulder competitions, the route setting recently changed to include more spectacular and acrobatic movements. The new boulder problems are a combination of classical climbing and bouldering with elements adapted from parkour sports or acrobatics. This evolution of climbing style demands different skills from the competition climbers compared to the climbers of the 1990s. Most climbing moves are dynamic, very three-dimensional, and heel hooks are frequently essential [59]. Reportedly, heel hooks have already caused an increase in knee and hip injuries in climbers [59, 60]. The increase in dynamic and spectacular moves, many of them encompassing larger heights, is subsequently increasing the injury rate and severity, as falls are more unpredictable and more difficult to control [14, 60]. Concerning injury prevention in the future, route setting, fall training for climbers, and compensatory training in beginners should be considered [22, 61].

A recent study by Meyers et al. [62] analyzed speed climbing training in adolescents and found that training regularly at practices on the speed wall was associated with a self-reported history of finger growth plate injuries among elite youth competition climbers. These preliminary data on speed climbing injuries need further investigation, as only few prior studies reported about speed climbing [17, 26].

To further evaluate these trends, especially since climbing's inclusion in the Tokyo Olympic program of 2021, we performed a single-center (Sportsmedicine Bamberg) injury surveillance in 436 climbing patients with a total number of 633 independent climbing-related injuries or complaints over the recent years of 2017–2018 [22]. It was found out that 77.1% of the injuries affected the upper extremities, 17.7% the lower extremities, and 5.2% other body regions. Injury severity was overall low. The most frequent injuries overall were finger pulley injuries (12.3%) and finger tenosynovitis (10.6%). Acute injuries were detected in 43.9% of all injuries and chronic overuse injuries in 56.1%. Bouldering accidents were the leading cause of acute injuries (60.4%). In comparison to our two prior study populations (1998–2001 and 2009–2012), we found (1) an overall decrease in upper extremity injuries, (2) an increase in lower extremity injuries, (3) a constant decrease of finger pulley injuries and epicondylitis, (4) an increase in knee injuries and shoulder dislocations, (5) an increase in adolescents' finger growth plate injuries, and (6) bouldering as the leading injury cause [22] (see Tables 2.5 and 2.6).

Table 2.5 shows the current trends within the ten most frequent climbing injuries and their dynamics over three different time spans in our database of climbing patients presented to our sports-medical center (Sportsmedicine Bamberg) [63].

Our current book focuses on sport climbing; nevertheless, a large number of sport climbers also practice other climbing disciplines, e.g., ice climbing. Insights and data on ice climbing epidemiology are also presented due to the overlap in training and outdoor execution of these disciplines (e.g., alpine climbing and ice climbing).

2.3.3 Ice Climbing

Ice climbing is experiencing an increased level of interest in the climbing community, especially in the form of water ice climbing on frozen waterfalls, glacier formations (ice caves), and also artificial outdoor ice climbing areas (e.g., Ouray Canyon, USA) (Fig. 2.8). Several years ago, Schöffl et al. had already evaluated 88 ice climbers with a retrospective questionnaire on accidents, injuries, risk management, and climbing frequency [34]. Sport activity time (ice climbing hours) was recorded, and the injury severity was analyzed using the NACA score. Mosimann [33] reported about 46 ice climbers who were rescued by the Swiss mountain rescue (Rega) after accidents. Runer et al. [64] evaluated 70 ice climbers prospectively over 1 year, and Maskovskiy et al. [65] analyzed the ice climbing demonstration competitions at the winter Olympics in Sochi.

Table 2.5 Distribution of diagnoses in our climbing clinic (ten most frequent injuries) (Lutter et al. [22])

Injuries 2017–2018 (*n* = 633)	*n*	%	Injuries 2009–2012 (*n* = 911)	*n*	%	Injuries 1998–2001 (*n* = 604)	*n*	%
Pulley injury (finger)	78	12.3	Pulley injury (finger)	140	15.4	Pulley injury (finger)	122	20.2
Tenosynovitis (finger)	67	10.6	Capsulitis (finger)	87	9.5	Epicondylitis	51	8.4
Capsulitis (finger)	49	7.7	Tenosynovitis (finger)	80	8.8	Tenosynovitis (finger)	42	7.0
Knee injury	45	7.1	SLAP tear	51	5.6	Strain finger joint capsule	37	6.1
SLAP tear (shoulder)	37	5.8	Epicondylitis	50	5.5	Skin abrasions	34	5.6
Impingement (shoulder)	34	5.4	Impingement (shoulder)	40	4.4	Back problems	24	4.0
Wrist strain	22	3.5	Strain finger flexor tendon	36	4.0	Knee injuries	14	2.3
Epicondylitis	21	3.3	Dupuytren disease	30	3.3	Fractures	14	2.3
Growth plate injuries (finger)	19	3.0	Strain finger joint capsule	25	2.7	Capsulitis (finger)	13	2.2
Spinal injuries	18	2.8	Ganglion finger flexor tendon	19	2.1	Ganglion finger flexor tendon	11	1.8

Table 2.6 Injury distribution according to body area as presented previously (data of trunk, spine, and pelvis merged). Values are n (%) (Lutter et al. [22])

Body area	2017–2018 (*n* = 633)	2009–2012 (*n* = 911)	1998–2001 (*n* = 604)
Finger	261 (41.2)	474 (52)	247 (41)
Shoulder	128 (20.2)	157 (17.2)	30 (5)
Hand	49 (7.7)	119 (13.1)	47 (7.8)
Forearm and elbow	49 (7.7)	83 (9.1)	81 (13.4)
Lower leg/foot	67 (10.6)	35 (3.8)	55 (9.1)
Knee	45 (7.1)	19 (2.1)	22 (3.6)
Trunk, spine, pelvis	34 (5.4)	21 (2.3)	43 (7.1)
Other	–	3 (0.3)	–

All of these studies found mostly minor graded injuries. Schöffl et al. [34] found that most of the acute injuries (61.3%) occurred while lead climbing and 23.8% while climbing second, and the rest was rare (6.3% belaying, 3.8% on return, and 2.5% on approach, other 2.5%). Most of the acute injuries (73.4%) happened in a waterfall, few in glacier ice walls (11.4%) and on arti-

Fig. 2.8 Water ice climbing with a high risk of objective danger (rock and ice fall, avalanche risk, changing weather conditions, etc.)

ficial ice walls (2.5%). Climber fall-related acute injuries amounted for 10.5%. Injuries were mainly open wounds (55.2%) and hematomas (21.9%), 71% were NACA 1, and no injury scored above NACA 3. The injury incidence was 4.07 per 1000 h for NACA 1–3 with 2.87 per 1000 h in NACA 1 and none in NACA 4–7. Overstrain injuries were mostly onto the shoulders and had an incidence of 0.77 per 1000 h. The injury risk correlated significantly with the body mass index ($p < 0.05$). Overstrain injuries correlated highly with the number of hours spent training ($p < 0.01$), the ice climbing level (grade) ($p < 0.01$), and the willingness to take risks in lead ($p < 0.01$).

Mosimann [33], reporting on 46 ice climbers rescued by the Swiss mountain rescue, found that 31% had no injury (NACA 0), 42% had NACA 2–3 injuries, 8% had NACA 4, 6% had NACA 5, and 13% (six climbers) had a fatal injury (NACA 7). The most frequent cause of an injury was a fall (55%), though falls caused no fatal injuries. The percentile death risk (fatality rate), which the author defined as the percentile portion of deaths in reference to the sum of all known emergencies, was reported as 13% for ice climbing. The author claimed that the fatality risk was higher for ice climbing than in mountaineering (8%), ski mountaineering (7.5%), and rock climbing (4%) but gave no reference for these data. Runer et al. [64] found an overall higher injury rate of 9.8 injuries per 1000 h, with this risk being the highest in the group of intermediate ice climbers. It was lower for beginners and elite ice climbers. The most frequent injuries were skin damage, contusions, and sprains. During the Winter Olympics in Sochi (16 days), more than 2500 beginners and 53 professional athletes climbed at a 17-m-high vertical artificial ice wall [65]. Injury incidence was overall low (0.83 per 1000 h). Again, the most frequent injuries which occurred were mostly minor, such as sprains, strains, and contusions.

While all of these studies show mostly minor graded trauma, serious injuries and fatal accidents can and do happen.

The general death numbers in ice climbing can be analyzed through the injury and fatality reports of the various Alpine Clubs [34]. The Canadian [66] and the American Alpine Club [67] have statistically recorded and analyzed all mountain accidents since 1951. In the USA, up to the year 2005, there were 6111 incidents with a total of 1373 (12%) fatalities [67]. Two hundred fifty-four (4%) of the accidents happened in ice, though no further evaluation of the ice climbing injuries was given. Nevertheless, if 4% of all injuries occur during ice climbing, then 4% of the deaths can be assumed to be related to ice climbing. This calculation yields a total of 55 fatally ice climbing injuries in 54 years, an average of one ice climbing fatality per year within the USA. The numbers for Canada are similar [66]. In 30 years, 92 mountaineers were injured while ice climbing, 30 fatally. Overall, the major ice climbing countries, Switzerland (up to 2006) and Canada (1951–2004), report approximately one death per year [33, 66]. Nevertheless, these number have increased in recent winters [34]. This is probably due to the fact that ice climbing itself has become much more popular.

2.4 Is Climbing a High-Risk Sport?

Meyers encyclopedia [68] defines extreme sports as the performance of exceptional sport disciplines in which the athlete deals with high mental and physical stress. If the sport contains an objective or subjective sensed risk of damage to health or life, it is considered a high-risk sport. Meyer's definition is accurate, but it does not define any real risk from the athlete's perspective. i.e., an experienced and highly skilled athlete is more likely to take and successfully manage higher perceived sporting risks compared to a novice [28]. Kajtna and Tusak [69] define high risk sports as any sport in which one has to accept the possibility of severe injury or death as an inherent part of the activity. In contradiction, Backx et al. [70] characterized high-risk sports as those performed mostly indoors with a high jump or contact rate, as present in volleyball or basketball.

Some authors substitute the terms "high-risk sport" and "extreme sports" or use these terms interchangeably. Young [71] included climbing under the term "extreme sports," together with inline skating, snowboarding, mountain biking, and others. Young stated that the category of extreme sports was fluid and the definition was inexact [71]. In support of Young's view, it is not known what the selection criteria were for an "extreme sport" to be included in the popular "X Games" media event. Climbing was repeatedly represented in this annual event.

Sport disciplines that are performed by a large population are subjectively considered harmless [72]. Therefore, more mainstream sports such as soccer, handball, or rugby are not perceived or characterized as high-risk sports, even though they have a much larger risk of injury compared to climbing [28]. Kite surfing reported a modest injury rate (7 per 1000 h) in a 6-month prospective study ($n = 235$) [73]. However, the injury incidence and severity rate was high and even recorded a fatality (124 injuries, 11 severe injuries, and 1 fatal injury) [73].

When assessing whether climbing should be considered a high-risk sport, it is obvious that each climbing subdiscipline is associated with different levels and types of risk of injury and fatality [28]. When climbing outdoors, there may be many objective dangers and physical hazards, for example, variable rock and ice quality, extreme weather conditions, weapon-like equipment (ice climbing), difficult approaches, and high mental and physical stress. In mountaineering, additional environmental factors can sometimes directly influence injuries and fatalities. These factors can include but are not limited to avalanches, crevasses, and altitude-induced illnesses with neurological dysfunction. However, these situations can still be avoided or sometimes successfully managed (i.e., weather forecasts, training in alpine climbing rescue skills, obtaining knowledge of local terrain, climbing permits, acclimatization and awareness of altitude-induced illnesses, access to helicopter mountain rescue, etc.). In contrast, the objective and external dangers are greatly reduced in indoor and sport climbing, but there is still a very low risk of a fatal injury [28].

The vast majority of climbers manage the risks with their climbing experience and skills, thereby avoiding serious injuries or fatality. Sport climbing, bouldering, and indoor climbing, including competitions, cannot be considered as high-risk sporting activities. Nevertheless, the risk of a possible fatal injury remains. Alpine, ice, and expedition climbing are all included in the broader definition of high-risk sports.

Nevertheless, based on various non-coherent definitions, many sport disciplines fulfill some criteria of a high-risk sport, even if these are often not proven through scientific numbers [41]. It is the task of sports medical science to further analyze this, which requires exact definitions of high-risk sports and injury scores which need to be developed and evaluated. For inter-study comparisons, risk evaluations per 1000 h as well as the implementation of common injury score are important tools to enable a more coherent comparison. For studies on climbing, we propose the use of the UIAA-MedCom score [30] or the IFSC MedCom score [17]).

2.5 Conclusion

In contrast to the general opinion, climbing sports show a minor injury risk and severity in comparison to many other sports. Nevertheless, some climbing subdisciplines such as alpine climbing have higher graded injuries and fatalities. The safest climbing activities are indoor, competition, sport climbing, and bouldering. Recently, there has been an increase in the reported number of injuries in indoor bouldering. This is based on the very dynamic movements of the sport and the vast increase in numbers of new beginner climbers.

References

1. Addiss DG, Baker SP. Mountaineering and rock-climbing injuries in US national parks. Ann Emerg Med. 1989;18(9):975–9.
2. Bowie WS, Hunt TK, Allen HA Jr. Rock-climbing injuries in Yosemite National Park. West J Med. 1988;149(2):172–7.

3. Schussmann LC, Lutz LJ, Shaw RR, Bohn CR. The epidemiology of mountaineering and rock climbing accidents. Wilderness Environ Med. 1990;1:235–48.
4. Paige TE, Fiore DC, Houston JD. Injury in traditional and sport rock climbing. Wilderness Environ Med. 1998;9(1):2–7.
5. Rooks MD, Johnston RB 3rd, Ensor CD, McIntosh B, James S. Injury patterns in recreational rock climbers. Am J Sports Med. 1995;23(6):683–5.
6. Gerdes EM, Hafner JW, Aldag JC. Injury patterns and safety practices of rock climbers. J Trauma. 2006;61(6):1517–25.
7. Smith LO. Alpine climbing: injuries and illness. Phys Med Rehabil Clin N Am. 2006;17(3):633–44.
8. Buzzacott P, Schöffl I, Chimiak J, Schöffl V. Rock climbing injuries treated in US emergency departments, 2008-2016. Wilderness Environ Med. 2019;30(2):121–8.
9. Jones G, Schöffl V, Johnson MI. Incidence, diagnosis, and management of injury in sport climbing and bouldering: a critical review. Curr Sports Med Rep. 2018;17(11):396–401.
10. Jones G, Johnson M. A critical review of the incidence and risk factors for finger injuries in rock climbing. Curr Sports Med Rep. 2016;15(6):400–9.
11. Jones G, Llewellyn D, Johnson MI. Previous injury as a risk factor for reinjury in rock climbing: a secondary analysis of data from a retrospective cross-sectional cohort survey of active rock climbers. BMJ Open Sport Exerc Med 2015;1(1).
12. Jones G, Asghar A, Llewellyn DJ. The epidemiology of rock climbing injuries. Br J Sports Med. 2007;2007:773–8.
13. Schöffl V, Lutter C, Wollings K, Schöffl I. Pediatric and adolescent injury in rock climbing. Res Sports Med. 2018;26(sup1):91–113.
14. Schöffl V, Lutter C. The "Newbie" syndrome. Wilderness Environ Med. 2017;28(4):377–80.
15. Schöffl V, Popp D, Küpper T, Schöffl I. Injury distribution in rock climbers—a prospective evaluation of 911 injuries between 2009-2012. Wilderness Environ Med. 2015;26(1):62–7.
16. Schöffl VR, Hoffmann G, Kupper T. Acute injury risk and severity in indoor climbing-a prospective analysis of 515,337 indoor climbing wall visits in 5 years. Wilderness Environ Med. 2013;24(3):187–94.
17. Schöffl V, Burtscher E, Coscia F. Injuries and medical incidences during the IFSC 2012 Climbing World Cup Series. Med Sport. 2013;17(4):168–70.
18. Lum ZC, Park L. Rock climbing injuries and time to return to sport in the recreational climber. J Orthop. 2019;16(4):361–3.
19. McDonald JW, Henrie AM, Teramoto M, Medina E, Willick SE. Descriptive epidemiology, medical evaluation, and outcomes of rock climbing injuries. Wilderness Environ Med. 2017;28(3):185–96.
20. Chang CY, Torriani M, Huang AJ. Rock climbing injuries: acute and chronic repetitive trauma. Curr Probl Diagn Radiol. 2016;45(3):205–14.
21. Woollings KY, McKay CD, Kang J, Meeuwisse WH, Emery CA. Incidence, mechanism and risk factors for injury in youth rock climbers. Br J Sports Med. 2015;49(1):44–50.
22. Lutter C, Tischer T, Hotfiel T, Frank L, Enz A, Simon M, et al. Current trends in sport climbing injuries after the inclusion into the Olympic program. Analysis of 633 injuries within the years 2017/18. Muscles Ligaments Tendons J. 2020;10(2):201–10.
23. Wright DM, Royle TJ, Marshall T. Indoor rock climbing: who gets injured? Br J Sports Med. 2001;35(3):181–5.
24. Schöffl V, Winkelmann HP. [Accident statistics at "indoor climbing walls"] Unfallstatistik an "Indoor-Kletteranlagen". Sportverletz Sportschaden. 1999;13(1):14–6.
25. Limb D. Injuries on British climbing walls. Br J Sports Med. 1995;29(3):168–70.
26. Schöffl V, Küpper T. Injuries at the 2005 World Championships in Rock Climbing. Wilderness Environ Med. 2006;17:187–90.
27. Hosaini SA, Atri AE, Kavosi A. Injuries at the Iranian championship in indoor rock climbing. Wilderness Environ Med. 2013;24(2):167–8.
28. Schöffl V, Morrison AB, Schwarz U, Schöffl I, Küpper T. Evaluation of injury and fatality risk in rock and ice climbing. Sport Med. 2010;40(8):657–79.
29. Schöffl V, Morrison A, Schöffl I, Küpper T. Epidemiology of injury in mountainering, rock and iceclimbing. In: Caine D, Heggie T, editors. Medicine and Sport Science - Epidemiology of Injury in Adventure and Extreme Sports. 58: Karger; 2012. p. 17–43.
30. Schöffl V, Morrison AB, Hefti U, Schwarz U, Küpper T. The UIAA medical commission injury classification for mountaineering and climbing sports. Wilderness Environ Med. 2011;22(1):46–51.
31. Valuri G, Stevenson M, Finch C, Hamer P, Elliott B. The validity of a four week self-recall of sports injuries. Inj Prev. 2005;11:135–7.
32. Veldman A, Fischer D, Brand J, Racky S, Klug P, Diefenbach M. Proposal for a new scoring system in international interhospital air transport. J Travel Med. 2001;8:154–7.
33. Mosimann U. Notfälle beim Eisklettern. Ther Ber. 2006;4:70–3.
34. Schöffl V, Schöffl I, Schwarz U, Hennig F, Küpper T. Injury-risk evaluation in water ice climbing. Med Sport. 2009;2(1–2):32–8.
35. Bein T. Scores—Hilfsmittel zur Risikoeinschätzung. In: Madler C, Jauch K, Werdan K, Siegrist J, Pajonk F, editors. Das NAW—Buch: Akutmedizin der ersten 24 Stunden. 3rd ed. München, Jena: Elvesier, Urban & Fischer; 2005. p. 193–9.
36. ÖGAN. Österreichische Gesellschaft für Qualitätssicherung und Ausbildung in der Notfallmedizin; 2008. www.oegan.at/jo/index.php?option=com_content&task=view&id=26&Itemid=24.

37. Woollings KY, McKay CD, Emery CA. Risk factors for injury in sport climbing and bouldering: a systematic review of the literature. Br J Sports Med. 2015;49(17):1094–9.
38. Jones G. Improving design characteristic to estimate load for future rock climbing studies. 4th International Rock Climbing Research Congres; Charmonix, France: IRCRA; 2018.
39. Schöffl V, Küpper T, Hartmann J, Schöffl I. Surgical repair of multiple pulley injuries—evaluation of a new combined pulley repair. J Hand Surg [Am]. 2012;37(2):224–30.
40. Schöffl V, Strohm P, Lutter C. Efficacy of corticosteroid injection in rock climber's tenosynovitis. Hand Surg Rehabil. 2019;38(5):317–22.
41. Schöffl V. Fels- und Eisklettern—Eine Risikoanalyse. Hamburg: Diplomica Verlag; 2009. p. 100.
42. Jakus PM, Shaw WD. An empirical analysis of rock climbers' response to hazard warnings. Risk Anal. 1996;16(4):581–6.
43. Nelson NG, McKenzie LB. Rock climbing injuries treated in emergency departments in the U.S., 1990-2007. Am J Prev Med. 2009;37(3):195–200.
44. Schöffl V, Hochholzer T, Winkelmann HP, Strecker W. Pulley injuries in rock climbers. Wilderness Environ Med. 2003;14(2):94–100.
45. Rooks MD. Rock climbing injuries. Sports Med. 1997;23(4):261–70.
46. Neuhof A, Hennig FF, et al. Injury risk evaluation in sport climbing. Int J Sports Med. 2011;32(10).
47. Josephsen G, Shinneman S, Tamayo-Sarver J, Josephsen K, Boulware D, Hunt M, et al. Injuries in bouldering: a prospective study. Wilderness Environ Med. 2007;18(4):271–80.
48. Mannering FL, Grodsky LL. Statistical analysis of motorcyclists' perceived accident risk. Accid Anal Prev. 1995;27(1):21–31.
49. Junge A, Dvorak J. Soccer injuries. Sports Med. 2004;34(13):929–38.
50. Wedderkopp N, Kaltoft M, Lundgaard B, Rosendahl M, Froberg K. Injuries in young female players in European team handball. Scand J Med Sci Sports. 1997;7(6):342–7.
51. Molsa J, Kujala U, Nasman O, Lehtipuu TP, Airaksinen O. Injury profile in ice hockey from the 1970s through the 1990s in Finland. Am J Sports Med. 2000;28(3):322–7.
52. Gabbett T. Incidence of injury in amateur rugby league sevens. Br J Sports Med. 2002;36(1):23–6.
53. Logan AJ, Makwana N, Mason G, Dias J. Acute hand and wrist injuries in experienced rock climbers. Br J Sports Med. 2004;38(5):545–8.
54. Gatterer H, Niedermeier M, Pococco E, Fruhauf A, Faulhaber M, Menz V, et al. Mortality in different mountain sports activities primarily practiced in the summer season-a narrative review. Int J Environ Res Public Health. 2019;16(20).
55. Jones G, Asghar A, Llewellyn DJ. The epidemiology of rock climbing injuries. Br J Sports Med. 2007;42(9):773–8.
56. Rohrbough JT, Mudge MK, Schilling RC. Overuse injuries in the elite rock climber. Med Sci Sports Exerc. 2000;32(8):1369–72.
57. Chitnavis JP, Gibbons CL, Hirigoyen M, Lloyd Parry J, Simpson AH. Accidents with horses: what has changed in 20 years? Injury. 1996;27(2):103–5.
58. DAV. Bergunfallstatistik 2004–2005. München; 2006.
59. Schöffl V, Lutter C, Popp D. The "heel hook"—a climbing-specific technique to injure the leg. Wilderness Environ Med. 2016;27(2):294–301.
60. Lutter C, Tischer T, Cooper C, Frank L, Hotfiel T, Lenz R, et al. Mechanisms of acute knee injuries in bouldering and rock climbing athletes. Am J Sports Med. 2020;48(3):730–8.
61. Schöffl V, Korb L, Matros P. ACT—Adjunct Compensatory Training. Hamburg: Tredition; 2020. p. 92.
62. Meyers RN, Howell DR, Provance AJ. The association of finger growth plate injury history and speed climbing in youth competition climbers. Wilderness Environ Med. 2020;31(4):394–9.
63. Schöffl V, Simon M, Lutter C. Finger- und Schulterverletzungen im Klettersport (Finger and shoulder injuries in rock climbing). Orthopade. 2019;48(12):1005–12.
64. Runer A, Lampl K, Neunhauserer D, Runer F, Frick N, Seitlinger G, et al. A 1-year prospective analysis of ice climbing injuries. Clin J Sport Med. 2017;27(2):161–7.
65. Mashkovskiy E, Beverly JM, Stocker U, Bychkovskiy S. Ice climbing festival in Sochi 2014 Winter Olympics: medical management and injury analysis. Wilderness Environ Med. 2016;27(1):117–24.
66. Canadian-Alpine-Club. 2007. www.alpineclub-edm.org/accidents/type.asp?type=Ice+Climbing.
67. American-Alpine-Club. Accidents in North American Mountaineering. Am Alpine Club. 2006;9(1):59.
68. Meyers. Meyers LexikonOnline 2.0 2008 [16.01.2008]. www.lexikon.meyers.de/meyers/extremsport.
69. Kajtna T, Tusak M. Some psychological studies of high risk sports. Kinesiologia Slovenica. 2004;10(1):96–105.
70. Backx FJG, Beijer HJM, Bol E, Erich WBM. Injuries in high-risk persons and high-risk sports. Am J Sports Med. 1991;19:124–30.
71. Young CC. Extreme sports: injuries and medical coverage. Curr Sports Med Rep. 2002;1(5):306–11.
72. Henke T, H. G. Die Risikobewertung der verschiedenen Sportarten—Epidemiologie von Sportverletzungen. In: Bergler R, editor. Irrationalität und Risiko Gesundheitliche Risikofaktoren und deren naturwissenschaftliche und psychologische Bewertung. Köln: Kölner Universitätsverlag; 2000. p. 300–18.
73. Nickel C, Zernial O, Musahl V, Hansen U, Zantop T, Petersen W. A prospective study of kitesurfing injuries. Am J Sports Med. 2004;32(4):921–7.

3 Anatomy and Biomechanics of the Hand

Volker Schöffl and Isabelle Schöffl

Isabelle Schöffl bouldering at Halfway Log Dump, Ontario, Canada, Photo Michael Simon

V. Schöffl (✉)
Department of Orthopedic and Trauma Surgery, Center of Sportsmedicine, Klinikum Bamberg, Bamberg, Germany

School of Clinical and Applied Sciences, Leeds Becket University, Leeds, UK

Section of Wilderness Medicine, Department of Emergency Medicine, University of Colorado School of Medicine, Denver, CO, USA

Department of Orthopedic and Trauma Surgery, Friedrich-Alexander Universität Erlangen-Nürnberg, Erlangen, Germany

I. Schöffl
Department of Pediatric Cardiology, Friedrich-Alexander Universität Erlangen-Nürnberg, Erlangen, Bayern, Germany

V. Schöffl et al. (eds.), *Climbing Medicine*, https://doi.org/10.1007/978-3-030-72184-8_3

3.1 Anatomy

The flexor and extensor tendons must be looked at separately, and the lumbricalis muscles are yet another entity [1]. The complex movements of the fingers and wrist are only possible through a combined effort of the extrinsic and intrinsic muscles [2]. The extrinsic muscle system describes the muscles originating from the elbow and forearm and inserting into structures within the hand. The intrinsic muscles are situated completely within the hand. They are divided into four groups: the thenar, hypothenar, lumbrical, and interosseous muscles. In addition many anatomical variants exist [3].

3.1.1 The Extensor Tendons

The long fingers have four common extensor tendons as well as two tendons which are dedicated to a single finger—extensor indicis for the second digit and extensor digiti minimi for the fifth digit. The tendon of the extensor digiti minimi runs through the fifth tendon compartment, while all other tendons run through the fourth compartment [1]. On the dorsum of the hand at the level of the metacarpophalangeal joints, there are many cross-connections known as the connexi intertendinei. At the proximal interphalangeal joint (PIP), the extensor tendons separate into two lateral reins and one central rein (tractus intermedius). The thumb has two extensor tendons, the extensor pollicis longus (third extensor compartment) and the extensor pollicis brevis (first extensor compartment). Together, these tendons form the so-called extrinsic system, tendons of muscles which originate proximal of the hand itself [1] but are moving the hand and fingers. The extrinsic system is supported by the intrinsic system, muscles originating within in the hand, the mm. lumbricales, the mm. interossei, and the thenar and hypothenar muscles [1]. The lumbricalis and interossei are exceptions as they originate from the flexor tendons themselves and end in the tendinous hood of the extensor tendons [4]. Their function is flexion in the metacarpophalangeal joint (MCP) and extension in the proximal (PIP) and distal interphalangeal joint (DIP) [1, 4, 5]. The muscle bodies of the lumbrical muscles have many anatomical variants and often have one common muscle body for the muscles of two adjoining fingers, mostly for the fourth and fifth finger [3, 4, 6]. If one of these fingers is extended and the adjoining one is flexed, the common muscle body receives shear force, which was reported by Schweizer as "quadriga" effect and leads to a "lumbrical shift" phenomenon [6] (Fig. 3.1).

3.1.2 The Flexor Tendons and Their Functional System with the Pulleys and Tendon Sheath

The flexor tendon apparatus has to be considered as a functional unit of tendons, tendon sheath, and pulley system [1]. The long fingers and the thumb differ with respect to the flexor tendons. The long fingers have two flexor tendons, the flexor digitorum profundus (FDP) and flexor digitorum superficialis (FDS) which run through the carpal tunnel and pulleys and intersect at the chiasm [1].

The one thumb flexor tendon (flexor pollicis longus) runs through the carpal tunnel, on the radial aspect of the forearm and then through an osteofibrous channel up to the base of the phalanx and is reinforced by two pulleys [1].

The annular ligaments and the cruciate ligaments represent a reinforcement of the tendon sheaths. Thus, the tendons are guided in an osteofibrous channel to the phalanges [7]. Five annular (A1–A5) and 3 weaker crucial ligaments (C1–C3) can be distinguished [7, 8] (Fig. 3.2). Pattern and arrangement of these ligaments vary with a high diversity for the A3 and C1/C2 pulleys [8, 9]. All pulleys have different functions in stabilizing the flexor tendons at the palmar sides of the phalanges [8, 10–12]. The main function of the flexor tendon pulley system is to hold the flexor tendons close to the bone, thus converting linear force into torque resulting in rotation at the interphalangeal (IP) and metacarpophalangeal (MCP) joints [13]. The A2 and A4 annular ligaments play the most important role in force transmis-

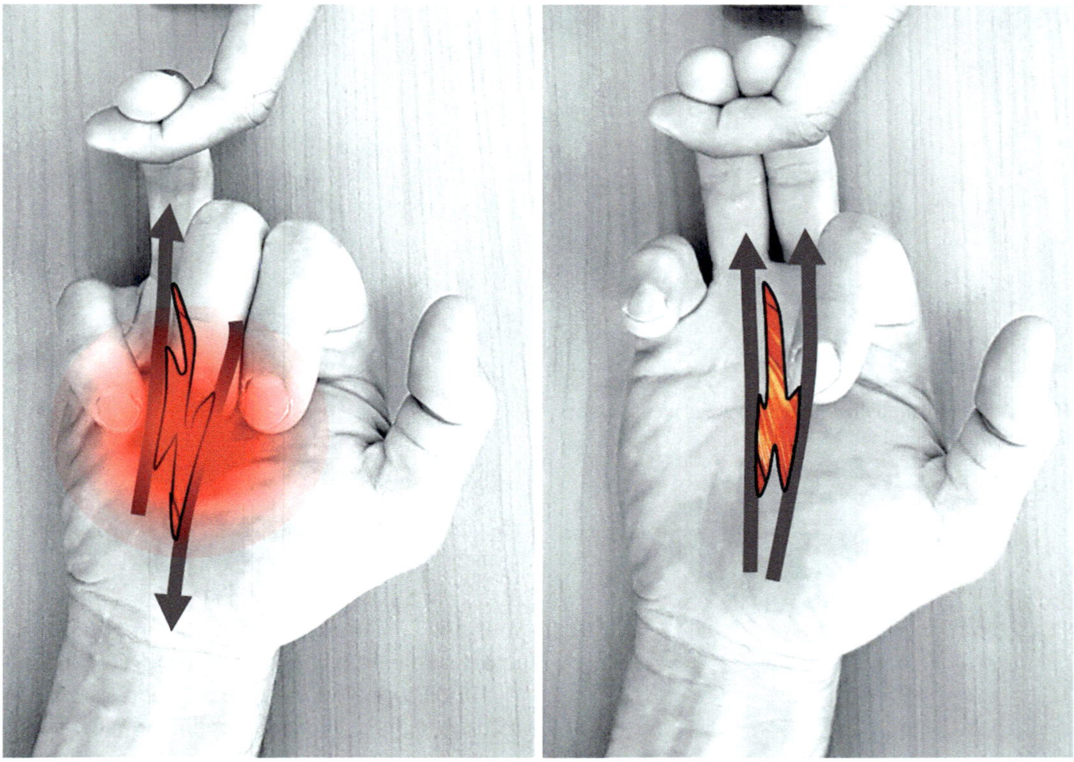

Fig. 3.1 The "quadriga" effect of the lumbrical muscles [6]

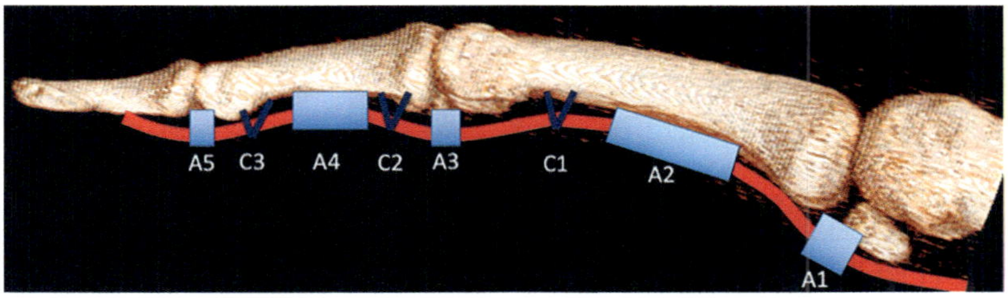

Fig. 3.2 The finger flexor tendons pulley system

sion and deflection of the flexor tendons [14–17]. A minor role is performed by the A1 and A5 pulleys, which are expandable [13]. However, the A3 pulley also controls the extent of tendon bowstringing as it is situated at the PIP joint and is thus more effective in maintaining the tendons close to the phalanges [18–21].

The pulley system keeps the tendons close to the bone. The force of the flexor tendons is transferred efficiently in flexion and hyperextension to reach the full range of motion [13]. A loss of one or several of the pulleys will cause an increased distance between the flexor tendons and the bone (tb = tendon-bone distance), which is called "bowstringing" and leads to a loss of strength and a decreased range of motion [7] (Figs. 3.3 and 3.4). These deficits are a function of the specific pulley or pulleys being ruptured. The bowstringing is minor, if only one pulley is ruptured and then can only be detected via ultrasound or MRI/CT imaging. If more than one pulley is ruptured, the bowstringing is visible and palpable and also leads to a decreased range of motion in the PIP joint. The A2 and A4 pulleys are believed to be the most important ones in preventing bowstringing and thus ensuring optimal force transmission [12, 18, 22,

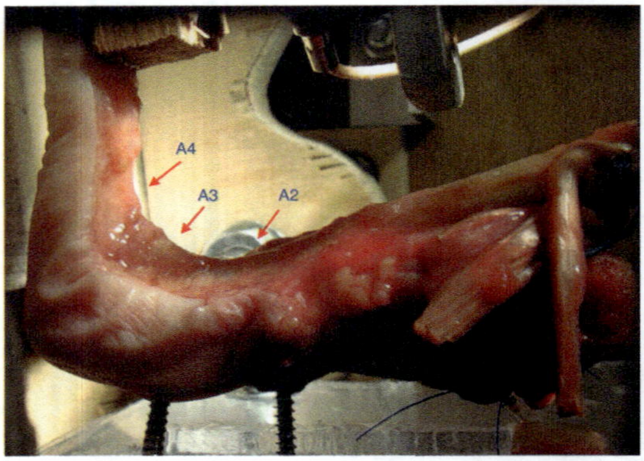

Fig. 3.3 Complete pulley system in a cadaver stress test in the biomechanical lab

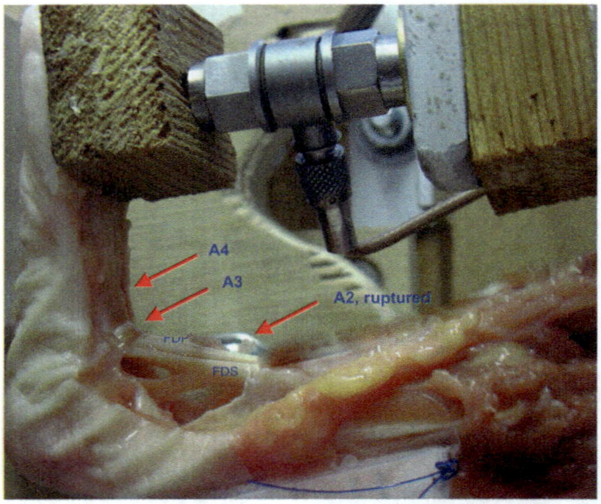

Fig. 3.4 Torn A2 pulley in a cadaver stress test in the biomechanical lab. The increased distance of the flexor tendons to the bone is clearly visible, the A3 and A4 pulley are intact

23]. Over time the increased tendon bone distance after multiple pulley ruptures performs as an increased lever arm onto the center of rotation of the PIP joint, causing contracture. In addition to their mechanical function, the pulleys themselves are also important for nutrition of the flexor tendons and actively secrete hyaluronic acid into the tendon sheath [10, 11] (Fig. 3.5).

The lumbricalis and interossei muscles are exceptions as they originate from the flexor tendons themselves and end in the tendinous hood of the extensor tendons [4]. They are part of the intrinsic muscle system. Their function is flexion

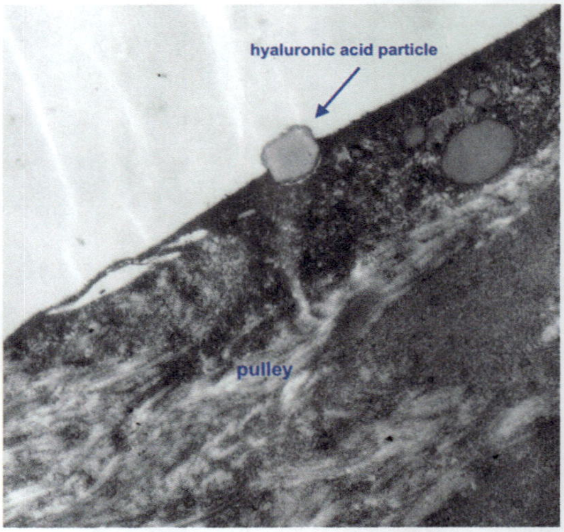

Fig. 3.5 Electron microscopy of a pulley which secretes hyaluronic acid [1]

in the MCP and extension in the proximal (PIP) and distal interphalangeal joint (DIP) [1, 4, 5].

Blood supply of the flexor tendons is guaranteed through the "vinculae tendinae" in the region of the osseous insertion of the tendon as well as in the osteofibrous channel [1, 4, 5]. Venous drainage is performed through the same system [1]. Verdan [24] divided the flexor and extensor tendons into different regions of interest regarding injuries, prognosis, and nutrition [25].

3.1.3 The Physis (Growth Plate)

The physis is situated between the epiphysis and the metaphysis of long bones. It enables the bone to grow and is only detectable in children and adolescents. When growth seizes after the end of puberty, the physes become fully calcified and cannot be detected any longer. The physis is the place of endochondral calcification. It is made up of chondrocytes, at different stages of differentiation [26]. There are basically three different stages of different differentiation, termed zones. The chondrocytes in the resting zone replicate at a slow rate and replenish the pool of proliferative chondrocytes [27]. The chondrocytes in the proliferative zone replicate at a high rate resulting in cells lining up along the long axis of the bone

[26]. Once the cells stop dividing, they differentiate into hypertrophic chondrocytes increasing their height six- to tenfold, making up the main part of longitudinal growth [28]. The hypertrophic chondrocytes then calcify the surrounding extracellular matrix and produce factors that attract the invading bone cells and blood vessels, including vascular endothelial growth factor [29], before undergoing apoptosis.

3.2 Tendon Healing

Histologically, tendons consist of extremely long collagenous fibers which are arranged into bundles. Similar to a rope ladder, elastic fibers and vessels are entangled in between those bundles [30]. To provide low friction during movement, tendons are covered by a paratendineum or a tendon sheath. Pulleys strengthen these sheaths along the phalanges. Friction plays an important role in the origin of injuries and chronic inflammatory diseases of the tendons, their sheaths and pulleys [10, 11, 31, 32]. Following an injury, healing processes emerge from the paratendineum and the paratendinous tissue, necessitating a distinction of an extrinsic and an intrinsic healing process [5, 30, 32, 33]. Characteristic for an extrinsic healing process is a distinctive inflammatory response followed by proliferation and remodeling [1]. Fibroblasts of the paratenon play an important role in migration which leads to adhesions. Immobilization supports the adhesion process [32–36]. The intrinsic healing process, supported by movement of the tendon, is characterized by immigration of "fibroblast like tenocytes," which produce the collagenous tissue and carry out the remodeling process [30, 32–36]. The inflammatory response is minimal, the clinical outcome better. This is the ratio of the widely recommended early passive movement therapy, which leads to a better nutrition and strength of the tendon [30, 32–36]. The following factors predict the tendon healing: age, overall health, scar formation disposition, motivation, injury risk based on Verdan's zones [24], injury type, synovial containment, as well as the surgical technique [1, 32]. Three phases of tendon healing are defined. First, a migration of peripheral cells and invasion of blood vessels occur; second, the tendon and surrounding tissues heal. Remodeling happens in the third phase of healing due to movement and function of the tendon [37].

The tendon gains its daily life-loading capacity after 12 weeks of healing; sportive activities are allowed earliest 4 months past injury. The remodeling process can last up to 12 months [1].

3.3 Biomechanics

3.3.1 Pulleys

In climbing two main finger positions are used for holding small edges, the hanging and the crimping position (Figs. 3.6 and 3.7). Additionally, many other hand positions exist for the various other shapes of holds, be it indoors or

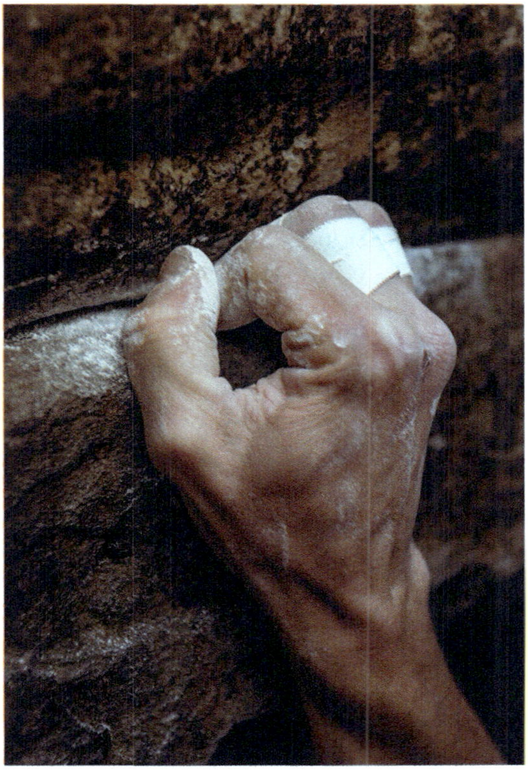

Fig. 3.6 The crimping position transfers a high stress load onto the finger flexor tendon pulleys and onto the joint cartilage. (Photo by Kilian Reil)

Fig. 3.7 In the hanging finger position, the passive structures, the joint capsule, the flexor tendons, and the ligaments receive the highest load. (Photo by Kilian Reil)

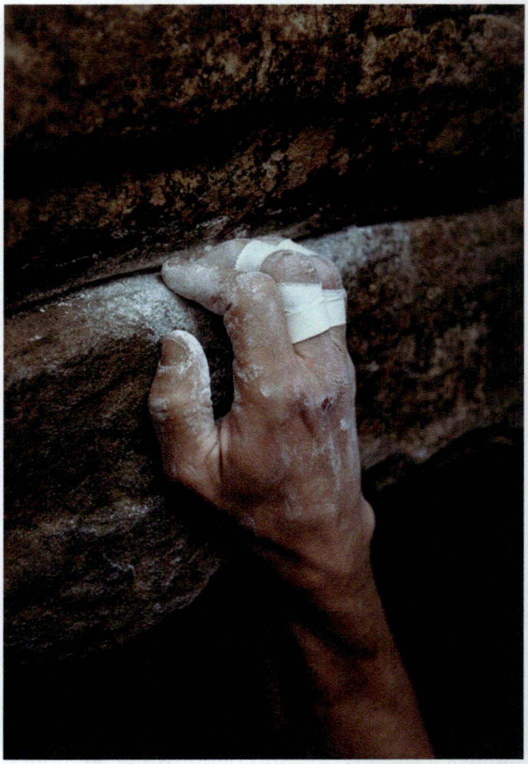

Fig. 3.8 The "half-crimp" reduces the stress onto the fingers. (Photo by Kilian Reil)

outdoors. Finger locks in crack climbing, underclinging, one- or two-finger pockets, pinch grip, stacked fingers grip, etc. are also frequently used hand and finger positions. Lately the "half-crimp" (Fig. 3.8) became popular as it is apparently placing less stress onto the pulleys and cartilage of the finger joints [3, 13].

However, the crimp grip and the hanging finger position are the most common especially for holding on to small edges. In the crimp position, the distal interphalangeal joints are hyper-extended, the proximal interphalangeal joints flexed, and the metacarpophalangeal joints extended. The wrist joint is extended to increase strength development for the finger flexion [38]. In the hanging finger position, the PIP, DIP, and MCP joints are flexed, while the wrist can be in any position from flexed, neutral, to extended. The crimping position helps to increase strength transfer in holding very small edges; thus, many climbers use it for these kinds of holds [39]. The crimping position leads to high pressure on the cartilage of the small finger joints and high stress on the pulley system. In the hanging position, most stress is transferred to the passive structures, the tendons, joint capsule, and its ligaments. The half-crimping position has recently been recommended more and more often as it decreases the high stress of the full crimp, while it still allows good strength [40]. Two other finger positions lead to biomechanically based stress, the "stacked" finger position (Fig. 3.9), and the "mono" (one finger pocket, additionally also two finger pocket) (Fig. 3.10). In the stacked finger position, a shear stress is forced onto the MCP and PIP joint as well as to the connexus intertendineus of the extensor tendons. The mono (one finger

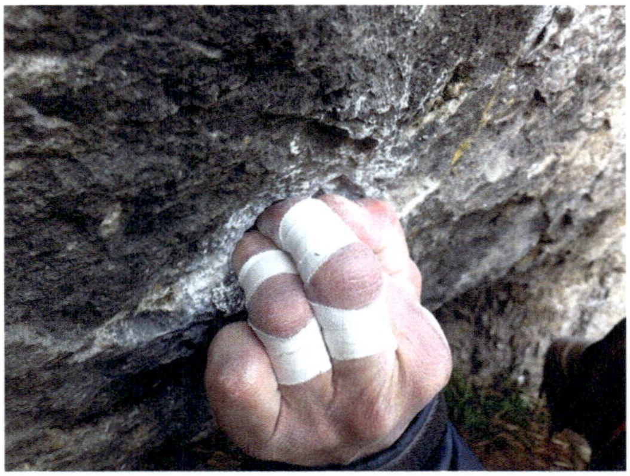

Fig. 3.9 A stacked finger position leads to shear forces onto the MCP and PIP joint as well as the connexus intertendineus

Fig. 3.10 The one finger pocket (mono) leads to a high stress onto the joint capsule and the flexor tendons and due to the fact that the adjoint fingers are flexed to a "quadriga" effect of the lumbrical muscles (Photo by Michael Simon)

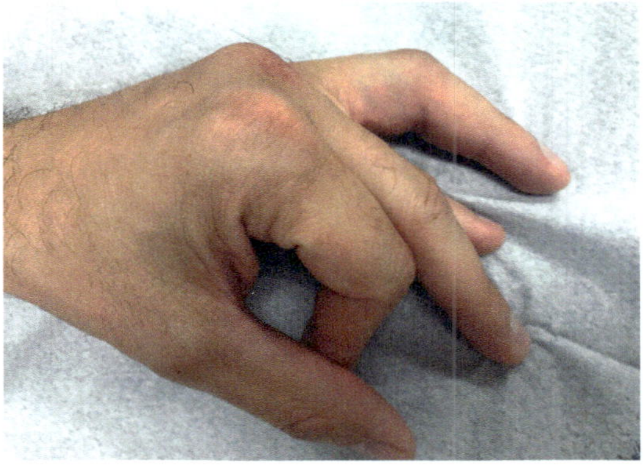

Fig. 3.11 Clinical bowstringing in a A2/3/4 rupture

pocket) and the two finger pocket positions lead to a high stress onto the joint capsule and the flexor tendons and due to the fact that the adjoint fingers are flexed, to develop more strength, to a "quadriga" effect of the lumbrical muscles [6, 40].

The main function of the flexor tendon pulley system is to hold the flexor tendons close to the bone, thus converting linear force into torque resulting in rotation at the interphalangeal and metacarpophalangeal joints [13]. A loss of one or several of the pulleys will cause an increased distance between the flexor tendons and the bone, a "bowstringing," and leads to a loss of strength and a decreased range of motion [7, 18]. Figure 3.11 shows a clinical bowstringing and Fig. 3.12 a cadaver specimen after iatrogenic A2/3/4 rupture in our biomechanical lab. The clinical picture of a loss of function is depending on the respective injured pulley. The A2 and the A4 pulley are considered as the most important pulleys to prevent a bowstringing, while the A3 pulley is weaker and more flexible, thus having a minor impact onto the flexor tendons course. It also does not insert onto the bone but on the palmar plate.

The forces acting on the pulleys during climbing can be estimated using mathematical models which incorporate anatomical observations into mathematical formulas [13, 41] (Fig. 3.13). The pulley deflects the tendon at the point of contact, and thus

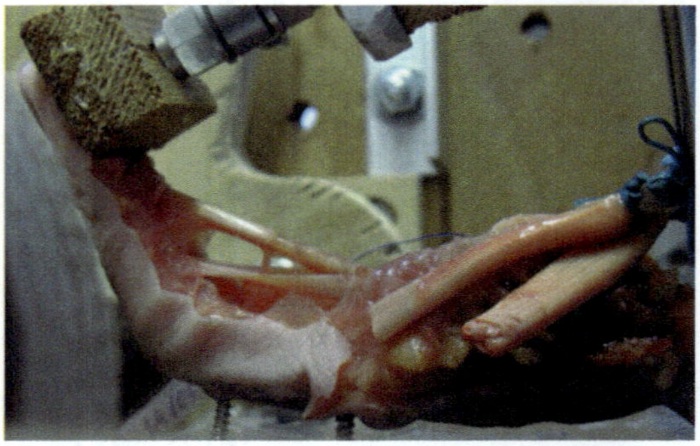

Fig. 3.12 Bowstringing in a cadaver finger in the biomechanical lab

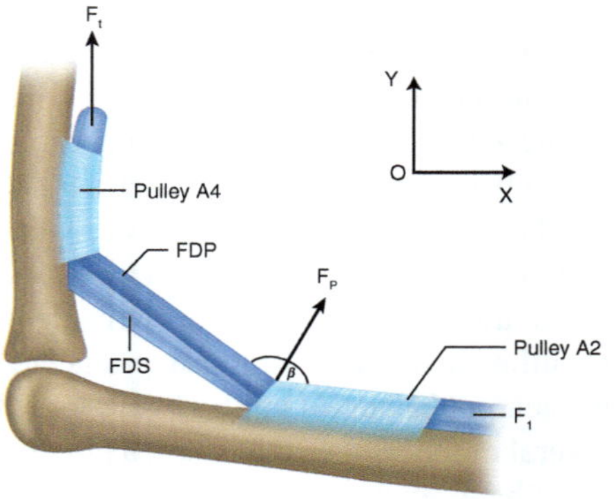

Fig. 3.13 Important for the force onto the A2 pulley (F_p) is the angle ß, which depends from the flexor angle of the PIP joint [13]

the force acting on the pulley (F_P) is a function of the angle between the tendon and the pulley (β) and the force developed in the tendon (F_T):

$$F_P = 2F_T \cos\left(\frac{\beta}{2}\right)$$

Furthermore, there is a relative mismatch between the width of the phalanx and the width of the tendons which leads to differing forces according to this mismatch.

$$F_i = F_P\left(1+\left(\frac{L_B - L_T}{2 * H_T}\right)^2\right)^{\frac{1}{2}}$$

F_i represents the force acting on the pulley insertion, L_B and L_T the width of the bone and tendon, and H_T the height of the tendon.

And finally, pulleys have a certain potential to stretch when loaded so that the angle between the tendon and the pulley will decrease once the pulley is loaded. Therefore, the stiffness of the pulley needs to be considered when modeling the forces acting on the pulley during climbing [15].

Simplified, it can be concluded that once the finger is flexed, the angle between the pulley and the tendon decreases, and the force acting on the pulley increases. This in turn signifies that a crimp grip position leads to higher forces acting on the pulley system than a hanging finger position. The higher the mismatch between the tendon and the bone width is, the higher the forces acting on the pulley.

Consequently, a maximum flexion in the PIP joint leads most frequently to a pulley rupture. These high forces are obviously happening during climbing but can also happen during other activities while stressing the fingers in a crimp grip position (e.g., lifting heavy items) [42]. The pulleys which are located around the PIP joint (A2, 3 and A4) are the ones which are mostly affected [31], but also singular c pulley injuries are reported [43].

The smaller this angle becomes, the higher the force is onto the respective pulley [13].

Butcher and Jenny [44] mathematically analyzed forces in various finger positions and found a force of 599 N on the pulley system with a PIP joint flexion of 135°. Biomechanical analyses of Bollen [22] estimated forces up to 450 N acting on the A2 pulley, and a model by Roloff et al. [13] calculated forces of 268 N for the A4 pulley in the crimp position, to be 268.8 N. Considering these high forces acting on the pulley system, the maximal tear load of pulleys needs to be considered. Lin et al. [45] found a maximal tear load of the A2 pulley of 400 N, while Widstrom et al. [46, 47] found 137 N. Even though these measurements show a wide variety, these results show that the forces estimated to act on the pulleys in the crimping position are very close to their biological tearing force in vitro. If, in addition to these high forces, the climbers hand slips off a hold or a foot comes loose, the biological strength of the pulley system is bound to fail.

Most cadaver studies on pulley failures used concentric forces onto these pulleys, mostly in behalf of pulling onto the pulleys until they tore [10]. Nevertheless, this may not be the real stress leading to rupture. Most injured climbers report that the movement leading to their pulley rupture can be described as an eccentric movement: passive opening of the hand, slipping off a foothold with a sudden increase of force on the pulley system, or a strength deficit while trying to hold onto a hold [39, 48, 49]. The analysis of a normal movement pattern in climbing shows that there is a pulling, concentric movement of the arm, with flexion of the elbow, but higher up the chain, there is an eccentric movement of the PIP joint; the fingers are slightly opening up. Many climbers report that the injury occurred at the end of a longer climbing day, when trying to hold an edge in a tired state with the consequence of not being able to maintain the crimping position. This in turn leads to an opening of the fingers [7]. In further cadaver studies, we therefore examined pulley ruptures during concentric versus eccentric movements, as well as in hanging versus crimping position [10, 11]. We found that pulley ruptures happened more easily during an eccentric movement than during a concentric movement (failure load: eccentric movement, 237 N on FDP, concentric movement 354 N on FDP) [10]. The reason for this discrepancy between concentric and eccentric loading is probably due to friction between tendon and tendon sheath/pulleys. Studies on birds and bats have shown that the inner coating of their pulleys, facing the flexor tendons, has an explicit rough surface. As a consequence of the high friction in between this rough inner surface and the flexor tendons, these animals can hang passively on their claws, without active muscular activation. Similar mechanisms could be shown for birds [50].

Examinations of human pulleys via electron microscopy show similar patterns in humans. Schweizer [31] was able to measure this increased friction in human pulleys and their possible correlation to pulley rupture in an in vivo study. In conclusion, it can be said that friction underneath the pulleys is an advantage for a climber, as this will increase flexor strength. However, an increased friction between the inner surface of the pulley and the tendons may be the reason why some climbers are more prone to pulley injuries than others. The obvious downside in "rough" pulleys and great finger strength is the higher incidence of pulley injuries in these people.

Even if the mathematical determined force values as well as the pulley tear forces in the cadaver models can only be partially transferred into the dynamic pulley flexor tendon system in vivo, these findings clearly show that in a crimping position the forces acting on the flexor pulleys are close to their tearing strength [7]. This explains the fact that a pulley rupture is one of the most common climbing-specific diagnoses. These high forces which are certainly occurring in climbing can also happen in daily live activities, e.g., lifting a heavy item [42]. The youngest patient with a pulley lesion was a 2.5-year-old girl, injuring her finger while playing in a sand box and her brother stepping onto the finger, while she was pulling it away.

The ring finger is most often concerned in pulley ruptures followed by the middle finger and the other fingers [7], which can possibly also be explained through a biomechanical analysis of the movement pattern.

In the crimp position, the carpal joint performs in addition to an overextension an ulnar abduction [7]. Caused by this ulnar abduction in combination with a mild supination while crimping, the ring finger receives the highest stress [51]. Especially as the ring finger is less powerful than the middle and the index finger, this rotational moment increases stress and injury potential to its pulleys [52]. Furthermore, the adjoint little finger is in most cases much shorter than the ring finger and therefore offers little support to the ring finger. Injuries to the index finger are rare, probably due to its decreased stress during the abovementioned supination. In crimping with the full hand, the middle finger is supported on both sides by the ring and index finger. The index finger is supported by the thumb which is often stacked on top of it, while nothing supports the ring finger. Pulley injuries to the middle finger are most frequent if the crimping is only performed with two fingers [53].

3.3.2 The Growth Plate

Fractures to the growth plate have previously been described in other sports on various anatomical locations. The first cases were observed in the proximal humeral growth plate of young baseball players and have been termed the Little Leaguer's shoulder [54–56]. The cause for this injury is believed to be a consequence of the whiplash action of the arm during a throwing movement [57]. As a consequence of repetitious throwing, a widening of the epiphyseal plate as in a Salter Harris I fracture [57] occurs. Therefore, most of the epiphyseal fractures observed in baseball are Salter Harris I fractures [55, 56]. After this first description of disruptions to the growth plate as chronic injury in baseball, stress fractures of the distal radius in gymnasts were being recognized on a regular basis [58–60]. The pathomechanism is believed to be complex rotational vaults [58, 61]. As a consequence of this loading mechanism, the injuries most often observed are Salter Harris II stress fractures with an epiphyseal widening and irregularity and cystic changes of the metaphyseal aspect of the growth plate [58, 59, 61]. The climbing community only started reporting epiphyseal stress fractures in the fingers in 1997 [62] and 1999 [63]. As a consequence of repetitive loading of the fingers, the fractures observed were always in the proximal interphalangeal joint [62]. Most often they were fractures of the Salter Harris III type with a fracture through the epiphysis of the middle phalanx [64].

All these injuries have repetitive loading as a common denominator. As there is no true blood supply to the physis but only an advance from the blood vessels of the epiphysis and metaphysis as well as from the perichondrial ring and vessels of the periosteum [65], repetitive loading can alter the metaphyseal perfusion and thus interfere with the mineralization of the hypertrophied chondrocytes [66]. The hypertrophic zone then widens as a consequence of constant growth in the germinal and proliferative zones. Usually, this widening of the epiphyseal plate is only temporary, as the resting and dividing cellular layer of the growth plate, and the attendant epiphyseal and metaphyseal blood supplies are essentially undisturbed. If the ischemic condition sets in as a consequence of repetitive loading with too much weight, an osseous necrosis and deformity within the developing ossification center can be caused, leading to growth irregularities in the physis. As these changes localized, asymmetric growth can be observed, or the entire physis may be involved leading to an overall slowdown of the rate of growth or even complete cessation of growth in that joint [67].

The growth plate seems to be exceptionally susceptible to injuries during periods of rapid growth [68–71] that is during the growth spurt during puberty. The strength of the physeal cartilage has been shown to decrease during pubescence in animal studies [72] as well as in humans [73]. This decrease in strength is a consequence of the structural changes during rapid growth leading to a thicker and more fragile plate as well as to the bone mineralization lagging behind the linear growth, rendering the bone more porous [74]. Furthermore, shear resistance is a function of the amount of matrix and collagen. Increased cell size in the hypertrophic zone reduces the amount of matrix available to resist shear, and an increase in weakness is thus predictable [75].

The process of longitudinal bone growth is governed by a complex network of endocrine signals, including growth hormone, insulin-like growth factor I, glucocorticoid, thyroid hormone, estrogen, androgen, vitamin D, and leptin [26]. However, how much the different hormone levels impact on the risk of growth plate fractures is questionable and has so far has not been examined.

Whether an increase in muscle-tendon tightness about the joints as a consequence to the growth spurt leads to an excessive muscular stress on the physis is controversial in the literature [71, 76, 77].

The mechanism behind this injury is believed to be a high repetitive stress on the proximal interphalangeal joint using a crimp grip position of the hand (Fig. 3.6) [64, 78]. In this finger position, the physis is subjected to a compressing force coming from the fingertip on one hand and from the flexor muscles on the other hand. As the finger is flexed to a maximum, this compressive

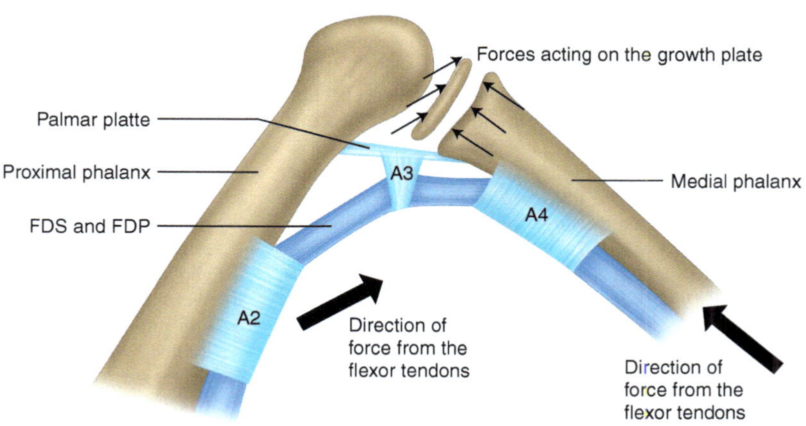

Fig. 3.14 Forces acting on the growth plate of the finger, in the crimp grip position, with a resulting compressing force on the growth plate coming from the fingertip on one side and from the flexor muscles on the other side

force is uneven, applying more force to the dorsal aspect of the growth plate (Fig. 3.14). This in turn leads to the fracture line going through the dorsal aspect of the growth plate as well as the epiphysis. This is exaggerated in campus board training in which the athlete performs climbing movements with the feet off the ground, thus subjecting the fingers to the full body weight. A study investigating radiographic changes in the fingers of top level climbers revealed that only those climbers who performed this kind of training had osteoarthrotic changes [79, 80].

3.4 Conclusion

The so-called crimp grip position, which is commonly used in climbers, is a very specific position of the fingers where the proximal interphalangeal joint is flexed more than 90°, and the distal interphalangeal joint is hyper-extended, resulting in very high loads onto the pulleys of the flexor tendon sheath, which are three to four times higher than the forces acting at the fingertip [81]. Also, the existence of friction between the flexor tendons and the pulleys [10, 31] adds to the understanding of the pathomechanism of the rupture of a pulley, which mostly happens during eccentric movement [10]. Adolescents during their growth spurt are most prone to injure the growth plate due to forces pushing the physis out on the dorsal aspect of the PIP joint. An injury to the growth plate can lead to serious impairment of the finger with malaligned growth or cessation of growth altogether. Campus boarding and the crimp grip position are the main causes for injuries to the growth plate.

References

1. Schöffl V, Heid A, Küpper T. Tendon injuries of the hand. World J Orthop. 2012;3(6):62–9.
2. Straub G. Anatomie und Biomechanik der Hand und Finger. In: Schöffl V und Hochholzer, T., editor. Sportklettern—Aktuelle sportmedizinische Aspekte. 1. Ebenhausen: Lochner Verlag; 2004. p. 134–144.
3. Schöffl V, Schöffl I, Frank L, Küpper T, Simon M, Lutter C. Tendon injuries in the hands in rock climbers: epidemiology, anatomy, biomechanics and treatment—an update. Muscles Ligaments Tendons J. 2020;10:233–43.
4. Lutter C, Schweizer A, Schöffl V, Römer F, Bayer T. Lumbrical muscle tear: clinical presentation, imaging findings and outcome. J Hand Surg Eur. 2018;43(7):767–75.
5. Schöffl V, Winkelmann HP. [Traumatic and degenerative tendon lesions of the hand]. Der Orthopäde 2010;39(12):1108–1116.
6. Schweizer A. Lumbrical tears in rock climbers. J Hand Surg (Br). 2003;28(2):187–9.
7. Schöffl V, Schöffl I. Injuries to the finger flexor pulley system in rock climbers—current concepts. J Hand Surg [Am]. 2006;31(4):647–54.
8. Doyle JR. Anatomy of the finger flexor tendon sheath and pulley system. J Hand Surg [Am]. 1988;13(4):473–84.
9. Hahn P, Lanz U. [Flexor pulleys of the fingers. Anatomy, biomechanics, reconstruction] Die Ringbänder der Fingerbeugesehnen. Anatmoie,

Biomechanik, Wiederherstellung. Handchir Mikrochir Plast Chir. 1996;28(5):265–70.

10. Schöffl I, Oppelt K, Jungert J, Schweizer A, Bayer T, Neuhuber W, et al. The influence of concentric and eccentric loading on the finger pulley system. J Biomech. 2009;42(13):2124–8.
11. Schöffl I, Oppelt K, Jungert J, Schweizer A, Neuhuber W, Schöffl V. The influence of the crimp and slope grip position on the finger pulley system. J Biomech. 2009;42(13):2183–7.
12. Manske PR, Lesker PA. Strength of human pulleys. Hand. 1977;9(2):147–52.
13. Roloff I, Schöffl VR, Vigouroux L, Quaine F. Biomechanical model for the determination of the forces acting on the finger pulley system. J Biomech. 2006;39(5):915–23.
14. An KN, Berglund L, Uchiyama S, Coert JH. Measurement of friction between pulley and flexor tendon. Biomed Sci Instrum. 1993;29:1–7.
15. Lin GT, Cooney WP, Amadio PC, An KN. Mechanical properties of human pulleys. J Hand Surg (Br). 1990;15(4):429–34.
16. Lin GT, Amadio PC, An KN, Cooney WP. Functional anatomy of the human digital flexor pulley system. J Hand Surg [Am]. 1989;14(6):949–56.
17. Doyle JR. Anatomy of the flexor tendon sheath and pulley system: a current review. J Hand Surg [Am]. 1989;14(2 Pt 2):349–51.
18. Marco RA, Sharkey NA, Smith TS, Zissimos AG. Pathomechanics of closed rupture of the flexor tendon pulleys in rock climbers. J Bone Joint Surg Am. 1998;80(7):1012–9.
19. Rispler D, Greenwald D, Shumway S, Allan C, Mass D. Efficiency of the flexor tendon pulley system in human cadaver hands. J Hand Surg [Am]. 1996;21(3):444–50.
20. Hamman J, Ali A, Phillips C, Cunningham B, Mass DP. A biomechanical study of the flexor digitorum superficialis: effects of digital pulley excision and loss of the flexor digitorum profundus. J Hand Surg [Am]. 1997;22(2):328–35.
21. Zao CF, Amadio PC, L.B.K. A. The A3 pulley. J Hand Surg [Am]. 2000;25:270–6.
22. Bollen SR. Injury to the A2 pulley in rock climbers. J Hand Surg (Br). 1990;15(2):268–70.
23. Bowers WH, Wolf JW Jr, Nehil JL, Bittinger S. The proximal interphalangeal joint volar plate. I. An anatomical and biomechanical study. J Hand Surg [Am]. 1980;5(1):79–88.
24. Verdan CE. Primary and secondary repair of flexor and extensor tendon injuries. In: Flynn JE, editor. Hand surgery. Baltimore: Williams & Willkins; 1966. p. 220–42.
25. Voigt C. Sehnenverletzungen an der Hand. Chirurg. 2002;73:744–67.
26. Nilsson O, Marino R, De Luca F, Phillip M, Baron J. Endocrine regulation of the growth plate. Horm Res. 2005;64(4):157–65.
27. Abad V, Meyers JL, Weise M, Gafni RI, Barnes KM, Nilsson O, et al. The role of the resting zone in growth plate chondrogenesis. Endocrinology. 2002;143(5):1851–7.
28. Hunziker EB. Mechanism of longitudinal bone growth and its regulation by growth plate chondrocytes. Microsc Res Tech. 1994;28(6):505–19.
29. Gerber HP, Vu TH, Ryan AM, Kowalski J, Werb Z, Ferrara N. VEGF couples hypertrophic cartilage remodeling, ossification and angiogenesis during endochondral bone formation. Nat Med. 1999;5(6):623–8.
30. Werdin F, Schaller HE. [Combined flexor tendon and nerve injury of the hand]. Der Orthopade. 2008;37(12):1202–1209.
31. Schweizer A, Frank O, Ochsner PE, Jacob HA. Friction between human finger flexor tendons and pulleys at high loads. J Biomech. 2003;36(1):63–71.
32. Taras JS, Gray RM, Culp RW. Complications of flexor tendon injuries. Hand Clin. 1994;10(1):93–109.
33. Boyer MI, Strickland JW, Engles D, Sachar K, Leversedge FJ. Flexor tendon repair and rehabilitation: state of the art in 2002. Instr Course Lect. 2003;52:137–61.
34. Strickland JW. The scientific basis for advances in flexor tendon surgery. J Hand Ther. 2005;18(2):94–110; quiz 1.
35. Choueka J, Heminger H, Mass DP. Cyclical testing of zone II flexor tendon repairs. J Hand Surg [Am]. 2000;25(6):1127–34.
36. Stein T, Ali A, Hamman J, Mass DP. A randomized biomechanical study of zone II human flexor tendon repairs analyzed in a linear model. J Hand Surg [Am]. 1998;23(6):1043–5.
37. Werber KD. Beugesehnenverletzungen der Hand. Unfallchirurg. 2005;108(10):873–82.
38. Schöffl V, Schweizer A. Sportclimbing related injuries and overuse syndromes. In: Seifert L, Wolf P, Schweizer A, editors. IRCRA handbook of climbing & mountaineering from science to performance. Research in sport and exercise science. London, New York: Routledge; 2016. p. 59–75.
39. Schöffl V, Hochholzer T, Winkelmann HP, Strecker W. Pulley injuries in rock climbers. Wilderness Environ Med. 2003;14(2):94–100.
40. Schöffl V, Schöffl I, Hochholzer T, Lutter C. Klettermedizin—Grundlagen, Unfälle. Verletzungen und Therapie: Springer; 2020.
41. Hume EL, Hutchinson DT, Jaeger SA, Hunter JM. Biomechanics of pulley reconstruction. J Hand Surg [Am]. 1991;16(4):722–30.
42. Schöffl V, Jüngert J. Closed flexor pulley injuries in non-climbing activities. J Hand Surg [Am]. 2006;31(5):806–10.
43. Schöffl V, Schöffl I. Isolated cruciate pulley injuries in rock climbers. J Hand Surg Eur. 2010;35(3):245–6.
44. Burtscher M, Jenny E. Häufigste trainingsbedingte Beschwerden und Verletzungen bei Sportkletterern. Prakt Sport Trauma Sportmed. 1987;2:15–21.

45. Lin GT, Amadio PC, An KN, Cooney WP, Chao EY. Biomechanical analysis of finger flexor pulley reconstruction. J Hand Surg (Br). 1989;14(3):278–82.
46. Widstrom CJ, Doyle JR, Johnson G, Manske PR, McGee R. A mechanical study of six digital pulley reconstruction techniques: Part II. Strength of individual reconstructions. J Hand Surg [Am]. 1989;14(5):826–9.
47. Widstrom CJ, Johnson G, Doyle JR, Manske PR, Inhofe P. A mechanical study of six digital pulley reconstruction techniques: Part I. Mechanical effectiveness. J Hand Surg [Am]. 1989;14(5):821–5.
48. Gabl M, Rangger C, Lutz M, Fink C, Rudisch A, Pechlaner S. Disruption of the finger flexor pulley system in elite rock climbers. Am J Sports Med. 1998;26(5):651–5.
49. Moutet F. [Flexor tendon pulley system: anatomy, pathology, treatment]. Chir Main. 2003;22(1):1–12.
50. Haffner M. A tendon-locking mechanism in two climbing rodents, Muscardinus avellanarius and Micromys minutus (Mammalia, Rodentia). J Morphol. 1996;229(2):219–27.
51. Schöffl V, Hochholzer T, Winkelmann HP, Roloff I, Strecker W. [Pulley injuries in sport climbers] Verletzungen des Ringbandapparates bei Sportkletterern. Handchir Mikrochir Plast Chir. 2004;36(4):224–30.
52. Quaine F, Vigouroux L, Martin L. Effect of simulated rock climbing finger postures on force sharing among the fingers. Clin Biomech (Bristol, Avon). 2003;18(5):385–8.
53. Schöffl V, Hochholzer T, Winkelmann HP, Strecker W. [Therapy of injuries of the pulley system in sport climbers] Zur Therapie von Ringbandverletzungen bei Sportkletterern. Handchir Mikrochir Plast Chir. 2004;36(4):231–6.
54. Dotter WE. Little leaguer's shoulder: a fracture of the proximal epiphysial cartilage of the humerus due to baseball pitching. Guthrie Clin Bull. 1953;23(1):68–72.
55. Carson WG Jr, Gasser SI. Little Leaguer's shoulder. A report of 23 cases. Am J Sports Med. 1998;26(4):575–80.
56. Rettig AC, Wurth TR, Mieling P. Nonunion of olecranon stress fractures in adolescent baseball pitchers: a case series of 5 athletes. Am J Sports Med. 2006;34(4):653–6.
57. Adams JE. Little league shoulder: osteochondrosis of the proximal humeral epiphysis in boy baseball pitchers. Calif Med. 1966;105(1):22–5.
58. Read MT. Stress fractures of the distal radius in adolescent gymnasts. Br J Sports Med. 1981;15(4):272–6.
59. Carter SR, Aldridge MJ, Fitzgerald R, Davies AM. Stress changes of the wrist in adolescent gymnasts. Br J Radiol. 1988;61(722):109–12.
60. Roy S, Caine D, Singer KM. Stress changes of the distal radial epiphysis in young gymnasts. A report of twenty-one cases and a review of the literature. Am J Sports Med. 1985;13(5):301–8.
61. Bak K, Boeckstyns M. Epiphysiodesis for bilateral irregular closure of the distal radial physis in a gymnast. Scand J Med Sci Sports. 1997;7(6):363–6.
62. Hochholzer T, Schöffl V, Krause R. Finger-Epiphysenverletzungen jugendlicher Sportkletterer. Sport Ortho Trauma. 1997;13(2):100–3.
63. Chell J, Stevens K, Preston B, Davis TR. Bilateral fractures of the middle phalanx of the middle finger in an adolescent climber. Am J Sports Med. 1999;27(6):817–9.
64. Hochholzer T, Schöffl V, Bischof B. [Epiphyseal fractures of the middle phalanx in sportclimbers] Epiphysenfrakturen der Fingermittelgelenke bei Sportkletterern. Sport Ortho Trauma. 2002;18:87–92.
65. Lipp EJ. Athletic physeal injury in children and adolescents. Orthop Nurs. 1998;17(2):17–22.
66. Caine D, DiFiori J, Maffulli N. Physeal injuries in children's and youth sports: reasons for concern? Br J Sports Med. 2006;40(9):749–60.
67. Laor T, Wall EJ, Vu LP. Physeal widening in the knee due to stress injury in child athletes. AJR. 2006;186(5):1260–4.
68. Flachsmann R, Broom ND, Hardy AE, Moltschaniwskyj G. Why is the adolescent joint particularly susceptible to osteochondral shear fracture? Clin Orthop Relat Res. 2000;381:212–21.
69. Bailey DA, McKay HA, Mirwald RL, Crocker PR, Faulkner RA. A six-year longitudinal study of the relationship of physical activity to bone mineral accrual in growing children: the university of Saskatchewan bone mineral accrual study. J Bone Miner Res. 1999;14(10):1672–9.
70. DiFiori JP, Mandelbaum BR. Wrist pain in a young gymnast: unusual radiographic findings and MRI evidence of growth plate injury. Med Sci Sports Exerc. 1996;28(12):1453–8.
71. Micheli LJ. Overuse injuries in children's sports: the growth factor. Orthop Clin North Am. 1983;14(2):337–60.
72. Bright RW, Burstein AH, Elmore SM. Epiphyseal-plate cartilage. A biomechanical and histological analysis of failure modes. J Bone Joint Surg. 1974;56(4):688–703.
73. Morscher E. Strength and morphology of growth cartilage under hormonal influence of puberty. Animal experiments and clinical study on the etiology of local growth disorders during puberty. Reconstr Surg Traumatol. 1968;10:3–104.
74. Bailey DA, Wedge JH, McCulloch RG, Martin AD, Bernhardson SC. Epidemiology of fractures of the distal end of the radius in children as associated with growth. J Bone Joint Surg. 1989;71(8):1225–31.
75. Alexander CJ. Effect of growth rate on the strength of the growth plate-shaft junction. Skelet Radiol. 1976;1:67–76.

76. Feldman D, Shrier I, Rossignol M, Abenhaim L. Adolescent growth is not associated with changes in flexibility. Clin J Sport Med. 1999;9(1):24–9.
77. Connolly SA, Connolly LP, Jaramillo D. Imaging of sports injuries in children and adolescents. Radiol Clin N Am. 2001;39(4):773–90.
78. Hochholzer T, Schöffl VR. Epiphyseal fractures of the finger middle joints in young sport climbers. Wilderness Environ Med. 2005;16(3):139–42.
79. Schöffl VR, Hoffmann PM, Imhoff A, Küpper T, Schöffl I, Hochholzer T, et al. Long-term radiographic adaptations to stress of high-level and recreational rock climbing in former adolescent athletes: an 11-year prospective longitudinal study. Orthop J Sports Med. 2018;6(9):2325967118792847.
80. Schöffl VR, Hochholzer T, Imhoff AB, Schöffl I. Radiographic adaptations to the stress of high-level rock climbing in junior athletes: a 5-year longitudinal study of the German Junior National Team and a group of recreational climbers. Am J Sports Med. 2007;35(1):86–92.
81. Schweizer A. Biomechanical properties of the crimp grip position in rock climbers. J Biomech. 2001;34(2):217–23.

Historical Development of a Physiological Model for Rock Climbing Performance

4

P. B. Watts

Luc Schöffl bouldering in Oberngrub, Frankenjura, Germany, Photo Michael Simon

P. B. Watts (✉)
School of Health and Human Performance, Northern Michigan University, Marquette, MI, USA
e-mail: pwatts@nmu.edu

V. Schöffl et al. (eds.), *Climbing Medicine*, https://doi.org/10.1007/978-3-030-72184-8_4

4.1 Background

Successful performance in rock climbing is physically demanding and involves the integration of many factors associated with production of the work required to ascend over specific terrain. Recreational climbers may find success through maintenance of a high level of general physical fitness however, performance at the highest levels likely requires physiological adaptations likened to that of high-performance athletes. This chapter will explore the more notable physiological aspects of high-level rock climbing. The objective is to provide a brief historical overview of the development of a theoretical physiological model for high-level climbing performance. The chapter is not intended as a comprehensive review of research to date. For a more complete exploration, the reader is referred to the published reviews of Watts [1] and Saul et al. [2].

Figure 4.1 presents a simplified conceptual diagram for development of a model for athletic-level performance via integration of measured characteristics of the performer and the specific stress demands of the activity performed. This model provides a basis upon which associated research results may be evaluated and integrated. In order to follow and apply this model-building concept for rock climbing, an objective measure of climber performance is necessary.

Description of the performance levels of participants in research studies has typically been relative to the most difficult rock route a participant could currently ascend without falling, referred to as red point (RP) ability. Attempts to describe climber performance ability have also used adjectives such as "expert" and "elite." However, these have been inconsistently applied.

Historically, climbers by nature tended toward a degree of competitiveness both internally with the self and externally with other climbers or the rock terrain. The expression of climber ability relative to a subjective rating of terrain difficulty naturally evolved. The first difficulty rating system for rock climbing was developed in the late 1800s and was a precursor to the *Union Internationale des Associations d'Alpinisme* (UIAA) scale established in the 1940s. Other localized numerical systems were developed in intervening years. The Yosemite Decimal System (YDS) for rating the difficulty of specific climbing routes appeared in the 1950s and became widely used in North America, while other systems developed in the United Kingdom, France, and other regions of the world.

Since no common difficulty scale is used worldwide and the existing scales are subjective and perhaps in constant evolution, there has been a need for a standard to use for scientific research. To provide a degree of standardization, a numerical and adjective scale for climber ability was developed by Draper et al. [3] of the International Rock Climbing Research Association (IRCRA). The IRCRA ability grouping scale is open ended and currently

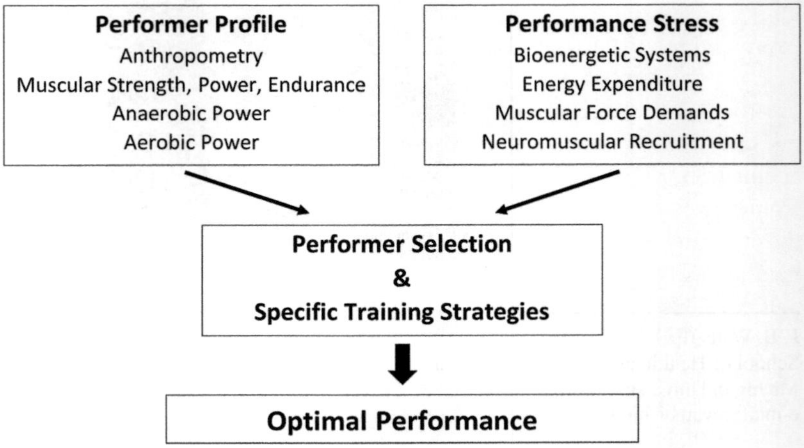

Fig. 4.1 Concept for development of a theoretical physiological model for optimal performance of a physically demanding task

Table 4.1 IRCRA group level and scale grades compared with the Yosemite Decimal System (YDS) and French system grades. Adapted from Draper, N, D Giles, V Schoffl, F Fuss, PB Watts, et al. Comparative grading scales, statistical analyses, climber descriptors and ability grouping: International Rock Climbing Research Association position statement. Sports Tech. 2016. https://doi.org/10.1080/19346182.2015.1107081

IRCRA performance group level		IRCRA scale	YDS	French
Level 2 male	Level 2 female	10	5.10a	5+
		11	5.10b	6a
		12	5.10c	6a+
		13	5.10d	6b
		14	5.11a	6b+
	Level 3 female	15	5.11b	6c
		16	5.11c	6c+
		17	5.11d	7a
Level 3 male		18	5.12a	7a+
		19	5.12b	7b
		20	5.12c	7b+
	Level 4 female	21	5.12d	7c
		22	5.13a	7c+
		23	5.13b	8a
Level 4 male		24	5.13c	8a+
		25	5.13d	8b
		26	5.14a	8b+
	Level 5 female	27	5.14b	8c
Level 5 male		28	5.14c	8c+
		29	5.14d	9a
		30	5.15a	9a+
		31	5.15b	9b
		32	5.15c	9b+
		33[a]	5.15d	9c

[a]Grade currently reported as of this publication but not included in the original published IRCRA table

extends from 1 (level I ability) to 33 (level 5 ability). Grade comparisons for IRCRA "intermediate" through "higher elite" levels relative to the YDS and French rating systems are presented in Table 4.1. In this chapter, all climber ability levels and route difficulty ratings will be expressed relative to the IRCRA scale.

Early researchers were challenged by the environment and movement nature of rock climbing and by instrumentation limitations for data acquisition. More recent research has benefitted from newer measurement technology; however, the changing nature of climbing, as more difficult natural terrain is discovered and attempted, presents challenges. The increasing interest in competition climbing on artificial structures and specializations into lead, bouldering, and speed events also affects the likely physiological determinants of high-level performance. The models upon which climber attributes, climbing stress and physical training strategies are based remain, for the most part, theoretical.

A look at how the "world's best" performance has evolved from 1960 to more recent years is available in Fig. 4.2. The fourth-order polynomial trendline indicates a relatively steady increase in ability relative to RP ascents with a possible plateau tendency since 2010. Although the trendline indicates a plateau or, at best, only slow progression of absolute route difficulty in recent years, the number of climbers who perform at the highest level has increased. For example, in 2014, only six ascents of IRCRA level 31 were made, while in 2018, there were 21 new RP first ascents at the IRCRA level 30 or higher, and 81 different climbers had performed RP ascents at level 30 or higher [4]. The relative distribution of climber abilities within a wider range for a given year is not known.

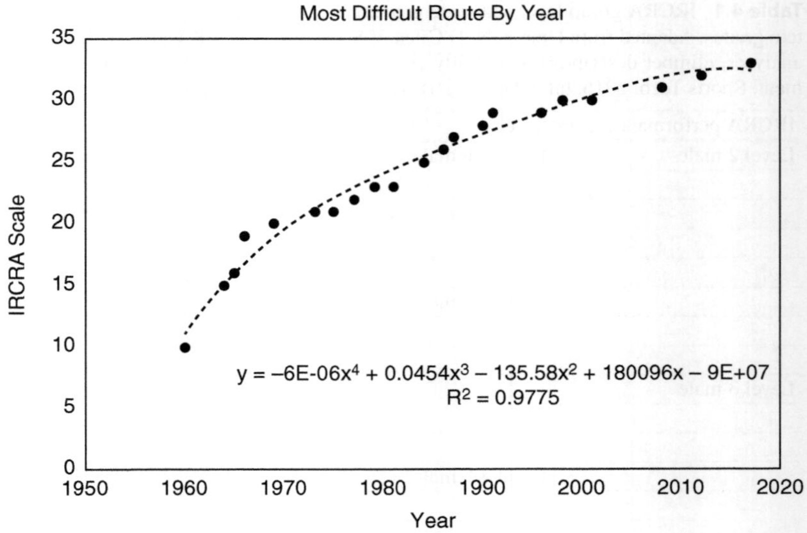

Fig. 4.2 Plot of world's most difficult route ascent by year, 1960–2017

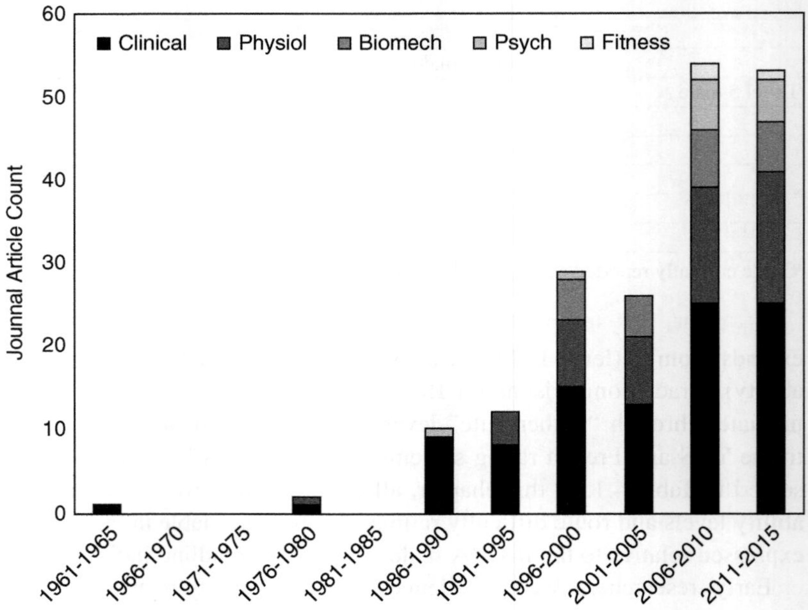

Fig. 4.3 Published research 1960–2019 via PubMed search on "rock climbing" and "rock climbers"

4.2 Physiological Study of Climbers and Climbing

Early scientific research on rock climbing was focused on the general nature of the activity and the physical aspects of climbing along with various injuries suffered by climbers. In the latter 1980s, as competition rock climbing grew in popularity, research models traditionally used for the study of athletes were applied to climbing and climbers. The specific areas of performance physiology, biomechanics, and sport psychology have evidenced increased climbing research to date. A simple online search of the USA's National Library of Medicine site (https://pubmed.ncbi.nlm.nih.gov/) with the terms "rock climbing" and "rock climbers" reveals the topical trends illustrated in Fig. 4.3. A PubMed search in

Table 4.2 Selected anthropometric measures reported for high-level rock climbers

Reference	Subject characteristics	Height (cm)	Mass (kg)	%fat
Males:				
Watts et al. [5]	7 males IRCRA 26[a]	179.3 ± 5.2	62.4 ± 4.5	4.8 ± 2.3[b]
Watts et al. [16]	11 males IRCRA 23	175.6 ± 8.9	65.9 ± 8.6	5.4 ± 1.5[b]
España-Romero et al. [12]	10 males IRCRA 20	172.0 ± 4.0	65.5 ± 4.5	7.6 ± 2.1[b] 13.3 ± 3.3[c]
Ozimek et al. [20]	6 males IRCRA 25–27	177.4 ± 5.6	66.9 ± 5.8	7.9 ± 4.8[b]
Females:				
Watts et al. [5]	6 females IRCRA 23	162.3 ± 4.6	46.8 ± 4.9	9.6 ± 1.9[b]
España-Romero et al. [12]	9 females IRCRA 15	162.0 ± 3.0	59.2 ± 3.6	18.0 ± 3.6[b] 25.2 ± 3.6[b]
Giles et al. [7]	14 females IRCRA 19.5	163.4 ± 4.7	56.5 ± 6.3	23.3 ± 4.8[b]

[a]*IRCRA* scale are means unless otherwise indicated
[b]Estimated via skinfold measures
[c]Estimated via DXA

July of 2020 found 255 studies published through 2019. Of the 255 papers, 71, or 28%, focused on physiological aspects of performance.

4.2.1 Anthropometry and Body Composition

The nature of climbing involves the work of moving the body along a relatively specific route against the negative force generated by the effect of gravity on body mass. Additionally, the force requirements for support and movement during climbing are often primarily imposed on the relatively small musculature of the upper body, in particular the muscles in control of positioning of the hands and fingers.

Anthropometry and body composition have long been of interest since body weight is a major factor for the work demand in climbing. In general, a low body mass reduces the work of climbing and reduces specific force requirements of the musculature. There is a compromise involved, however, since one positive factor of strength is muscle mass which, in turn, adds to the overall work of climbing.

Assessment of relative fat mass, expressed as percent body fat, is of particular interest, since fat mass can be a negative factor as it increases the work of climbing without directly contributing to support and movement. A summary of anthropometry results from selected research studies is presented in Table 4.2.

Watts et al. [5] published the first anthropometric study of competitive male and female sport rock climbers in 1993. Data were recorded at a 1989 international competition with 21 of 29 male and 18 of 21 female semifinalists participating. Seven of the ten male finalists and all six female finalists participated. Climbers were found to be small in stature and low in body mass with low sum of skinfold measures and estimated body fat percentage. Stature, body mass, and percent body fat averaged 179.3 ± 5.2 and 162.3 ± 4.6 cm, 62.4 ± 4.5 and 46.8 ± 4.9 kg, and 4.8 ± 2.3% and 9.6 ± 1.9% for male and female finalists, respectively. The body fat measures would be considered very low and, for both males and females, were at or below estimated essential body fat levels.

More recently, España-Romero et al. (2009) [6] reported similar height and weight means for male and female climbers of IRCRA level 3; however, percent fat means were higher when estimated via skinfold measures or via dual-energy X-ray absorptiometry (DXA). Giles et al. [7] also reported higher percent fat values when calculated from skinfold measures in 14 female climbers than the earlier study of Watts et al. [5].

Although the more recent studies have indicated higher percent fat values in some elite climbers, the best climbers still tend to be relatively small in stature with low body mass. Aside from the early study of Watts et al. [5], research has, for the most part, studied climber samples that were convenient to the geographical location of the research. This consideration may contribute to the differences between historical and more recent data. In addition, a variety of measurement methodologies and estimation procedures have been employed. The degrees to which these factors may have influenced the specific results are not known.

Other anthropometric characteristics, such as finger length, arm length and the ratio of arm span to height, or "ape index," have been of interest but have not consistently shown relationships with climbing ability ratings [8].

4.2.2 Bioenergetic Power

Bioenergetics involves the conversion of stored chemical energy into the mechanical energy of performing work. Generally, three primary bioenergetic systems may be described. A phosphagen-based system involves stored adenosine triphosphate (ATP) and creatine phosphate (PCr), which can provide energy rapidly, without the involvement of oxygen, for fast powerful muscle contractions. This phosphagen or ATP-PCr system is relatively limited in capacity, however, and may be nearly depleted within a few seconds, typically 5–8 s of total-body intense effort. The ATP-PCr system is limited primarily by the level of PCr stored within the muscle. Depleted PCr is restored relatively quickly between muscular contractions if blood circulation and oxygen delivery are adequate.

A carbohydrate-fueled metabolic system may also convert stored chemical energy quickly though more slowly than the phosphagen system. This system is usually termed glycolysis when the substrate is glucose and glycogenolysis when the initial substrate is stored muscle glycogen. When the rate of energy conversion is high in this pathway, the addition of oxygen is not required, and the process is considered to proceed anaerobically but with a consequent accumulation of lactate in the muscle and blood. This accumulation of lactate and a resulting dissociation of hydrogen ion (H^+) have been associated with fatigue. With the presence of oxygen, lactate may be removed as an aerobic substrate or, with adequate blood flow, shuttled to other, less active, muscle for metabolic removal. Collectively, whether the substrate is glycogen, glucose, or both, this system is typically termed anaerobic or fast glycolysis.

When the rate of energy conversion is slower and adequate blood flow and oxygen are available to the muscle, the glycolytic process may proceed through a number of additional steps aerobically, without lactate accumulation. This aerobic oxidative metabolic system may also utilize fats and protein substrates and help spare the limited stores of carbohydrate within the muscle.

Table 4.3 provides a summary of the three bioenergetic systems. Along with the rate of energy expenditure demand, a climber's ability to take in, circulate, and utilize oxygen during a performance generally controls which bioenergetic system is primarily involved.

Table 4.3 Summary of bioenergetic system support of muscular work in climbing

Bioenergetic system	Substrate	Oxygen requirement	Power (energy/time)	Capacity (total energy available)
Phosphagen	ATP & PCr	None	Highest	Limited by muscle PCr stores. Typically enables 5–8 sec of maximal total-body exertion
Anaerobic (fast) glycolysis	Glycogen and glucose	None	High	Moderate capacity limited primarily by unbuffered H^+ ions dissociated from lactic acid. Requires adequate blood flow and oxygen for recovery
Oxidative	Glycogen, glucose, fatty acids, proteins	Required	Moderate	High capacity limited by substrate availability

Assessment of the power output capability of the aerobic, or oxidative, bioenergetic system is accomplished by measuring a climber's maximum ability to uptake and utilize oxygen (VO_2max). Since an individual's VO_2max can differ among different activities and with the degree of active muscle mass, the highest VO_2 attained during a specific activity mode, such as climbing, is usually referred to as a peak oxygen uptake or VO_2pk.

Studies have reported VO_2pk values for rock climbers of 54.8 ± 5.0 mL kg^{-1} min^{-1} [9] and 55.2 ± 3.6 mL kg^{-1} min^{-1} [10] for treadmill running. Billat et al. [9] reported a VO_2pk of 22.3 ± 2.6 mL kg^{-1} min^{-1} for an arm pulling test, and Booth et al. [11] described a mean of 43.8 ± 2.2 mL kg^{-1} min^{-1} during fast climbing in seven highly skilled climbers. España-Romero et al. [12] utilized a special climbing treadmill (treadwall) to observe climbing specific VO_2 peaks of 53.6 ± 3.7 and 49.2 ± 3.5 in male and female climbers, respectively, during fast climbing.

The VO_2pk levels observed in climbers would be considered "excellent" for general aerobic fitness but are well below the typical levels of 70–90 mL kg^{-1} min^{-1} observed in elite aerobic athletes such as distance runners and cross-country skiers. The lower values for climbers likely reflect a smaller intensely activated muscle mass during the work of climbing compared with the more total-body involvement of other endurance-type competitive activities.

Regardless of the lower total-body VO_2pk for climbers, the specific aerobic power and capacity of localized muscle groups have been found to be high. Fryer et al. [13] estimated an oxidative capacity index for forearm musculature in climbers through near-infrared spectroscopy (NIRS). This oxidative capacity index and maximal hemoglobin-myoglobin desaturation and VO_2pk during treadwall climbing were significant predictors of climbing ability and explained over 67% of the variance in RP climbing ability.

In addition to describing the maximal and peak bioenergetic characteristics of climbers, early research began to observe physiological responses and demands of actual performance during climbing. Billat et al. [9] and Mermier et al. [14] used Douglas bags to collect expired air for VO_2 analysis during route climbing. Watts and Drobish [10] employed a nonmotorized climbing treadmill (Brewer's Ledge Treadwall®) to record the first continuous VO_2 measurements during climbing at different angles. These studies found the average climbing VO_2 to range between 24 and 32 mL kg^{-1} min^{-1} regardless of terrain angle.

As portable expired air analysis systems became available in the 1990s, researchers began to observe physiological responses continuously during actual climbing on indoor artificial walls and outdoors on real rock. Watts et al. [15] found average and peak VO_2 of 24.7 ± 4.3 and 31.9 ± 5.3 mL kg^{-1} min^{-1}, respectively, during ascents of a competition style indoor route rated IRCRA 19. Booth et al. [11] recorded a mean VO_2 of 32.8 mL kg^{-1} min^{-1} during ascents of an outdoor route. In a more recent study, España-Romero et al. [6] recorded peak VO_2 levels of 36.9 ± 4.9 mL kg^{-1} min^{-1} during on-sight ascents of a moderate grade indoor route rated as 10 on the IRCRA scale.

In studies involving route ascents, VO_2 tends to plateau after 1.5–2.0 min of climbing, which suggests a steady-state condition is attained. A general integration of results from studies that assessed VO_2pk during maximal testing and those that measured VO_2 during typical route climbing tasks indicates the stress of actual climbing likely requires approximately 70% of VO_2pk, which would be compatible with a steady-state primarily aerobic condition.

The idea of a steady-state condition during route ascents has been challenged by results that indicated significantly elevated post-climbing VO_2 [15] and elevated blood lactate concentration [9, 15, 16]. Although blood lactate is elevated with climbing, post-ascent levels have been relatively low with observed means of 3.2–7.0 mmol L^{-1}. Furthermore, Watts et al. [15] found pre-climb values of 3.5 ± 1.9 mmol L^{-1} in climbers who felt "warmed-up" and ready to climb a competition style route.

Bertuzzi et al. [17] have attempted to rate the relative contributions of the three bioenergetic systems for energy expenditure during routes of different difficulty levels. By their estimation, the contributions of the ATP-PC phosphagen, anaerobic glycolytic, and aerobic oxidative systems for elite climbers were 35.8 ± 6.7%, 22.3 ± 7.2%, and 41.9 ± 7.4% for easy (IRCRA 10), moderate (IRCRA 15), and difficult (IRCRA 19) routes, respectively. Although climbing demands the full range of bioenergetic support, the phosphagen and aerobic systems appear to be dominant.

4.2.3 Muscular Strength, Endurance, and Power

Factors of muscular performance have intuitively been of interest to climbers. These factors may be categorized into strength, the ability of muscle to generate high force; endurance, the ability of muscle to perform repeated contractions or to sustain a contraction level over time; and power, the level of force production relative to time and/or velocity of movement.

Force production by muscle is a function of the activation of motor units within a muscle. A motor unit consists of a specific neuron and the number of muscle fibers it activates. The number of motor units within a given muscle and the number of fibers within a motor unit vary across different muscles and for different individuals and are considered to be primarily determined by genetic factors. Conversely, the size of muscle fibers within a motor unit can be increased through specific training.

Relative to climbing, muscle contractions may be categorized as isotonic, where the muscle changes length either concentrically, by shortening, or eccentrically, by lengthening under load, or as isometric where muscle length remains static during force production.

Dynamic muscular force is expressed by a force-velocity relationship. On a fiber basis, the absolute contraction force declines with increasing velocity of contraction. The highest velocity of muscle fiber contraction is typically generated at 30–40% of the fiber's peak force capability, and the highest force is produced at zero velocity during isometric or static contraction. Since establishing and maintaining contact with a specific hold in climbing is primarily a static task, isometric strength of the muscle controlling the fingers of climbers was of early interest.

The most obvious target for descriptive strength research with elite climbers has been isometric handgrip force, typically measured with a grip dynamometer, Fig. 4.4a. Watts et al. [10, 16] found absolute handgrip scores to be rather "average" in elite climbers; however, when handgrip strength was expressed as strength–body mass ratio, males placed at the 80th percentile and females placed at the 90th percentile for age and gender matched North American population norms. This finding of a high strength–mass ratio for accomplished climbers has been consistent across other studies as well.

In the 1990s, researchers began to question the specificity of handgrip dynamometry for measuring finger and hand strength in climbers. Simple observation of the hand-to-rock contact interface in climbing revealed minimal use of finger-thumb opposition during ascents of the steep edging routes popular at the time. In response to the potential limitation of handgrip dynamometry, Grant et al. [18] and Watts and Jensen [19] were among the first to construct strain gauge or force sensor devices to enable measurement of finger-curl force without thumb opposition, Fig. 4.4b, though these researchers did not measure values in elite-level climbers.

During actual pull-type movements in climbing, the finger-curl force required to maintain contact with holds is less likely to be of a concentric nature as an eccentric nature as the fingers work to resist extension. Ozimek et al. [20] measured finger force from a more eccentric contraction perspective as the total force, body weight plus added weight, a climber could support while hanging from the fingers on a 2.5 cm edge. Much higher force, 1266.3 ± 147.2 N in this study, may be generated during such hangs than during standard concentric format dynamometry.

A summary of selected research relative to hand and finger force measures is presented in Table 4.4.

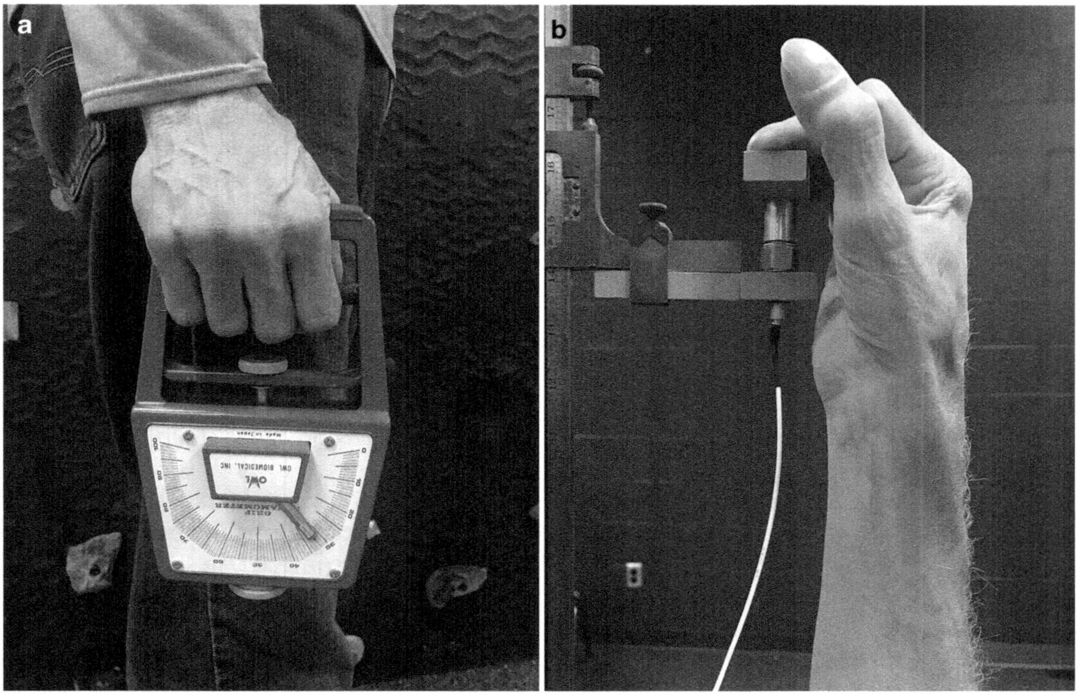

Fig. 4.4 Historical methods of measuring "grip" strength in climbers. (**a**) standard handgrip dynamometer and (**b**) plate and force sensor to measure finger curl force without opposition of the thumb

Table 4.4 Selected measures of hand and finger strength reported for high-level rock climbers

Reference	Subject characteristics	Measurement technique	Strength (N)	Strength–mass ratio
Males:				
Watts et al. [5]	7 males IRCRA 26 (mean RP[a])	Handgrip dynamometry	477.7 ± 89.3	0.78 ± 0.13
Watts et al. [16]	11 males IRCRA 23 (mean RP)	Handgrip dynamometry	581.6 ± 69.6	Not reported
Watts et al. [15]	15 males IRCRA 20–27 (range)	Handgrip dynamometry	507.2 ± 73.6	0.77 ± 0.07
Levernier and Laffaye [26]	14 males IRCRA 25 (≥)	"Power grip" sensor—slope crimp position	476.6 ± 74.4 521.6 ± 36.2	0.72 ± 0.05 0.83 ± 0.05
Ozimek et al. [20]	6 males IRCRA 25–27 (range)	Weighted hang on 2.5 cm edge	1266.3 ± 147.2	1.93 ± 0.20
Females:				
Watts et al. [5]	6 females IRCRA 23 (mean RP)	Handgrip dynamometry	297.2 ± 30.4	0.65 ± 0.04
Giles et al. [7]	14 females IRCRA 19.5 (mean RP)	Finger force on 20 mm edge	408.4 ± 62.3	Not reported

[a]RP is redpoint ascent

The high strength values in climbers may also be significant due to the relationship of strength to submaximal muscular contraction endurance. As specific maximum voluntary contraction force increases, a given absolute contraction force will be at a lower percentage of the new maximum voluntary contraction force, and an increase in endurance measures would be expected. The noted cardiovascular factors previously discussed would also contribute to increased muscular endurance, particularly for repeated contractions.

Most research has assessed muscular endurance as isometric contraction holding time at a given submaximal percentage of maximum force or as the time to maintain repeated timed contractions at a given percentage of maximum force. Limited early studies consistently found hand and upper body muscular endurance to be higher in accomplished climbers than recreational climbers or non-climbers [18, 21].

Watts et al. [16] found handgrip endurance to be impacted greater than handgrip strength with climbing to the point of failure. This group studied grip strength and endurance before and after climbers performed continuous climbing on difficult terrain, rated 18 on the IRCRA scale, to the point of a fall. In this study, maximum voluntary contraction (MVC) force decreased by 22% and holding time at 70% of the pre-climb MVC force decreased by 57%.

Ferguson and Brown [22] found experienced climbers to have double the endurance time for 40% maximum handgrip during rhythmic maximum contractions than sedentary non-climbers. Furthermore, a significantly enhanced forearm vasodilator capacity was found in the climbers in this study, which would enhance the ability to recover between contractions via aerobic metabolism.

Muscular power has also been of interest, and most assessments in climbers have looked at high velocity explosive power with body weight load during dynamic moves. Laffaye et al. [23] used accelerometers to assess upper body power in boulderers and lead climbers during an explosive pull and release vertical dynamic move off large "jug" holds. Relative power, in Watts per kilogram body mass, was 28.4 ± 7.55 and 23.4 ± 3.7 W/kg in boulderers and lead climbers, respectively.

Giles et al. [7] have recently described muscular power characteristics in high-level female climbers. This group measured lower body power via a counter movement vertical jump test and upper body power via a power slap test, a type of explosive pull-up movement to a one-hand slap of a high point. The mean power slap score was significantly higher for an elite group (n = 14, ability = IRCRA 19.5 ± 3.1) than for an advanced lower ability group (n = 13, ability = IRCRA 15.9 ± 1.4). Lower body power was not significantly different between the ability groups.

4.3 Prediction of Climber Performance Via Physiological Characteristics

Often, research on rock climbers has associated various anthropometric and physiological measures with self-reported RP climbing ability of the research participants. Significant predictors of performance have been described; however, many studies have involved participants of a wide range of abilities. General physical fitness characteristics can be very predictive within groups with a range of abilities from novice to expert. As the ability range becomes more homogeneous, such as with a group of elite level climbers, the physiological predictors of performance likely become more narrow and specific. This phenomenon has been observed for other athlete groups. In a classic study of marathon runners, Sjödin and Svedenhag [24] found VO_2max to significantly correlate with marathon race pace with a correlation of $r = 0.78$ in 75 runners with marathon paces between 3.0 and 5.4 m/s (42 km race times of 4:00 and 2:16 h:min); however, the correlation was reduced to a nonsignificant $r = 0.01$ for fast runners with paces between 4.7 and 5.4 m/s (race times between 2:30 and 2:16).

In the study of elite competition climbers of Watts et al. [5], stepwise regression analysis was employed with stature, body mass, stature to mass ratio, sum of seven skinfolds, estimated body fat percentage, grip strength, strength to mass ratio, and arm volume as independent variables and self-reported red point ability as the dependent variable. Ability in 39 climbers with an ability range of approximately IRCRA 21–27 was significantly predictable with strength to mass ratio and percent body fat accounting for 33% of the variance in ability.

Baláš et al. [25] studied 205 sport climbers, 136 males and 69 females, within an ability range of IRCRA 5–29, with an array of anthropometric and strength measures along with experience and training volume. Across this broad ability range, grip strength to weight ratio and hand/arm endurance along with estimated percent body fat were good predictors of performance.

Mermier et al. [14] published the first attempt at associating climber characteristics with actual measured climbing performance. Forty-four climbers attempted two artificial routes as on-sight top-roped climbing. Handhold contacts, or *moves*, on each route progressively increased in difficulty and maintenance of contact with successive handholds scored points. Together, the routes involved 63 *moves* up to IRCRA 22 level. A principal components analysis reduced the large number of measured variables to three components: training, anthropometry, and flexibility. Subsequent regression analysis found 39% of performance variance to be explained by the trainable factors, 15% by anthropometry, and 10% by flexibility. This study also involved climbers of a broad range of abilities, IRCRA 6–24, and did not look at a more homogeneous sample at an elite ability level.

In a more recent study, MacKinzie et al. [8] assessed 47 variable scores for 44 males (IRCRA 9–23) and 33 females (IRCRA 9–20). Test areas included were anthropometry; balance; muscular strength, endurance, and power; aerobic power (as running VO_2max and arm-cranking VO_2peak); and hand-eye and foot-eye spatial coordination. For this broad range of climbing ability levels, shoulder power and endurance assessed as maximum pull-ups, arm-crank power, and timed bent-arm hang were the best predictors of performance.

Based on the integration schematic of Fig. 4.1, Watts [1] (2004) proposed an initial performance model for difficult climbing performance with the following components:

- Small stature and high strength–mass ratio. This would be especially important for strength of the musculature that controls hand and finger positions. A low body fat percentage may also contribute to strength–mass ratio, although extreme reductions in body fat can have negative consequences for health.
- High level of isometric muscle contraction endurance, particularly for repeated contractions in the musculature that controls hand and finger positions.
- Moderately high total-body aerobic power (VO_2max ≈ 50–55 mL kg^{-1} min^{-1} for males).

Subsequent research has suggested additional components for the model:

- High capacity of phosphagen, ATP and PCr, energy yield.
- High vasodilator capacity and high oxidative power and capacity of the forearm musculature.
- High upper body explosive power.

References

1. Watts PB. Physiology of difficult rock climbing. Eur J Appl Physiol. 2004;91(4):361–72.
2. Saul D, Steinmetz G, Lehmann W, Schilling AF. Determinants for success in climbing: a systematic review. J Exerc Sci Fit. 2019;17(3):91–100.
3. Draper N, Giles D, Schöffl V, Fuss K, Watts P, Wolf P, et al. Comparative grading scales, statistical analyses, climbers descriptors and ability grouping: International Rock Climbing Research Association position statement. Sports Technol. 2016;8(3–4):88–94.
4. The Hardest Sport Climbs in The World, Rock and Ice. [Internet]. 2020 [cited Accessed 9 August 2020]. https://rockandice/inside-beta/the-hardest-sport-climbs-in-the-world/.
5. Watts PB, Martin DT, Durtschi S. Anthropometric profiles of elite male and female competitive sport rock climbers. J Sports Sci. 1993;11(2):113–7.
6. España-Romero V, Jensen RL, Sanchez X, Ostrowski ML, Szekely JE, Watts PB. Physiological responses in

rock climbing with repeated ascents over a 10-week period. Eur J Appl Physiol. 2012;112(3):821–8.
7. Giles D, Barnes K, Taylor NF, Chidley J, Mitchell K, Torr O, et al. Anthropometry and performance characteristics of recreational advanced to elite female rock climbers. J Sport Sci. 2020;18:1–9.
8. MacKenzie R, Monaghan L, Masson RA, Werner AK, Caprez TS, Johnston L, et al. Physical and physiological determinants of rock climbing. Int J Sports Physiol Perform. 2019:1–12.
9. Billat V, Palleja P, Charlaix T, Rizzardo P, Janel N. Energy specificity of rock climbing and aerobic capacity in competitive sport rock climbers. J Sports Med Phys Fitness. 1995;35(1):20–4.
10. Watts PB, Drobish KM. Physiological responses to simulated rock climbing at different angles. Med Sci Sports Exerc. 1998;30(7):1118–22.
11. Booth J, Marino F, Hill C, Gwinn T. Energy cost of sport rock climbing in elite performers. Br J Sports Med. 1999;33(1):14–8.
12. España-Romero V, Ortega Porcel FB, Artero EG, Jimenez-Pavon D, Sainz G, Castillo Garzon MJ, et al. Climbing time to exhaustion is a determinant of climbing performance in high-level sport climbers. Eur J Appl Physiol. 2009;107(5):517–25.
13. Fryer SM, Giles D, Palomino IG, de la OPA, Espana-Romero V. Hemodynamic and cardiorespiratory predictors of sport rock climbing performance. J Strength Cond Res. 2018;32(12):3534–41.
14. Mermier CM, Janot JM, Parker DL, Swan JG. Physiological and anthropometric determinants of sport climbing performance. Br J Sports Med. 2000;34(5):359–65; discussion 66.
15. Watts PB, Daggett M, Gallagher P, Wilkins B. Metabolic response during sport rock climbing and the effects of active versus passive recovery. Int J Sports Med. 2000;21(3):185–90.
16. Watts P, Newbury V, Sulentic J. Acute changes in handgrip strength, endurance, and blood lactate with sustained sport rock climbing. J Sports Med Phys Fitness. 1996;36(4):255–60.
17. Bertuzzi RC, Franchini E, Kokubun E, Kiss MA. Energy system contributions in indoor rock climbing. Eur J Appl Physiol. 2007;
18. Grant S, Hynes V, Whittaker A, Aitchison T. Anthropometric, strength, endurance and flexibility characteristics of elite and recreational climbers. J Sports Sci. 1996;14(4):301–9.
19. Watts PB, Jensen RL. Reliability of peak forces during a finger curl motion common in rock climbing. Meas Physical Ed Exer Sci. 2003;7(4):263–7.
20. Ozimek M, Rokowski R, Draga P, Ljakh V, Ambroży T, Krawczyk M, et al. The role of physique, strength and endurance in the achievements of elite climbers. PLoS One. 2017;12(8):e0182026.
21. Cutts A, Bollen SR. Grip strength and endurance in rock climbers. Proc Inst Mech Eng H. 1993;207(2):87–92.
22. Ferguson RA, Brown MD. Arterial blood pressure and forearm vascular conductance responses to sustained and rhythmic isometric exercise and arterial occlusion in trained rock climbers and untrained sedentary subjects. Eur J Appl Physiol Occup Physiol. 1997;76(2):174–80.
23. Laffaye G, Collin JM, Levernier G, Padulo J. Upper-limb power test in rock-climbing. Int J Sports Med. 2014;35(8):670–5.
24. Sjödin B, Svedenhag J. Applied physiology of marathon running. Sports Med. 1985;2:83–99.
25. Baláš J, Pecha O, Martin AJ, Cochrane D. Hand–arm strength and endurance as predictors of climbing performance. Eur J Sport Sci. 2012;1:16–25.
26. Levenier G, G Laffaye. Four weeks of finger grip training increases the rate of force development and the maximal force in elite and top world-ranking climbers. J Strength Cond Res. 2019;33(9):2471–2480.

5 Imaging of Climbing Injuries

T. Bayer

Jens Liße in „Stalingrad direkt", Frankenjura, Germany, Photo Michael Simon

T. Bayer (✉)
Department of Radiology and Neuroradiology, Klinikum Fürth, Fürth, Germany
e-mail: thomas.bayer@klinikum-fuerth.de

V. Schöffl et al. (eds.), *Climbing Medicine*, https://doi.org/10.1007/978-3-030-72184-8_5

Radiological imaging is essential for correct diagnosis of climbing injuries. Classical X-ray procedures and computed tomography (CT) as cross-sectional imaging are excellent for detecting bony changes, such as degeneration, fracture, or avulsion. However, they have limited soft tissue contrast and are associated with exposure of the patient to ionizing radiation. Sonography is widely applied, is easy to learn, and can be performed functionally during the patient's movement. It is suited for the primary diagnosis of soft tissue injuries caused by climbing, such as the pulley rupture. Magnetic resonance imaging (MRI) offers better soft tissue contrast than any other cross-sectional imaging and is established for further diagnosis of numerous musculoskeletal issues in climbing.

5.1 X-ray

5.1.1 X-ray Procedures for the Diagnosis of Climbing Injuries

The X-ray procedures used in diagnostics today include classical projection radiography (synonymous with "conventional X-ray" or "classic X-ray") and other procedures derived from it, such as fluoroscopy or computed tomography (CT). All procedures are suited for the diagnosis of bone pathologies but with limited soft tissue contrast compared to sonography or MRI. Regarding climbing, X-ray and CT are particularly important; fluoroscopy is less so.

In all acute trauma, e.g., a fall or a major distortion, the primary goal is to detect an acute bony pathology (i.e., fracture). In these cases, conventional X-ray is the primary imaging method of choice. However, some anatomical regions may be difficult to assess with X-ray because of the superimposition of many small or complex bony structures, for example, the wrist or the tarsus. In these regions, additional X-ray projections with special settings (e.g., so-called scaphoid targeting at the wrist, Figs. 5.12, 5.13, 5.14, and 5.15) may be useful. When X-ray diagnosis remains unclear and the patient is still suspicious for fracture, additional cross-sectional imaging should be performed. CT is primarily indicated, allowing exact fracture diagnosis in more difficult anatomical regions.

In patients without acute trauma and with gradual developing complaints, X-ray and CT are less important, as these issues are mostly caused by soft tissue injuries. In most cases, a fracture may already be excluded clinically. Nevertheless, X-ray is part of the basic examination. The primary aim is to rule out chronic bone pathology, such as degeneration, osteophyte formation, and soft tissue calcification, but also bone tumors, bony cysts/ganglia, or systemic changes.

One major exception are stress fractures, which may result from chronic repetitive microtrauma, such as in the hamate hook [1, 2] or the middle finger growth plate of teenagers [3]. Stress fractures of the hand are a unique characteristic in climbing and are best detected with MRI. Trabecular microfracturing is usually too small to be visible with X-rays or CT but can be detected due to an accompanying bone marrow edema with improved soft tissue contrast during MRI.

5.1.2 Tips for X-ray Procedures

X-ray examinations should be performed by trained technicians. Standard images are defined for almost every joint and every region of the body and are a basic requirement for reliable imaging. Long-term expertise is recommended to learn proper radiological assessment. An open approach to X-ray imaging in the everyday clinical routine is useful. We encourage a regular discussion of findings with partners in radiology, orthopedy, and trauma surgery. X-ray image quality should be adequate, and standards of adjustment techniques are necessary for generating proper imaging. Typically, X-ray in two planes consists of a correct projection in an anterior-posterior and lateral plane with clear identification of joint surfaces. Anatomically complex regions, such as the wrist, require special imaging, such as the scaphoid target image or the ball image. If the bone findings are unclear, a supplementary CT is usually required.

5.1.3 Fundamentals of X-rays

All X-ray procedures rely on weakening of radiation in the human body. This is greater in thicker and denser objects and increases with the charge of atomic nucleus of tissue. With different degrees of attenuation, an image of the inside of the body can be taken, and thus, for example, bones can be distinguished from soft tissue. In conventional X-rays, three-dimensional objects are summed up in the direction of the projection plane. Thus, the result of X-ray imaging is two-dimensional. In consequence, injuries are often not accessible in only one projection plane. A dislocated fracture, for example, may not be distinguished in anterior-posterior projection, whereas this would be visible in lateral projection.

CT is also based on X-ray weakening, with the difference that an X-ray tube rotates around the human body during imaging, thus producing cross-sectional images without any overlap. Based on the acquired radiation attenuation per angle of rotation, density values can be assigned to each volume unit (voxel) by a computer calculation. These density values can be displayed in grayscale images along any slice plane without overlapping, including 3D techniques (e.g., with so-called volume rendering views, Fig. 5.18; maximum intensity projections, Fig. 5.17; or multiplanar reconstructions, Fig. 5.16). Due to the exposure of the patient to ionizing radiation, the operation of all X-ray-based systems is usually subject to strict safety regulations.

5.2 Sonography

5.2.1 Sonography for the Diagnosis of Climbing Injuries

Sonography (ultrasound or "US") is the standard first-line imaging modality for diagnosing climbing-associated soft tissue injuries. The major advantage is that it can be performed functionally, i.e., during movement or when the muscles and tendons are stressed. It is particularly suitable for diagnosing hand and finger pathologies due to the low tissue penetration depth required. At the same time, it enables high detail resolution. Regarding climbing injuries, US is also important for elbow and shoulder diagnostics. In principle, it can be applied to all joints and regions of the human body.

Functional US examination has become a well-established standard in the detection of pulley ruptures [4]. Bowstringing sign under forced finger flexion (Figs. 5.1 and 5.2) is a strong indicator for pulley rupture. By measuring the distance between bone and tendon, this sign allows definition of the severity and location of a pulley rupture. In trigger finger inflammation (tenosynovitis stenosans), increased friction or chronic inflammatory thickening of the pulleys and tendons may be imaged functionally during movement. Functional examination may be designed

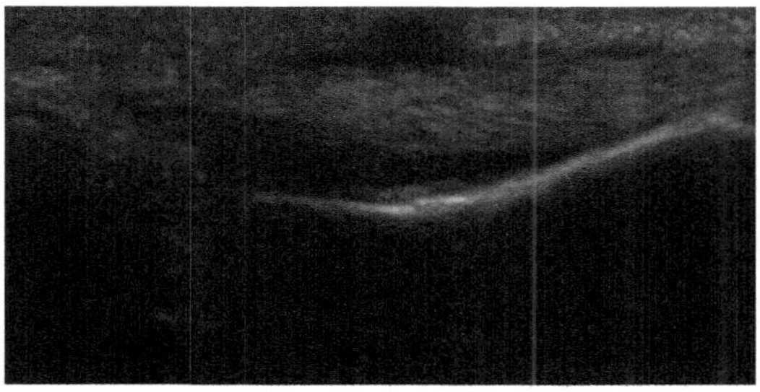

Fig. 5.1 Sonography under provocation, massive bowstringing over the A2 pulley

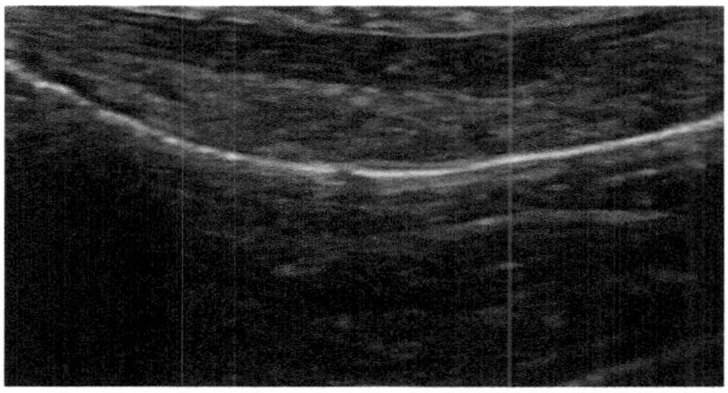

Fig. 5.2 Sonography under provocation, massive bowstringing over the A4 pulley

individually, dependent on the joint and the patient clinics. For shoulder rotator cuff injuries, a wide range of functional sonographic examination techniques has been described.

Fluid collections, tenosynovitis, and ganglion cysts can be identified by the characteristics of echo-free (black) water with dorsal sound amplification (white in the image). Bones and calcifications can be identified by total reflection. This means that the surface of the corresponding structure appears white, whereas the zone behind the object appears black. Ultrasound allows the detection of muscular injuries, such as fiber tear, bleeding, or discontinuity with definition of the resulting severity. The intramuscular viscoelastic tissue characteristic may be examined with the so-called elastography. Small image shifts during pressure application with the transducer probe are evaluated in a computer analysis. As a result, the stretching ability of the muscles can be displayed in spatial resolute image. A particularly well-established sonographic application is so-called duplex sonography, in which the blood flow within vessels or tissue is displayed in color-coded images using the Doppler effect.

5.2.2 Tips for Sonography

Ultrasound devices may be compact and easily portable resembling laptop design. This allows acute trauma examination, for example, during a competition. We recommend modern ultrasound devices in combination with high-resolution linear transducer probes. High megahertz transducers (typically 10–20 MHz) are particularly suitable for hand and finger examinations. Sonography depends on the individual expertise of the examiner; thus, proper training is helpful. Finger examinations benefit from short linear transducers with a small contact surface, as this hardly impairs the patient's freedom of movement during functional examinations. Mechanical sound wave coupling is crucial for good results, and air between the patient and the transducer probe must be avoided. In order to ensure sound coupling, ultrasound gel can be used, whereas functional hand and finger examinations can also be performed elegantly in a water bath. We recommend a standardized setup, in which the patient and examiner sit opposite each other with the patient's hand on a stable examination table.

5.2.3 Fundamentals of Sonography

Ultrasonic waves are mechanical vibrations that propagate in the human body in a longitudinal direction in the form of compression and decompression. When passing through tissue, the sound waves are weakened by various effects, such as absorption, refraction, scattering, divergence, and reflection. The latter effect is particularly noticeable at tissue interfaces with different acoustic properties. The sound waves are particularly strongly reflected at locations with so-called high acoustic impedance. The reflected sound waves are converted into electrical signals in a specially arranged system of piezoelectric elements in the transducer probe. Based on these signals, the location-coded image is reproduced by a computer in real time as a high-contrast image corresponding to the acoustic tissue properties. The power of ultrasound waves increases with increasing frequency, while, at the same time, the penetration depth decreases.

5.3 Magnetic Resonance Imaging (MRI)

5.3.1 MRI for the Diagnosis of Climbing Injuries

Magnetic resonance imaging is particularly important for further diagnostics of climbing injuries. It has a better potential to depict pathological and post-traumatic soft tissue conditions with higher spatial resolution and contrast than any other imaging modality. The slice direction can be chosen freely in any plane and/or in three dimensions. Numerous special MRI techniques are available, including application of intravenously or intra-articular contrast agent. MRI is suitable for all joints and regions of the body. Dedicated examination protocols with specially designed anatomical surface-receiving coils are available for almost every joint.

The MRI detectability of all climber-specific injuries of the hand and fingers is excellent. Among these are all flexor tendon injuries, extensor tendon injuries, tenosynovitis, capsular lesions, volar plate injuries, lumbrical muscle injuries, Dupuytren's disease, bony pathologies such as epiphyseal joint injuries, and many more.

Various MRI techniques for further diagnostics of the pulley system have been described. Pulley ruptures can be identified indirectly by bowstringing with the same approach as in sonography. In addition, however, the morphology and localization of the individual ruptures can be displayed directly. Direct pulley injury imaging requires particularly powerful MR scanners with strong magnetic field strength, enabling high spatial resolution [5, 6]. It allows definition of the exact location and shape of the rupture, which is advantageous for presurgical evaluation. Complications, such as a dislocation of a ruptured pulley ligament between flexor tendons and bone, are easy to detect. This has been described in literature as the so-called FLIP phenomenon (flap irritation phenomenon, Fig. 5.4).

The standard MRI is performed in a neutral position. However, functional examination is possible, requiring additional time (Fig. 5.5). For example, a hand may also be examined in the crimp-grip position [7]. Dynamic examinations of the fingers may be performed with MR cinematography [8]. This allows evaluation of further indirect physiological and pathological movement patterns of the tendons and the joints. This may be particularly useful regarding the volar plate and the A3 pulley. Despite not being standard, functional examinations may also contribute to better diagnostics in other joints, such as the shoulder for diagnosing instability (e.g., ABER positioning; abduction elevation external rotation).

MR arthrography (Fig. 5.21) requires administration of intra-articular contrast agent and enables a better assessment of injuries to the capsular ligaments. This technique is regularly performed for climbers. It is particularly useful for detecting injuries to the biceps tendon anchor and the glenoid labrum in the context of SLAP injury (superior labrum anterior to posterior). In the hip joint, it improves diagnostics of an femoroacetabular impingement (FAI).

Many other modern MR techniques are available, allowing a more profound and specific joint diagnostic. One example is cartilage imaging, in which the quality of cartilage is assessed compositionally, using mapping techniques with analysis of the cartilage quality. The wide range of all possible modern MR applications would go beyond the scope of this chapter.

5.3.2 Tips for MRI

A basic understanding of the different MRI sequences is useful. The specific image contrast depends on the sequence weighting (see Sect. 4.2.3). In T2-weighted sequences, fat and water are particularly rich in signal (hyperintense, bright on the image), whereas in T1-weighted sequences, only fat is rich in signal (note: T2, fat and water; T1 only, fat). Other tissues, such as tendons, capsules, connective tissue, and compact bone, are usually low in signal (hypointense, dark). A precise diagnosis of these structures requires improved techniques, such as proton density (PD)-weighted sequences. In PD imaging, both fat and water are rich in signal, but the detail recognition of the previously mentioned low-signal connective tissue structures is better. For musculoskeletal MRI and thus for diagnostics of climbing injuries, PD sequences should always be included in the examination protocol.

One common characteristic of all of these MRI sequences is a relatively bright image, representing hyperintense fat and water. In contrast, sequences can also appear almost completely dark. In these images, a fat saturation (fs) technique was probably applied. Fat saturation (fs) techniques can be combined with the previously mentioned bright standard sequences (corresponding sequence names, e.g., T1 fs, T2 fs, PD fs), which will increase the acquisition time slightly. The main advantage is the higher sensitivity for injuries as a bright area in the overall dark image. With some practice, even an inexperienced MRI viewer may be able to diagnose relevant musculoskeletal injuries on PD fs sequence (Fig. 5.8 and 5.9).

It is beneficial to include approximately the same number of sequences with and without fs in an MRI. The choice of correct slice plane (e.g., axial, coronal, sagittal, or 3D image) and slice thickness according to the expected injury is of utmost importance. The slice thickness should match the size of the joint. For example, a slice thickness of 2 mm is recommended for finger and wrist examinations, whereas a slice thickness of 3–4 mm is usually enough for larger joints, such as the hip or knee. As sport-specific injuries often result in subtle joint changes, application of modern MRI scanners with dedicated multichannel surface coils in combination with high magnetic field strength is recommended whenever possible. For instance, a 3 Tesla scanner is potentially superior to a 1.5 Tesla scanner, resulting in better image quality.

5.3.3 Basics of MRI

MRI is the result of nuclear physical effects in the permanently strong magnetic field of a superconducting magnet. Outside the MRI scanner, the spins of hydrogen atoms in the human body are mostly disordered. Under the influence of the strong magnetic field in the MRI, the nuclear spins are aligned, resulting in a measurable magnetization along the main magnetic field axis. The nuclear spins perform a directed "gyroscopic motion" along this axis; this is called "precession" in physics. The precession frequency increases proportionally with the magnetic field strength; it is 42.48 MHz/Tesla. For imaging, artificial attenuations of the magnetic field strength are performed with gradient coils in different spatial directions, which allows an exact location encoding of the MR signals. In addition, high-frequency (HF) pulses are transmitted with transmitter coils under resonance conditions. High-frequency pulses lead to a targeted "excitation" of the nuclear spins, resulting in longitudinal and transverse magnetization. After termination of an RF pulse, spins lose the absorbed energy by emitting the MR signal; this "relaxation process" can be detected with receiving coils.

MRI imaging is performed by repeated application of HF pulses with intermediate detection of MR signals. Such a sequence of excitation and relaxation of the spins is called an MR sequence. A spatial assignment of the received MR signal to a corresponding coordinate in the later slice image is done in complex computer calculations. Due to the different distribution of water and fat in the body and the different relaxation characteristics of different tissues, a particularly high-contrast image of soft tissues can be obtained with correspondingly differently designed MR sequences.

5.4 Typical Differential Diagnosis of Climbing Injuries

5.4.1 Pulley Rupture

A 23-year-old climber noticed a popping sound with acute severe pain in his right ring finger while climbing. This is a severe injury with the rupture of the A2, A3, and A4 ring ligaments (i.e., triple pulley rupture). The patient received a surgical pulley reconstruction due to the massive functional loss with risk of subsequent flexion contracture and the evidence of a flap irritation phenomenon (FLIP) (Figs. 5.1, 5.2, 5.3, 5.4, and 5.5).

5.4.2 Stress Fracture of the Growth Plate

A 14-year-old ambitious climbing athlete presented with pain and swelling over the PIP joint of the left middle finger, which gradually appeared over the course of several weeks. The diagnosis was a typical finding of an epiphyseal stress fracture of the middle phalanx base. A training stop for 3 months and cast splinting was performed. Since X-ray diagnostics could not show the stress fracture in this case, a follow-up

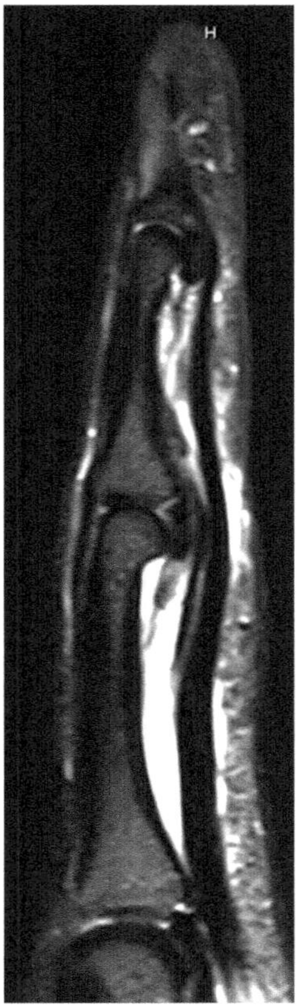

Fig. 5.3 Sagittal PD fs: triple rupture of the ligament with hematoma in the tendon sheath and bowstringing over the A2 and A4 pulleys. In the neutral position, no clear bowstring over the A3 pulley

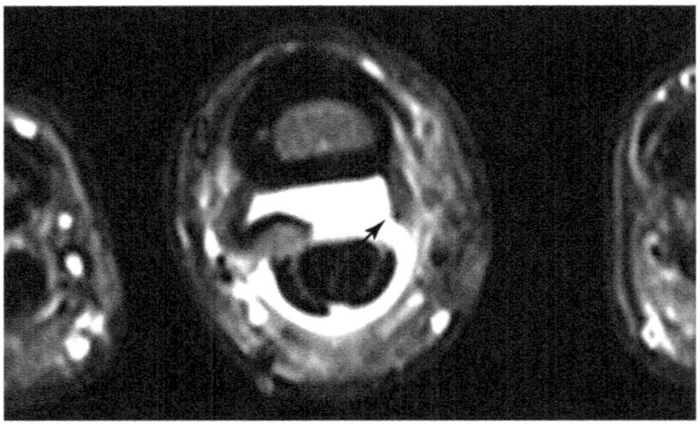

Fig. 5.4 Axial PD fs: A2 pulley rupture with dislocation of the ligamental stump between the flexor tendon and phalanx in the sense of a flap irritation phenomenon (FLIP)

MRI was performed to confirm fracture healing after 3 months (Figs. 5.6, 5.7, 5.8, and 5.9).

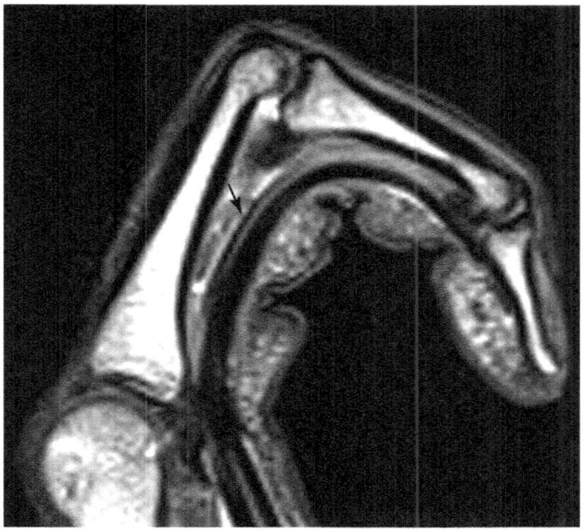

Fig. 5.5 Maximum finger flexion achievable by the patient in the MR functional examination with sagittal PD: triple (A2-A3-A4) pulley rupture with increased bowstringing over the phalanges and also over the A3 pulley (note the increased distance of the flexor tendons to the volar plate of the PIP joint)

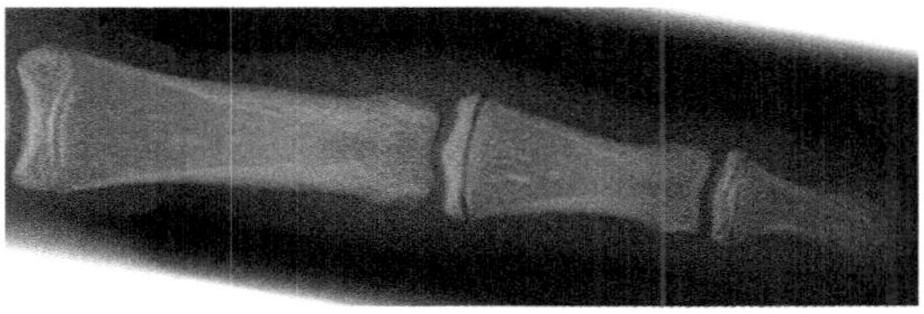

Fig. 5.6 X-ray in a.p. projection: soft tissue swelling over the PIP joint without clear evidence of fracture

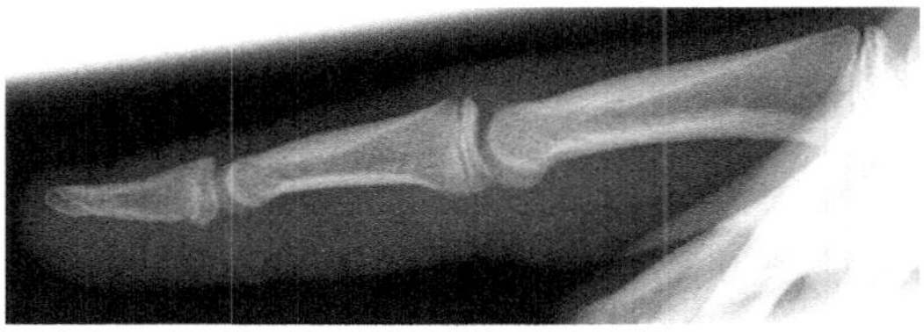

Fig. 5.7 Lateral X-ray projection: soft tissue swelling over the PIP joint without clear evidence of fracture

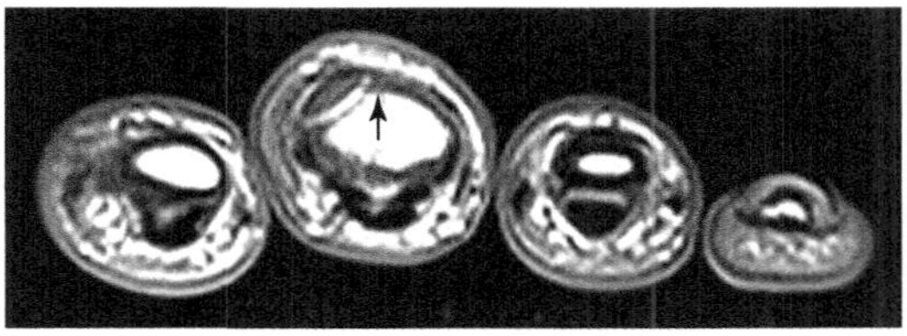

Fig. 5.8 Axial T1-weighed MRI: hyperintense fracture line in the middle phalanx base Dig. 3

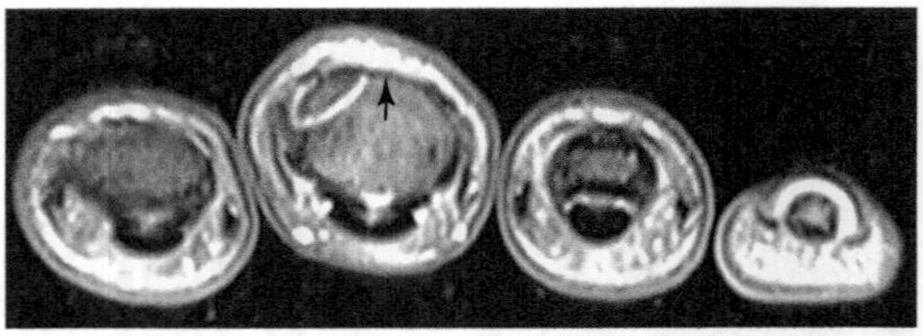

Fig. 5.9 Axial PD weighted fs MRI: hyperintense fracture line in the middle phalanx base Dig. 3, the fracture is easier to detect compared to Fig. 5.8 due to fat suppression (fs)

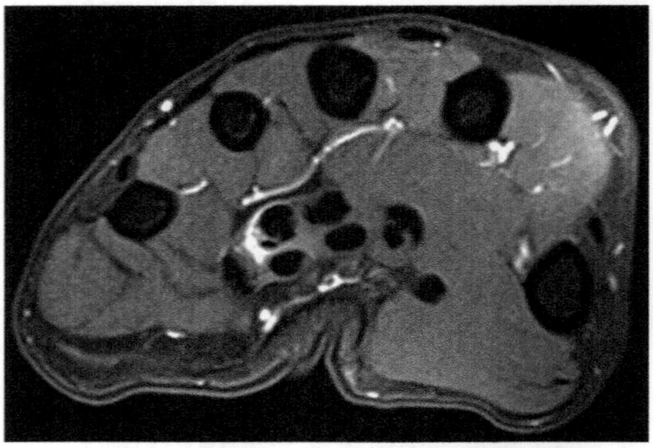

Fig. 5.10 Axial PDfs MRI: hyperintense hematoma surrounding the fourth flexor digitorum profundus tendon with tear of the lumbrical muscle tendon at the origin in the sense of a grade III lumbrical shift syndrome

5.4.3 Lumbrical Shift Syndrome

A 37-year-old female climber presented with acute pain and swelling in the palm of her hand on the right side above the flexor tendon of the ring finger. The injury healed conservatively within 2 months without functional sequel (Figs. 5.10 and 5.11).

5.4.4 Trabecular Microfracture of the Scaphoid

A 16-year-old climber fell on his wrist while bouldering. Unlike MRI, the scaphoid series and CT did not show this microfracture. The injury healed conservatively after 8 weeks in a cast (Figs. 5.12, 5.13, 5.14, 5.15, 5.16, 5.17, 5.18, and 5.19).

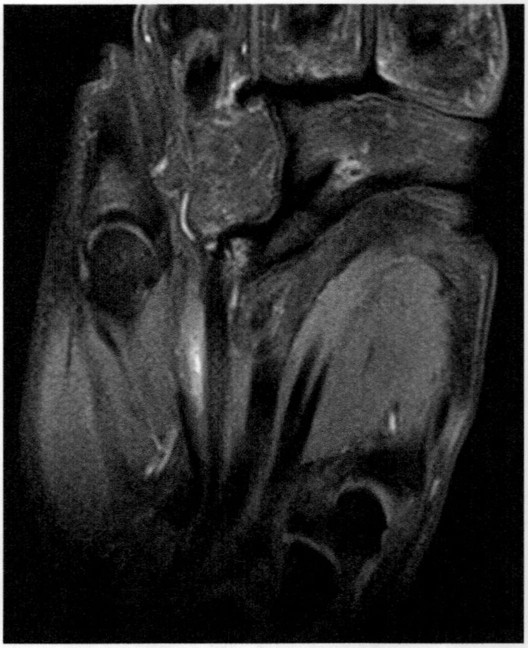

Fig. 5.11 Coronary PDfs MRI: hyperintense hematoma along the flexor digitorum profundus tendon Dig. 4 and the third lumbrical muscle

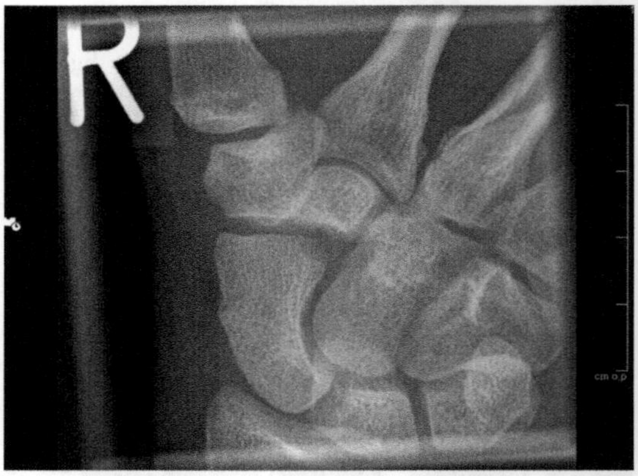

Fig. 5.12 Inconspicuous scaphoid series

5.4.5 MR Arthrography for Diagnosing Labrum Tear

A 33-year-old climber presented with chronic shoulder problems, pain during the night at rest, recurrent subluxation events, and a feeling of

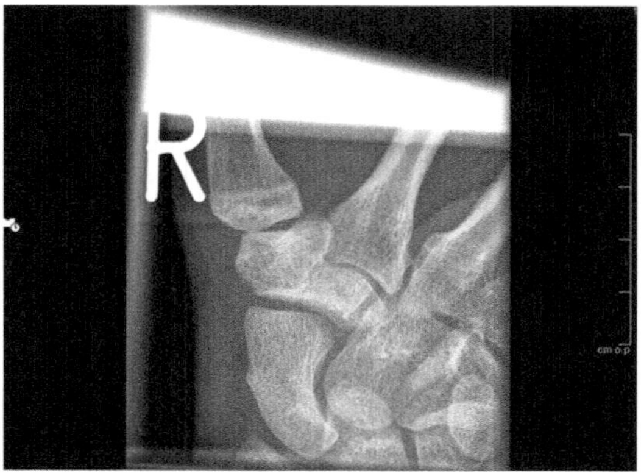

Fig. 5.13 Inconspicuous scaphoid series

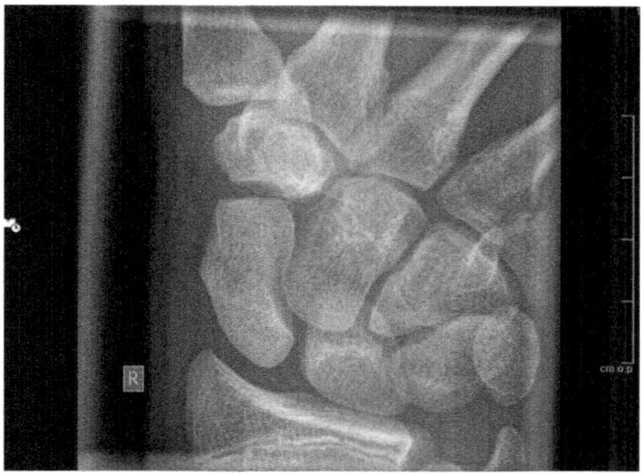

Fig. 5.15 Inconspicuous scaphoid series

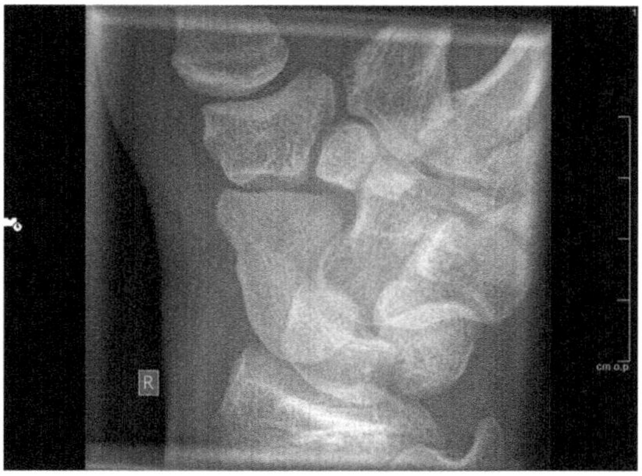

Fig. 5.14 Inconspicuous scaphoid series

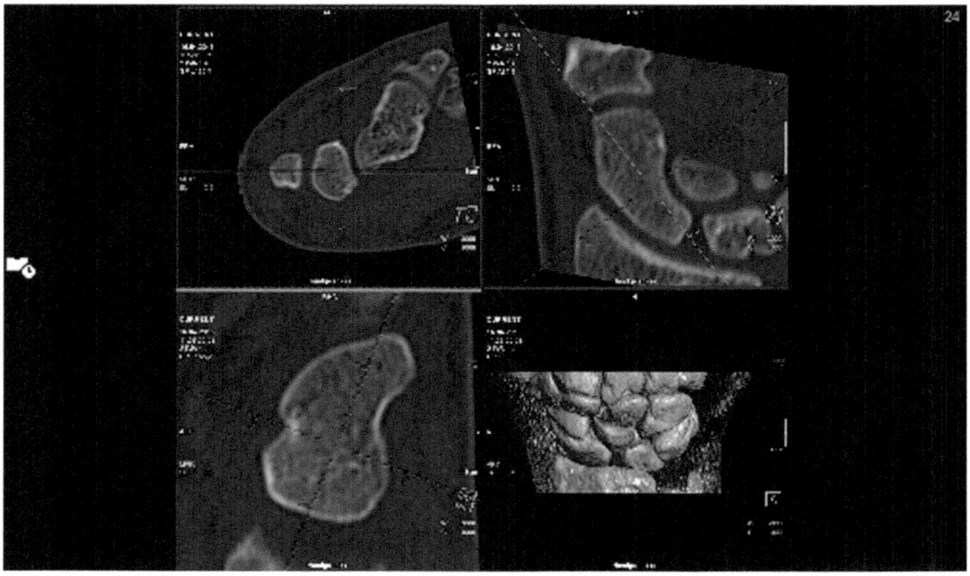

Fig. 5.16 Inconspicuous computed tomography: multiplanar reconstruction (MPR)

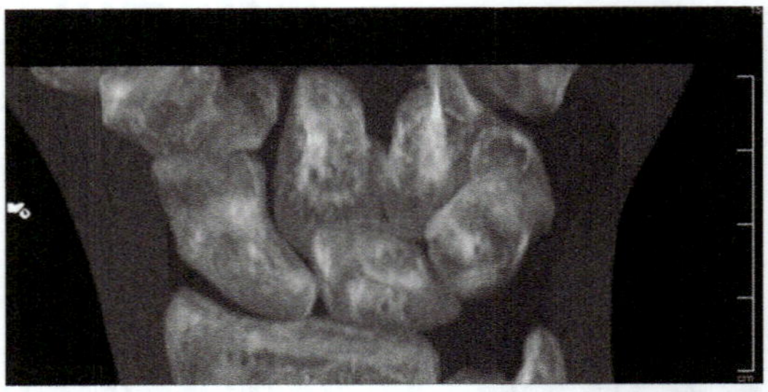

Fig. 5.17 Inconspicuous computed tomography: maximum intensity projection (MIP)

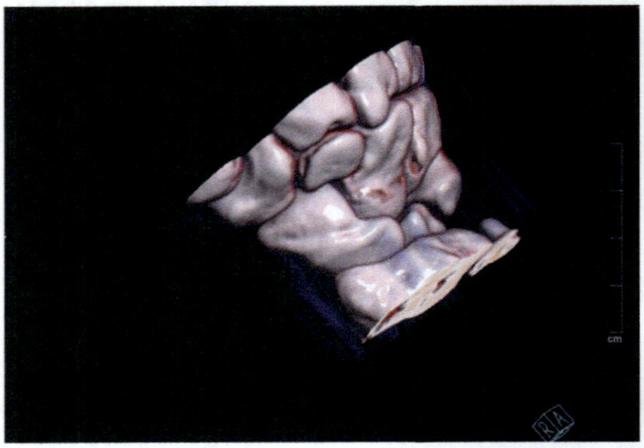

Fig. 5.18 Inconspicuous computed tomography: volume rendering

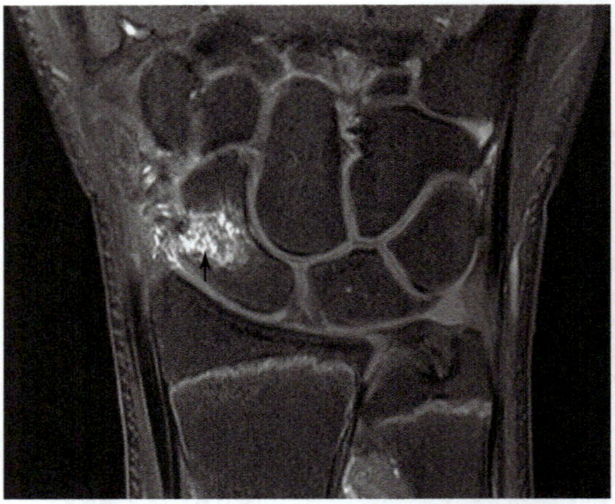

Fig. 5.19 Axial PDfs MRI: detection of a trabecular microfracture of the scaphoid with hyperintense blurred fracture zone

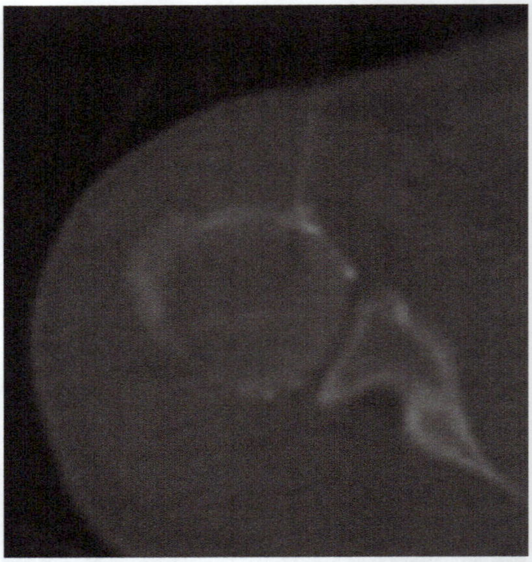

Fig. 5.20 CT-guided puncture of the right shoulder joint for application of MRI contrast agent

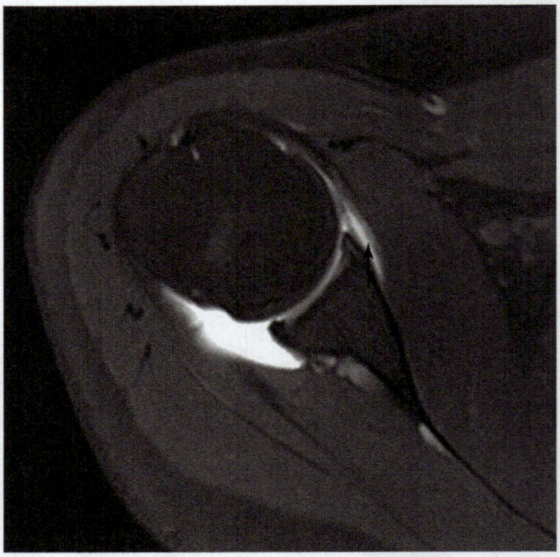

Fig. 5.21 MR arthrography with evidence of a tear in the ventral labrum without lifting of the periosteum, a so-called Perthes lesion

instability while climbing. The suspected labrum tear could be visualized particularly well in MR arthrography, where contrast agent had been applied intra-articularly previously. A surgical labrum repair was performed (Figs. 5.20 and 5.21).

References

1. Lutter C, Schweizer A, Hochholzer T, Bayer T, Schoffl V. Pulling harder than the hamate tolerates: evaluation of hamate injuries in rock climbing and bouldering. Wilderness Environ Med. 2016;27(4):492–9.
2. Bayer T, Schweizer A. Stress fracture of the hook of the hamate as a result of intensive climbing. J Hand Surg Eur. 2009;34(2):276–7.
3. Bayer T, Schoffl VR, Lenhart M, Herold T. Epiphyseal stress fractures of finger phalanges in adolescent climbing athletes: a 3.0-Tesla magnetic resonance imaging evaluation. Skelet Radiol. 2013;42(11):1521–5.
4. Klauser A, Frauscher F, Bodner G, Halpern EJ, Schocke MF, Springer P, et al. Finger pulley injuries in extreme rock climbers: depiction with dynamic US. Radiology. 2002;222(3):755–61.
5. Guntern D, Goncalves-Matoso V, Gray A, Picht C, Schnyder P, Theumann N. Finger A2 pulley lesions in rock climbers: detection and characterization with magnetic resonance imaging at 3 Tesla—initial results. Investig Radiol. 2007;42(6):435–41.
6. Goncalves-Matoso V, Guntern D, Gray A, Schnyder P, Picht C, Theumann N. Optimal 3-T MRI for depiction of the finger A2 pulley: comparison between T1-weighted, fat-saturated T2-weighted and gadolinium-enhanced fat-saturated T1-weighted sequences. Skelet Radiol. 2008;37(4):307–12.
7. Bayer T, Fries S, Schweizer A, Schöffl I, Janka R, Bongartz G. Stress examination of flexor tendon pulley rupture in the crimp grip position: a 1.5-Tesla MRI cadaver study. Skelet Radiol. 2014;
8. Bayer T, Adler W, Janka R, Uder M, Roemer F. Magnetic resonance cinematography of the fingers: a 3.0 Tesla feasibility study with comparison of incremental and continuous dynamic protocols. Skelet Radiol. 2017;46(12):1721–8.

Part II

Orthopedics of the Upper Extremity

Hand and Fingers

6

Volker Schöffl, Thomas Hochholzer, Y. El-Sheikh, and Christoph Lutter

"Half crimping" in Halfway Log Dump, Ontario, Canada, Photo Michael Simon

V. Schöffl (✉)
Department of Orthopedic and Trauma Surgery, Center of Sportsmedicine, Klinikum Bamberg, Bamberg, Germany

School of Clinical and Applied Sciences, Leeds Becket University, Leeds, UK

Section of Wilderness Medicine, Department of Emergency Medicine, University of Colorado School of Medicine, Denver, CO, USA

Department of Orthopedic and Trauma Surgery, Friedrich-Alexander Universität Erlangen-Nürnberg, Erlangen, Germany

T. Hochholzer
Private Clinic Hochrum, Innsbruck, Austria

Y. El-Sheikh
Department of Surgery, North York General Hospital, Toronto, ON, Canada

Division of Plastic & Reconstructive Surgery, Department of Surgery, University of Toronto, Toronto, ON, Canada

Climbing Escalade Canada, Ottawa, ON, Canada

C. Lutter
Department of Orthopedic Surgery, University Hospital Rostock, Rostock, Germany

V. Schöffl et al. (eds.), *Climbing Medicine*, https://doi.org/10.1007/978-3-030-72184-8_6

6.1 Clinical Examination of the Hand

After taking a thorough history of the presenting injury, a detailed clinical examination is crucial to arriving at an accurate diagnosis. The history should focus on what kind of climbing move and finger position lead to the trauma and clarify whether the pain arose before, during or after climbing (e.g., tenosynovitis of the finger is usually painful before and after climbing; this injury may be pain-free during climbing if warmed up properly. A pulley strain is usually painful only under stress when climbing using the crimp position of the fingers).

The clinical examination both actively and passively tests the range of motion (ROM) of all three joints of the respective finger: the metacarpophalangeal (MCP), the proximal interphalangeal (PIP), and the distal interphalangeal (DIP). For the thumb, the ROM is tested for the basal joint, MCP joint, and IP joint. The function of the flexor tendons, flexor digitorum superficialis (FDS), and flexor digitorum profundus (FDP) needs to be tested separately (Figs. 6.1 and 6.2). The collateral ligaments of the MCP, PIP, and DIP joints need to be tested both in extension and 30° flexion in order to assess the proper and accessory components of each ligament (Fig. 6.3). The palmar plate of the PIP joint is assessed through the so-called "translation test," which is similar to the Lachman test in the knee joint (Fig. 6.4). For the detection of flexor tendon bowstring phenomenon in a flexor pulley injury, the respective finger undergoes flexion against resistance. The patient is asked to flex the injured finger against their own thumb, while the clinician palpates and observes for clinical flexor

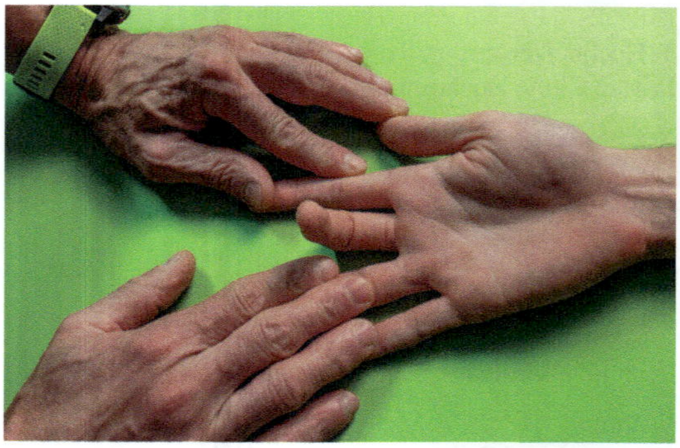

Fig. 6.1 Examination of the superficial flexor tendon (m. flexor digitorum superficialis) (FDS) (Photo M. Simon)

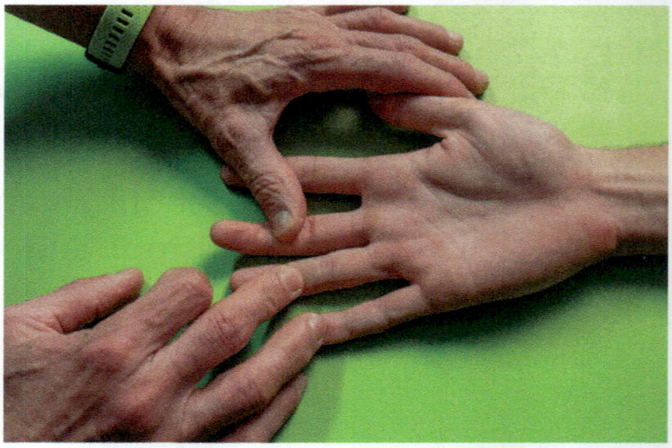

Fig. 6.2 Examination of the deep flexor tendon (m. flexor digitorum profundus) (FDP) (Photo M. Simon)

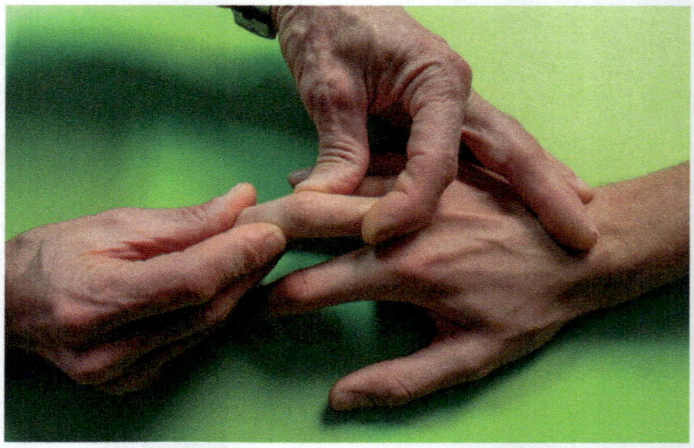

Fig. 6.3 Examination of the collateral ligaments of the PIP in 30° flexion (Photo M. Simon)

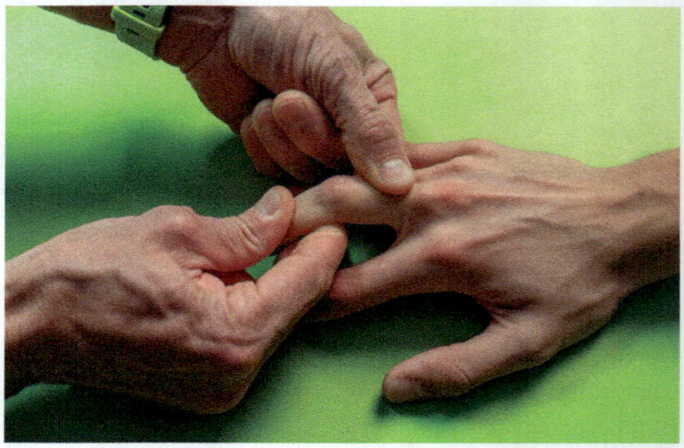

Fig. 6.4 Translation test of the palmar plate (Photo M. Simon)

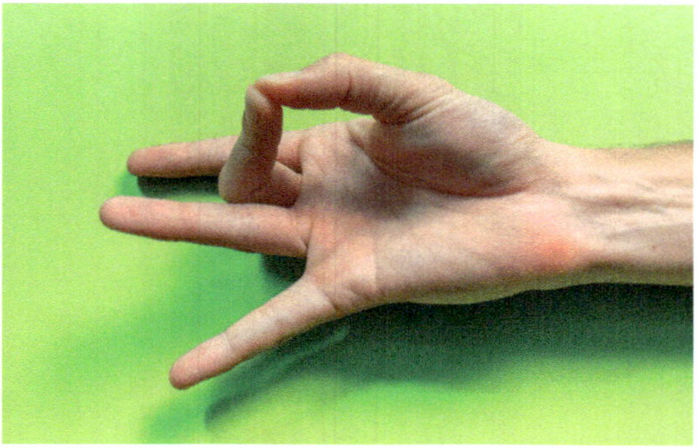

Fig. 6.5 Examination of a bowstring phenomenon (Photo M. Simon)

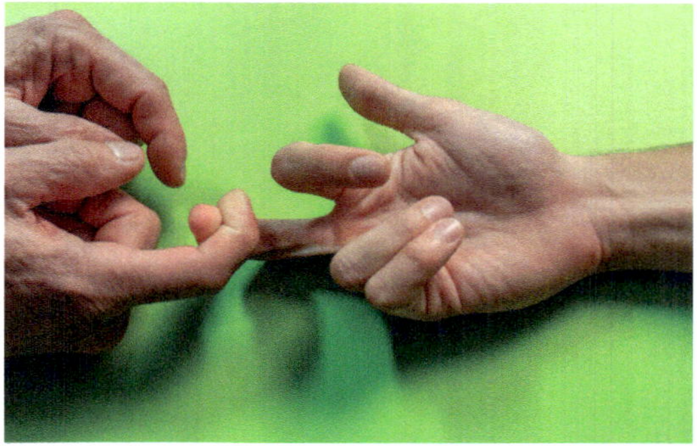

Fig. 6.6 Lumbrical shift test: an extended middle finger is stressed while the neighboring fingers are bent, thus leading to a stress of the lumbrical muscles common body (Photo M. Simon)

tendon bowstringing in the palmar direction (Fig. 6.5). A palmar-sided tenderness to palpation at the finger's proximal phalanx is suggestive of an A2 pulley injury or inflammatory tenosynovitis. A laterally located palpation tenderness at the same phalanx is suggestive of a lumbrical muscle injury. The "lumbrical shift test" is performed for a specific examination of the lumbrical tendons (Fig. 6.6). A dorsally located point tenderness at the PIP joint is frequently found in epiphyseal fractures in adolescent climbers or joint capsulitis (articular synovitis) in skeletally mature climbers. To detect hamulus fractures a specific stress test is performed (Fig. 6.7). Finger flexion in ulnar abduction causes shear forces onto the hamulus, provoking pain.

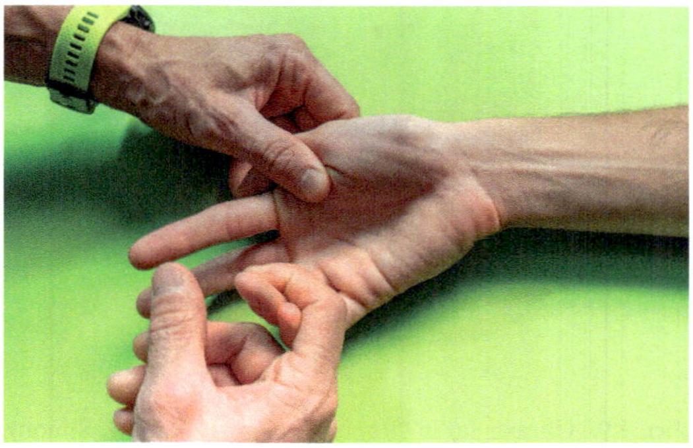

Fig. 6.7 Hamulus stress test: finger flexion in ulnar abduction causes shear forces onto the hamulus, provoking pain (Photo M. Simon)

6.2 Diagnostic Imaging

Many acute and chronic finger conditions require plain radiographs to confirm the diagnosis and/or rule out other possible diagnoses. However, in recent years, dynamic musculoskeletal ultrasound has become the "workhorse" imaging technique for climbers' finger injuries. Ultrasound assessment is very user-dependent and therefore may not be easily available to all healthcare providers. In many European countries, ultrasound examinations are performed routinely by orthopedic surgeons or sports medical doctors themselves. However, in other countries such as the UK, Canada, and certain regions in the USA, these exams are performed by either ultrasound technicians or radiologists. Consequently, some of the recommendations given within this book, especially in the chapter on hand injuries, may be difficult to implement in other countries. In the diagnosis of pulley injuries, tenosynovitis, or ganglion cysts, ultrasounds are very cost-effective and accurate with high sensitivity and specificity [1–4]. The technique of finger ultrasound examination is easy to learn and very efficient [5]. We predominantly use small, linear transducers ("hockey stick") with 10–18 MHz in longitudinal and transversal planes (Fig. 6.8). For signal enhancement, either a gel standoff pad or normal ultrasound gel is used. The ultrasound is performed statically and dynamically; for proper detection of inflammation, a color-coded duplex ultrasound evaluation is used. In diagnosing pulley injuries, the dis-

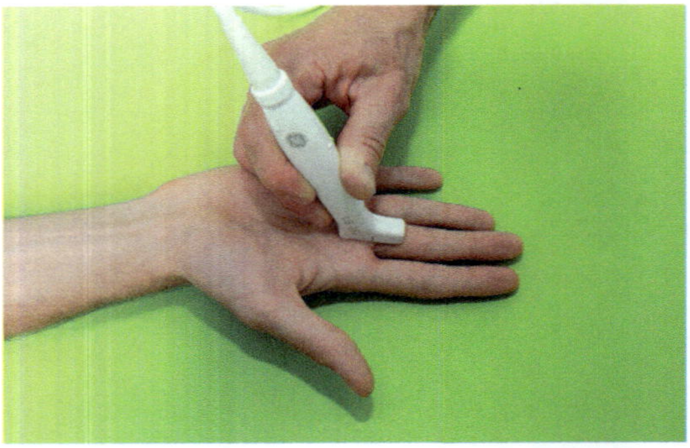

Fig. 6.8 Ultrasound examination with an 18 Mhz linear transducer (Hockey stick) (Photo M. Simon)

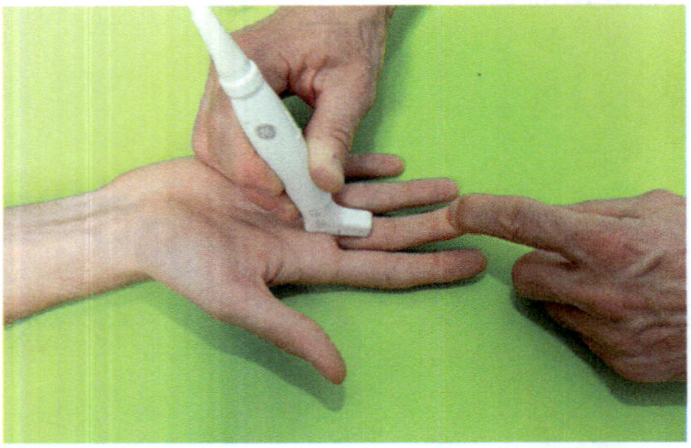

Fig. 6.9 Ultrasound in "forced flexion" (Photo M. Simon)

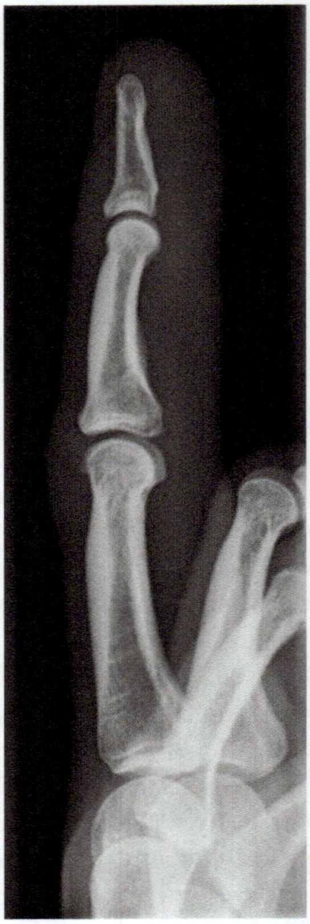

Fig. 6.10 Cortical hypertrophy of a 28-year-old climber as a stress adaptation (normal finding)

tance between the flexor tendons and the bone (TB = tendon-bone distance) is measured under "forced flexion" [6] (Fig. 6.9). The forced flexion is achieved by asking the patient to use a healthy finger to press against the fingertip of the injured finger, which undergoes active flexion during examination. This causes stress and tension on the flexor tendons of the injured finger, increasing a possible TB in an injury. The TB is measured at the midportion of the base phalanx for the A2, at the midportion of the middle phalanx of for the A4, and at the level of the palmar plate for the A3 pulley. In the last case, the palmar plate-tendon distance is measured, as the A3 pulley inserts into the palmar plate, not the bone [7, 8]. An additional MRI or CT scan is only required in rare cases [9–18].

6.2.1 Normal Findings

The high impact of rock climbing onto the fingers leads to physiological adaptations over the years. These adaptations must be strictly separated from pathological changes [19]. Through radiographic and MRI analyses, Hochholzer et al. [15] and Heuck et al. [20] identified these adaptations. The authors found an adaptive hypertrophy of the joint capsules, thickening of the collateral ligaments, cortical hypertrophy, and a hypertrophy of up to 50% of the flexor tendons themselves (Fig. 6.10). Hochholzer and Heuck also showed that certain reactions, such as cortical hypertrophy, are likely protective adaptations to the high stress of the sport rather than early pathological osteoarthritic changes. Therefore, clinical and radiographical findings must be interpreted carefully. In an analysis of radiographs in young high-level climbers as well as 140 experienced climbers, Schöffl et al. [21] classified the following findings as stress reactions, using a modified Kellgren Lawrence classification [22]: cortical hypertrophy, subchondral sclerosis/increased thickness of epiphysis, calcifications of the insertion of the flexor digitorum superficialis or the flexor digitorum profundus tendon, broadened proximal, and/or distal interphalangeal joint base.

6.2.2 Distribution of Finger Injuries

Table 6.1 summarizes the distribution of diagnoses in our own patients presenting with finger pain over three different time periods.

In the following sections, we discuss various specific finger injuries. In general, distinction should be made between acute traumatic finger injuries (e.g., pulley injury, tendon strain, ligamental injury, soft tissue injury, etc.) and chronic overstrain injuries (e.g., tenosynovitis, osteoarthritis, epiphyseal growth plate injury, lumbrical shift syndrome, etc.).

6.3 Finger Injuries

6.3.1 Pulley Injuries

An injury to the finger flexor tendon pulleys is the most common climbing-specific injury. Closed traumatic ruptures of finger flexor tendon pulleys arose as a new complex trauma in the middle of the 1980s. Bollen [23–25] and Tropet et al. [26] were the first to observe this type of injury; subsequently, much research has been carried out. In the past, the clinical management recommendations varied widely for these injuries. Currently, there is more consensus with little variation in treatment regimens between various authors [27]. Since this injury has become more well-known in medical literature, it has also been described in other sports [28] as well as in nonathletes and children [29]. Pulley injuries occur not only in the well-known annular (A) pulleys but also as isolated injuries of the cruciate (C) pulleys [30].

6.3.1.1 Clinical Presentation

Most climbers report an acute onset of pain while performing a hard move with the finger in crimp position. There may be a slip of the foot from a foothold, resulting in sudden loading of the finger. In some cases, a loud popping noise is heard. Occasionally, the rupture can occur due to chronic repetitive trauma and fatigue failure. In cases of prior chronic tenosynovitis treated with steroid injections, the pulley can rupture even in cases of minor stress if the appropriate rest after the last injection was neglected. Biomechanical analyses showed that pulley ruptures are mostly caused by eccentric finger movement [31].

Physical examination reveals tenderness over the palmar level of the injured pulley accompanied by swelling and occasionally hematoma. A visible bowstring only occurs with multiple pulley ruptures (Fig. 6.11). In cases of A4 or A2 pulley rupture, increased distance of the flexor tendon to the bone can be palpated occasionally if the injured finger is opposed to the thumb in a crimping position (i.e., palpable bowstring) [23]. Since a definitive diagnosis cannot be made based on clinical examination exclusively, we always perform a dynamic ultrasound assessment. In cases of special questions or diagnostic uncertainty, an MRI can be helpful [7, 17, 32].

Table 6.1 Most frequent finger injuries 2017–2018 ($n = 261$), 2009–2012 ($n = 474$), and 1998–2001 ($n = 247$) [109]

Finger injuries 2017–2018 ($n = 261$) (2 years)	n	%	Finger injuries 2009–2012 ($n = 474$) (4 years)	n	%	Finger injuries 1998–2001 ($n = 247$) (4 years)	n	%
Pulley injury	78	29.9	Pulley injury	140	29.5	Pulley injury	122	49.4
Tenosynovitis flexor tendon	67	25.7	Capsulitis	87	18.4	Tenosynovitis	42	17.0
Capsulitis	49	18.8	Tenosynovitis flexor tendon	80	16.9	Strain finger joint capsule	37	15.0
Epiphyseal fracture	19	7.3	Strain flexor tendon	36	7.6	Capsulitis	13	5.3
Lumbrical tear/strain	12	4.6	Strain finger joint capsule	25	5.3	Ganglion	11	4.5
Strain finger joint capsule	10	3.8	Ganglion finger flexor tendon	19	4.0	Strain flexor tendon	7	2.8
Osteoarthritis	5	1.9	Lumbrical shift syndrome	19	4.0	Fracture	7	2.8
Strain flexor tendon	4	1.5	Collateral ligament injury	17	3.6	Osteoarthritis	7	2.8
Ganglion finger flexor tendon	3	1.2	Epiphyseal fracture	16	3.4	Soft tissue injury	5	2.0
Contusion	3	1.2	Osteoarthritis	14	3.0	Tendon rupture	4	1.6
Phlegmonia/cellulitis	2	0.7	Extensor hood syndrome	7	1.5	Collateral ligament injury	3	1.2
Collateral ligament injury	2	0.7	Lumbrical tear/strain	4	0.8	Osseous tear fibrocartilago palmaris	2	0.8
Distorsion thumb	1	0.3	Snap finger	3	0.6	Epiphyseal fracture	2	0.8
Disruption volar plate	1	0.3	Cartilage injury	2	0.4	Lumbrical tear/strain	2	0.8
Cartilage injury	1	0.3	Flip phenomena	2	0.4	Phlegmonia/cellulitis	1	0.4
Neuropraxia	1	0.3	Broken osteophyte	1	0.2	Finger amputation	1	0.4
PIP joint dislocation	1	0.3	Avulsion fracture	1	0.2	–	–	–
Snap finger	1	0.3	Flexor contraction	1	0.2	–	–	–
Contracture finger flexor tendon	1	0.3	Rupture connexus intertend.	1	0.2	–	–	–
–	–	–	Enchondroma	1	0.2	–	–	–
–	–	–	Contusion	1	0.2	–	–	–
–	–	–	Tendon rupture	1	0.2	–	–	–

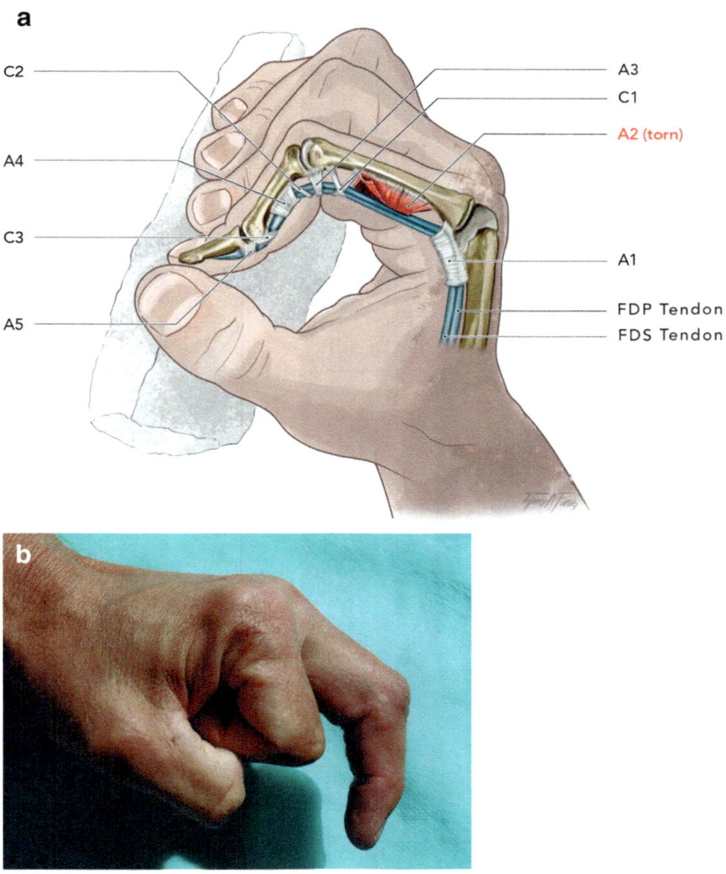

Fig. 6.11 Bowstringing in a A2 pulley rupture. (**a**) biomechanical drawing, (**b**) clinical presentation (Pic by Tiffany Fung)

6.3.1.2 Diagnostic

The diagnostic follows our algorithm [33]:

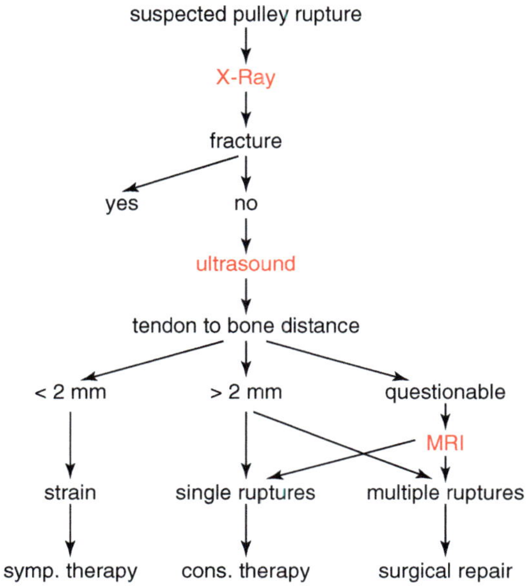

According to the algorithm as well as the reports of Gabl et al. [34], radiographs should be performed to exclude fractures or osseous tears of the palmar plate as well as chronic repetitive stress epiphyseal fractures in adolescent climbers, if warranted by the clinical picture. For further evaluation, the "gold standard" is the dynamic ultrasound. MRI is only performed if the X-rays and ultrasound do not provide a conclusive diagnosis (Fig. 6.12).

The advantage of the ultrasound is the fact that it can be performed dynamically, which is only possible with an MRI in very specific studies [7, 16]. A direct visualization of the various pulleys is often possible with a high-resolution ultrasound probe (15–18 MHz, "hockey stick" probe) (Fig. 6.13). Valid detection of the tendon-bone distance (TB) under "forced flexion" can also always be performed with a lower resolution probe, even with some of the new handheld ultrasound devices [1] (Fig. 6.14). Regarding the TB

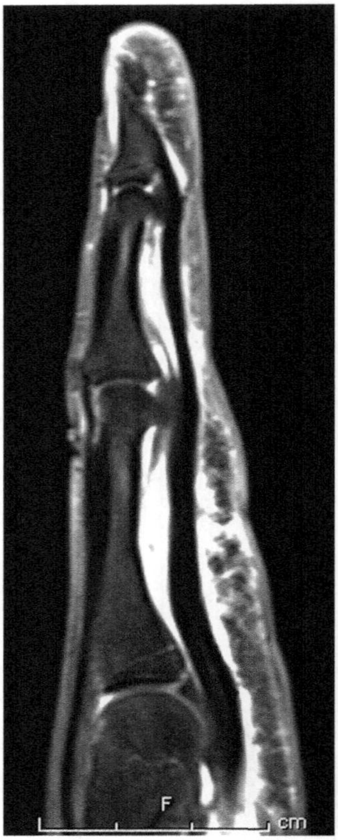

Fig. 6.12 MRI in a A2,3,4 pulley rupture

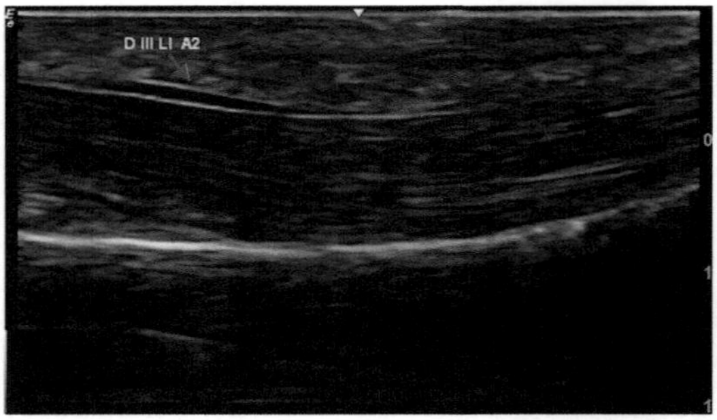

Fig. 6.13 Normal A2 pulley in the ultrasound (the flexor tendon follows the course of the bone; a direct visualization of the pulley is possible)

distance in a noninjured population, Bassemir et al. [35] described the normal distances in an analysis of 200 non-climbing individuals.

The following cutoffs could be evaluated in our cadaver study for pulley injury detection under "forced flexion" of 10 N [7] and are our clinical guidelines:

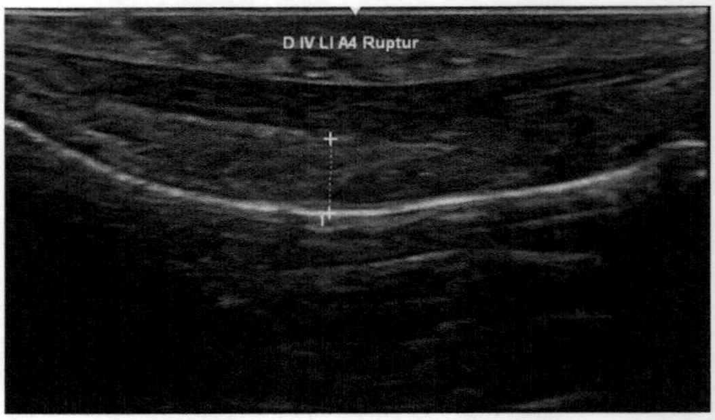

Fig. 6.14 A4 pulley rupture (the flexor tendons are distant to the bone)

Ultrasound criteria for pulley tear under forced flexion of 10 N [7]:

- **A2: TB > 2 mm: sensitivity 94%, specificity 100%**
- **A3: VP > 0.9 mm: sensitivity 76%, specificity 94%**
- **A4: TB > 2 mm: sensitivity 90%, specificity 97%**

(TB = tendon-bone distance, VP = volar plate distance)

For the A2 and the A4 pulley, the TB distance is measured at the midportion of the respective phalange; for the A3, the distance tendon to volar plate is measured. To get the correct measuring point of the volar plate distance (VP) distance, the length of the VP is measured from its most proximal to its most distal point. A point is then defined on the proximal phalanx where the curvature of the condyle transitions from concave to convex. Then, the distance is measured between this point and the middle of the volar aspect of the VP (VP bone). Next, the distance is determined between this point (measuring base point) and the tendon in extension to obtain the VP tendon distance [7] (Fig. 6.15).

Clear TB distance cutoffs for single C pulley injuries do not exist, as the TB distance is only slightly increased. For the diagnosis of C pulley injuries, a combination of the detection of hematoma, a change of the tendons course in the dynamic ultrasound examination, and a minor increase in TB distance lead to the final diagnosis.

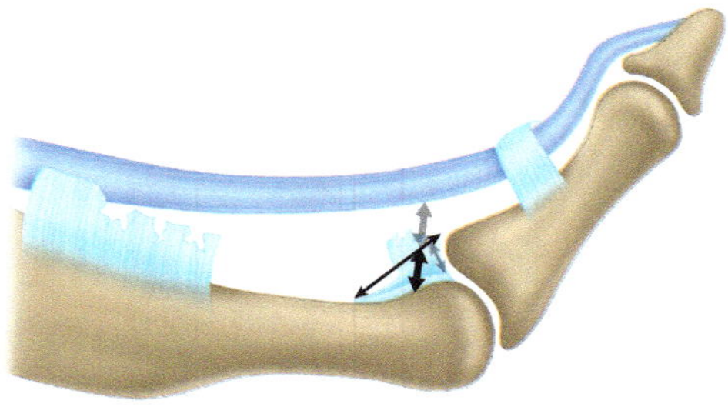

Fig. 6.15 Stylized drawing of a finger after pulley injury of the A2 and A3 pulleys with the measurements taken during the study (thick black arrow = VP bone distance, thick grey arrow = VP tendon distance, thin black arrow = VP length, thin grey arrow = VP thickness) [7]

Multiple complete pulley ruptures, especially triple-pulley ruptures (i.e., A2/3/4) can cause clinically visible bowstringing and a functional deficit, as the longitudinal force of the flexor tendons can no longer be transferred into a rotation of the finger joints (DIP, PIP). Without treatment, this will often lead to chronic flexion contractures, usually at the PIP joint (Fig. 6.16).

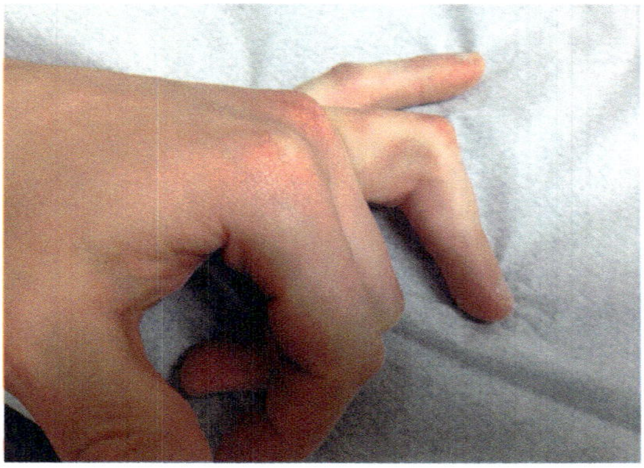

Fig. 6.16 Contracture as a result of an untreated A2,3,4 pulley rupture

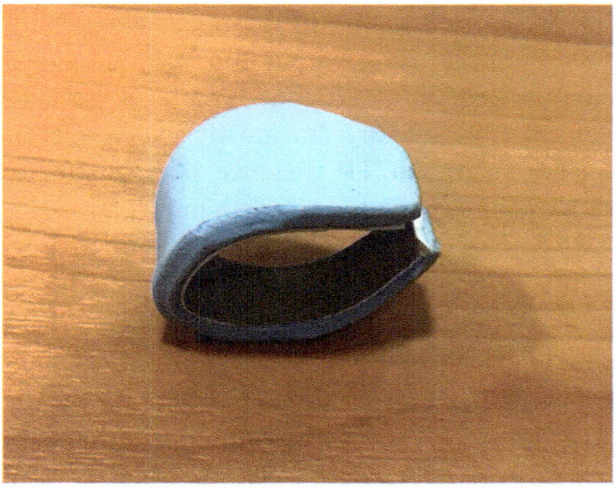

Fig. 6.17 Pulley support ring (fixed with a circular tape)

6.3.1.3 Treatment

Single pulley injuries (grade I–III) receive conservative therapy with an optional initial short period of immobilization (depending on the grade of swelling and collateral damage) and early functional therapy with a thermoplastic pulley support ring (Figs. 6.17 and 6.18) [36]. The design for these pulley support rings presented by Schneeberger et al. [37] is the best, as these are the only ones, which allow for anatomic repositioning of the flexor tendons without compression of the neurovascular bundle. Circular rings cannot relocate the tendon close enough to the bone without compromising the digital neurovascular structures. With anatomic repositioning of the tendons, the injured pulley, which generally does not tear in the middle but from one side of its insertion [38, 39], can heal while forming a stable scar (neo-pulley). An ultrasound evaluation is essential to exclude a FLIP-phenomenon (flap irritation phenomenon), in which the torn pulley slips underneath the flexor tendons. This prevents relocation of the tendons and produces chronic irritation of the tendons, resulting in tenosynovitis. Occasionally, a FLIP-phenomenon requires a secondary surgical pulley repair [40, 41]. The initial immobilization is only rarely necessary; the pulley support ring should be worn for 6–8 weeks in a single pulley tear [37]. For a grade 1 injury, pulley strain taping is usually sufficient. When wearing the pulley support ring, it is crucial to maintain proper skincare and daily application of nourishing ointments

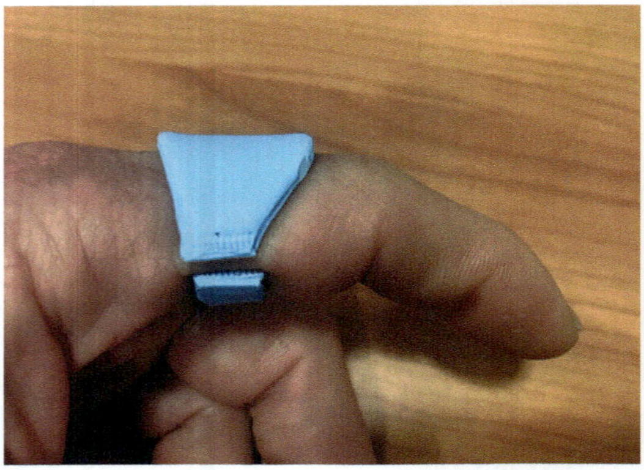

Fig. 6.18 Pulley support ring (fixed with a circular tape)

(e.g., panthenol) to prevent pressure sores (Fig. 6.19). A coating of the inner layer of the pulley support rings with a thin layer of leather can reduce skin ulcerations. The results of this conservative therapy are mostly excellent; the former increased TB distance is permanently reduced and the initial strength in finger flexion will resolve as the climbers regain their pre-injury strength level [42]. Following the initial pulley protection ring, H-taping of the finger should be applied during the "return-to-sport" period for up to 6–12 months [43].

In grade IV injuries (multiple pulley injuries), the general recommendation is a surgical repair [36]. Otherwise, the permanently increased TB distance and scarring will cause flexion contracture [27]. If this contracture is already present, a later, secondary surgical reduction is challenging. In recent years, we have modified our original therapeutic and grading system [36, 44], as recent studies also showed a good outcome of conservative management in certain types of multiple pulley ruptures without clinical bowstringing [27, 37]. Thus, we divided grade IV injuries into the subgroups of grade IVa and grade IVb injuries. In some cases, grade IVa injuries can be treated with conservative therapy [36]. Grade IVa injuries represent A2/A3 or A3/A4 injuries with no major clinical bowstringing, an ultrasound-proven possibility of repositioning the flexor tendons to the bone, a therapy start of less than 10 days after the injury and no contracture. A2/A3 or A3/A4 injuries with obvious bowstringing are categorized as

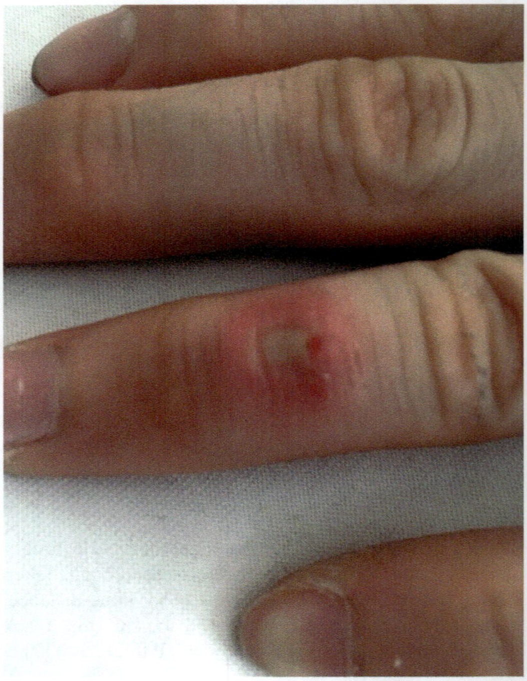

Fig. 6.19 Pressure skin ulceration caused by a pulley support ring

grade IVb injuries, as are all other multiple pulley injuries (see Table 6.2) [36].

Nevertheless, it is important to mention that the decision toward a conservative management in a multiple tear should always be made specifically for each individual case. This decision should be made in consultation with the patient and based on careful analysis of the clinical findings, including a proper ultrasound diagnosis. Additionally, close patient follow-up is necessary (including ultrasound controls), and surgical repair must immediately be performed if contracture begins to develop. In many ways, conservative therapy for a multiple pulley injury is actually more challenging, than surgical repair, and should only be undertaken by experienced clinical teams at high-volume centers. Conservative therapy of double a pulley rupture is only possible if the therapy starts promptly after the injury (ideally within one to 2 weeks), if flexor tendons can be anatomically repositioned under ultrasound monitoring with no FLIP-phenomenon [41], and if no flexion contracture has developed

Table 6.2 Therapeutic guidelines of pulley injuries in rock climbers (Consensus Statement of Expert Commission Sportsmedicine Bamberg (Volker Schöffl, Isabelle Schöffl, Christoph Lutter, Michael Simon, Yasser El-Sheikh), Bamberg 2019 [36], modified after (Schöffl et al. 2004c))

	Grade I	Grade II	Grade III	Grade IV a	Grad IV b
	Pulley strain	Complete tear of A3 or A4, partial tear of A2	Complete tear of A2	Multiple ruptures: – A2/A3 or A3/A4 rupture if: – No major clinical bowstring – Ultrasound-prooven possibility of reposition of the flexor tendon to the bone – Therapy starting <10 d after injury – No contracture	Multiple ruptures: – A2/A3 or A3/4 with obvious clinical bowsrting – A2/A3/A4 rupture – Singular pulley rupture with FLIP phenomen – Singular rupture with increasing contracture – Singular rupture with secondary, therapy-resistant, tenosynovitis
Therapy	Conservative	Conservative	Conservative	Conservative, if secondary onset of PIP contracture >20° secondary surgical	Surgical
Immobilization	None	optional, <5 days	Optional, <5 days	Optional, <5 days	Postsurgical 14 days
Functional therapy with pulley protection (defined)	2–4 weeks H-tape (during day time) or thermoplastic ring	6 weeks thermoplastic pulley ring	6–8 weeks thermoplastic pulley ring	8 weeks thermoplastic pulley ring	4 weeks thermoplastic ring (after 2 weeks of immobilization)
Easy sport-specific activities	After 4 weeks	After 6 weeks	After 8 weeks	After 10 weeks	After 4 months
Full sport-specific activities	After 6 weeks	After 8–10 weeks	After 3 months	After 4 months	After 6 months
H-taping during climbing	3 months	3 months	3 months	>12 months	>12 months

and a proper pulley ring is applied, keeping the tendons close to the bone [27]. The pulley support ring is worn for 23 h per day and is only removed once daily for skin care.

If a significant secondary flexion contracture of more than 20° extension deficit occurs, a surgical revision and pulley reconstruction is recommended. A minor contracture of up to 5° often develops and can be accommodated. In complex rupture situations (multiple pulleys (grade IVb) or after failure of a conservative therapy in grade IVa injuries, surgical repair in the form of a pulley reconstruction is necessary. In these cases, the clinical presentation is majorly compromised movement and reduced flexion strength of the DIP joint. If untreated, secondary contracture at the PIP will follow [41].

Suture repair of the torn pulley at any stage is rarely possible, due to primary and secondary contracture of the ligament, with shortening and friability of the tissues preventing reliable end-to-end approximation, so pulley reconstruction is necessary. Out of the many various pulley reconstruction techniques (including the "Kleinert and Bennet" repair based on Weilby's idea [45], Karev's "belt-loop" technique [46], "single-loop"

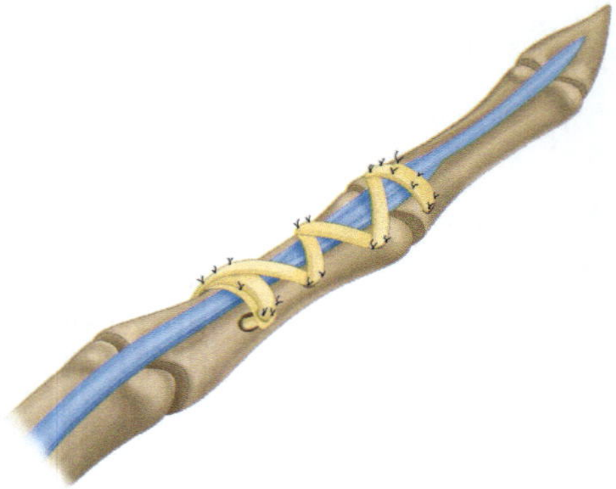

Fig. 6.20 Transosseous pulley repair [57]

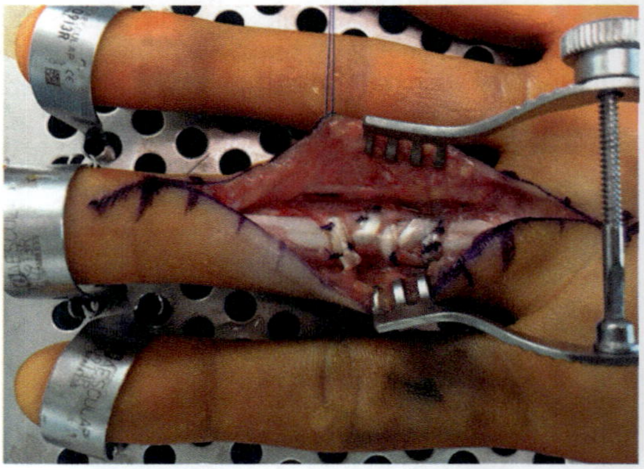

Fig. 6.21 Transosseous pulley repair with a free palmaris longus graft during surgery

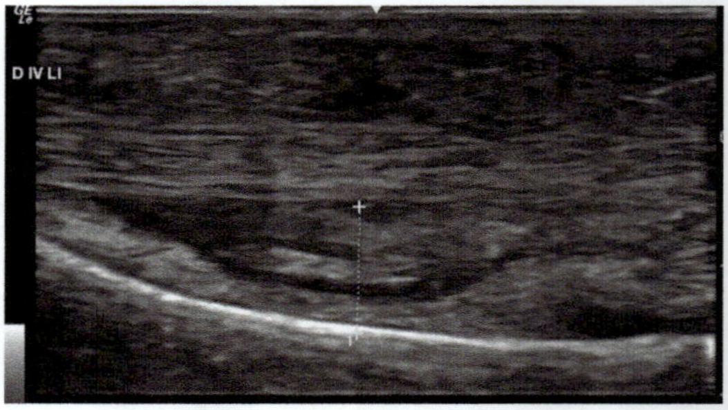

Fig. 6.22 A2 pulley rupture with an increased tendon-bone distance (see the line in the figure) and the trunk of the former pulley squeezed underneath the flexor tendons as a "FLIP"-phenomenon

technique according to Bunnel [47], "Lister repair" with retinaculum flexorum [48], palmaris longus tendon transplantation through the volar plate according to Doyle [49], Widstrom's "loop-and-a-half" technique [50], "triple-loop technique" by Okutsu [51], or "extensor retinaculum graft" by Gabl [52] and Moutet [53]), we favor a modified "loop-and-a-half" technique with a distal continuation to the A3 pulley [54]. This combines the advantages of the "loop-and-a-half" as the strongest and the Weilby repair as the most functional repair [50, 54, 55]. We recently noticed that some osseous necrosis of the phalanx can occur following this repair, likely due to the high pressure of the circulation onto the bone and its blood vessels [56]. Therefore, we have modified our procedure into a transosseous repair (Figs. 6.20 and 6.21) [57]. A FLIP-phenomenon, an interposition of the pulleys stump underneath the tendons and the bone, as shown in Fig. 6.22 (Fig. 6.22) can render a conservative therapy impossible and require surgery. This can either be performed using the palmaris longus or a retinaculum extensorum flap [40].

Postsurgically, the finger is immobilized until completion of skin healing, using a palmar-sided cast splint for a maximum of 2 weeks. Thereafter, a thermoplastic pulley support ring is implemented and gentle active ROM exercises commence. Six weeks after surgery, free movement without the support ring is allowed, and the functional therapy can be intensified. Specific rehabilitation exercises are presented in "One Move Too Many" (Schöffl V, Hochholzer T, Lightner S Jr) [58]. General exercise (e.g., light running) can be resumed after 2 weeks, once the surgical site edema has completely resolved. Easy climbing and climbing-specific training are allowed after 4 months (with protective H-taping) (Figs. 6.23, 6.24, 6.25, and 6.26). Taping should be continued through climbing for another 6 months. The full, sport-specific load can be assumed after 6–9 months.

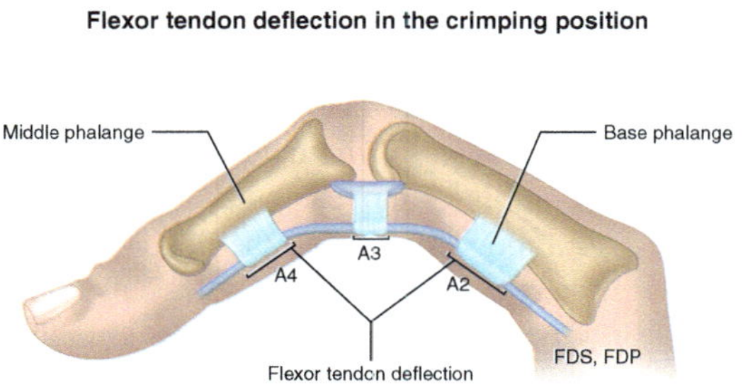

Fig. 6.23 Biomechanics of the H-tape. Based on the support of the A3 pulley, the stress onto the distal rim of the A2 pulley and the proximal rim of the A4 pulley is reduced, as the angulation of the flexor tendons is decreased. This helps in a tenosynovitis (see Sect. 5.3.2) and in a pulley injury [43]

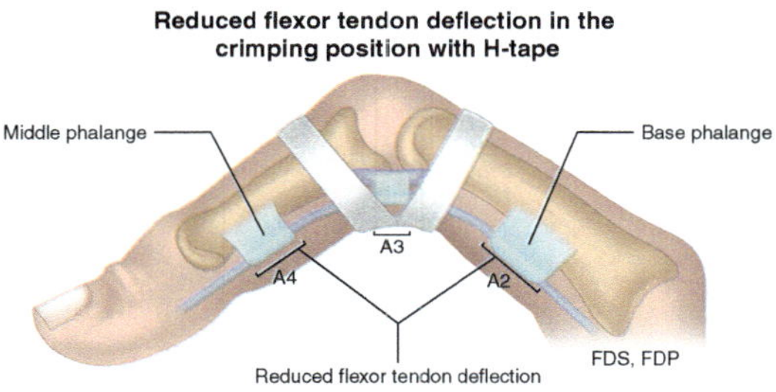

Fig. 6.24 Biomechanics of the H-tape. Based on the support of the A3 pulley, the stress onto the distal rim of the A2 pulley and the proximal rim of the A4 pulley is reduced, as the angulation of the flexor tendons is decreased. This helps in a tenosynovitis (see Sect. 5.3.2) and in a pulley injury [43]

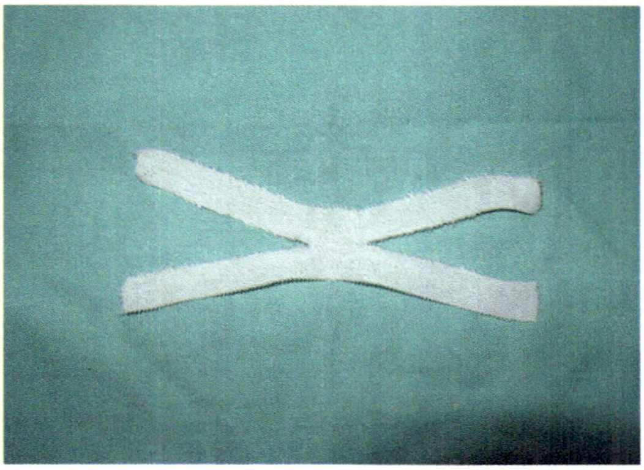

Fig. 6.25 H-taping

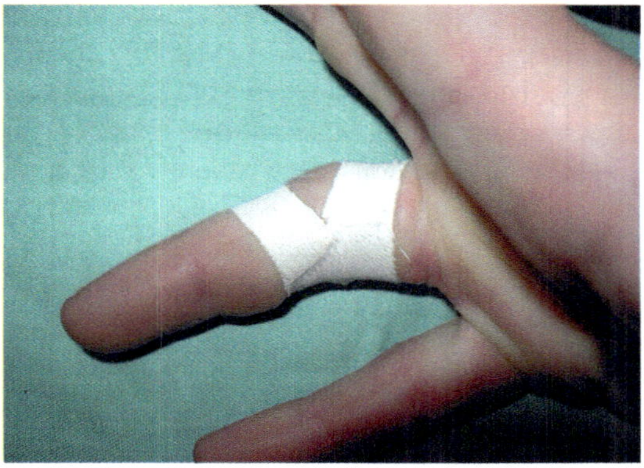

Fig. 6.26 H-taping

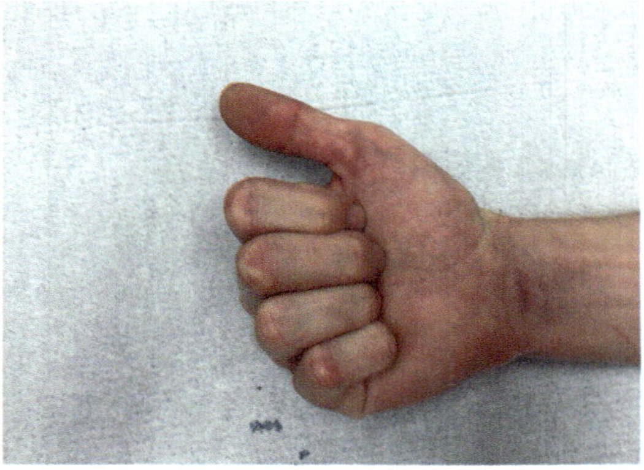

Fig. 6.27 Three-month postsurgical result of the patient in Fig. 6.21

6.3.1.4 Outcome

Schneeberger et al. [37] showed that through consistent use of the pulley support ring, the increased TB distance can be reduced. Also, the surgical management of multiple pulley ruptures showed good to excellent sport-specific results in our sport-specific pulley injury outcome score (see Chap. 2) [54] (Figs. 6.27 and 6.28). This is in agreement with the current literature, which shows good outcomes after pulley reconstruction in general, with no significant difference in outcomes between techniques [32, 52, 54, 59]. In an outcome analysis, Schöffl et al. found a "return to climbing" in all of the 81 conservatively treated patients (grade I–III); only one patient needed a

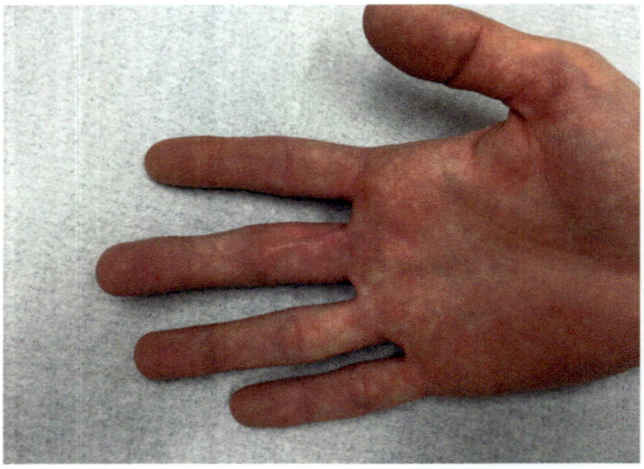

Fig. 6.28 Three-month postsurgical result of the patient in Fig. 6.21

secondary pulley reconstruction due to a chronic tenosynovitis [44]. In another study concerning finger strength after conservative treatment of single pulley injuries, no patients showed a remaining strength deficit [42]. Nevertheless, in both groups (conservative and surgical), a minor extension deficit (5°–7° contracture) at the PIP joint frequently remained. However, this usually has no clinical impact. If patients present for a pulley repair with a preexisting contracture due to an older pulley injury, we start with conservative dynamic extension splint therapy and aim to reduce the contracture to 20° prior to surgery. If this is not possible, we release the joint capsule surgically in addition to the pulley repair; however, minor permanent contracture of up to 20° must be expected.

6.3.1.5 Prevention

Many climbers try to protect their flexor tendon pulleys by applying circumferential tape around the proximal phalanx [27]. However, while some reduction of the bowstring is achieved via tape in cases of a disrupted pulley, no effect can be expected from protective pulley tape around an intact pulley [60, 61]. Taping over healthy pulleys did not show any biomechanical effect in a cadaver study [62]. Woollings et al. [63] even reported that preventive taping increases the risk for finger injuries. The main positive effect of prophylactic taping may be that the PIP joint is not flexed more than 80°–90° if the tape is applied close or even over the PIP joint itself [27]. A correct warming-up procedure and the avoidance of a pronounced crimp grip position seem to be more important factors in pulley injury prevention [27]. It has been shown that over the first 100 to 120 climbing moves, the amount of physiological bowstringing of the flexor tendons shows an increase of up to 30% [64]. This effect could only be shown with performing climbing movements and was not observed with other warming-up techniques [64]. Therefore, it is recommended to warm-up by climbing about 3–4 routes with 40 moves or 8–12 boulder problems [64] with increasing intensity. After a pulley injury, tape should be applied at either the distal end of the respective injured pulley [64] or as an H-tape at the level of the PIP joint [43] for 6–9 months (see Chap. 22. Taping) (Fig. 6.29). Another potential preventive measure for finger pulley injuries is controlled low-intensity eccentric training of the finger flexor muscle-tendon units, which is shown to help in tendon healing [65] and may also have some effect on the pulleys [66]. Nevertheless, scientific evidence for this is still pending.

- **History:** Mostly one hard move in a crimping position
- **Onset:** Sudden
- **Clinical:** Pressure tenderness palmar-sided at the pulley, often a "popping" sound during the event
- **Diagnostic:** Clinical findings, ultrasound, optional MRI
- **Therapy:** Singular pulley injuries conservative, multiple mostly surgical
- **Outcome:** Good, no consistent strength deficit

6.3.2 Finger Flexor Tenosynovitis

Inflammatory flexor tenosynovitis (aka: tendonitis, tendovaginitis) is inflammation of the flexor tendon sheath and is the most frequent overuse syndrome in climbers' fingers [67]. An inflammatory response occurs after repetitive stress, and its onset can be acute after one exceptionally hard training or climbing day, or creeping up

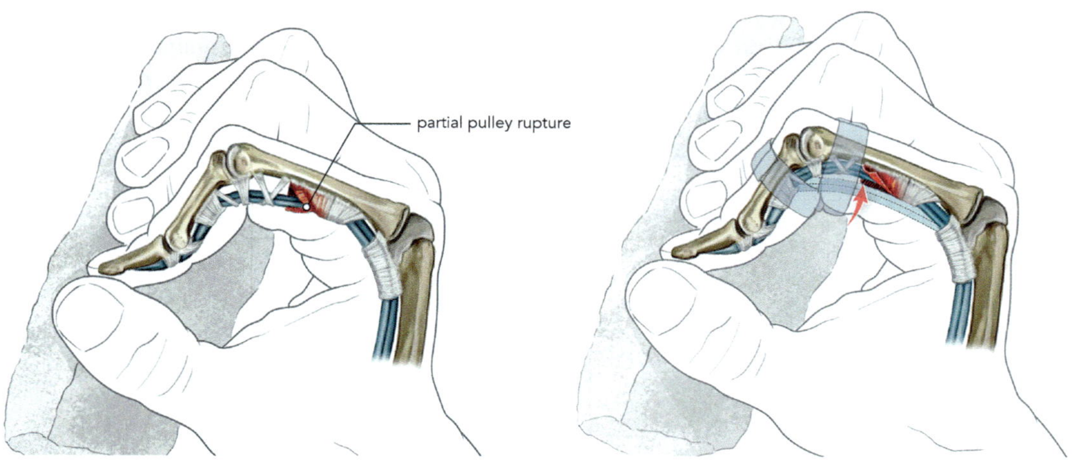

Fig. 6.29 H-taping in partial A2 tear. (Pic by Tiffany Fung)

slowly over days [19]. The latter condition is commonly referred to by laypeople and climbers as tendonitis but in fact represents an inflammation of the tendon sheath. Thus, the correct term is "tenosynovitis" [6, 19, 67].

The etiology of inflammatory tenosynovitis in climbers is usually repetitive high stress while performing a so-called "crimping" position [19]. In a crimping position, the flexor tendons are curved at the distal end of the A2 finger flexor tendon pulley, as well as the proximal end of the A4 pulley [67–69]. This causes an increased flexion of the tendons and thus increased friction onto the rim of the pulley [43, 69]. Repetition of this position can lead to a chronic inflammatory reaction of the tendon sheath [19, 67]. Even though no studies have been performed on a histopathological level, ultrasound examination of climbers with this condition, including duplex sonography, revealed an inflammatory reaction within the flexor sheath and its lining [1, 70, 71]. While there are generally multiple extrinsic and intrinsic factors in the origin of tenosynovitis, the main pathological stimulus for this condition is considered to be the excessive loading of tendons during vigorous training as performed in rock climbing [72]. Tendons respond to repetitive overload beyond the physiological threshold either with inflammation of the sheath, degeneration of the tendons, or both [73]. Tendon damage may even occur from stress within the physiological limits, as frequent cumulative microtrauma may not allow enough time for the tendons to heal [73], a case that is highly probable in ambitious climbers [67]. In general, no other sport than climbing places such an immense load onto the fingers. The frequent use of the crimping position in climbing is increasing this load additionally [68, 69].

6.3.3 Clinical Presentation and Diagnostic

The climber suffers from pain, occasionally accompanied by a minor swelling along the palmar surface of the finger, just around the same area as in a pulley injury. The pain can extend into the palm or the forearm [67] (Fig. 6.30). The clinical examination usually presents point tenderness at the level of the inflammation, mostly at the A2 or A4 pulley. Sometimes, the swelling of the tendon sheath is palpable due to the increased fluid in the area. A diagnosis can be made using ultrasound, with the hallmark finding of a "halo" phenomenon around the tendon [1, 6, 19, 71]. This increased accumulation of liquid around the tendon is most clearly visible in the transverse plane [5]. As climbers tend to have more liquid in their flexor tendon sheets after high stress on various ranges, no clear information can be given regarding the normal range [19, 67]. It is recommended to compare the ultrasound finding of the injured finger to the same finger on the contralateral side [19, 67, 74]. A "halo" of more than 2 millimeters in a cross-sectional plane is strongly indicative of tenosynovitis [19] (Figs. 6.31 and 6.32). Color duplex ultrasound can subsequently show the inflammation. In addition, an MRI examination can assist in the diagnosis but is usually not necessary (Figs. 6.33 and 6.34) [67].

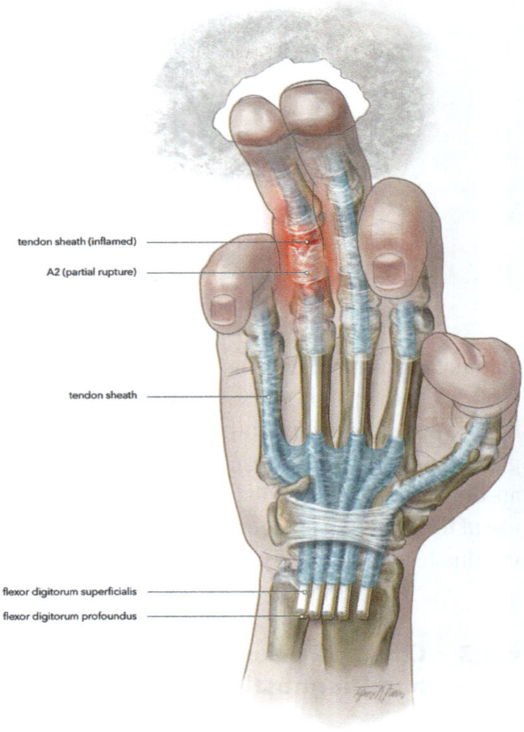

Fig. 6.30 Tenosynovitis after partial A2 tear (Pic by Tiffany Fung)

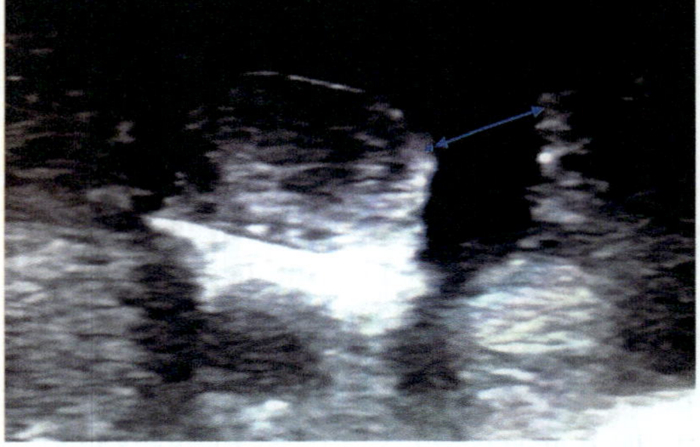

Fig. 6.31 "Halo" phenomenon in the ultrasound (the arrow shows the thickened tendon sheath)

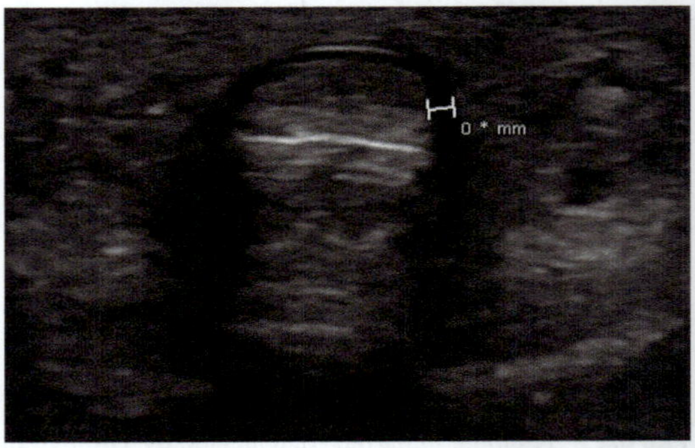

Fig. 6.32 Normal finding for comparison

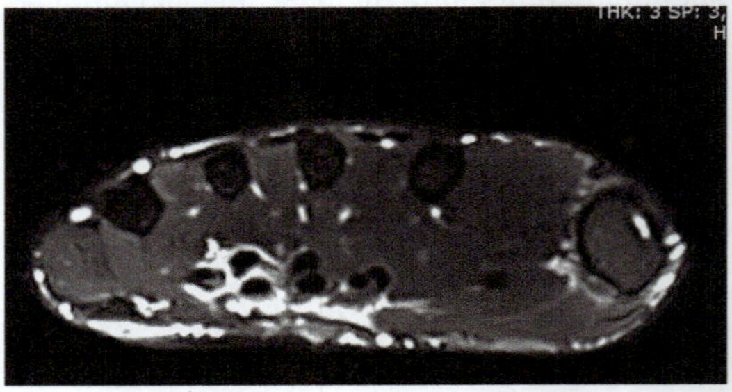

Fig. 6.33 Tenosynovitis of the palm in an MRI imaging

Fig. 6.34 Acupressure ring

6.3.4 Therapy

The initial therapy consists of anti-inflammatory medication, optional resting on a splint for several days, externals, brush massages, ice therapy and, in persisting conditions, local cortisone injections [19, 75]. These injections are not always avoidable, as chronic tenosynovitis can be stubborn. Surgical revision with tenosynovectomy is only required in rare cases. In mild cases, a stress reduction together with H-taping during sports will already decrease symptoms. If osteoarthrotic bone spurs are causing irritations to the tendons, removal of the bone spurs may be necessary. Alternative treatment options include radial shock wave therapy [76] or medicinal leech therapy [67].

For specific guidance in the treatment inflammatory flexor tenosynovitis, we developed a stage-related treatment regimen (Table 6.3) [67].

Stage 1 cases receive conservative therapy of self-massage with an acupressure ring (Fig. 6.34) or brush massages (with a toothbrush) to stimulate blood flow [67]. In addition, local overnight ointment dressings with Ichtolan® 20% (Ammonium bituminosulfonate, Ichtyol®, Hamburg, Germany) are applied for 10 days. Local ice therapy and nonsteroid anti-inflammatories can also be used [67]. The climbers are advised to either rest or to implement H-taping during climbing, depending on the extent of complaints or professional ability (e.g., during World Cup season, rest may not be possible). H-taping biomechanically deflects the angulation of the tendon and thus decreases friction at the distal rim of the A2 pulley and the proximal rim of the A4 pulley [43].

If the tenosynovitis persists beyond 6 weeks, the condition is considered chronic (stage 2), and injections into the tendon sheath are recommended. Various substances such as corticosteroids (triamcinolone, dexamethasone), hyaluronic acid, platelet-rich plasma (PRP), or NSAIDs are used for injection therapy [77–81]. We mostly perform steroid or PRP injections. Although these injections have been reported to show beneficial effects [19], they are controversial since inadvertent intratendinous injection (instead of a peritendinous injection) of corticosteroids may cause tendon rupture [75, 82]. Therefore, correct injection technique is essential. Based on our own experiences, we favor a crystalloid corticosteroid (dexamethasone) in a prefabricated mixture with lidocaine (Supertendin®, Carinopharm, Elze, FRG), as this is the least viscous injection mixture and can be administered through a thin hypodermic (insulin) needle (27Gx0.5). Per injection, 0.1–0.2 mL is administered. The injection localization is defined as the point of the largest halo ring detected via ultrasound. The injection is performed in two areas. Following proper skin disinfection, the needle is inserted at a 60° angle pointing to the fingertip until it reaches the flexor tendons. The patient is then directed to flex the finger, causing the needle to move with the flexor tendon. The needle is retracted until the tendon can move freely, and only the tip of the needle remains in contact with the tendon, at

Table 6.3 Stage-related treatment regimen for tenosynovitis of the finger flexor tendons in climbers (Modified after Schöffl et al. 2019b)

Stage	Time frame since onset of pain	Therapy	Climbing rest
1	< 6 weeks	Conservative, icing, local therapy with acupressure ring	0–14 days, then stepwise climbing load increase with H-taping [18]
2	>6 weeks (or after failed conservative therapy >4 weeks)	Local infiltration with "Supertendin®" (corticosteroid) Reinjection after 7–10 days or radial shock wave therapy	No climbing or hand-related sports in between the injections and at least for 10 days after the second injection, then stepwise climbing load increase with H-taping [18]
3	Persistent pain >6 weeks after second injection	Medicinal leech therapy	14 days rest, then stepwise climbing load increase with H-taping [18]
4	Persistent pain >6 weeks after failed leech therapy	Surgical tenosynovectomy	6 weeks rest, then stepwise climbing load increase with H-taping [18]

which point 0.1 mL is injected (Fig. 6.35). In the second step, the needle is inserted further through the flexor tendons until it reaches the bone. This guarantees that the needle tip's open lumen is in the space between the deep surface of the flexor tendons and the bone, where the second part of the injection (0.1 mL) is administered (Fig. 6.36) [67]. We always aim to perform two injections 7–10 days apart and prescribe climbing rest for at least 10 days after the second injection.

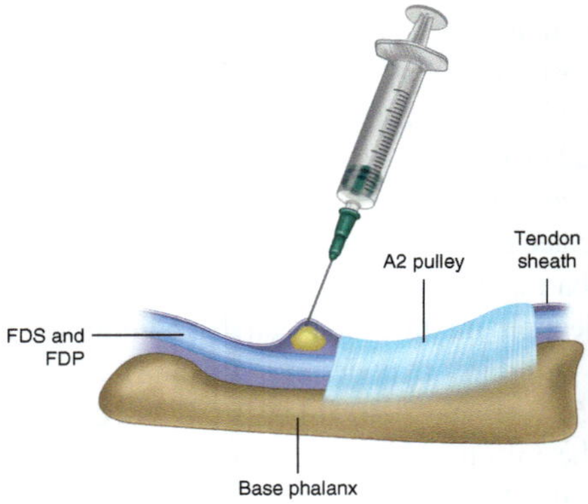

Fig. 6.35 Technique of steroid injection into the tendon sheath (Schöffl et al. 2019)

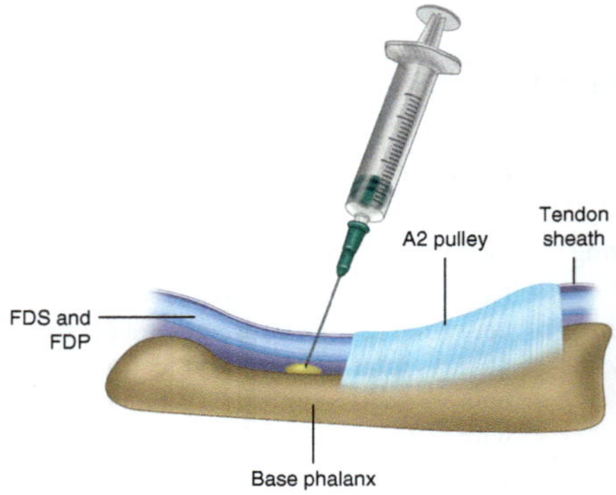

Fig. 6.36 Technique of steroid injection into the tendon sheath (Schöffl et al. 2019)

In an outcome analysis, we recently reported on 42 climbers with tenosynovitis, of which 31 climbers were free of pain (73.8%) after two steroid injections and a mean time frame of 21 days [67]. Only nine patients required another ongoing therapy. All athletes could restart their climbing, and all patients except one regained their preinjury climbing level. No complication based on the injections occurred during the study [67].

Radial shock wave therapy can be used instead of injections at certain stages [73, 76]. Radial shock wave therapy has demonstrated effectiveness in the therapy of chronic, non-inflammatory tendinosis, based on an increased neovascularization [73]. Before administering radial shock wave therapy, the stage of the tenosynovitis needs to be evaluated carefully, as this technique is contraindicated in acute inflammatory tendonitis. Thus, looking at our strict regimen, it may be applied in stage 2 instead of injections.

In patients who do not respond to injection or radial shock wave therapy, we administer medicinal leech (hirudo medicinalis) therapy (stage 3) (Fig. 6.37). In this treatment, a medicinal leech is applied at the tenosynovitis site after needle puncture and left there until it lets go. Fourteen days of climbing rest are prescribed afterwards. Leeches act by withdrawing blood and injecting active substances through their saliva into the host tissue [83]. During this process, more than 100 substances are injected [83–85]. Active substances identified within leech saliva today include hirudin, factor Xa inhibitor, and hyaluronidase [83]. These substances have anticoagulant, thrombolytic, anti-inflammatory, and pain-relieving effects [83–85]. With these described effects as well as the positive effect on symptomatic osteoarthritis at the first carpometacarpal joint, we started using medicinal leeches as an alternative before surgery [86]. After positive preliminary results in our patients with tenosynovitis and capsulitis and in literature [83–85], we added this into our therapeutic regimen (stage 3).

A tenosynovectomy (stage 4 therapy) is only necessary in very rare conditions [19, 67]. In the aforementioned study, no tenosynovectomy was necessary in any patient [67].

Fig. 6.37 Medicinal leech therapy in a chronic tenosynovitis

- **History:** Overstrain, chronic, extensive crimping
- **Onset:** Acute, sometimes chronic
- **Clinical:** Point tenderness of the tendon sheath at the rim of the pulley, sometimes palpable swelling
- **Diagnostic:** Clinical finding, ultrasound, MRI optional
- **Therapy:** Conservative, in persistent pain >6 weeks local steroid injections (twice)
- **Outcome:** Good, surgical revision rarely necessary

6.3.5 Joint Capsule Damage and Collateral Ligament Injury

Pulling on one-finger pockets causes high stress upon the fingers' passive joint structures, the articular capsule, and the ligaments (Fig. 6.38). The climber also tries to reduce the strength needed for pulling by jamming the finger in the pocket (Fig. 6.39). This can strain or rupture both the joint capsule or collateral ligaments.

Fig. 6.38 Passive joint capsule stress in hanging onto a "mono" (Photo: E.Haase)

An injury to the palmar plate typically occurs when the finger is jammed in a one- or two-finger pocket and while the climber tries to continue with the next move, resulting in a hyperextension force to the joint [19]. The diagnosis must include radiographic examination to exclude an associated avulsion fracture at any of the ligamentous attachments (Fig. 6.40). During a hyperextension injury of the PIP joint, the palmar plate can torn off the bone, sometimes with a little osseous fragment (Fig. 6.41). The collateral ligaments are assessed through ultrasound or fluoroscopy. Often, fractures can be visualized using ultrasound as well (Fig. 6.42). If joint capsuloligamentous injury is neglected, it can lead to flexion contracture or deviation deformity of

Fig. 6.39 Shear stress onto the fingers joint capsule in a one-finger pocket

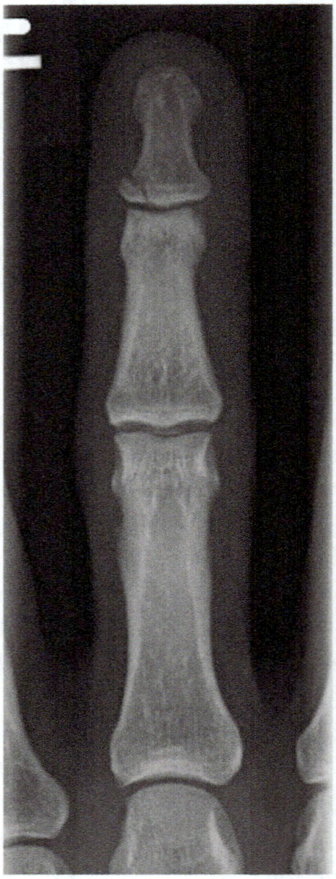

Fig. 6.40 Osseous collateral ligament avulsion after jamming in a pocket

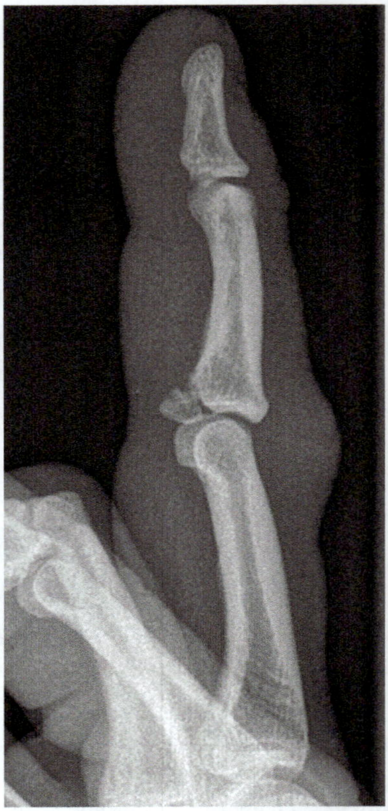

Fig. 6.41 Fracture of the palmar plate caused by overextension and a stuck finger in a pocket

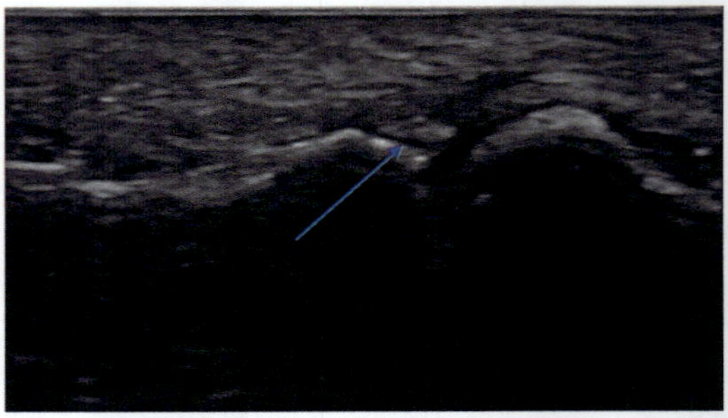

Fig. 6.42 Fracture detection in the ultrasound

the finger (Fig. 6.43) with chronic decreased range of motion and dysfunction (Figs. 6.44 and 6.45). Chronic joint effusions and synovitis can also cause joint contractures.

6.3.5.1 Therapy

Depending on the stability of the joint, treatment is usually early functional range of motion exercises facilitated by buddy taping. A short period of immobilization may be necessary in the case of joint instability or an associated unstable fracture. Overall, the goal is as little immobilization as possible because prolonged immobilization leads to a

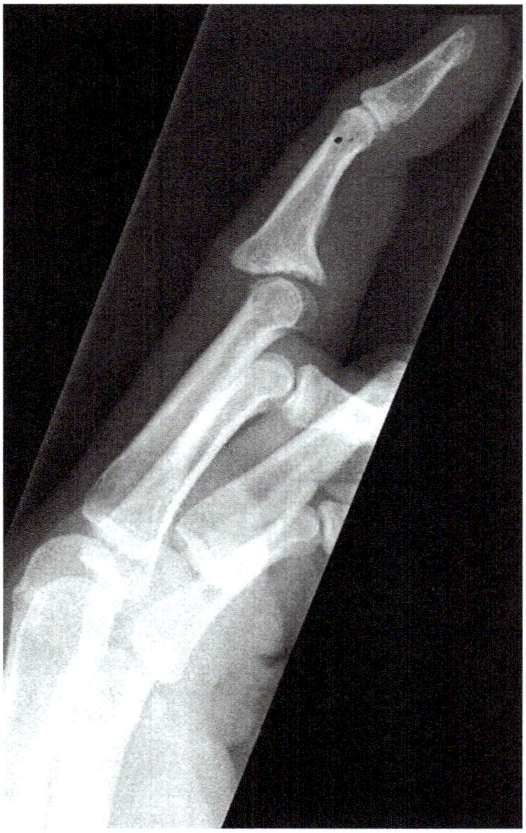

Fig. 6.43 In dislocation fused avulsion fracture of the palmar plate

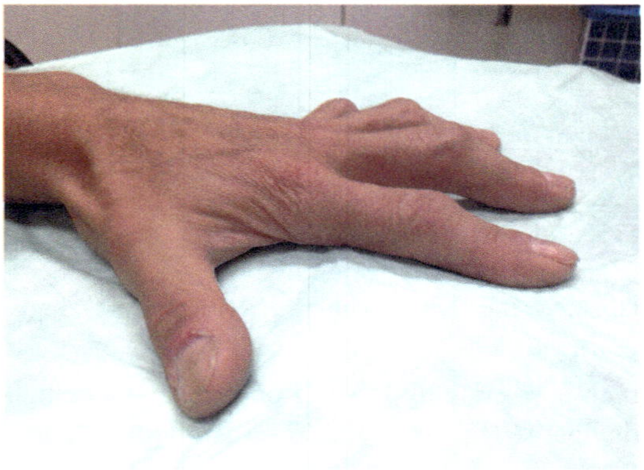

Fig. 6.44 Contracture of multiple fingers in a long-time climber

higher rate of contracture. Only chronic instabilities of a high-grade require surgical treatment. It is important to remember that untreated joint capsule injuries frequently lead to these contractures, which

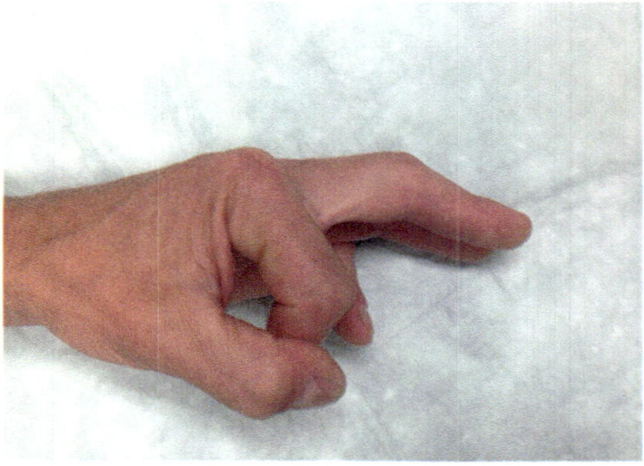

Fig. 6.45 Contracture of multiple fingers in a long-time climber

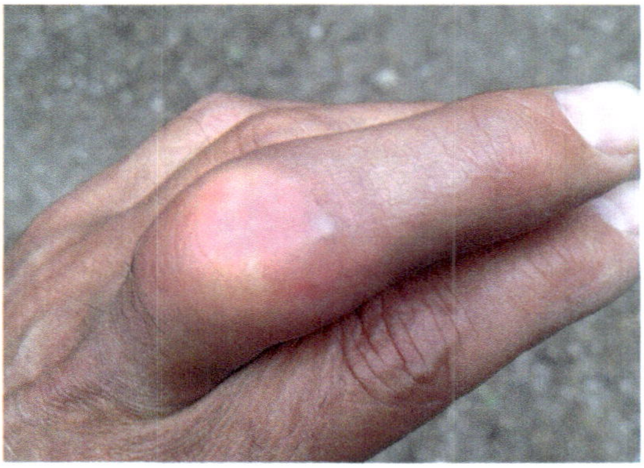

Fig. 6.46 Excessive inflammatory reaction after joint capsule injury. This case could be finally treated well with radiotherapy, but a persisting reduced range of motion resulted

are reported in a high number of older climbers (Figs. 6.44 and 6.45) [87–89]. In few cases, a prolonged inflammatory response can occur, which can be difficult to treat and will also lead to permanent articular damage (Figs. 6.46 and 6.47).

- **History:** Pulling on a one-finger pocket, finger jams
- **Onset:** Acute
- **Clinical:** Joint capsule swelling, painful reduced range of motion, joint instability, pain while pulling in the open grip position
- **Diagnostic:** Clinical finding, X-ray, ultrasound, optional fluoroscope
- **Therapy:** Early functional, rarely complete immobilization
- **Outcome:** Can lead to contracture

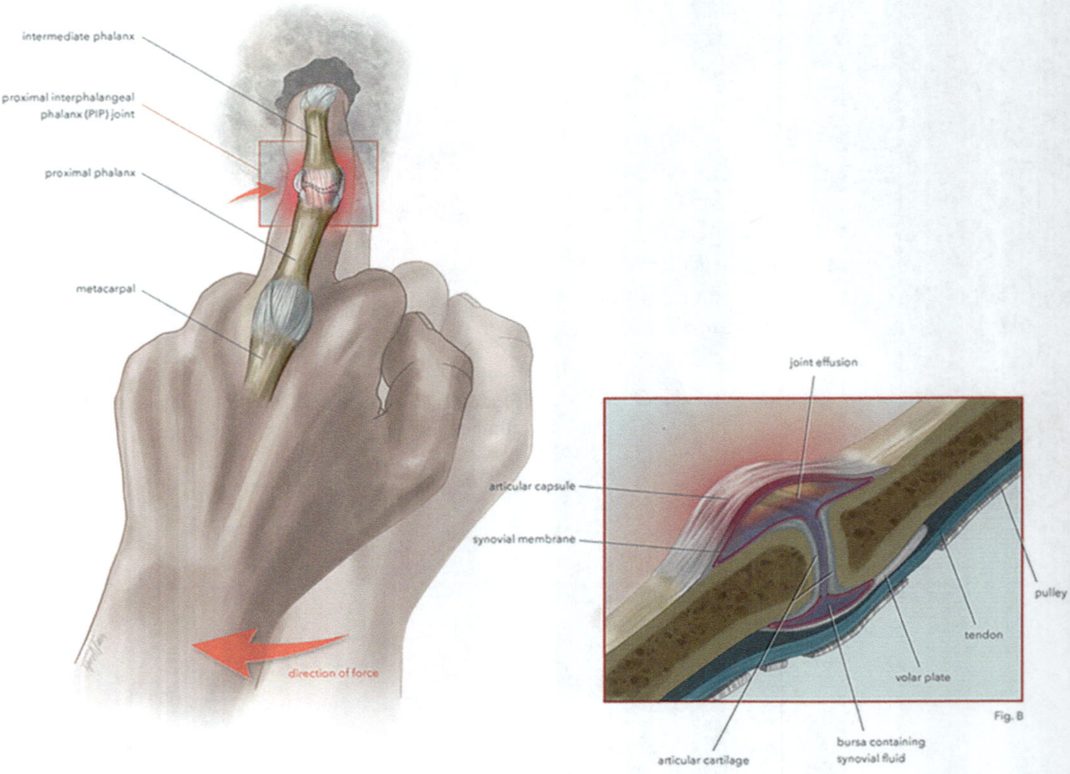

Fig. 6.47 Collateral ligament tear. (Pic by Tiffany Fung)

6.3.6 Chronic Capsulitis (Intra-Articular Synovitis)

High peak pressure within the finger IP joints during the crimp position causes the release of aggressive enzymes that can lead to a chronic inflammation of the joint capsule. If left untreated, chronic capsulitis is a precursor to chronic osteoarthritis (Figs. 6.48 and 6.49).

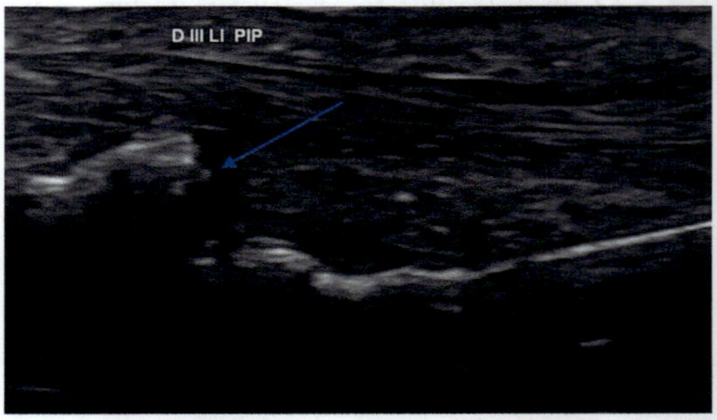

Fig. 6.48 Sonographic finding in a capsulitis

The term "capsulitis" is used to describe chronic synovitis of the finger (or toe) joints in climbing medicine. These can be caused by either repetitive stress or a single traumatic event with reactive effusion within the joint. The climber complains of early morning stiffness and swelling of the finger joints, a reduced range of motion, and pain, which often improves after activity (Fig. 6.50). The clinical examination shows a joint capsule effusion and typically dorsally sided pain under pressure and palpation. Overall, the increased thickness of the small finger joints, as seen in many long-term climbers, is not a pathology per se (Fig. 6.51) but is rather thought to be a functional adaptation of the joint capsule and collateral ligaments to the stresses of climbing [87, 89]. About 40% of climbers have these thickenings of the finger middle and end joints [87, 89]; some cases are just soft tissue adaptations and others are chronic joint effusions [20]. Frequently, there is a correlation with extensive use of the

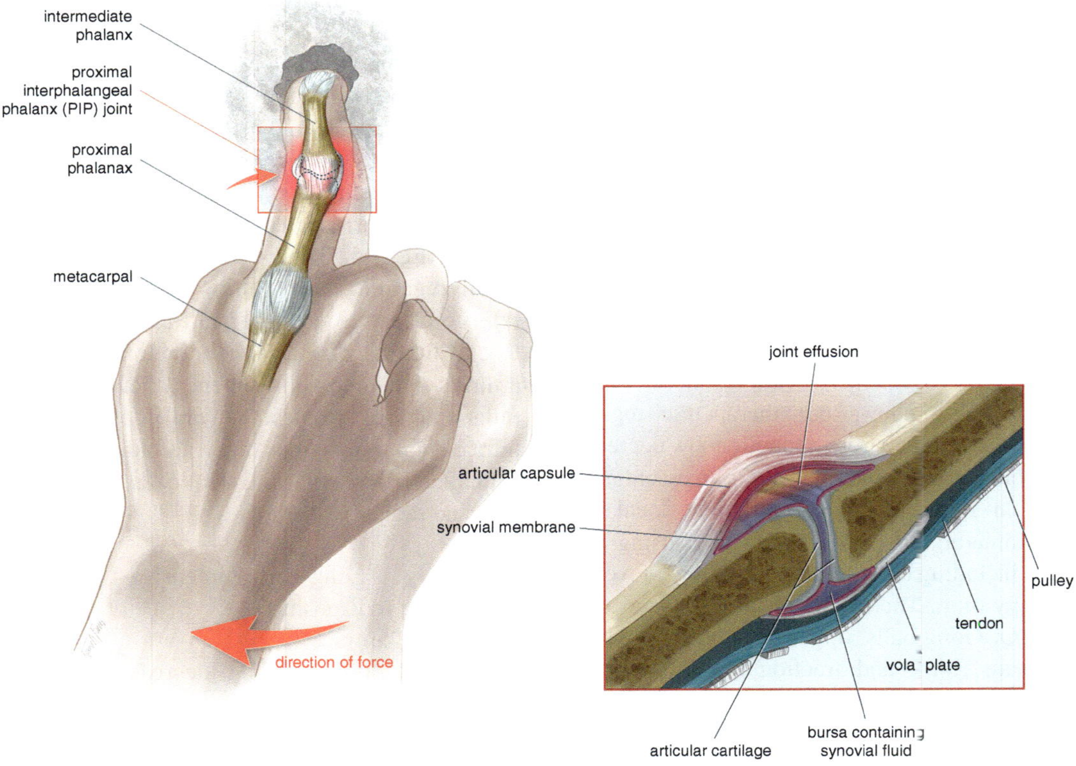

Fig. 6.49 Capsulitis following a collateral ligament tear (pic by Tiffany Fung)

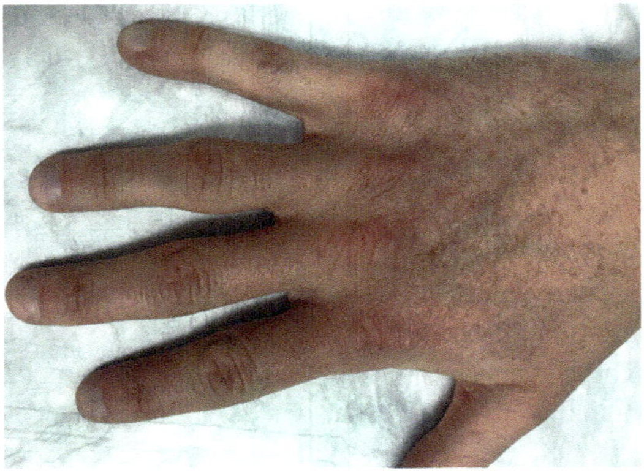

Fig. 6.50 Finger middle joint swelling at the ring finger in capsulitis

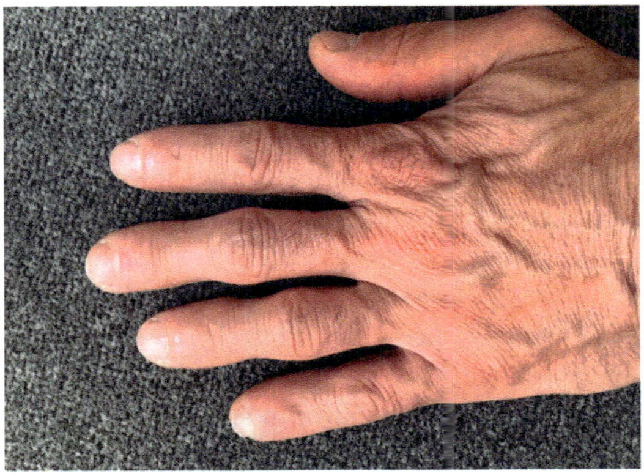

Fig. 6.51 Stress adaptations of the joint capsule after more than 35 years of climbing

crimping position. In this position, the finger PIP joint is almost completely flexed, and the DIP joint is hyperextended. The stress on the cartilage is concentrated onto certain areas of the joint and is not evenly distributed over the whole articular surface. These high-pressure peaks cause micro-injuries and tears of the cartilage and a release of aggressive enzymes into the joint. This can result in a reactive inflammatory reaction of the synovial membrane leading to hyperemia and effusion, which results in swollen joints in the clinical presentation and shows a joint space widening in X-ray examination ("X-ray sign 1") [90]. Also the X-ray may show an inflammatory-edematous synovial membrane in respect of a thickened soft tissue presentation ("X-ray sign 2") [90]. This effusion itself stresses the synovia and can be the start of a vicious circle of overstrain, effusion, and synovitis, leading to a chronic capsulitis and, if untreated, to osteoarthritis and joint destruction [58, 90–93].

The chronic effusion is mostly visible on the dorsal aspect of the respective joint. It leads to a restricted range of motion, reduced fine-motor coordination, and dull pain. The dull pain is very characteristic and based on the chronic irritation of the joint capsule nerve endings.

Table 6.4 shows our therapeutic regimen. Immediate joint stress reduction by decreasing climbing volume and intensity is an essential first step in management. In addition, local anti-inflammatory therapy is performed using practices including but not limited to icing, movement therapy (e.g., TheraBand Hand Exerciser®, softball, Qi-Gong balls, finger massage ring, or acupressure ring), and traction therapy. In more chronic conditions, hand baths with a medical sulfur solution, ointment dressings (e.g., Ichtolan® 20% (Ammoniumbituminosulfonate, Ichtyol®, Hamburg, Germany), and exercising with therapy putty are used. If all of these fail, we perform fluoroscopy-guided intra-articular injections of steroid, hyaluronic acid, or platelet-rich plasma (PRP) (Fig. 6.52). Combinations of treatments are used frequently (e.g., initial treatment with an anti-inflammatory steroid injection followed by a PRP injection 1 week later). In cases of persistent symptoms, a radio-synoviorthesis with erbium shows good results [87]. Alternatively, medicinal leech therapy can be an effective treatment. We recommend an additional substitution with chondroitin and glucosamine.

- **History:** Extensive crimping, high stress load
- **Onset:** Chronic
- **Clinical:** Joint swelling, dull pain, pain while moving
- **Diagnostic:** Clinical finding, ultrasound
- **Therapy:** Difficult, often long-term, intra-articular injections, sometimes radiosynoviorthesis
- **Outcome:** Good, but may take a long time (1–2 years)

6.3.7 Osteoarthritis

The question of whether long-term high-level climbing causes osteoarthrosis of the finger joints is widely discussed and also a topic of many scientific analyses. Studies report a positive correlation between climbing years and climbing level with osteoarthritic changes of the fingers [21, 94–97] (Figs. 6.53, 6.54, and 6.55). For further details regarding osteoarthritis in climbers, please refer to Chap. 13: Long-Term Effects of Intense Rock Climbing. The treatment options for osteoarthritis include movement therapy, anti-inflammatories (topical or oral), cryotherapy, intra-articular injections (e.g., steroid, hyaluronic acid, platelet-rich plasma, stem cells, autologous fat), and radiation therapy.

- **History:** Long-term intensive climbing, genetic predisposition.
- **Onset:** Chronic.
- **Clinical:** Joint swelling, decreased range of motion, pain.
- **Diagnostic:** Clinical finding, radiography.
- **Therapy:** Multiple options—conservative, injectable, radiation therapy, surgery.
- **Outcome:** The goal is to decrease symptoms and slow progression, with response to treatment and rates of degeneration varying widely.

6.3.8 Extensor Hood Syndrome

In athletes with a long history of climbing activity, progressive osteoarthritic changes of the small finger joints have been observed [88, 95, 98–100]. These changes can present as large bone spurs on both the flexor and extensor sides of the phalanges [65, 98] (Fig. 6.56). With repetitive use of the crimping position during climbing, these bone spurs can produce irritations to the extensor tendons [98] (Fig. 6.57). We reported on 13 rock climbers in a 3-year period complaining of dorsally sided pain of the proximal and/or the distal interphalangeal (PIP/DIP) joints [98]. Plain radiographs revealed dorsal bone spurs (osteophytes) on the PIP joint in all climbers and on the DIP joint in three climbers [98]. According to the Kellgren-

Table 6.4 Stage-related treatment regimen for capsulitis of the finger and toe joints in climbers

Stage	Time frame since onset of pain	Therapy	Climbing rest
1	<6 weeks	Conservative: icing, movement therapy, traction, hand baths with anti-inflammatory solution (e.g., medical sulfur solution), exercise therapy with therapy putty, acupressure ring, lymphatic drainage	• 0–14 days • Then stepwise climbing load increase in accordance to the clinical symptoms • Taping during climbing restart with a figure of eight tape
2	>6 weeks (or after failed conservative therapy >4 weeks)	*2a:* local infiltration with corticosteroid, reinjection after 7–10 days *2b:* alternative local infiltration with PRP (platelet-rich plasma), mesenchymal stem cells (MSC), or autologous fat micrograft (if applicable combination of corticosteroid or PRP)	• *2a:* no climbing or hand-related sports in between the injections and at least for 10 days after the second injection • *2b:* no climbing or hand-related sports 48 h after the injection • Then stepwise climbing load increase • Taping during climbing restart with a figure of eight tape
3	Persistent pain and swelling/effusion >6 weeks after second injection	RSO (radiosynoviorthesis)	• Immobilization of the respective joint for 48 h in a splint • Afterward light movement exercises; gradually restart of climbing activities after 6 weeks, with full load bearing after 10 weeks • Taping during climbing restart with a figure of eight tape
4	Persistent pain and swelling/effusion >6 month after first RSO	Second RSO if applicable in combination with simultaneous instillation of a corticoid or Medicinal leech therapy or Local radiation therapy (last resort: surgical synovectomy)	• Immobilization of the respective joint for 48 h in a splint • 6 weeks rest, then stepwise climbing load increase with full load bearing after 10 weeks • Taping during climbing restart with a figure of eight tape • Personalized follow-up regime in case of medicinal leech therapy, radiation, or synovectomy (prolonged immobilization until assured wound healing)

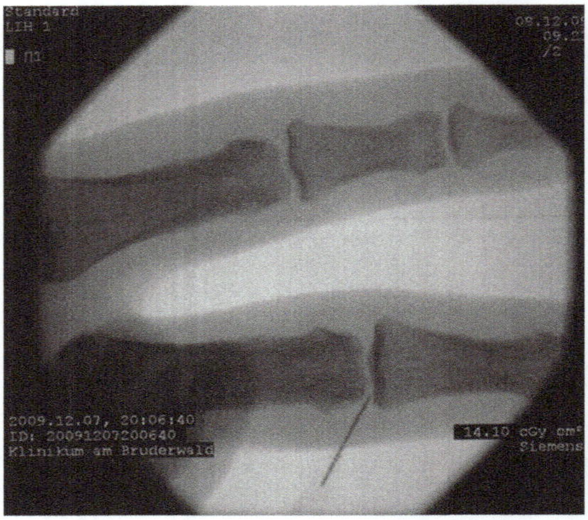

Fig. 6.52 Injection under fluoroscope monitoring

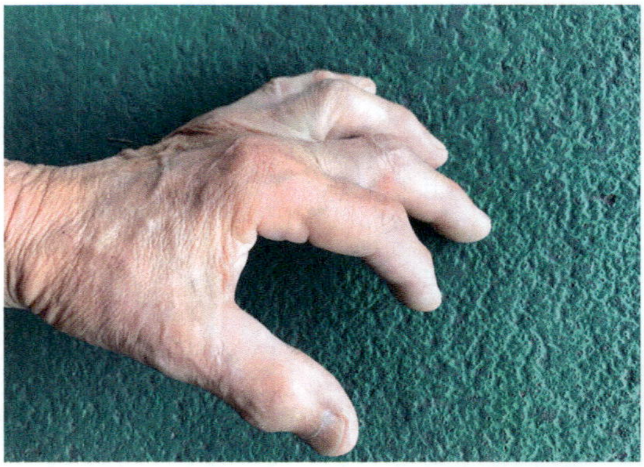

Fig. 6.53 Osteoarthritis in many finger joints in an 60-year-old climber who still climbs at a UIAA level 9 (YDS 5.12.c, Fr 7b+)

Lawrence scale [22], the radiographs (found bilaterally in seven cases) revealed 5 grade 2, 12 grade 3, and 3 grade 4 osteoarthritis. Each of these dorsal bone spurs was causing irritation of the extensor mechanism, resulting in fluid accumulation and tendonitis. In two cases, the dorsal osteophyte was fractured at the time of presentation [98].

Treatment for extensor hood mechanism is primarily conservative with anti-inflammatory ointment dressings, anti-inflammatory hand baths, or local steroid injections. In rare cases, however, a surgical excision of the dorsal-sided bone spurs is necessary to alleviate the extensor tendon irritation [98].

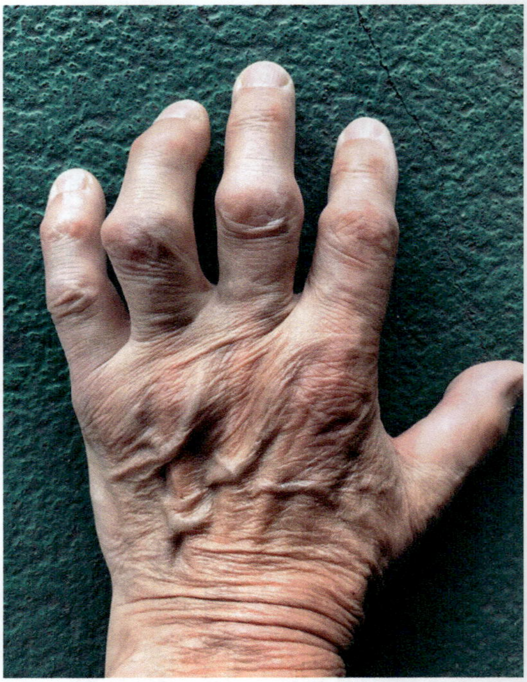

Fig. 6.54 Osteoarthritis in many finger joints in an 60-year-old climber who still climbs at a UIAA level 9 (YDS 5.12.c, Fr 7b+)

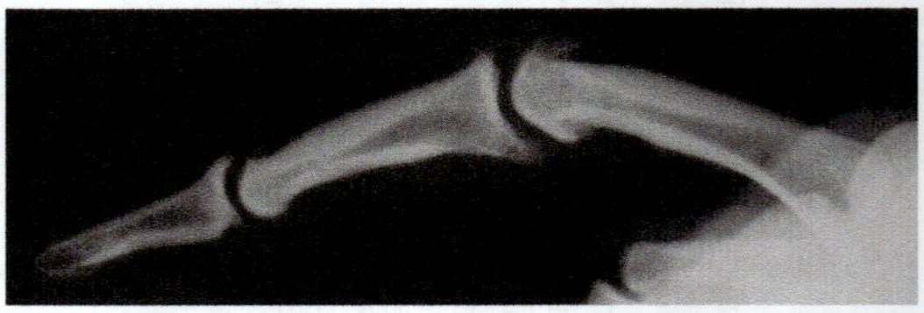

Fig. 6.55 Radiographic finding of osteoarthritis in a long-time climber (over 35 years of climbing). Surprisingly, the climber is almost pain-free!

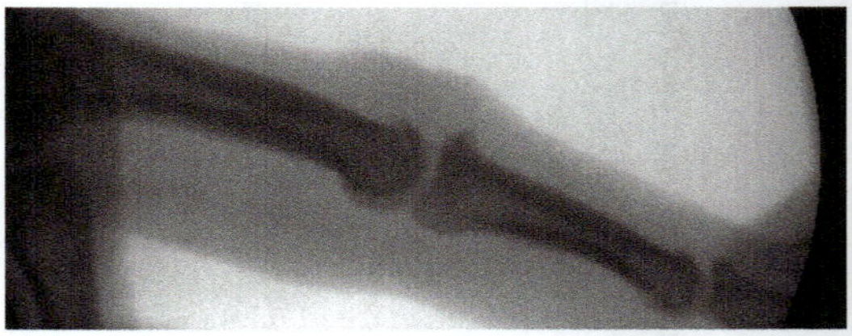

Fig. 6.56 Extensor hood syndrome in fluoroscope imaging

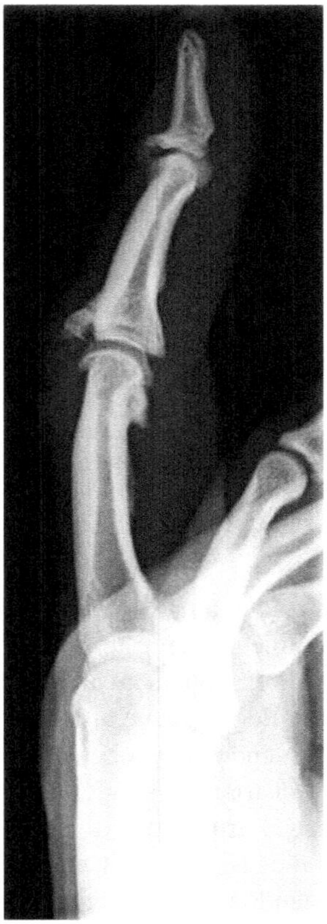

Fig. 6.57 Extensor hood syndrome

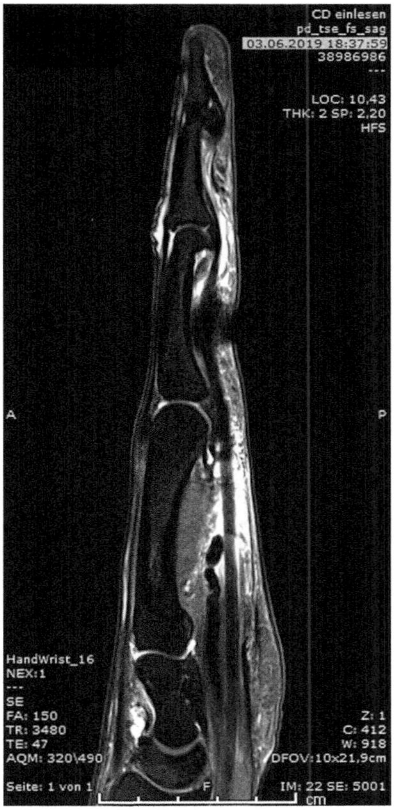

Fig. 6.58 Complete rupture of the deep flexor tendon after a one-finger pocket move

- **History:** Many years of climbing, with dorsal osteophytosis of the IP joints and gradual onset of dorsal-sided finger pain
- **Onset:** Chronic
- **Clinical:** Dorsally sided pain over the fingers' middle or end joint
- **Diagnostic:** Clinical finding, radiographic, ultrasound
- **Therapy:** Mostly conservative
- **Outcome:** Good with conservative treatment, though sometimes surgical treatment is necessary

6.3.9 Flexor Tendon Strains and Tears

Directly injured flexor tendons were observed in relatively few cases, most often caused by a sudden stress on a hand or finger in a hanging position (e.g., the foot slipping off a foothold) [19, 74] (Figs. 6.58 and 6.59). Tendon strains occur most frequently, whereas direct tears, complete or partial, are rare. Patients with flexor tendon strains present with pain running along the flexor tendon [58]. This pain increases in the open finger grip position and can sometimes be absent in a crimping position. Often, a strain of the lumbrical muscle is also present, and lumbrical shift sign will be positive. Physical examination should include isolated testing of the flexor digitorum profundus and superficialis, with the deep flexor tendon more commonly injured. The deep flexor tendons (FDP) are mostly affected. For further diagnoses, ultrasound and MRI are utilized. In flexor tendon strains, the recovery can be prolonged, and the reoccurrence rate is high. Therapy is conservative combined with therapeutic ultrasound, radial shock wave ther-

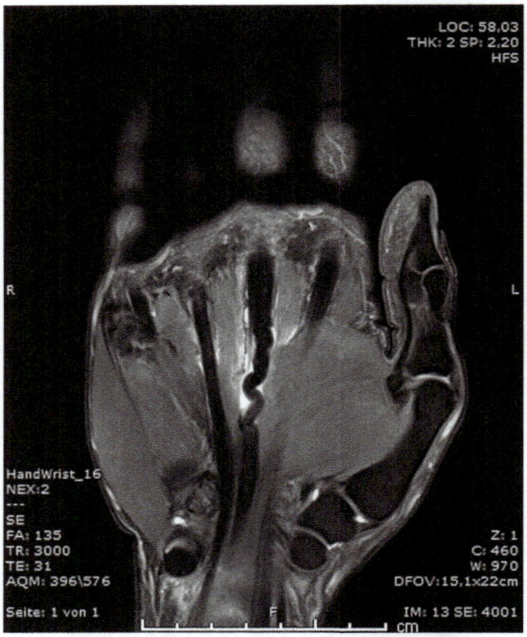

Fig. 6.59 Complete rupture of the deep flexor tendon after a one-finger pocket move

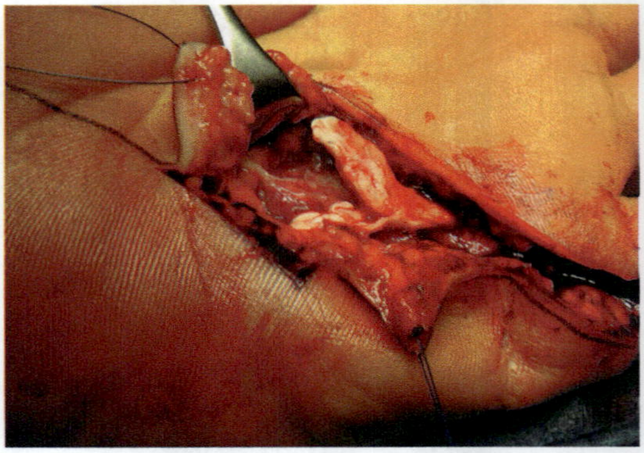

Fig. 6.60 Degenerative spontaneous rupture of both flexor tendons of the small finger while climbing

apy, and light eccentric training in a later stage of the rehabilitation. In rare cases, a partial tear of the tendon occurs, which can lead to tendon nodules and triggering [101, 102].

Open tendon injuries are common findings in general trauma and orthopedic patients and require primary surgical treatment. Closed lesions are more difficult to detect. A differentiation between complete or partial rupture must be made. A lesion of less than 60% of the tendons diameter should be treated conservatively [103]. Multiple biomechanical in vivo and ex vivo studies have demonstrated that conservatively treated partial tendon ruptures have a higher tear resistance ($p < 0.05$) compared to those which are surgically treated [103]. Chronic degenerative tendon ruptures such as those seen in rheumatoid arthritis patients must be distinguished from these traumatic ruptures. In rock climbers, we see a combined entity, with a chronic tendinosis and degeneration and, finally, an acute tear. High intensive stress combined with chronic tenosynovitis and tendinosis can lead to a rupture of the tendon, even after minor mechanical stress [19]. Additionally, the poor blood supply of the tendon, microtrauma, and microscopic structural damage can lead to a modest prognosis.

These complete tendon tears are surprisingly rare and require surgical treatment (Fig. 6.60). Usually, these degenerative attrition tendon ruptures leave the tendon in poor condition for repair, so other reconstructive surgical techniques are required, such as arthrodesis or tendon transfer. If the tear is at the level of the distal interphalangeal joint, flexor tendon reconstruction using free tendon grafting or DIP joint arthrodesis are the best options [19, 58, 104].

Flexor tendon closed avulsion injuries occur at the insertion of the FDP tendon to the distal phalanx. This injury is seen most frequently on the fourth finger, where the FDP-tendon is embedded in between the double-sided lumbrical tendons as shown by Manske and Lesker [105] in cadaver dissection. We have verified that this injury in rock climbers occurs in the setting of chronic degenerative damage to the tendon [104, 106].

These injuries often result in poor postoperative outcomes [19]. Four factors determine the prognosis of avulsion injuries: the extent of the retraction of the tendon, the remaining blood supply, the time interval between trauma and surgery, and the presence and size of an osseous fragment [107]. Recently, Schweizer et al. [108] reported on closed disruption of a single flexor digitorum superficialis tendon in three cases, two in climbing, one in judo. With conservative and functional treatment, all three patients regained normal function and return to sport. The authors concluded that the disruption of one FDS tendon slip is a differential diagnosis for a popping incident in the finger, which is typically associated with a closed flexor tendon pulley injury [108].

In rare cases, osteophyte irritation of the flexor tendons can occur and may require surgical removal to relieve pain and prevent future attrition rupture of the tendon [65, 74] (Fig. 6.61).

- **History:** Mostly degenerative, due to tendinosis. In rare cases, acute blunt disruption.
- **Onset:** Acute.
- **Clinical:** Loss of function, pain in a hanging finger position. Frequently in combination with lumbrical muscle strain or tear.
- **Diagnostic:** Clinical findings, ultrasound, MRI optional.

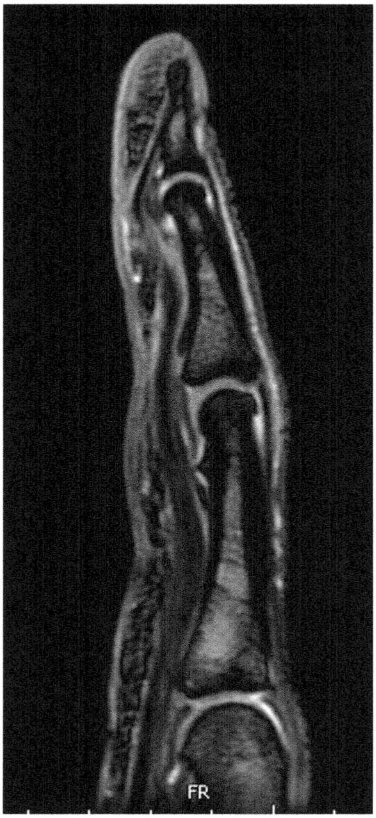

Fig. 6.61 Flexor tendon irritation based on a palmar-sided bone spur, which later needed surgical removal. The climber later regained international top climbing level

- **Therapy:** Conservative mostly, surgical in complete tears.
- **Outcome:** Recovery can be prolonged. Sometimes a joint fusion or tendon reconstruction is necessary.

6.3.10 Ganglion Cysts

Flexor tendon sheath and pulley ganglion cysts are frequent in rock climbers [74, 109, 110]. They are caused by high local pressure onto the flexor tendon sheath from sharp edges of finger holds, leading to microinjuries and tears which eventually form a ganglion cyst. However, flexor tendon sheath and pulley ganglia are not exclusive to climbing [111]. In climbers, they mostly occur on the palmar side but sometimes also on the extensor side, often in combination with a joint effusion (Figs. 6.62 and 6.63). Clinically, they are easy to detect; otherwise, ultrasonography can confirm the diagnosis (Figs. 6.64 and 6.65) [6, 19, 70, 112]. The treatment is conservative; sometimes, the cysts dissolve by themselves through direct pressure applied either while climbing or from gentle self-massage. Sometimes, a puncture and local steroid instillation are necessary. If these conservative options fail, surgical excision can be performed. An open surgery technique is used in order to visualize and protect the digital nerves during the ganglionectomy procedure.

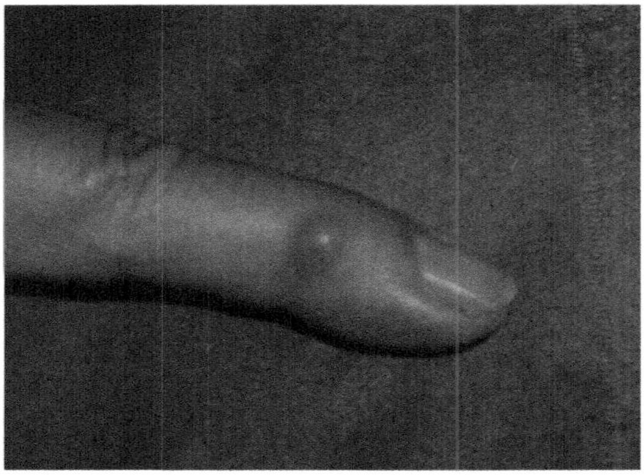

Fig. 6.62 Extensor sided ganglion

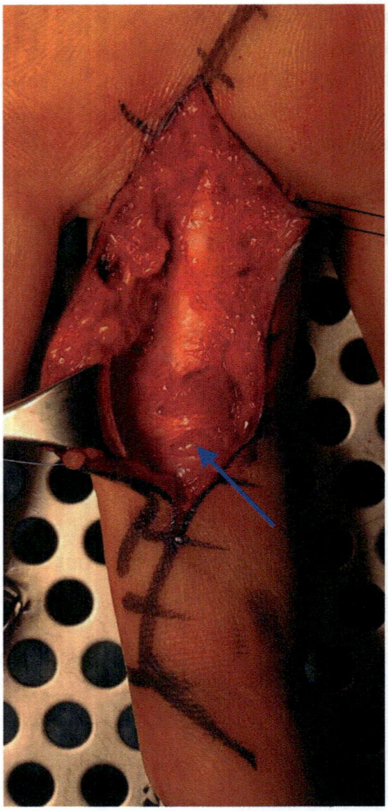

Fig. 6.63 Pulley ganglion as a random finding during surgery

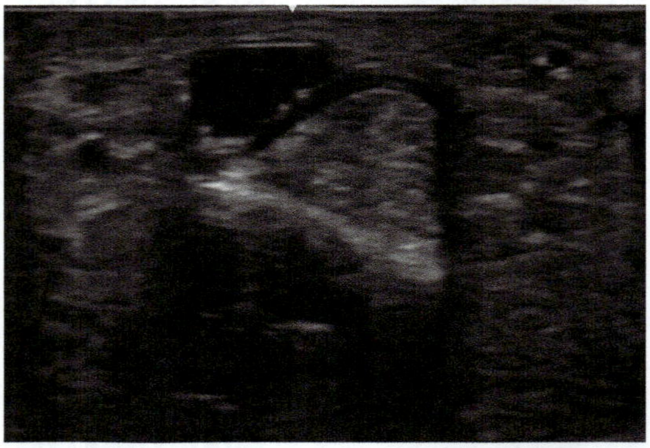

Fig. 6.64 A2 pulley ganglion in a transverse ultrasound plane

- **History:** Microtrauma
- **Onset:** Slow
- **Clinical:** Pressure tenderness, palpable nodule
- **Diagnostic:** Clinical finding, ultrasound

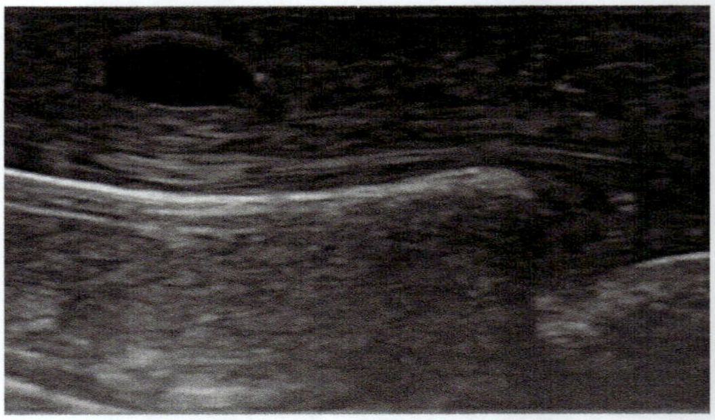

Fig. 6.65 A2 RB pulley ganglion in a longitudinal ultrasound plane

- **Therapy:** Conservative, sometimes steroid instillation
- **Outcome:** Good

6.3.11 Epiphyseal Growth Plate Fractures in Adolescent Climbers

Epiphyseal fracture of the finger middle phalanx base is an injury specific to high-level adolescent sport climbers [113]. Since the first reported case in 1997, there have been more than 100 of these fractures reported in literature [113–117], and recent reports show an increase in incidence [110, 118, 119]. Presently, they are reported to be the most frequent climbing-specific injury in adolescent climbers [113]. Most recently they have also been related to an increase in speed climbing training [120]. With the inclusion of climbing in the Olympics, a continuously increasing number of these injuries are to be expected, as overall training time and intensity will increase and the age of high level climbers will continue to decrease [121]. Patients are typically between 13 and 15 years old and in the phase of the pubertal growth spurt. The fractures are mostly "transient" fractures, meaning fractures during the phase of growth plate closure, where the palmar part of the growth plate is already fused and the dorsal part is still widely open [122] (Fig. 6.66) The middle or ring fingers are most commonly affected, and these fractures are believed to be caused from repetitive stress and microtrauma rather than from a single acute event [113].

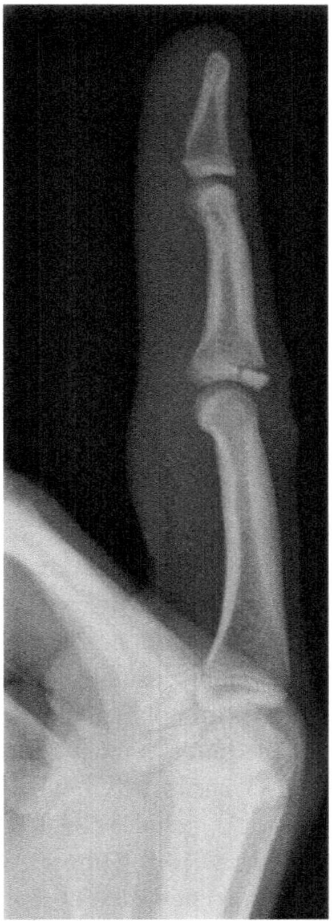

Fig. 6.66 Epiphyseal growth plate fracture as a transient fracture (age: 13 years)

Nevertheless, in a higher number of cases, the patients refer to a final, evolving event, e.g., one hard campus board workout or one exceptionally intense bouldering session.

These fractures are caused by the high pressure onto the dorsal aspect of the growth plate in a crimping position [113]. In addition, the pulling forces of the extensor tendon central slip is at the dorsal growth plate [115, 122, 123]. Bartschi et al. [124] also report a translatory shift of the fingers' middle phalanx base on the proximal phalanx head as a possible additional cause. The pulling force of the flexor tendons shifts the dorsal cortex in a crimping position dorsally, while the pressure from the palmar side onto the fingertips during crimping also pushes the cortex dorsally. Thus, two forces are applied to the dorsal growth plate fragment from two sides, shifting it dorsally.

We found that out of 16 youth climbers with finger pain, 14 revealed an epiphyseal fracture on magnetic resonance imaging and one climber showed bilateral fractures [110]. Of 20 injured young climbers up to the age of 14, 14 (70%) had an epiphyseal fracture [110]. Thus, an epiphyseal growth plate fracture is by far the most frequent injury in young climbers [125], with the overall incidence still increasing [95].

6.3.11.1 Diagnosis

The growth plates of the finger bones are not closed until the age of 17–19 years. Growth plates are the weakest structure of the finger in adolescents and are most susceptible to injuries [115]. As soon as pain without an obvious trauma in the PIP joints is apparent in an adolescent, the crimp grip position should not be allowed at all until the pain disappears or the growth plates have closed at the age of 17–19 years. In the clinical presentation, the climbers report dorsally sided pain at the PIP joint, mostly at the base of the middle phalanx. There is pressure tenderness at that area, sometimes accompanied by minor swelling and a palpable protruding bone fragment (Fig. 6.67).

After clinical suspicion, a standard radiograph can be performed to diagnose a fracture (Fig. 6.66). It is important to note that the mostly obliquely oriented fracture lines are obscured and may not be visible in conventional radiographs. If the plain radiographs are negative and there is a high index of suspicion for fracture, CT scan and/or MRI should be performed to rule out an occult fracture [123] (Fig. 6.68). A standardized diagnostic-therapeutic algorithm is currently published [126]. We recommend an MRI in all adolescent climbers with persistent finger pain for 2 weeks after sufficient rest and in the absence of acute finger trauma.

The fracture pattern is usually Salter-Harris type III (81%) [113, 114] on mostly the middle finger but sometimes the ring finger [63, 113, 114, 125]. Sonographic detection of these injuries is far more difficult. While many of them are detectable, some are not; thus, a normal ultrasound does not exclude an epiphyseal fracture. Nevertheless, as described in prior chapters, all

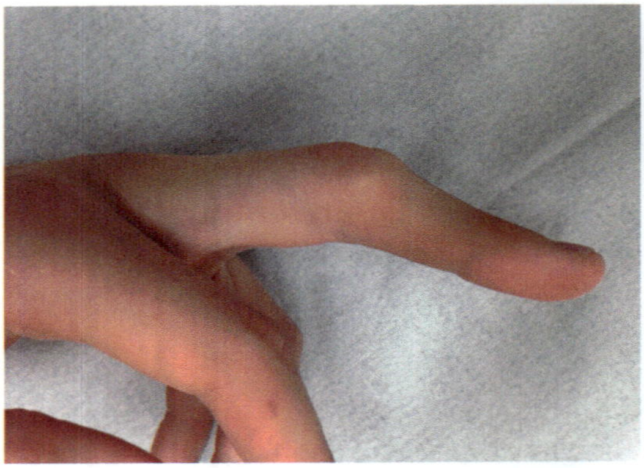

Fig. 6.67 Clinical finding in an epiphyseal growth plate fracture; note the protruding dorsal bone

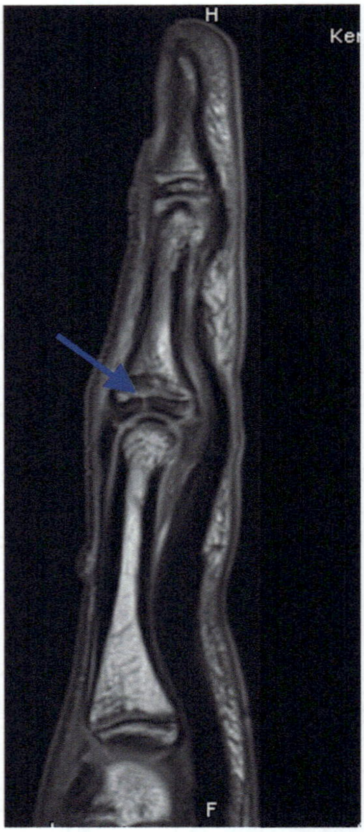

Fig. 6.68 Epiphyseal growth plate fracture in the MRI. The normal radiographs could not detect a fracture (age: 13 years)! MRI diagnostics are mandatory in clinical suspicion if the normal radiographs are negative

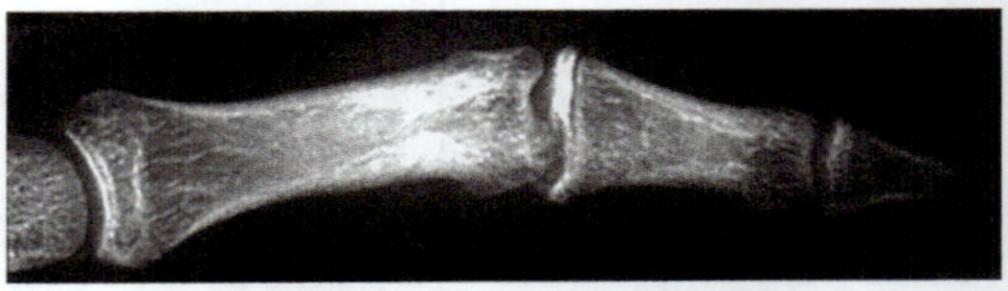

Fig. 6.69 Malunion of a growth plate fracture. No therapy was performed

finger injuries receive a diagnostic ultrasound evaluation in our clinic. If growth plate injuries are not diagnosed and neglected, they lead to permanent PIP joint osteoarthritis, often with stiffness, pain, and deformities (Fig. 6.69).

6.3.11.2 Therapy

Non-displaced fractures receive conservative therapy, including complete rest from climbing and gentle finger range of motion exercises until fracture healing, which usually occurs within 2–4 months [115]. Immobilization is usually not necessary, but the climber must cease all finger-stressing exercises until fracture healing has been confirmed. This also includes isolated hang board training with the neighboring fingers, as the extensor tendons are interconnected at the level of the metacarpophalangeal joint. Due to this connection, a training-related stress onto the ring finger also leads to a stress to the extensor tendons of the middle finger. This results in increased stress of the middle finger's growth plate, even if the middle finger is not directly involved in gripping during the exercise. A stepwise return to climb regimen was recently published by Meyers et al. [127]. Results of conservative management are excellent in patients who present with acute, non-displaced fractures. Patients presenting with chronic or displaced fractures may heal with conservative treatment, but some have incongruity of the middle phalanx base at the fracture site. Patients who continue to climb without allowing their fractures to heal are likely to develop osteonecrosis of the epiphyseal fracture fragment, with resultant permanent angular deformity and stiffness of the affected joint. Acute fractures with significant displacement are treated with closed

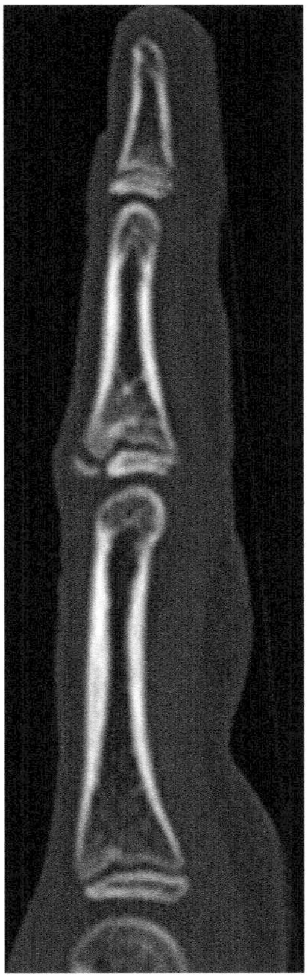

Fig. 6.70 Therapy-resistant sclerosed epiphyseal growth plate fracture (age; 15 years, female)

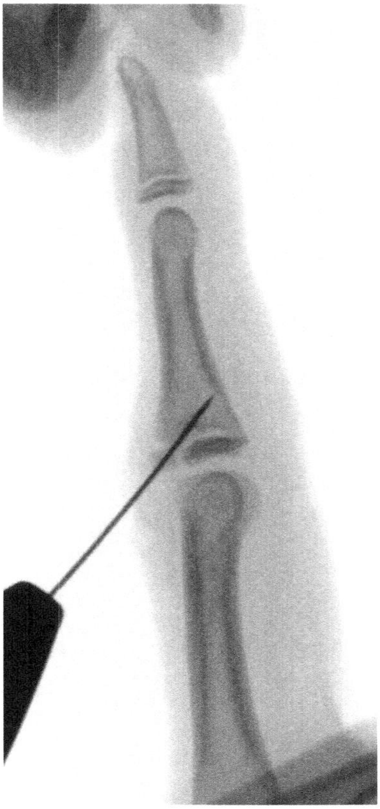

Fig. 6.71 Surgical spot drilling in patient of Fig. 6.70

reduction +/− percutaneous k-wire fixation. Rarely, an open reduction is considered if closed treatment fails; however, risk of an open procedure to the growth plate must be considered. In the case of mild displacement of the fracture, we still treat conservatively but immobilize for 2–3 weeks. In cases with a delayed osseous healing, we perform a CT scan to evaluate for sclerosis at the fracture margins, a sign of established nonunion, which will prevent bone healing without surgical manipulation (Fig. 6.70). If a CT scan shows a displaced fracture and sclerosis, we treat the injury with surgical epiphysiodesis via spot drilling [122] (Figs. 6.71 and 6.72). Recently a therapeutic algorithm was established and evaluated [128].

6.3.11.3 Results

In our series of 16 youth climbers with finger epiphyseal fractures, we found that nine (56%) showed complete healing, five (31%) had incongruence of the joint space, and two (13%) cases resulted in a deformity. In another study, we found that 70% of the youth climbers with epiphyseal fractures had good outcomes, whereas 23% were still pending and one case (4.5%) had permanent damage [113]. Lastly, we reported that intensive finger training (campus boarding, a previously known risk factor for epiphyseal fractures) during adolescence can lead to early-onset osteoarthritis of the hand up to a decade later [95]. We also described good outcome of patients treated with surgical spot drilling epiphysiodesis [122]. To date, we have performed these spot drillings in about ten cases, all with good results and complete fusion and fracture healing within 8–10 weeks.

The most important intervention factors for prevention of epiphyseal fractures are sports

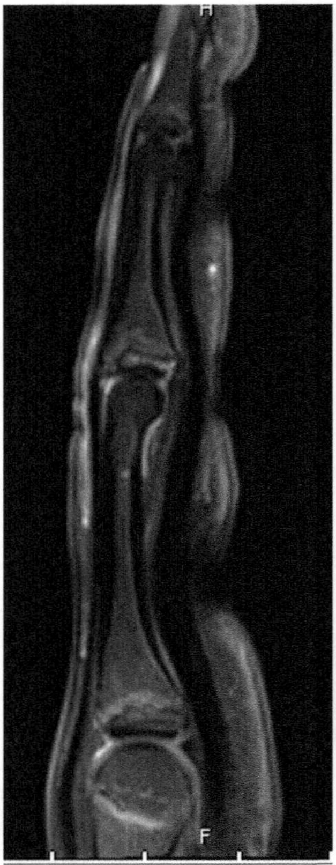

Fig. 6.72 MRI with complete fusion after 4 months

medical education and close sports medical surveillance of competitive adolescent climbers, with the avoidance of certain high-stress training exercises in this age group (e.g., campus board, training board with extra weight). The use of the campus board was shown in several studies to be a significant risk factor for the development of an early onset of finger joint osteoarthritis in adolescent high-level climbers [95]. Some eager trainers argue against this, stating that the junior climbers are also crimping during climbing, which is also provoking the injury. Nevertheless, we think that the extra stress of an additional campus board training or hang board training with extra weights certainly adds to the overall training load and may not even be necessary in this age group. Even though the pubertal growth spurt is considered the high-risk phase for these fatigue fractures, we have seen them in younger athletes without any signs of puberty onset (Tanner stage 0) and in athletes over the age of 18. Close training monitoring with analysis of the overall training load and avoidance of overly intensive load peaks is important. Additionally, sports medical supervision and annual clinical examination of the hands should be performed in youth climbers. Overall, finger pain in an adolescent after or during climbing that does not abate after 1 week of rest should be managed as an epiphyseal fracture until proven otherwise [113].

In conclusion, early detection of a growth plate fracture is critical to ensure high likelihood of healing without long-term consequences. Thus, educating sports medical doctors, trainers, athletes, and parents is crucial in order to instill acute clinical awareness for this injury in youth climbers with finger pain at the PIP joint level.

- **History:** Chronic, mostly crimping position
- **Onset:** Slow, sometimes acute initiating event
- **Clinical:** Dorsal-sided PIP joint tenderness
- **Diagnostic:** Clinical finding, MRI
- **Therapy:** Mostly conservative, sometimes surgical spot drilling epiphysiodesis
- **Outcome:** Good if treated, very bad if neglected (permanent damage)

6.3.12 Lumbrical Shift Syndrome and Lumbrical Muscle Tears

Injuries to the lumbrical muscles have been reported to be rare climbing-specific injuries by Schweizer et al. [129]. In recent years, we have seen an increasing number of these injuries in our patients as well as in international, top-level climbers [109, 121, 130].

The pathomechanics of lumbrical muscle injury are related to the "quadriga effect" (Fig. 6.73), which describes a shear stress on the bipennate third or fourth lumbrical muscles to the ring and small fingers, respectively. These two lumbricals frequently have a dual origin from adjacent flexor digitorum profundus tendons, so shear stress injury to the muscle can occur [129–131] from gripping positions in which one or two fingers are extended while the neighboring fin-

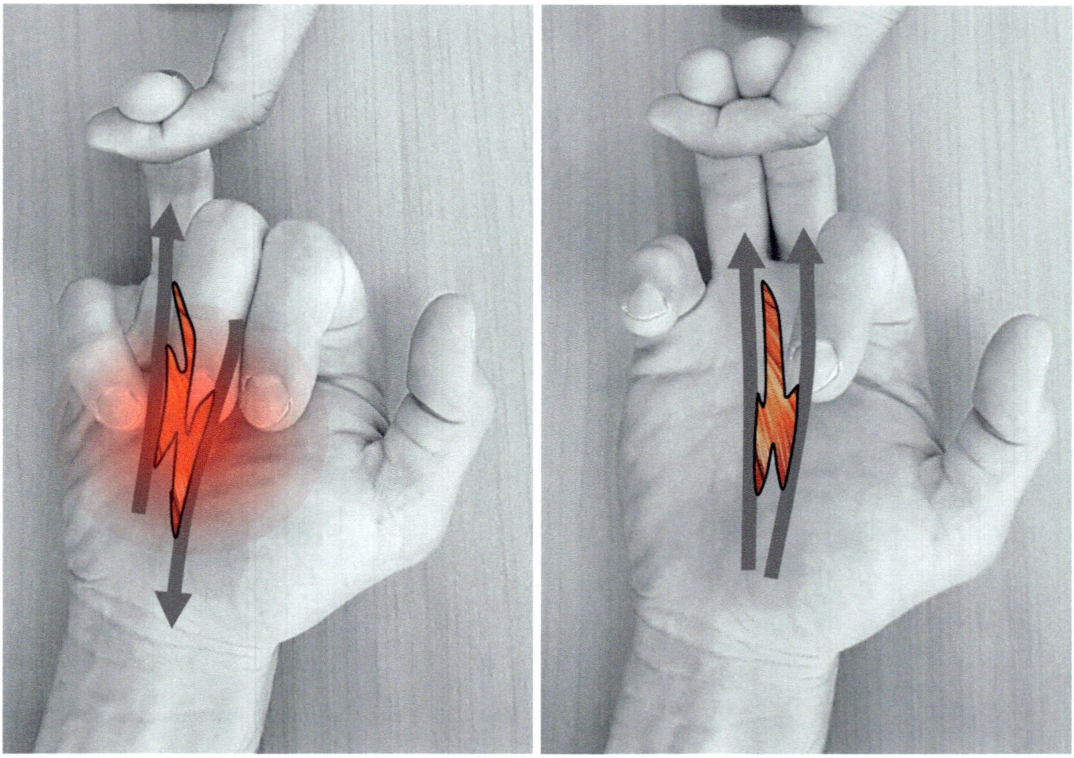

Fig. 6.73 Lumbrical shift test

gers are flexed by force [129, 131]. Typically, this occurs in one- or two-finger pockets.

This finger position increases the maximum strength by up to 50% but causes a shift of the FDP tendons and the common muscle body of the various fingers against each other, leading to muscle strains or partial tears [130] (Fig. 6.74). Injuries typically occur within the third or fourth lumbrical muscles because these are the only ones that have a bipennate origin [130, 132–134].

6.3.12.1 Diagnosis

The diagnosis is made using the lumbrical stress test (Fig. 6.1). This test is positive when pain is elicited while loading the injured finger in extension, with the adjacent fingers flexed. The pain is at the palm, sometimes radiating into the forearm along the course of the FDP tendon of the respective finger (Fig. 6.75). If the climber pulls with all fingers in extension, there is no pain. Also, crimping is usually pain-free. A detailed overview shows the three different grades of injuries (Fig. 6.5). Ultrasound shows an edema of the lumbrical muscle (grade II injury) or even a hematoma (grade III injury). In cases of hematoma (grade III injury), an MRI should be performed (Figs. 6.76 and 6.77).

6.3.12.2 Therapy and Outcome

Therapy consists of symptomatic treatment, taping, and carefully stretching of the muscle [129] and follows our guidelines (Fig. 6.78) [135]. Initially, a rest from climbing and careful movement exercises are recommended. Immobilization is only required in high-grade injuries. Immobilization of the injured lumbrical muscle is initially performed through buddy taping and later through specific lumbrical taping (see Chap. 22 "Taping"). It is important to start with stretching exercises early, which is done in the same way that the injury was provoked but with much less load [27]. In addition, radial shock wave therapy has been shown to provide good pain

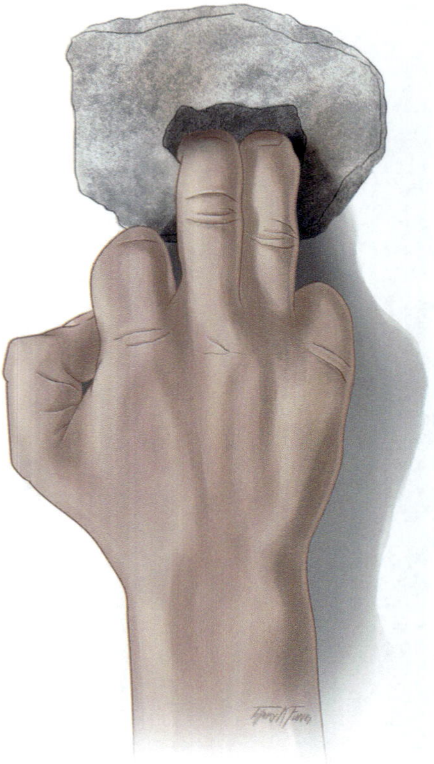

Fig. 6.74 Typical two-finger pocket position with lumbrical stress. (Pic by Tiffany Fung)

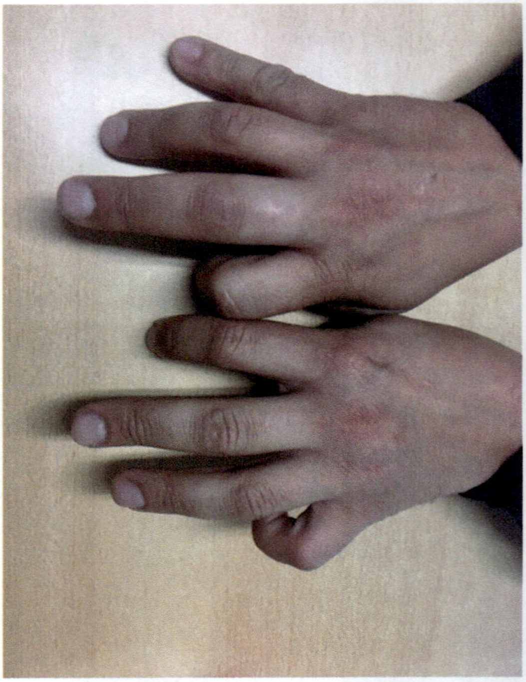

Fig. 6.75 Lumbrical muscle tear with clear swelling of the injured finger

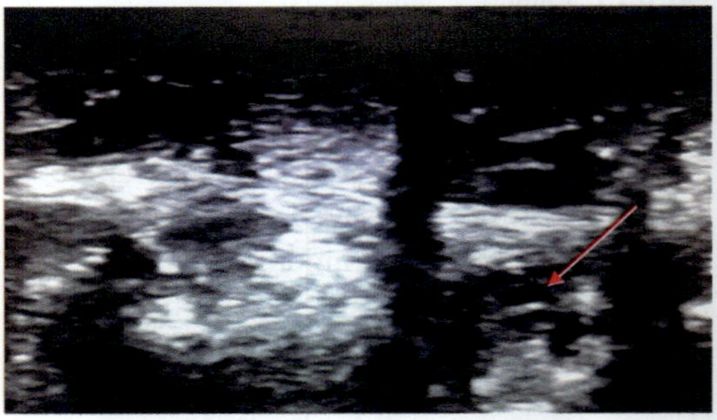

Fig. 6.76 Ultrasound of the lumbrical rupture in the same patient

relief [74]. Healing time is often prolonged with a high reoccurrence rate if exposed to similar stress. Lumbrical injuries are often correlated to flexor tendons strains.

We reviewed data from 60 consecutive patients with a positive lumbrical stress test, including clinical examination (n = 60/60), ultrasound (n = 60/60), magnetic resonance imaging (n = 12/60), and outcome (n = 60/60) [130]. Lumbrical muscle tears were graded as mentioned above (grade I–III injuries) [130]. The therapy consisted of adapted functional therapy [130]. Thirty percent of patients had grade I injuries (micro-trauma), 53% had grade II injuries (muscle fibre disruption), and 16% had grade III injuries (musculotendinous disruption) [130]. All patients had an uncomplicated outcome with complete recovery and unaffected return to climbing [130]. The healing period in grade III injuries was significantly longer than in the other two groups ($p < 0.001$).

- **History:** One or two-finger pocket
- **Onset:** Acute
- **Clinical:** Positive lumbrical stress test
- **Diagnostic:** Clinical finding, ultrasound, MRI
- **Therapy:** Conservative
- **Outcome:** Often prolonged healing time, high rate of reoccurrence

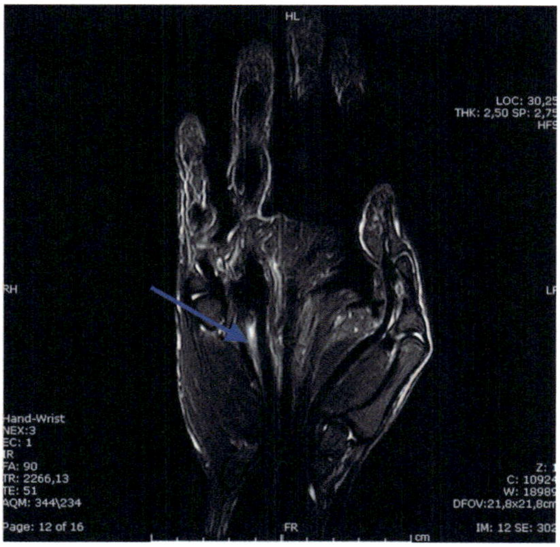

Fig. 6.77 Lumbrical rupture in the MRI

6.3.13 Cellulitis and Other Soft Tissue Infections

Climbing-related minor cuts of the skin can cause bacterial infections of the skin (cellulitis), which can spread to the deep soft tissues of the finger and the flexor tendon sheath (suppurative tenosynovitis) (Fig. 6.79). The classic clinical findings of suppurative tenosynovitis are Kanavel's cardinal signs (fusiform finger swelling, tenderness of the flexor sheath, pain with passive extension, and slight flexed position of the finger). The ultrasound can help in the detection of an edema or pus in the tendon sheath. In the early stages (24–72 h), suppurative tenosynovitis can sometimes be treated conservatively, with splinting, elevation, and oral or IV antibiotics. However, after this initial period, urgent surgical incision

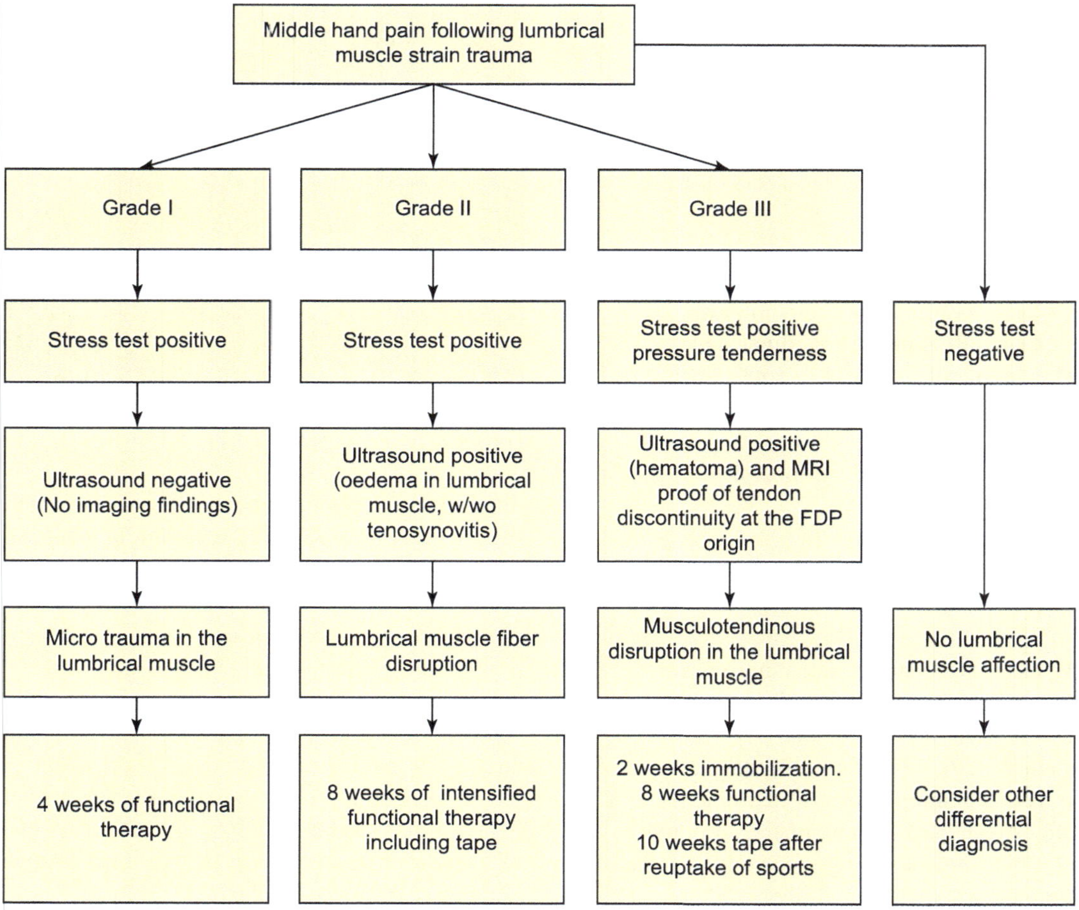

Fig. 6.78 Lumbrical rupture in the MRI

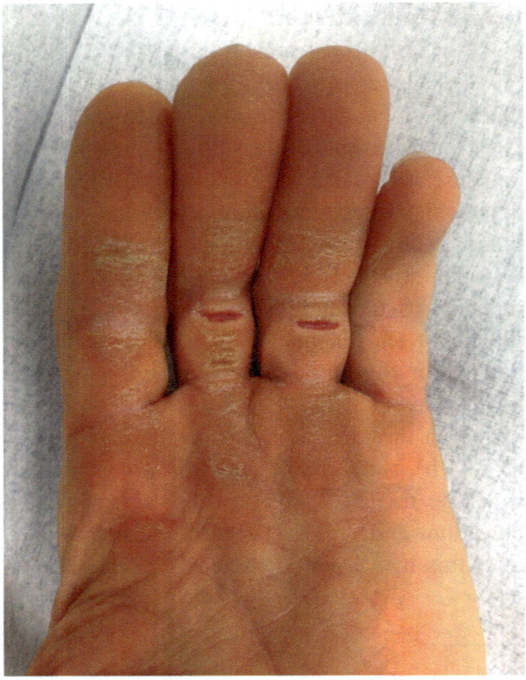

Fig. 6.79 Skin cuts are frequent

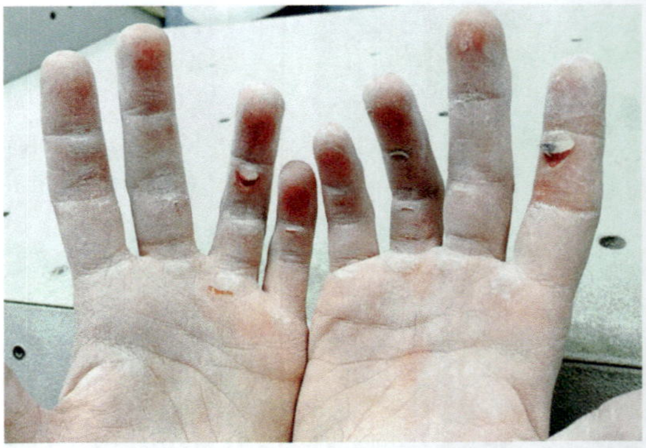

Fig. 6.80 Skin injuries can easily lead to infections (cellulitis)

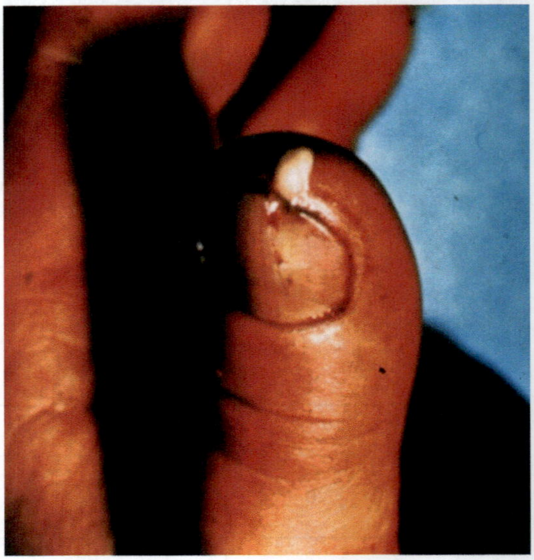

Fig. 6.81 Nailbed infection after a minor injury

and drainage +/− debridement are necessary to clear the infection before necrosis of the flexor tendons occurs (Figs. 6.80 and 6.81).

- **History:** Small skin breaks (cuts)
- **Onset:** Over a few hours to days
- **Clinical:** Kanavel's cardinal signs
- **Diagnostic:** Clinical findings, ultrasound
- **Therapy:** Initially conservative, otherwise surgical
- **Outcome:** Dependent on early diagnosis and treatment

6.3.14 Irritations of Finger Nerves

While nerve compression syndromes in climbers (e.g., carpal tunnel syndrome, supinator syndrome) are described and explored in several studies [136, 137], finger digital nerve irritation is a less commonly discussed problem and occurs almost exclusively in climbers, particularly when climbing on sharp finger pockets or jamming in finger cracks. If the climber uses one- or two-finger pockets, the fingertip is often jammed into the pocket (Fig. 6.82) [58], resulting in localized pressure on one or both digital nerves at the level of the pocket edge. If this pressure is high-load, prolonged, or repeated, it can result in a neuropraxic injury to the digital nerve, which causes a temporary nerve conduction block (i.e., mild acute compression neuropathy), with resultant numbness, paresthesias, or neuropathic pain in the corresponding finger digital nerve territory. It is difficult to predict how long the recovery time of the respective nerve will be, but most resolve within a few days to weeks.

Fig. 6.82 Jamming in pockets can lead to finger nerve damage

Therapy is conservative, consisting of regular finger massage, range of motion, and nerve gliding exercises as well as explanation of the cause of the problem to reassure the climber and help prevent further injury.

- **History:** High localized pressure onto the fingers digital nerve (s)
- **Onset:** Acute
- **Clinical:** Numb fingertip
- **Diagnostic:** Clinical finding
- **Therapy:** Conservative
- **Outcome:** Excellent

6.3.15 Dupuytren Contracture

Historically, this is a disease that usually occurs in men between the age of 40 and 60 and is caused by genetic predisposition, with other risk factors including diabetes, hypothyroidism, epilepsy, and chronic microtraumas associated with hard labor [138–140]. However, this condition is now on the rise in climbers, also those younger than 40. In a study published by Logan et al. in 2005, an increased prevalence of 20% was described for Dupuytren's disease in a relatively young rock climber sample (mean age 54 years) [141]. We have even seen several cases of early Dupuytren's contracture in climbers under 18 [19]. This condition is so rare among younger

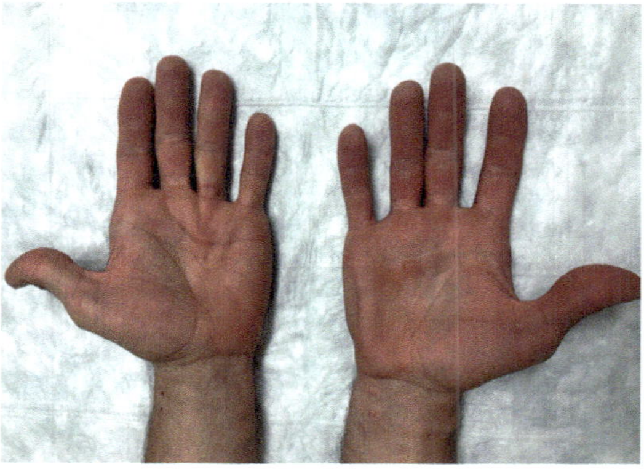

Fig. 6.83 Dupuytren's contracture in a climber

people that a number of young climbers who have suffered from it have actually been misdiagnosed with a malignant tumor [19]. In climbers, it is most likely caused by the repetitive stress and microtrauma to the palmar aponeurosis, which triggers the disease [19, 141]. Pervulesko et al. developed a self-diagnostic tool for rock climbers which was also used to evaluate Dupuytren's contracture frequency in climbers [142].

The therapy is prescribed according to severity and is mostly conservative. Most climbers only present with early stages, which rarely progress extensively (Fig. 6.83). In higher-grade stages (grade 3–4), various treatment techniques are available, including needle fasciotomy, Xiaflex® collagenase injection, radiation therapy, or selective fasciectomy surgery.

- **History:** Eecurrent mechanical stress onto the palmar aponeurosis
- **Onset:** Slow
- **Clinical:** Palpable and visible nodule
- **Diagnostic:** Clinical finding
- **Therapy:** Mostly conservative
- **Outcome:** Stage-related

6.3.16 Rope-Tangling Injuries, Avulsion Amputations

Rope-tangling injuries are an established pathology in climbing accidents and were first reported in 2003 [143], followed by subsequent reports which had different causative mechanisms [144, 145]. There are three major mechanisms involved with rope-entanglement or rope-catching injuries:

1. A sport climber takes a minor fall, just above the protection. While falling, he or she may grab the ascending rope, mostly as a reflex borne out by fear. This grabbing action is unnecessary, as "catching a fall" is the responsibility of the belayer at the other end of the rope. However, by instinctively grabbing the rope close to the protection point, the following injuries are possible. The climber may (1) catch the fall by him- or herself and get severe skin and soft tissue injuries in the palm, (2) get tangled with the belay point, (3) unwillingly unclip the belay device which may create the potential for more serious injuries, or (4) tangle the rope around his or her hand or fingers. All of these causative mechanisms have been reported [144].
2. A climber takes a long fall and unintentionally grabs the rope that comes from the harness. If the fall is long, more loose rope is involved. This longer rope can tangle or loop around the climber's hand or fingers, and, once the rope tightens at the moment of the fall impact, the resulting tension causes avulsion amputations of one or more digits, fractures, or severe contusions/bruises [144, 145].
3. A climber takes a fall on a steep or overhanging route. As he or she is dangling by the rope beneath the last point of protection, the standard technique in sport climbing to get back onto the wall is to perform one or two pull-ups on the rope and then let go. With each pull-up, the belayer takes in the excess rope or runs backward to reestablish rope tension. Thus, the climber can slowly make his way up the rope. However, we have one case study of a very experienced high-level climber who entangled his hand with the rope while performing this described technique which caused severe bruises, soft tissue damage, and metacarpal fractures.

The consequences are either avulsion amputations [144, 145] (Figs. 6.84 and 6.85), fractures or extensive soft tissue injuries (Fig. 6.86).

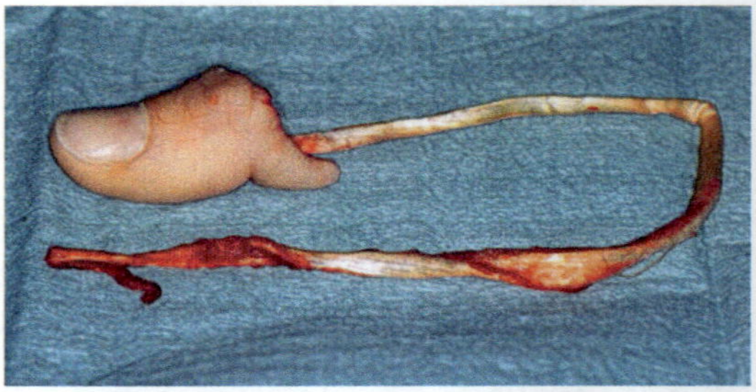

Fig. 6.84 Avulsion amputation through rope tangling in a climbers fall

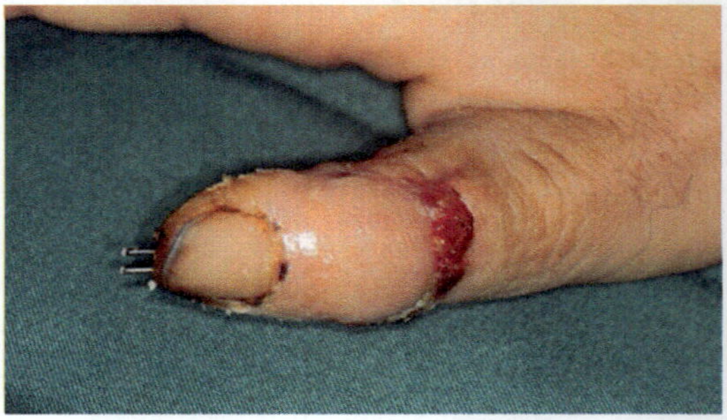

Fig. 6.85 The finger could be replanted with interphalangeal fusion

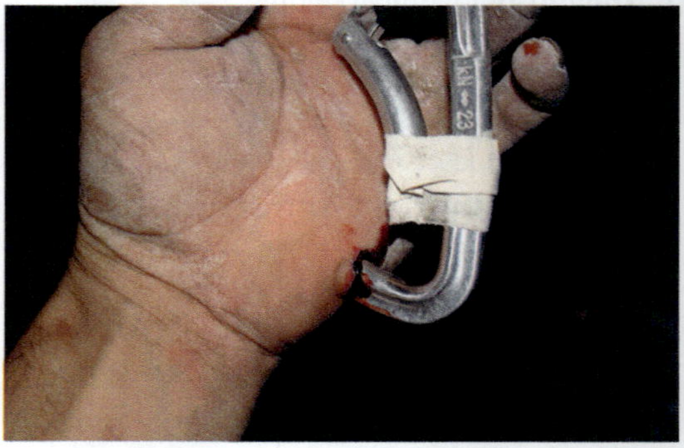

Fig. 6.86 Climber fall injury caused by getting caught in a carabiner

The best treatment for these injuries is prevention, so climbers are advised to avoid grabbing the rope or carabiners during a fall [144]. In correct fall position, the hands should be away from the rope and used instead for stabilizing the body in the air during a fall [144] (Fig. 6.87) The therapy

Fig. 6.87 During a fall, the hands should be off the rope and used to stabilize the body in the air to avoid rope-tangling injury

of these rope-tangling injuries depends on their severity and follows the general orthopedic trauma guidelines. In the most minor cases, only rest and gentle range of motion exercises are required. In the most severe cases in which there is avulsion amputation of one or more digits, emergency surgical reconstruction in the form of digital replantation is required. Unfortunately, due to the wide zone of soft tissue injury in avulsion amputations, replanted digits often require further surgery, ranging from revision surgery to address severe stiffness to revision amputation surgery, if the digit necroses due to vascular compromise.

- **History:** Rope tangling during a fall or finger/hand getting crushed in a carabiner
- **Onset:** Acute
- **Clinical:** Dependent on trauma
- **Diagnostic:** Physical examination and x-ray
- **Therapy:** Dependent on extend of injury
- **Outcome:** Dependent on extend of injury

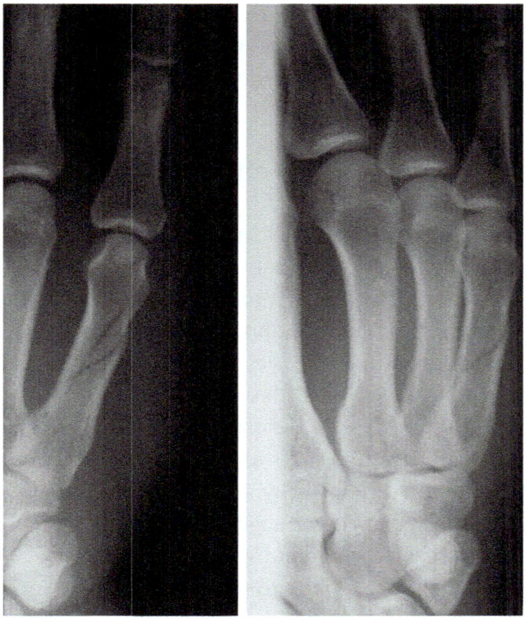

Fig. 6.88 Fracture of the fifth metacarpal in a climber caused by a fall while the hand was stuck in a crack in a long alpine climb. He finished the climb and a cast was applied the next day

6.3.17 Fractures

Most finger fractures in climbers are caused by direct trauma during rock fall or when hitting the rock during a fall. Another injury mechanism is an indirect trauma caused by a jammed hand or finger in a crack or pocket, causing non-axial forces or bending of the finger. Note that some minor fractures can present with the clinical symptoms of a pulley rupture; thus, in suspicion of a pulley rupture, an X-ray to exclude a fracture should be performed [32].

Finger fractures must be treated according to trauma/orthopedic surgical standards [19, 146, 147] (Fig. 6.88) specific to the fractured bone and the fracture pattern. These treatments vary from a short period of rest and gentle range of motion exercises for 4–6 weeks to surgery for reduction and hardware fixation of the fracture.

- **History:** Mmostly direct trauma
- **Onset:** Acute
- **Clinical:** Deformity, pain while movement
- **Diagnostic:** Physical examination and X-ray, CT scan

- **Therapy:** Dependent on the kind of fracture either conservative or surgical
- **Outcome:** Dependent on extend of injury

6.3.18 Cartilage Compression Injury

Cartilage compression syndrome, or "over-crimping" injury, is an entity that has not been previously described in climbing literature but is now seen more frequently, probably because of the increasing difficulty level of climbing in recent years. We have found this condition to be associated with overuse of the crimping position. In this position, the PIP joint is almost completely flexed while the DIP joint is hyperextended. This causes stress points on the cartilage concentrated at certain small areas of the joint. Many pockets and small edges additionally require some tilting and twisting of the fingers, leading to shear forces onto the joints' cartilage surface. These high-pressure peaks compress the cartilage and cause microinjuries. This causes a reaction of the joints' inner synovial membrane which secrets fluid, leading to an articular effusion and pain. Long term, this injury could be a precursor to capsulitis and chronic synovitis of the IP joint. In its acute state, this condition causes pain when crimping, usually at the DIP joint. Open finger position moves are usually pain-free. Diagnosis is mostly based on the history and clinical examination, though MRI can also detect minor bone bruising and edema. The therapy consists mostly of avoidance of the triggering movement or complete rest from climbing until pain subsides completely. Additionally, icing, ointment dressings, and sometimes intra-articular injections with platelet-rich plasma (PRP) are performed. Oral supplementation with chondroitin and glucosamine sulfate also helps to nourish the cartilage and may be beneficial (Figs. 6.89 and 6.90).

- **History:** Hard crimping move
- **Onset:** Acute
- **Clinical:** Painful crimping test with sharp pain
- **Diagnostic:** History and clinical finding, maybe MRI
- **Therapy:** Conservative
- **Outcome:** Can result in chronic capsulitis

Fig. 6.89 Hard crimping puts high pressure onto the cartilage of the IP joints (Photo Michael Simon)

6.4 Conclusion

Modern climbing places tremendous unnatural strain on the joints and soft tissues of the fingers. Therefore, it is not surprising that the fingers are the most commonly injured body part in climbers, with more than 20 different climbing-related finger pathologies described to date. Several of these finger conditions, such as an acute closed rupture of the flexor pulley, repetitive stress epiphyseal fractures, and closed lumbrical muscle shear injuries, are interesting in that they are found almost exclusively in climbers.

Diagnosis of finger pain in climbers can usually be made with a detailed history of the injury, including the timeframe of injury, mechanism of injury, and aggravating and alleviating factors, such as specific gripping positions. Thereafter, a detailed examination of the hand, with injury-specific physical examination maneuvers, should help narrow down the diagnosis, which can then be confirmed with imaging techniques such as X-ray or dynamic ultrasound. MRI or CT scans

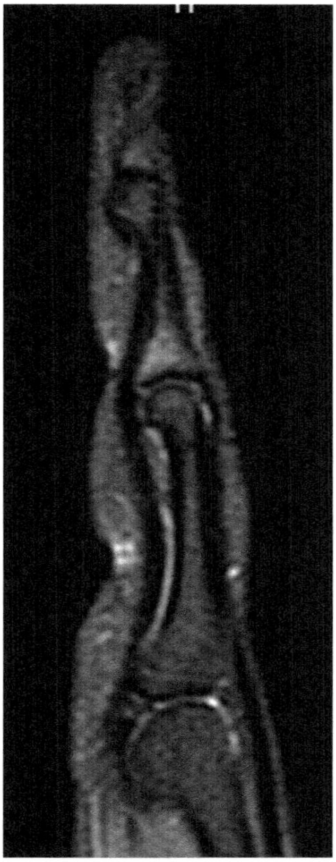

Fig. 6.90 Bone edema after hard crimping in the middle phalanx shown in the MRI

may be required in cases in which the diagnosis is still uncertain.

Treatment of climbing-related finger injuries is usually conservative, including a brief period of rest and anti-inflammatory measures, followed by focused physiotherapy and gradual return to sport-specific loading of the injured structure(s) with supportive taping as required. When conservative treatment fails, injectable and other minimally invasive treatments, such as steroid, PRP, stem cells, fat grafting, radial shock wave therapy, RSO, or leech therapy, may be beneficial. Surgical treatment is usually reserved for severe or specific conditions such as:

- Rope entanglement injuries
- Grossly displaced unstable fractures
- Unstable ligamentous injuries
- Complete combined closed flexor pulley ruptures
- Complete flexor or extensor tendon ruptures
- Non-united repetitive stress epiphyseal fractures
- Severe acute infections
- Chronic inflammatory conditions not resolved after exhausting nonsurgical treatment options
- Other non-climbing-specific hand surgical conditions such as arthritis, compressive neuropathies, ganglion cysts, tumors, and Dupuytren's contractures

References

1. Klauser A, Frauscher F, Bodner G, Halpern EJ, Schocke MF, Springer P, et al. Finger pulley injuries in extreme rock climbers: depiction with dynamic US. Radiology. 2002;222(3):755–61.
2. Garcia K, Jaramillo D, Rubesova E. Ultrasound evaluation of stress injuries and physiological adaptations in the fingers of adolescent competitive rock climbers. Pediatr Radiol. 2017;48(3):366–73.
3. Martinoli C, Bianchi S, Cotten A. Imaging of rock climbing injuries. Semin Musculoskelet Radiol. 2005;9(4):334–45.
4. McGeorge DD, McGeorge S. Diagnostic medical ultrasound in the management of hand injuries. J Hand Surg (Br). 1990;15:256–61.
5. Klauser A, Stadlbauer KH, Frauscher F, Herold M, Klima G, Schirmer M, et al. Value of transducer positions in the measurement of finger flexor tendon thickness by sonography. J Ultrasound Med. 2004;23(3):331–7.
6. Klauser A, Bodner G, Frauscher F, Gabl M, Zur Nedden D. Finger injuries in extreme rock climbers. Assessment of high-resolution ultrasonography. Am J Sports Med. 1999;27(6):733–7.
7. Schöffl I, Deeg J, Lutter C, Bayer T, Schöffl V. Diagnosis of A3 pulley injuries using ultrasound. Sportverletz Sportschaden. 2018;32(4):251–9.
8. Schöffl I, Hugel A, Schöffl V, Rascher W, Jüngert J. Diagnosis of complex pulley ruptures using ultrasound in cadaver models. Ultrasound Med Biol. 2017;43(3):662–9.
9. Bodner G, Rudisch A, Gabl M, Judmaier W, Springer P, Klauser A. Diagnosis of digital flexor tendon annular pulley disruption: comparison of high frequency ultrasound and MRI. Ultraschall Med. 1999;20(4):131–6.
10. Gabl M, Lener M, Pechlaner S, Lutz M, Rudisch A. [Rupture or stress injury of the flexor tendon pulleys? Early diagnosis with MRI]. Handchir Mikrochir Plast Chir 1996;28(6):317–321.

11. Martinoli C, Bianchi S, Nebiolo M, Derchi LE, Garcia JF. Sonographic evaluation of digital annular pulley tears. Skelet Radiol. 2000;29(7):387–91.
12. Schöffl V, Hochholzer T, Winkelmann HP, Roloff I, Strecker W. [Pulley injuries in sport climbers]. Handchir Mikrochir Plast Chir. 2004;36(4):224–30.
13. Schöffl V, Hochholzer T, Winkelmann HP, Strecker W. [Injuries of the pulley system in sport climbers]. Handchir Mikrochir Plast Chir. 2004;36(4):231–236.
14. Le Viet D, Rousselin B, Roulot E, Lantieri L, Godefroy D. Diagnosis of digital pulley rupture by computed tomography. J Hand Surg [Am]. 1996;21(2):245–8.
15. Hochholzer T, Heuk A, Hawe W, Bernett P. MRT der Hand und Finger bei Verletzungen und Überlastunsgbeschwerden von Sportkletterern. In: Bernett P, Jeschke D, editors. Sport und Medizin—Pro und Contra, Kongressband des Deutschen Sportärztekongesses in München. München: W.Zuckschwerdt Verlag; 1990. p. 380–3.
16. Bayer T, Adler W, Janka R, Uder M, Roemer F. Magnetic resonance cinematography of the fingers: a 3.0 Tesla feasibility study with comparison of incremental and continuous dynamic protocols. Skelet Radiol. 2017;46(12):1721–8.
17. Bayer T, Fries S, Schweizer A, Schöffl I, Janka R, Bongartz G. Stress examination of flexor tendon pulley rupture in the crimp grip position: a 1.5-Tesla MRI cadaver study. Skelet Radiol. 2015;44(1):77–84.
18. Bayer T, Adler W, Schweizer A, Schöffl I, Uder M, Janka R. Evaluation of finger A3 pulley rupture in the crimp grip position-a magnetic resonance imaging cadaver study. Skelet Radiol. 2015;44(9):1279–85.
19. Schöffl VR, Schöffl I. Finger pain in rock climbers: reaching the right differential diagnosis and therapy. J Sports Med Phys Fitness. 2007;47(1):70–8.
20. Heuck A, Hochholzer T, Keinath C. Die MRT von Hand und Handgelenk bei Sportkletterern: Darstellung von Verletzungen und Überlastungsfolgen. Radiologe. 1992;32:248–54.
21. Schöffl V, Hochholzer T, Imhoff A. Radiographic changes in the hands and fingers of young, high-level climbers. Am J Sports Med. 2004;32(7):1688–94.
22. Kellgren JH, Lawrence JS. Radiological assessment of osteo-arthrosis. Ann Rheum Dis. 1957;16:494–502.
23. Bollen SR. Injury to the A2 pulley in rock climbers. J Hand Surg (Br). 1990;15(2):268–70.
24. Bollen SR. Soft tissue injury in extreme rock climbers. Br J Sports Med. 1988;22(4):145–7.
25. Bollen SR, Gunson CK. Hand injuries in competition climbers. Br J Sports Med. 1990;24(1):16–8.
26. Tropet Y, Menez D, Balmat P, Pem R, Vichard P. Closed traumatic rupture of the ring finger flexor tendon pulley. J Hand Surg [Am]. 1990;15(5):745–7.
27. Schöffl V, Schweizer A. Sportclimbing related injuries and overuse syndromes. In: Seifert L, Wolf P, Schweizer A, editors. IRCRA handbook of climbing & mountaineering from science to performance. Research in sport and exercise science. London, New York: Routledge; 2016. p. 59–75.
28. Lourie GM, Hamby Z, Raasch WG, Chandler JB, Porter JL. Annular flexor pulley injuries in professional baseball pitchers: a case series. Am J Sports Med. 2011;39(2):421–4.
29. Schöffl V, Jüngert J. Closed flexor pulley injuries in non-climbing activities. J Hand Surg [Am]. 2006;31(5):806–10.
30. Schöffl V, Schöffl I. Isolated cruciate pulley ruptures in rock climbers. J Hand Surgery (Eur). 2010;35:245–6.
31. Schöffl I, Oppelt K, Jungert J, Schweizer A, Bayer T, Neuhuber W, et al. The influence of concentric and eccentric loading on the finger pulley system. J Biomech. 2009;
32. Schöffl V, Hochholzer T, Winkelmann HP, Strecker W. Pulley injuries in rock climbers. Wilderness Environ Med. 2003;14(2):94–100.
33. Schöffl V, Hochholzer T, Winkelmann HP, Roloff I, Strecker W. [Pulley injuries in sport climbers] Verletzungen des Ringbandapparates bei Sportkletterern. Handchir Mikrochir Plast Chir. 2004;36(4):224–30.
34. Gabl M, Rangger C, Lutz M, Fink C, Rudisch A, Pechlaner S. Disruption of the finger flexor pulley system in elite rock climbers. Am J Sports Med. 1998;26(5):651–5.
35. Bassemir D, Unglaub F, Hahn P, Muller LP, Bruckner T, Spies CK. Sonographical parameters of the finger pulley system in healthy adults. Arch Orthop Trauma Surg. 2015;135(11):1615–22.
36. Lutter C, Tischer T, Schöffl VR. Olympic competition climbing: the beginning of a new era-a narrative review. Br J Sports Med. 2020;
37. Schneeberger M, Schweizer A. Pulley ruptures in rock climbers: outcome of conservative treatment with the pulley-protection splint; a series of 47 cases. Wilderness Environ Med. 2016;27(2):211–8.
38. Schöffl I, Oppelt K, Jüngert J, Schweizer A, Neuhuber W, Schöffl V. The influence of the crimp and slope grip position on the finger pulley system. J Biomech. 2009;42(13):2183–7.
39. Schoffl I, Oppelt K, Jungert J, Schweizer A, Bayer T, Neuhuber W, et al. The influence of concentric and eccentric loading on the finger pulley system. J Biomech. 2009;42(13):2124–8.
40. Schöffl I, Baier T, Schöffl V. Flap irritation phenomenon (FLIP): etiology of chronic tenosynovitis after finger pulley rupture. J Appl Biomech. 2011;27(4):291–6.
41. Schöffl V, Schöffl I. Injuries to the finger flexor pulley system in rock climbers—current concepts. J Hand Surg [Am]. 2006;31(4):647–54.
42. Schöffl V, Einwag F, Strecker W, Schöffl I. Strength measurement after conservatively treated pulley ruptures in climbers. Med Sci Sports Exerc. 2006;38(4):637–43.
43. Schöffl I, Einwag F, Strecker W, Hennig F, Schöffl V. Impact of taping after finger flexor tendon pul-

ley ruptures in rock climbers. J Apl Biomech. 2007;23:52–62.
44. Schöffl V, Hochholzer T, Winkelmann HP, Strecker W. [Therapy of injuries of the pulley system in sport climbers] Zur Therapie von Ringbandverletzungen bei Sportkletterern. Handchir Mikrochir Plast Chir. 2004;36(4):231–6.
45. Kleinert HE, Bennett JB. Digital pulley reconstruction employing the always present rim of the previous pulley. J Hand Surg [Am]. 1978;3(3):297–8.
46. Karev A, Stahl S, Taran A. The mechanical efficiency of the pulley system in normal digits compared with a reconstructed system using the "belt loop" technique. J Hand Surg [Am]. 1987:596–601.
47. Bunnell S. Surgery of the hand. Philadelphia: JB Lippincott; 1944. p. 315.
48. Lister G. Indications and techniques for the repair of the finger flexor tendon sheath. Hand Clin. 1985;1:85–95.
49. Doyle JR, Blythe W. The finger tendon flexor sheath and pulleys: anatomy and reconstruction. In: AAOS symposium of tendon surgery in the hand. St. Louis: Mosby; 1975. p. 81–87.
50. Widstrom CJ, Doyle JR, Johnson G, Manske PR, McGee R. A mechanical study of six digital pulley reconstruction techniques: Part II. Strength of individual reconstructions. J Hand Surg [Am]. 1989;14(5):826–9.
51. Okutsu I, Ninomiya S, Hiraki S, Inanami H, Kuroshima N. Three-loup technique for A2 pulley reconstruction. J Hand Surg [Am]. 1987;12:790–4.
52. Gabl M, Reinhart C, Lutz M, Bodner G, Angermann P, Pechlaner S. The use of a graft from the second extensor compartment to reconstruct the A2 flexor pulley in the long finger. J Hand Surg (Br). 2000;25(1):98–101.
53. Moutet F. [Flexor tendon pulley system: anatomy, pathology, treatment]. Chir Main 2003;22(1):1–12.
54. Schöffl V, Küpper T, Hartmann J, Schöffl I. Surgical repair of multiple pulley injuries—evaluation of a new combined pulley repair. J Hand Surg [Am]. 2012;37(2):224–30.
55. Widstrom CJ, Johnson G, Doyle JR, Manske PR, Inhofe P. A mechanical study of six digital pulley reconstruction techniques: Part I. Mechanical effectiveness. J Hand Surg [Am]. 1989;14(5):821–5.
56. Lutter C, Schöffl V. Intermittent unspecific osteitis and cortex atrophy of the proximal phalanx after surgical pulley repair. BMJ Case Rep. 2015;2015:bcr2015213109.
57. Schöffl I, Meisel J, Lutter C, Schöffl V. Feasibility of a new pulley repair: a cadaver study. J Hand Surgery. 2018;43(4):380.e1–7.
58. Schöffl V, Hochholzer T, Lightner S. One move too many. Boulder, CO: Sharp End Publishing; 2016.
59. Arora R, Fritz D, Zimmermann R, Lutz M, Kamelger F, Klauser AS, et al. Reconstruction of the digital flexor pulley system: a retrospective comparison of two methods of treatment. J Hand Surg (Br). 2007;32(1):60–6.
60. Schweizer A. Biomechanical effectiveness of taping the A2 pulley in rock climbers. J Hand Surg (Br). 2000;25(1):102–7.
61. Warme WJ, Brooks D. The effect of circumferential taping on flexor tendon pulley failure in rock climbers. Am J Sports Med. 2000;28(5):674–8.
62. Schöffl I, Einwag F, Strecker W, Hennig F, Schöffl V. Impact of taping after finger flexor tendon pulley ruptures in rock climbers. J Appl Biomech. 2007;23(1):52–62.
63. Woollings KY, McKay CD, Emery CA. Risk factors for injury in sport climbing and bouldering: a systematic review of the literature. Br J Sports Med. 2015;49(17):1094–9.
64. Schweizer A. Biomechanical properties of the crimp grip position in rock climbers. J Biomech. 2001;34(2):217–23.
65. Schöffl V, Heid A, Küpper T. Tendon injuries of the hand. World J Orthop. 2012;3(6):62–9.
66. Schöffl V, Korb L, Matros P. ACT—Adjunct Compensatory Training. Hamburg: Tredition; 2020. p. 92.
67. Schöffl V, Strohm P, Lutter C. Efficacy of corticosteroid injection in rock climber's tenosynovitis. Hand Surg Rehabil. 2019.
68. Schöffl I, Oppelt K, Jungert J, Schweizer A, Neuhuber W, Schöffl V. The influence of the crimp and slope grip position on the finger pulley system. J Biomech. 2009;
69. An KN, Berglund L, Uchiyama S, Coert JH. Measurement of friction between pulley and flexor tendon. Biomed Sci Instrum. 1993;29:1–7.
70. Klauser A, Frauscher F, Gabl M, Smekal V. Highfrequency Sonography in the detection of finger injuries in Sports climbing [Hochauflösende Sonograohie zur Erkennung von Fingerverletzungen beim Sportklettern]. Sport Ortho Sport Trauma. 2005;21:24–30.
71. Danielsen MA. Ultrasonography for diagnosis, monitoring and treatment of tenosynovitis in patients with rheumatoid arthritis. Dan Med J. 2018;65(3):B5474.
72. Selvanetti A, Cipolla M, Puddu G. Overuse tendon injuries: basic sciences and classifications. Oper Tech Sports Med. 1997;5:110–7.
73. Sharma P, Maffulli N. Tendon injury and tendinopathy: healing and repair. J Bone Joint Surg Am. 2005;87(1):187–202.
74. Schöffl V, Schöffl I, Frank L, Küpper T, Simon M, Lutter C. Tendon injuries in the hands in rock climbers: epidemiology, anatomy, biomechanics and treatment—an update. Muscles Ligaments Tendons J. 2020;10(2):233–43.
75. Jones G, Johnson MI. A critical review of the incidence and risk factors for finger injuries in rock climbing. Curr Sports Med Rep. 2016;15(6):400–9.
76. Lutter C, Schöffl V, Hotfiel T, Schmitz C, Milz S, Strohm P. Radial extracorporeal shock wave therapy in flexor tendon pathology of the hand: a feasibility study. Technol Health Care. 2019;28(1):77–83.

77. Fleisch SB, Spindler KP, Lee DH. Corticosteroid injections in the treatment of trigger finger: a level I and II systematic review. J Am Acad Orthop Surg. 2007;15(3):166–71.
78. Callegari L, Spano E, Bini A, Valli F, Genovese E, Fugazzola C. Ultrasound-guided injection of a corticosteroid and hyaluronic acid. A potential new approach to the treatment of trigger finger. Drugs R D. 2011;11(2):137–45.
79. Shakeel H, Ahmad TS. Steroid injection versus NSAID injection for trigger finger: a comparative study of early outcomes. J Hand Surg [Am]. 2012;37(7):1319–23.
80. Ring D, Lozano-Calderon S, Shin R, Bastian P, Mudgal C, Jupiter J. A prospective randomized controlled trial of injection of dexamethasone versus triamcinolone for idiopathic trigger finger. J Hand Surg [Am]. 2008;33(4):516–22.
81. Salim N, Abdullah S, Haflah NH. Outcome of corticosteroid injection versus physiotherapy in the treatment of mild trigger fingers. J Hand Surg Eur. 2012;37(1):27.
82. Tanaka T, Ogawa T, Yanai T, Okano E, Kohyama S, Ochiai N. Disruption of flexor tendon after intrasheath triamcinolone acetonide injection for trigger digits: two case reports. J Hand Surg Asian Pac. 2017;22(3):380–3.
83. Koeppen D, Aurich M, Rampp T. Medicinal leech therapy in pain syndromes: a narrative review. Wien Med Wochenschr. 2013;164(5–6):95–102.
84. Sig AK, Guney M, Uskudar Guclu A, Ozmen E. Medicinal leech therapy-an overall perspective. Integr Med Res. 2017;6(4):337–43.
85. Bäcker M, Lüdtke R, Afra D, Cesur Ö, Langhorst J, Fink M, et al. Effectiveness of leech therapy in chronic lateral epicondylitis. Clin J Pain. 2011;27(5):442–7.
86. Michalsen A, Lüdtke R, Cesur Ö, Afra D, Musial F, Baecker M, et al. Effectiveness of leech therapy in women with symptomatic arthrosis of the first carpometacarpal joint: a randomized controlled trial. Pain. 2008;15(137(2)):452–9.
87. Hochholzer T, Schöffl V. Soweit die Hände greifen. 4te ed. Ebenhausen: Lochner Verlag; 2014.
88. Hochholzer T, Schöffl V. [Osteoarthrosis in fingerjoints of rockclimbers] Degenerative Veränderungen der Fingergelenke bei Sportkletterern. D Z Sportmed. 2009;60(6):145–9.
89. Rotmann I. Überlastungssyndrome der Hand bei tschechoslowakischen Spitzenkletterern. Prakt Sport Trauma Sportmed. 1987;2:41–3.
90. Freyschmidt J. Skeletterkrankungen. Springer; 2018.
91. Scanzallo CR, Goldring SR. The role of synovitis in osteoarthritis patheogenesis. Bone. 2012;51(2):249–57.
92. Mathiessen A, Conaghan PG. Synovitis in osteoarthritis: current understandin with therapeutic implications. Arthitis Res Therapy. 2017;19(18):2–9.
93. Schöffl V, Hochholzer T, El-Sheikh Y, Lutter C. Hand and fingers. In: Schöffl V, Schöffl I, Hocholzer T, Lutter C, editors. Climbing medicine—a practical guide. 1. Heidelberg: Springer; 2022.
94. Schöffl VR, Hochholzer T, Imhoff AB, Schoffl I. Radiographic adaptations to the stress of high-level rock climbing in junior athletes: a 5-year longitudinal study of the German junior national team and a group of recreational climbers. Am J Sports Med. 2007;35(1):86–92.
95. Schöffl VR, Hoffmann PM, Imhoff A, Küpper T, Schöffl I, Hochholzer T, et al. Long-term radiographic adaptations to stress of high-level and recreational rock climbing in former adolescent athletes: an 11-year prospective longitudinal study. Orthop J Sports Med. 2018;6(9):2325967118792847.
96. Sylvester AD, Christensen AM, Kramer PA. Factors influencing osteological changes in the hands and fingers of rock climbers. J Anat. 2006;209(5):597–609.
97. Rohrbough JT, Mudge MK, Schilling RC, Jansen C. Radiographic osteoarthritis in the hands of rock climbers. Am J Orthop. 1998;27(11):734–8.
98. Schöffl V, Hochholzer T, Schöffl I. Extensor hood syndrome—osteophytic irritation of digital extensor tendons in rock climbers. Wilderness Environ Med. 2011;21(3):253–6.
99. Pastor T, Fröhlich S, Spörri J, Schreiber T, Schweizer A. Cartilage abnormalities and osteophytes in the fingers of elite sport climbers: an ultrasonography-based cross-sectional study. E J Sport Science. 2019;20(2):269–76.
100. Schreiber T, Allenspach P, Seifert B, Schweizer A. Connective tissue adaptations in the fingers of performance sport climbers. Eur J Sport Sci. 2015;15(8):696–702.
101. Amis AA, Jones MM. The interior of the flexor tendon sheath of the finger. The functional significance of its structure. J Bone Joint Surg Br. 1988;70(4):583–7.
102. Serafini G, Derchi LE, Quadri P, Martinoli C, Orio O, Cavallo A, et al. High resolution sonography of the flexor tendons in trigger fingers. J Ultrasound Med. 1996;15(3):213–9.
103. Boyer MI, Strickland JW, Engles D, Sachar K, Leversedge FJ. Flexor tendon repair and rehabilitation: state of the art in 2002. Instr Course Lect. 2003;52:137–61.
104. Schöffl V, Winkelmann HP. [Traumatic and degenerative tendon lesions of the hand]. Orthopade. 2010;39(12):1108–16.
105. Manske PR, Lesker PA. Avulsion of the ring finger flexor digitorum profundus tendon: an experimental study. Hand. 1978;10(1):52–5.
106. Schöffl V. Rock climbing. In: Engelhardt M, editor. Sports injuries. 1. Munich-Jena: Elsevier; 2011.
107. Aronowitz ER, Leddy JP. Closed tendon injuries of the hand and wrist in athletes. Clin Sports Med. 1998;17(3):449–67.
108. Schweizer A, Bayer T. Closed disruption of a single flexor digitorum superficialis tendon slip: 3 cases. Hand Surg Rehabil. 2019;38(2):121–4.

109. Lutter C, Tischer T, Hotfiel T, Frank L, Enz A, Simon M, et al. Current trends in sport climbing injuries after the inclusion into the Olympic program. Analysis of 633 injuries within the years 2017/18. Muscle Tendons Ligament J. 2020;10(2):201–10.
110. Schöffl V, Popp D, Küpper T, Schöffl I. Injury distribution in rock climbers—a prospective evaluation of 911 injuries between 2009–2012. Wilderness Environ Med. 2015;26(1):62–7.
111. Abe Y, Watson HK, Renaud S. Flexor tendon sheath ganglion: analysis of 128 cases. Hand Surg. 2004;9(1):1–4.
112. Klauser A, Frauscher F, Bodner G, Cihak C, Gabl M, Schocke M, et al. [Value of high-resolution ultrasound in the evaluation of finger injuries in extreme sport climbers]. Ultraschall Med 2000;21(2):73–78.
113. Schöffl I, Schöffl V. Epiphyseal stress fractures in the fingers of adolescents: biomechanics, pathomechanism, and risk factors. Eur J Sports Med. 2015;3(1):27–37.
114. Schöffl V, Lutter C, Woollings K, Schöffl I. Pediatric and adolescent injury in rock climbing. Res Sports Med. 2018;26(sup1):91–113.
115. Hochholzer T, Schöffl VR. Epiphyseal fractures of the finger middle joints in young sport climbers. Wilderness Environ Med. 2005;16(3):139–42.
116. Hochholzer T, Schöffl V, Krause R. Finger-Epiphysenverletzungen jugendlicher Sportkletterer. Sport Ortho Trauma. 1997;13(2):100–3.
117. Chell J, Stevens K, Preston B, Davis TR. Bilateral fractures of the middle phalanx of the middle finger in an adolescent climber. Am J Sports Med. 1999;27(6):817–9.
118. Schöffl V, Lutter C, Wollings K, Schöffl I. Pediatric and adolescent injury in rock climbing. Res Sports Med. 2018;26(sup1):91–113.
119. Caine D, Meyers RN, Schöffl V, Mafulli N. Periphyseal stress injuries in young athletes: a systematic review. Sports Med. 2021.
120. Meyers RN, Howell DR, Provance AJ. The association of finger growth plate injury history and speed climbing in youth competition climbers. Wilderness Environ Med. 2020;31(4):394–9.
121. Lutter C, El-Sheikh Y, Schöffl I, Schöffl V. Sport climbing: medical considerations for this new Olympic discipline. Br J Sports Med. 2017;51(1):2–3.
122. El-Sheikh Y, Lutter C, Schöffl I, Schöffl V, Flohe S. Surgical management of proximal interphalangeal joint repetitive stress epiphyseal fracture nonunion in elite sport climbers. J Hand Surg [Am]. 2018;43(6):572.e1–5.
123. Bayer T, Schöffl VR, Lenhart M, Herold T. Epiphyseal stress fractures of finger phalanges in adolescent climbing athletes: a 3.0-Tesla magnetic resonance imaging evaluation. Skelet Radiol. 2013;42(11):1521–5.
124. Bärtschi N, Scheibler A, Schweizer A. Symptomatic epiphyseal sprains and stress fractures of the finger phalanges in adolescent sport climbers. Hand Surg Rehabil. 2019;38(4):251–6.
125. Schweizer A, Göhner SK. Sport climbing, bouldering and associated injuries in childhood and adolescence. Orthopade 2019;48L998–1004.
126. Schöffl V, Schöffl I, Flohe S, El-Sheikh Y, Lutter C. Evaluation of a diagnostic-therapeutic algorithm for epiphyseal growth plate stress injuries in adolescent climbers. Am J Sports Med. 2021.
127. Meyers RN, Schöffl VR, Mei-Dan O, Povance AJ. Returning to climb after epiphyseal finger stress fracture. Curr Sports Med Rep. 2020; 19(11):457–62.
128. Schöffl V, Schöffl I, Flohe S, El-Sheikh Y, Lutter C. Evaluation of a diagnostic-therapeutic algorithm for finger epiphyseal growth plate stress injuries in adolescent climbers. Am J Sports Med. 2022;50(1):229–37. https://doi.org/10.1177/03635465211056956.
129. Schweizer A. Lumbrical tears in rock climbers. J Hand Surg (Br). 2003;28(2):187–9.
130. Lutter C, Schweizer A, Schöffl V, Romer F, Bayer T. Lumbrical muscle tear: clinical presentation, imaging findings and outcome. J Hand Surg Eur. 2018;43(7):767–75.
131. Verdan C, Poulenas I. Anatomic and functional relations between the tendons of the long palmar muscle and the long flexor muscle of the thumb at their crossing in the carpus. Ann Chir Plast. 1975;20(2):191–6.
132. Wang EH, Loftus WK, Bird SJ, Sampson MJ. Ring finger lumbrical origin strain: a case series with imaging findings. Skelet Radiol. 2016;45(12):1729–34.
133. Parminder K. Morphological study of lumbricals—a cadaveric study. J Clin Diagn Res. 2013;7(8):1558–60.
134. Palti R, Vigler M. Anatomy and function of lumbrical muscles. Hand Clin. 2012;28(1):13–7.
135. Lutter C, Schweizer A, Schöffl V, Römer F, Bayer T. Lumbrical muscle tear: clinical presentation, imaging findings and outcome. J Hand Surg (Eur). 2018;
136. Peters P. Nerve compression syndromes in sport climbers. Int J Sports Med. 2001;22(8):611–7.
137. Hochholzer T, Krause R, Heuk A. [Nerve compression syndromes in sports climbers]. Sportverletz Sportschaden. 1993;7(2):84–87.
138. Geoghegan JM, Forbes J, Clark DI, Smith C, Hubbard R. Dupuytren's disease risk factors. J Hand Surg (Br). 2004;29(5):423–6.
139. Millesi H, Reihsner R, Hamilton G, Mallinger R, Menzel EJ. Biomechanical properties of normal tendons, normal palmar aponeuroses and palmar aponeuroses from patients with Dupuytren's disease subjected to elastase and chondroitinase treatment. Connect Tissue Res. 1995;31(2):109–15.
140. Straub G. [Musculoskeletal problems in rock climbing—a review]. Wien Med Wochenschr. 2000;150(8–9):186–90.
141. Logan AJ, Mason G, Dias J, Makwana N. Can rock climbing lead to Dupuytren's disease? Br J Sports Med. 2005;39(9):639–44.

142. Pervulesko N, Schöffl V, Gormasz C. Evaluation of a self-diagnostic tool for Dupuytren's disease in rock climbers. Hand Therapy. 2011;16:45–8.
143. Hochholzer T, Schöffl V. One move too many. Ebenhausen: Lochner Verlag; 2003.
144. Schöffl V, Küpper T. Rope tangling injuries—how should a climber fall? Wilderness Environ Med. 2008;19(2):146–9.
145. Jakubietz RG, Jakubietz MG, Gruenert JG. Digital amputation caused by climbing-rope entanglement. Wilderness Environ Med. 2006;17:178–9.
146. Hochholzer T, Schöffl V. One move to many. Ebenhausen: Lochner Verlag; 2003.
147. Schöffl V, Hochholzer T. Sportklettern—Aktuelle sportmedizinische Aspekte. Ebenhausen: Lochner Verlag; 2004.

Wrist Injuries

7

Christoph Lutter and Volker Schöffl

Isabelle Schöffl in „North Star", Frankenjura, Germany, Photo Kilian Reil

C. Lutter (✉)
Department of Orthopedic Surgery, University Hospital Rostock, Rostock, Germany

V. Schöffl
Department of Orthopedic and Trauma Surgery, Center of Sportsmedicine, Klinikum Bamberg, Bamberg, Germany

School of Clinical and Applied Sciences, Leeds Becket University, Leeds, UK

Section of Wilderness Medicine, Department of Emergency Medicine, University of Colorado School of Medicine, Denver, CO, USA

Department of Orthopedic and Trauma Surgery, Friedrich-Alexander Universität Erlangen-Nürnberg, Erlangen, Germany

7.1 Overview (Statistics)

Depending on the classification of injuries and overuse damage to the wrist region, complaints in this area account for approximately 7–10% of all climbing-related injury patterns [1–3]. This is the fourth most common injury region after finger, shoulder, lower extremity, and foot injuries and must therefore be considered when treating climbers and boulderers [1–3]. According to current figures, the most common symptom is the ulnar-sided overuse reaction of the wrist [3]. In addition, Dupuytren's contracture of the palm accounts for a

V. Schöffl et al. (eds.), *Climbing Medicine*, https://doi.org/10.1007/978-3-030-72184-8_7

large proportion of chronic complaints of the metacarpus and can often be diagnosed, especially in long-time (older) athletes [4] (see also Chap. 15, "Climbing in Older Athletes.").

7.2 Diagnosis

The initial diagnostic procedure for complaints of the wrist, metacarpus, and distal forearm area follows the usual criteria and should be structured as follows:

1. *Patient history*: How long have the complaints existed? After which strain did they first appear? Can the complaints be triggered by certain grip/pull shapes? Do the complaints radiate? Have the pain characteristics changed? Which treatments have been administered so far?
2. *Clinical examination*: Inspectorial abnormalities (swelling, color changes of the skin, etc.). Palpation of the corresponding region (pressure tenderness, etc.). Stress tests depending on the presumed diagnosis (hamate stress test, forced radial/ulnar duction, etc.). This should always be accompanied by a comparative examination of the opposite side!
3. *Imaging procedures*: Sonographic examination of the corresponding region, X-ray, CT, MRI diagnostics, depending on the suspected diagnosis.

Detailed anamnesis and a proper clinical examination are of utmost importance in the clarification of wrist pain and pathologies in the metacarpus and distal forearm. With appropriate knowledge of the various common clinical patterns and the respective stress tests, correct diagnoses can often be made with a relatively high degree of certainty. Unnecessary imaging procedures can thus be avoided. Regarding the clarification of pain in the area of the metacarpals, also refer to Chap. 9, "Finger Injuries."

7.3 Bone Edema

Bone marrow edema (BME) of the carpus, the metacarpus, and the distal forearm is usually the result of chronic overloading, usually triggered by repetitive, unphysiological movements [5–9]. Pathophysiologically, bone marrow edema corresponds to an increased accumulation of tissue fluid and blood within the respective bone as well as increased blood flow in the affected area [10]. It should be noted that the climbing-related stress-induced bone marrow edema is based on a different pathogenesis than transient bone marrow edema, ischemic osteonecrosis, vascular necrosis, or bone contusions [11–14]. The diagnosis of corresponding overuse damage has become more frequent in recent years due to the increasing use of MRI diagnostics [5, 6, 10].

Bone marrow edema can be divided into the following categories: mechanical bone edema (chronic overload, fractures, malpositions, degenerative), ischemic bone edema (osteonecrosis), inflammatory bone edema (rheumatoid arthritis, infectious), idiopathic (transient) bone edema, or bone edema due to malignancies. The most frequently affected areas for stress-related BME are usually static areas of the lower extremities. However, BME of the upper extremities, especially in the hand, are very rarely described in literature [8, 14–18].

In climbing, however, bone marrow edema is more frequent due to the continuous tensile load and the resulting shear forces in the hand, wrist, and forearm [5, 7]. Based on the complex anatomy of the carpus, overload reactions can affect all carpal bones as well as in the metacarpals and forearms. However, BME are frequently described in the following regions: lunate, capitate, hamate, bases of the metacarpals, distal radius, and distal ulna [5–7]. It should also be noted that the osseous reactions described differ from osteochondroses such as Kienböck's disease and others. The differentiation to these disease patterns can sometimes be difficult (Fig. 7.1).

Affected climbing athletes commonly complain of a slow-onset and increasing pain around the affected bone, which usually improves to a certain extent when the load is reduced (rest). During examination, the pain can usually be triggered or intensified by palpation of the affected region. Specific stress tests (see hamate fractures, ulnocarpal pain reactions, and Chap. 9, "Finger Injuries") can also be helpful [5]. Bone marrow edema generally remains inconspicuous in images produced by sonography, X-ray, and CT scans [10, 14, 19].

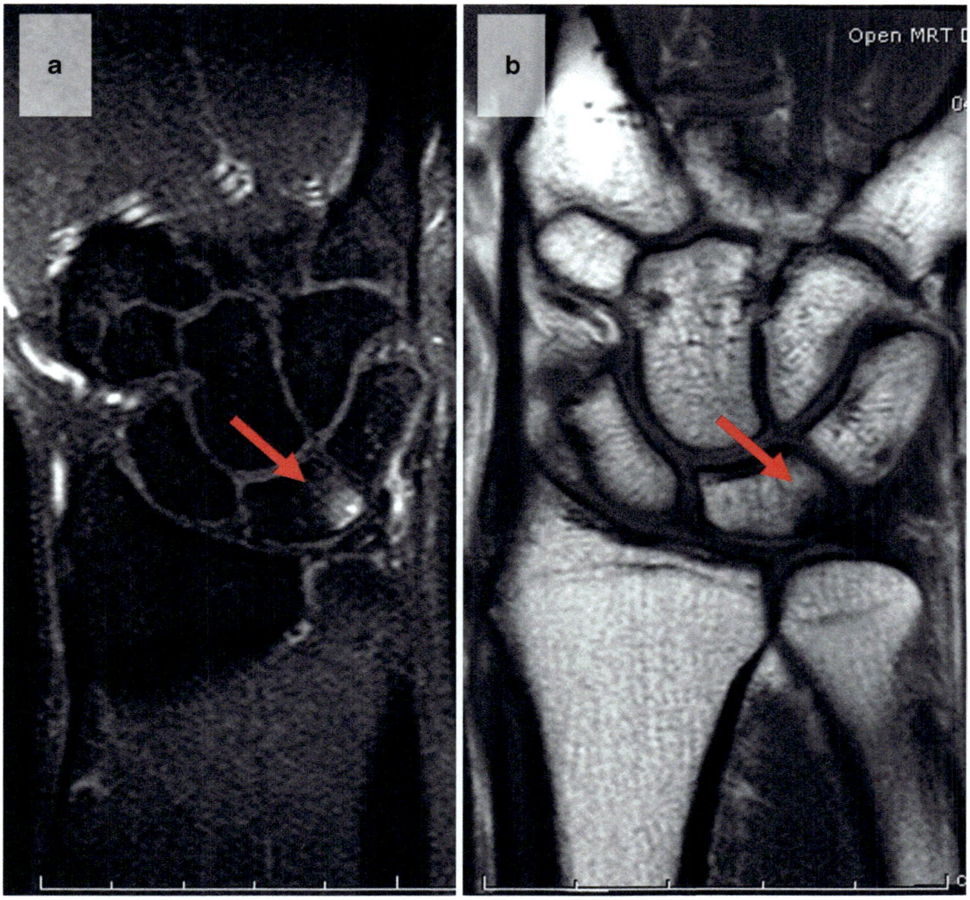

Fig. 7.1 MRI of bone marrow edema within the lunate (**a**, T2 weighting; **b**, T1 weighting)

However, MRI depicts the stress reaction within the affected bone. The therapeutic approach consists of immediate and long-term stress reduction with subsequent adjustment of training. Supplementary shockwave therapy can also help to promote healing; however, there is no corresponding evidence of this in the literature to date (Fig. 7.2).

After the symptoms have subsided, supporting circular wrist tape can be applied; however, this should only be used if it gives the athlete a feeling of stabilization [20].

It can be assumed that the bone marrow edema of the wrist will become increasingly important due to the increasing popularity of bouldering worldwide [21]. An early and correct diagnosis and sufficient treatment can prevent serious secondary conditions, such as stress fractures caused by overstrain [5–7, 19].

- **Etiology:** Bone marrow edema within the wrist is based on chronic overuse.
- **Onset:** Usually slow-onset and load-dependent complaints.
- **Clinic:** Circumscribed pain symptoms around the affected bone.
- **Diagnosis:** Clinical and magnetic resonance imaging (MRI).
- **Therapy:** Immediate stress reduction and long-term training change. Mostly long healing process.
- **Result:** Mostly complete healing possible.

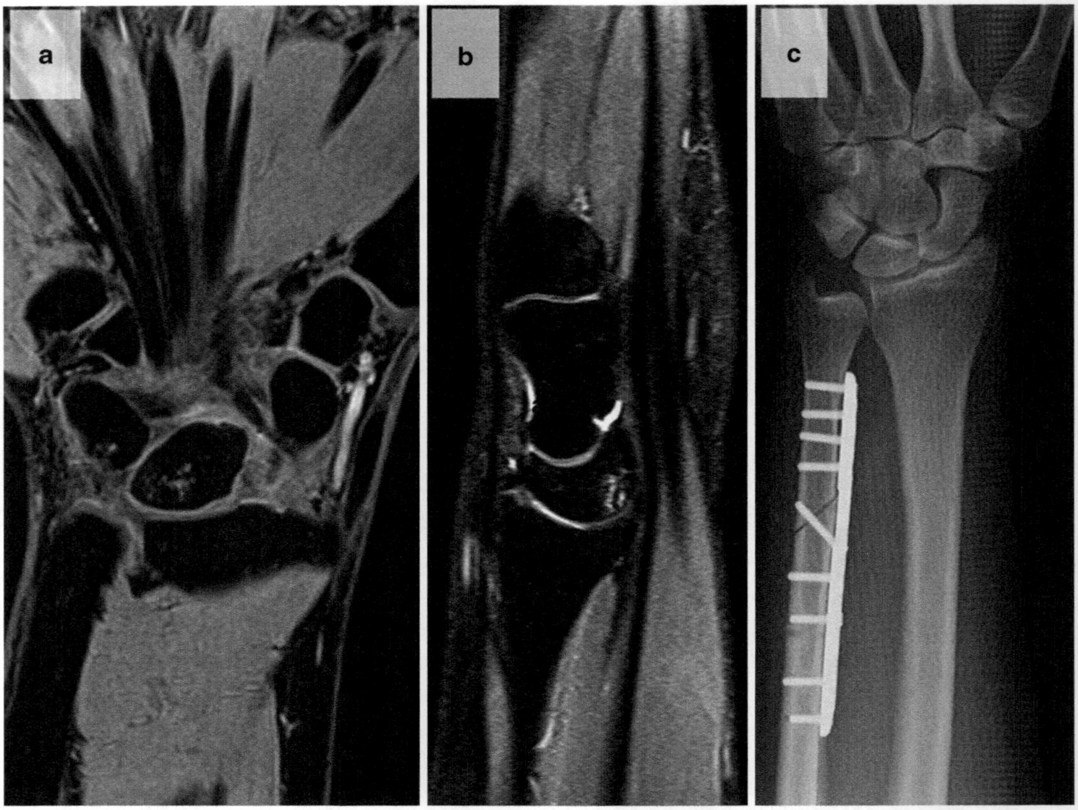

Fig. 7.2 MRI (**a**, **b**) and X-ray (**c**, left wrist) of a 30-year-old female bouldering athlete who had reported long-lasting and consistent pain even after having received arthroscopic debridement of the TFCC 2 years earlier (positive ulna variance). An extended shockwave therapy of the wrist (lunate) was performed in addition to long rest. However, follow-up MRI and CT scans did not show sufficient bony healing within the lunate. Finally, an ulna shortening procedure was performed (**c**)

7.4 Hamate Fractures

Sport-related fractures of the hamate are usually caused by direct impact trauma (e.g., direct impact from handlebars (e.g., mountain biking) or from a club (e.g., baseball or golf)) [22–28]. These direct impact traumas are rare in climbing [5, 19], in which the cause is usually indirect overloading of the hamate [5, 22]. In particular, the hamate hook, the bony deflection pulley of the flexor tendons of the ring and little finger, is subjected to enormous stress due to the high contact pressure [5]. This high contact pressure increases in ulnar-abducted hand positions (e.g., under grips) and under high tensile forces on the ring and little finger. Hamate fractures in climbing almost exclusively affect the hook and are divided depending on their location into type I(I) fractures (avulsion fractures at the tip of the hook), type I(II) fractures (center part of the hook), and type I(III) fractures (base of the hook) (classification according to Milk and Xiong) (Fig. 7.3) [5, 22].

The patient's medical history generally reports a slow-onset, load-dependent, and progressive pain on the volar and ulnar side of the

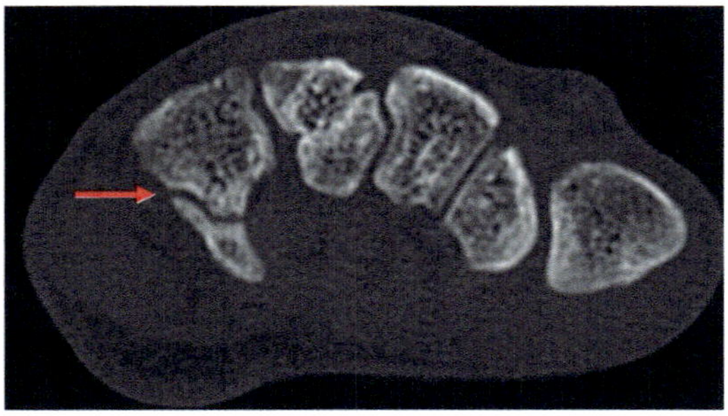

Fig. 7.3 CT scan of a hamate hook base fracture (see arrowhead)

wrist. It can also be helpful to administer a so-called hamate stress test, in which the patient is asked to flex his ring and little finger strongly against resistance (e.g., the examiner's finger) with an ulnar-abducted wrist (see Fig. 5.1. Fig. 7) [29]. An overload reaction or a fracture in the area of the hamate is indicated by significant pain in the affected area, which is less pronounced in a subsequent retest in the radially abducted wrist under the same load on the ring and little finger [29]. To confirm the diagnosis, immediate magnetic resonance imaging or computed tomography should be performed; conventional radiological examination can sometimes be misleading [5, 19, 30, 31].

The authors recommend that a conservative therapy attempt (initial consistent immobilization!) should always be carried out first [5, 22]. Since the hamate hook plays an essential role as a redirection of the flexor tendons (Dig. IV/V), immediate resection of the bone portion leads to a loss of strength; this should be avoided in climbers [32]. Conservative therapy often promises sufficient healing of the fracture if the patient is not participating in sports. In the case of insufficient bone healing, either spot-drilling of the fracture site (Fig. 7.4) or compression osteosynthesis using a Herbert screw should be considered (Fig. 7.5). A resection of the hamulus should only be performed in exceptional cases as it may lead to strength deficit (Fig. 7.6) [22, 30–33]. Regardless of the form of therapy, the training must be adjusted following the injury to avoid refractures or other injuries. A circular wrist tape can be considered after resuming sport [20].

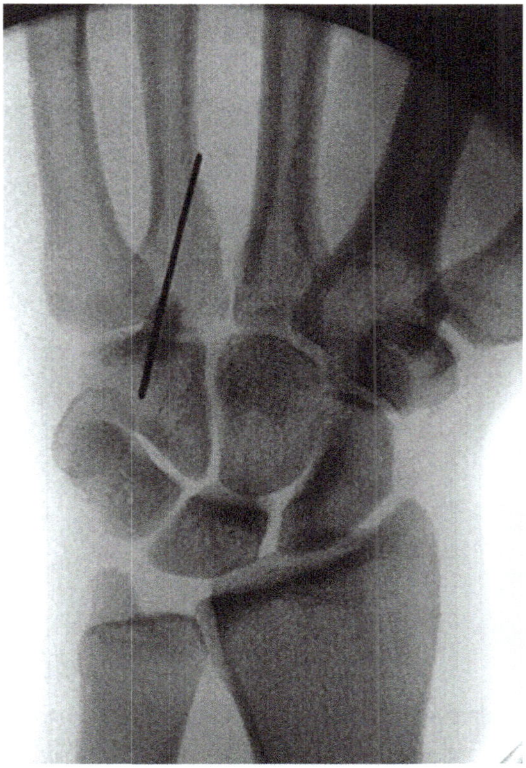

Fig. 7.4 Spot-drilling of nonunion of the hamate using a Kirschner wire

- **Etiology:** Chronic overload reaction due to contact pressure of the flexor tendons. Initially bone edema in the hamate (hook); continuous strain can cause fracture.
- **Onset:** Slow onset of pain, in case of fracture sudden pain exacerbation.
- **Clinic:** Pain on the ulnar and volar aspect of the wrist, especially during ulnar duction and loading of Dig. IV/V.
- **Diagnosis:** Clinically and CT/MRI. Positive hamate stress test (see Chap. 5).
- **Therapy:** Initially conservative therapy, surgical therapy only in cases of long-term therapy failure.
- **Result:** Mostly complete healing under conservative therapy, occasionally surgical procedure (e.g., osteosynthesis, drilling, Herbert screw osteosynthesis, or resection of the hamate hook) necessary.

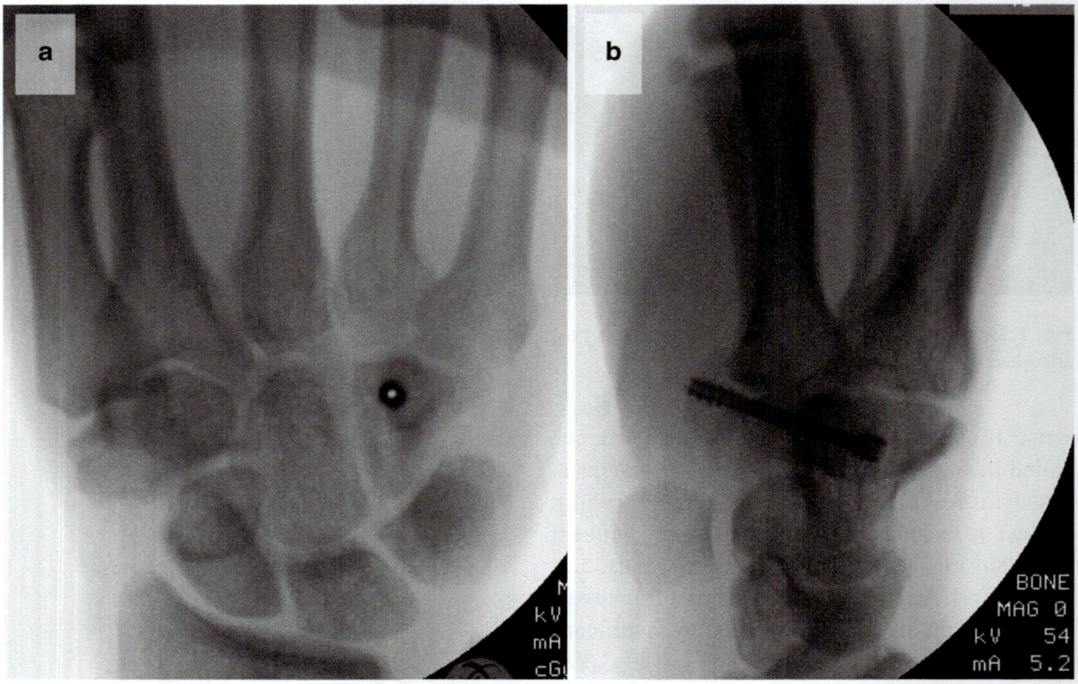

Fig. 7.5 Osteosynthesis of a nonunion of the hamate using a Herbert screw (**a**, a.p. projection; **b**, lateral projection)

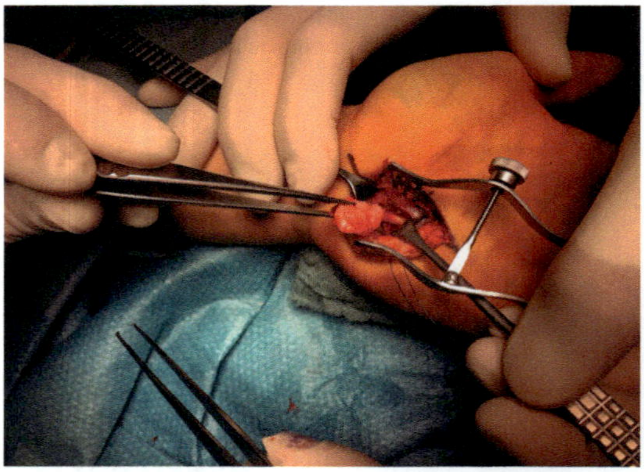

Fig. 7.6 Resection of the hamate hook following insufficient bony healing

7.5 Dupuytren's Contracture

A nodular remodeling reaction in the palm of the hand can be diagnosed frequently in climbers, without a known direct connection between climbing and the clinical picture of Dupuytren's contracture (see Fig. 7.7) [34–36]. However, it is assumed that recurrent microtrauma in the palmar aponeurosis area caused by climbing could cause or accelerate corresponding remodeling reactions [1–3, 36].

In early stages, the disease manifests itself as a nodular hardening in the palm (usually Dig. IV/V); in advanced stages, it leads to retraction and limited extension of the affected fingers [34, 35].

Although individual cases have been described in young climbers (see Fig. 7.7), this disease is usually found in older athletes. Current figures show that in older climbing athletes (>35 years), this change in the palm is one of the ten most frequently diagnosed symptoms [4]. It is not known whether or to what extent climbing contributes to the genesis and the progress of the alteration reactions. In our own patient collective, the proportion of Dupuytren's contracture symptoms in athletes >35 years is 3%. The diagnosis (clinical diagnosis) and therapy of Dupuytren's contracture in climbers does not differ significantly from that in non-climbers [23, 34, 35].

Fortunately, the contracture is usually relatively painless for climbers and hinders only very few climbers. Additionally, the affected athletes

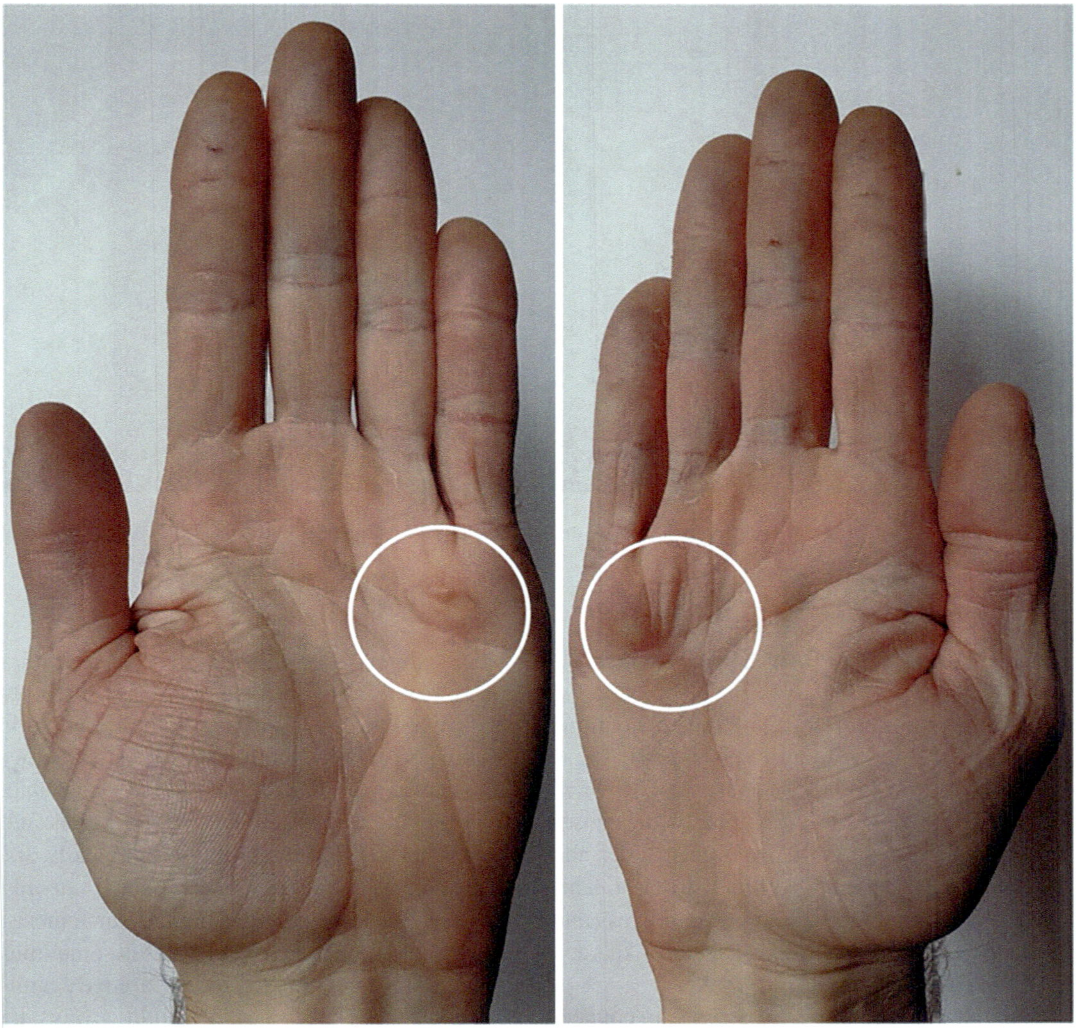

Fig. 7.7 A 34-year-old climber (former IFSC Boulder World cup athlete) who had developed an early stage Dupuytren's contracture on both hands over the last 10 years (Dig. IV/V)

rarely progress beyond the initial stage of the disease. In this initial stage, a "wait-and-see" or conservative approach can usually be taken. In case of (pressure) pain or disruptive movement restrictions, the surgical solution and resection of the scar strands is performed [34, 35].

- **Etiology:** Multifactorial, possible etiology due to climbing-related microtrauma of the palmar aponeurosis

- **Onset:** Creeping hardening in the palm area
- **Clinic:** Nodular hardening of the palm (mainly Dig. IV/V)
- **Diagnosis:** Clinical
- **Therapy:** Conservative, in case of progress surgical resection of the scar strands
- **Result:** Initially often asymptomatic, in advanced stages limited climbing ability

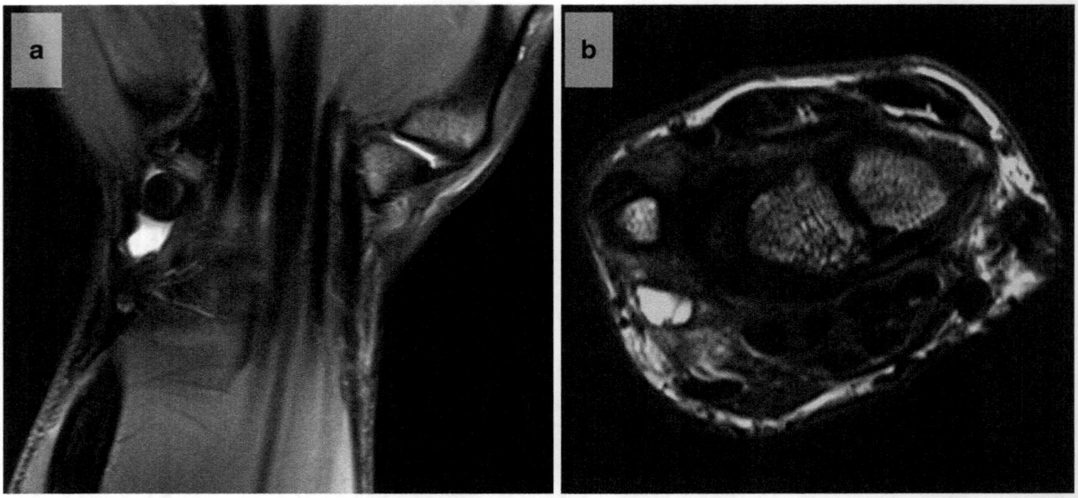

Fig. 7.8 MRI of an effusion with synovialitis in the distal radioulnar joint (**a**, coronary MRI image; **b**, axial projection)

7.6 Ulnocarpal Pain Syndromes

Ulnocarpal pain syndromes in climbers continue to be a challenge for treating physicians. These complaints can have various causes, all of which are difficult to diagnose. Additionally, the causes are difficult to distinguish from each other and are sometimes extremely difficult to treat. This often results in a lengthy diagnostic process and an often slow and frustrating course of therapy for affected athletes.

The following causes are discussed in connection with ulnar-sided wrist pain in climbers:

1. Lesions of the triangular fibrocartilaginous complex (TFCC lesions)
2. Instabilities or subluxations of the distal radioulnar joint (DRUJ)
3. Lesions of the lunotriquetral ligament [37]
4. Carpal instabilities (midcarpal instability) [38]
5. Triquetral impingement ligament tear (TILT) syndrome [39]
6. Overloading reactions around the extensor carpi ulnaris tendon
7. Synovitis with plica syndrome

For correct diagnosis, the physical examination is usually more precise than magnetic resonance imaging. The latter usually shows synovitis as a sign of stress reaction (Fig. 7.8). Clear structural causes for this are less often detectable. However, the reproducibility of the forces and loads occurring during climbing is clinically difficult and often, the patient's complaints can only be determined by taking a precise anamnesis. Reproduction of the pain symptoms is sometimes not possible. Nevertheless, a comparative examination of the opposite side should always be performed.

Climbers often describe a painful feeling of instability in the distal radioulnar joint (DRUJ). During stress in climbing, the force is usually transmitted via the carpus to the radius and ulna. The ligamentous apparatus of the TFCC forms one of the main stabilizers of the DRUJ. A corresponding load can therefore lead to synovitis in this area. The DRUJ also has little muscular stabilization, which favors the high load on the DRUJ. In the event of instability or subluxation of the DRUJ, overloading of the extensor carpi ulnaris tendon may also occur. Anatomical variations (especially ulna-plus/ulna-minus variations) can

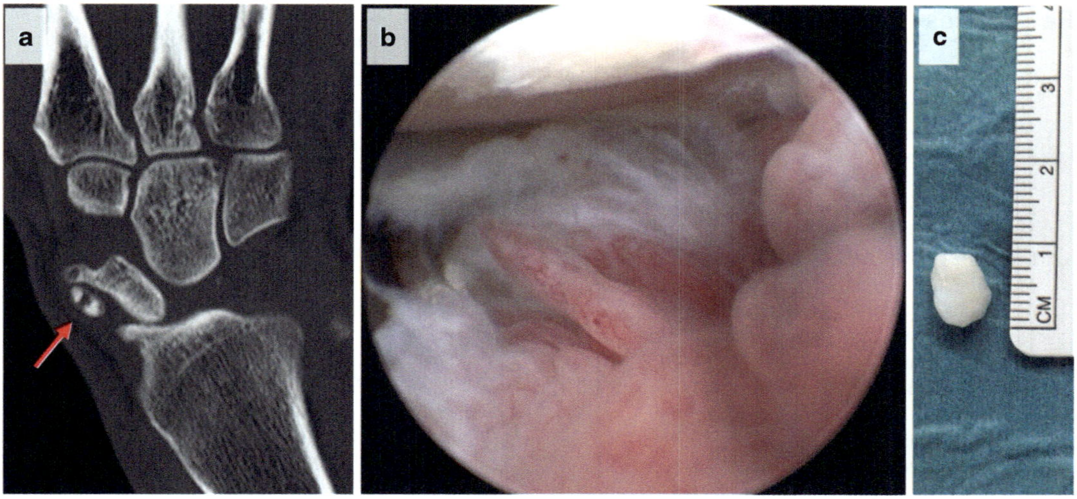

Fig. 7.9 A 43-year-old male climbing athlete with ulnar-sided wrist pain. CT scan shows an intra-articular body in the wrist joint (**a**, arrow). Arthroscopy of the wrist shows synovialitis (**b**) and confirms the loose body within the joint (**c**, after resection)

promote the symptoms. Therapeutically, an initial conservative approach is usually recommended. The aim should be to reduce the load by adapting the training and avoiding the triggering hand positions. In case of persistent pain reactions and especially in case of clear synovitis confirmed by MRI, corticosteroid injections may be considered. Circular wrist tape often provides a feeling of stability for affected athletes. If therapy continues to fail and clinical/MRI suspicion of soft-tissue instability exists, wrist arthroscopy with synovectomy and, if necessary, stabilization of injured structures can be performed (Fig. 7.9).

- **Etiology:** Largely unclear. Suspected—TFCC lesions, DRUJ subluxations, lesions of the lunotriquetral ligament stump, midcarpal instabilities, "triquetral impingement ligament tear (TILT)" syndrome and overloading reactions around the extensor carpi ulnaris tendon
- **Onset:** Usually a slow start
- **Clinic:** Stress-dependent ulnocarpal pain
- **Diagnosis:** Clinical, MRT
- **Therapy:** Conservative (taping), if necessary, injection therapy. Arthroscopy in case of therapy failure
- **Result:** Often moderate response to therapy

7.7 Conclusion

In summary, complaints around the metacarpus, wrist, and distal forearm form a large group of clinical pictures in climbers. With the appropriate level of knowledge about the underlying injuries and overuse damage, most complaints can be quickly and correctly diagnosed. Except for the ulnocarpal pain syndromes, most forms of injury respond well to the appropriate forms of therapy; however, healing processes are generally protracted, and stress reactions and adjustment of training after resumption of sport are unavoidable.

References

1. Schöffl V, Hochholzer T, Winkelmann HP, Strecker W. Pulley injuries in rock climbers. Wilderness Environ Med. 2003;14(2):94–100.
2. Schöffl V, Popp D, Küpper T, Schöffl I. Injury trends in rock climbers: evaluation of a case series of 911 injuries between 2009 and 2012. Wilderness Environ Med. 2015;26(1):62–7.
3. Lutter C, Tischer T, Hotfiel T, Frank L, Enz A, Simon M, et al. Current trends in sport climbing injuries after the inclusion into the Olympic program. Analysis of 633 injuries within the years 2017/18. Muscles Ligaments Tendons J. 2020;10(2):201–10.
4. Lutter C, Hotfiel T, Tischer T, Lenz R, Schöffl V. Evaluation of rock climbing related injuries in older athletes. Wilderness Environ Med. 2019;30(4):362–8.
5. Lutter C, Schweizer A, Hochholzer T, Bayer T, Schöffl V. Pulling harder than the hamate tolerates: evaluation of hamate injuries in rock climbing and bouldering. Wilderness Environ Med. 2016;27(4):492–9.
6. Lutter CH, Bayer T, Schöffl V. Rock climbing related bone marrow edema of the hand: a follow-up study. Clin J Sport Med. 2018;28(4):382–8.
7. Hochholzer TBT, Straub G, Schöffl V. Overuse bone marrow edema of the hands in sport climbers. Sport Orthop Traumatol. 2013;29:219–24.
8. Korompilias AV, Karantanas AH, Lykissas MG, Beris AE. Bone marrow edema syndrome. Skelet Radiol. 2009;38(5):425–36.
9. Patel S. Primary bone marrow oedema syndromes. Rheumatology. 2014;53(5):785–92.
10. Meizer R, Radda C, Stolz G, Kotsaris S, Petje G, Krasny C, et al. MRI-controlled analysis of 104 patients with painful bone marrow edema in different joint localizations treated with the prostacyclin analogue iloprost. Wien Klin Wochenschr. 2005;117(7–8):278–86.
11. Bennell KL, Malcolm SA, Thomas SA, Wark JD, Brukner PD. The incidence and distribution of stress fractures in competitive track and field athletes. A twelve-month prospective study. Am J Sports Med. 1996;24(2):211–7.
12. Biedert MRBM, Gal I. Bone bruise—radiologische und morphologische Aspekte. Sportortopädie Traumatologie. 2007;23(1):36–41.
13. Doury P. Bone-marrow oedema, transient osteoporosis, and algodystrophy. J Bone Joint Surg. 1994;76(6):993–4.
14. Grampp S, Henk CB, Mostbeck GH. Overuse edema in the bone marrow of the hand: demonstration with MRI. J Comput Assist Tomogr. 1998;22(1):25–7.
15. Aigner N, Petje G, Schneider W, Krasny C, Grill F, Landsiedl F. Juvenile bone-marrow oedema of the acetabulum treated by iloprost. J Bone Joint Surg. 2002;84(7):1050–2.
16. Aigner N, Petje G, Schneider W, Meizer R, Wlk M, Kotsaris S, et al. Bone marrow edema syndrome of the femoral head: treatment with the prostacyclin analogue iloprost vs. core decompression: an MRI-controlled study. Wien Klin Wochenschr. 2005;117(4):130–5.
17. Bartl C, Imhoff A, Bartl R. Treatment of bone marrow edema syndrome with intravenous ibandronate. Arch Orthop Trauma Surg. 2012;132(12):1781–8.
18. Hofmann SKJ, Vakil-Adli A. Painful bone marrow oedema of the knee: differential diagnosis and therapeutic concepts. Orthop Clin N Am. 2004;35:321–33.
19. Bayer T, Schweizer A. Stress fracture of the hook of the hamate as a result of intensive climbing. J Hand Surg Eur Vol. 2009;34(2):276–7.
20. Lutter C, Schöffl V. Circular wrist tape—how to correctly use this technique in climbing and bouldering. Sports Orthop Traumatol. 2017;33(4):410–3.
21. Lutter C, El-Sheikh Y, Schöffl I, Schöffl V. Sport climbing: medical considerations for this new Olympic discipline. Br J Sports Med. 2017;51(1):2–3.
22. Aldridge JM 3rd, Mallon WJ. Hook of the hamate fractures in competitive golfers: results of treatment by excision of the fractured hook of the hamate. Orthopedics. 2003;26(7):717–9.
23. Bishop AT, Beckenbaugh RD. Fracture of the hamate hook. J Hand Surg Am. 1988;13(1):135–9.
24. Boulas HJ, Milek MA. Hook of the hamate fractures. Diagnosis, treatment, and complications. Orthop Rev. 1990;19(6):518–29.
25. Egawa M, Asai T. Fracture of the hook of the hamate: report of six cases and the suitability of computerized tomography. J Hand Surg Am. 1983;8(4):393–8.
26. Foucher G, Schuind F, Merle M, Brunelli F. Fractures of the hook of the hamate. J Hand Surgery. 1985;10(2):205–10.
27. Guha AR, Marynissen H. Stress fracture of the hook of the hamate. Br J Sports Med. 2002;36(3):224–5.
28. Hirano K, Inoue G. Classification and treatment of hamate fractures. Hand Surg. 2005;10(2–3):151–7.
29. Wright TW, Moser MW, Sahajpal DT. Hook of hamate pull test. J Hand Surg Am. 2010;35(11):1887–9.
30. Walsh JJt, Bishop AT. Diagnosis and management of hamate hook fractures. Hand Clin. 2000;16(3):397–403, viii.
31. Xiong G, Dai L, Zheng W, Sun Y, Tian G. Clinical classification and treatment strategy of hamate hook fracture. J Huazhong Univ Sci Technolog Med Sci. 2010;30(6):762–6.
32. Demirkan F, Calandruccio JH, Diangelo D. Biomechanical evaluation of flexor tendon function after hamate hook excision. J Hand Surg Am. 2003;28(1):138–43.
33. Milek MA, Boulas HJ. Flexor tendon ruptures secondary to hamate hook fractures. J Hand Surg Am. 1990;15(5):740–4.
34. Ball C, Izadi D, Verjee LS, Chan J, Nanchahal J. Systematic review of non-surgical treatments

for early dupuytren's disease. BMC Musculoskelet Disord. 2016;17(1):345.
35. Soreide E, Murad MH, Denbeigh JM, Lewallen EA, Dudakovic A, Nordsletten L, et al. Treatment of Dupuytren's contracture: a systematic review. Bone Joint J. 2018;100-b(9):1138–45.
36. Schöffl V, Hochholzer S., Lightner S. One move too many. Ebenhausen: Lochner; 2016.
37. Tay SC, Berger RA, Parker WL. Longitudinal split tears of the ulnotriquetral ligament. Hand Clin. 2010;26(4):495–501.
38. von Schroeder HP. Dorsal wrist plication for midcarpal instability. J Hand Surg Am. 2018;43(4):354–9.
39. Watson HK, Weinzweig J. Triquetral impingement ligament tear (tilt). J Hand Surgery. 1999;24(3):321–4.

8 Elbow and Forearm

M. Simon, Christoph Lutter, and Volker Schöffl

Michael Simon climbing in Kalymnos, Greece, Photo Archive Michael Simon

M. Simon (✉)
Department of Orthopedic and Trauma Surgery, Center of Sportsmedicine, Klinikum Bamberg, Bamberg, Germany
e-mail: info@michaelsimon.org

C. Lutter
Department of Orthopedic Surgery, University Hospital Rostock, Rostock, Germany

V. Schöffl
Department of Orthopedic and Trauma Surgery, Center of Sportsmedicine, Klinikum Bamberg, Bamberg, Germany

School of Clinical and Applied Sciences, Leeds Becket University, Leeds, UK

Section of Wilderness Medicine, Department of Emergency Medicine, University of Colorado School of Medicine, Denver, CO, USA

Department of Orthopedic and Trauma Surgery, Friedrich-Alexander Universität Erlangen-Nürnberg, Erlangen, Germany

V. Schöffl et al. (eds.), *Climbing Medicine*, https://doi.org/10.1007/978-3-030-72184-8_8

8.1 Overview

Pathologies in the area of the elbow and forearm are the fourth most common type of injury in climbing after finger, shoulder, and hand injuries, accounting for 7.7% of all pathologies in our own patient sample [1]. Furthermore, a study by Gronhaug et al., which included 667 active climbers with complaints around the elbow, found an incidence of 17.7% of all climbing-related injuries, making this the third most common pathology in climbers [2]. Hereby, mainly chronic pathologies caused by overload are found. Epicondylitis humeri, with a share of up to 8.4%, is one of the ten most common diagnoses in climbers. Elbow pain can also be caused by brachialis, biceps, and triceps muscle tendinopathies [3]. Traumatic injuries in the area of the elbow and forearm, such as ruptures of the distal biceps tendon as well as higher-value muscle or vascular injuries, occur occasionally in climbing and require a quick, correct diagnosis and prompt surgical treatment. If load-dependent pain in the forearm persists for a long time after climbing, the differential diagnosis of the rarely occurring functional compartment syndrome of the forearm musculature must also be considered [4]. In the following section, the specifics of the main climbing-related injuries around the elbow, their diagnosis, and their treatment are discussed.

8.2 Lateral and Medial Epicondylitis

Epicondylitis humeri is one of the most common complaints among climbing athletes, although, in recent years, a slight decrease in incidence has been observed in our own patient sample. Nevertheless, with 3.3% of all complaints, it is still among the ten most common climbing-related pathologies [1]. A distinction is made between lateral (radial) and medial (ulnar) epicondylitis, with medial epicondylitis occurring more often in climbers [5, 6]. Patients report local pressure pain at the epicondyle and, sometimes, pain radiating into the forearm muscles, a decrease in strength, and pain at rest. The complaints are caused by repetitive overstrain of the origin of the forearm extensors (lateral side) or flexors (medial side), whereby chronic microtrauma leads to degenerative changes and ultimately to tendinopathy [7]. Muscular imbalance between the forearm flexors and extensors can also be one of the triggers for epicondylitis, since training the antagonists (extensor group) is often neglected by many climbers. The diagnosis is made by clinical examination, including provocation tests (lateral, isometric extension of the wrist as well as the middle finger when the elbow is extended; medial, isometric flexion in the wrist as well as the fingers when the elbow is extended) and sonography, if necessary. Further diagnostics in the form of X-ray and MRI should only be performed in the case of refractory complaints under conservative therapy to exclude any other underlying pathology (e.g., arthrosis, osteochondritis dissecans, loose bodies inside the joint) (Fig. 8.1).

At first, the therapy is always conservative, and there are a variety of treatment options available: nonsteroidal anti-inflammatory drugs (NSAID), platelet-rich plasma (PRP) injections, physical therapy, manual therapy, extracorporeal shockwave therapy, acupuncture, and bandages. Often, only a combination of several methods is successful for treatment, since the complaints can be very persistent [7–9]. Eccentric strength train-

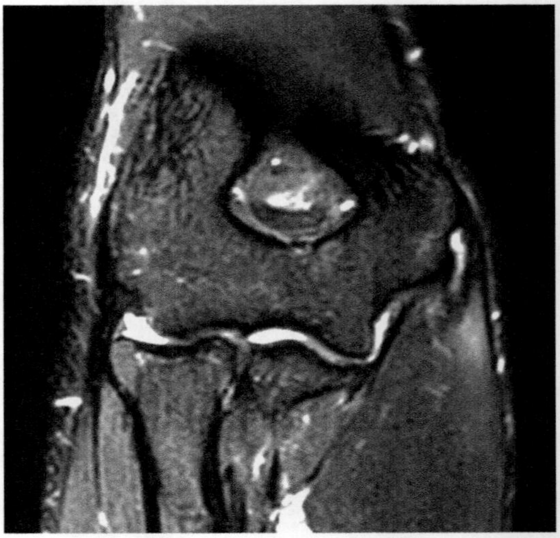

Fig. 8.1 MRI of an elbow with a medial epicondylitis humeri and a perifocal edema

ing, stretching measures, and antagonist training often relieve symptoms and should be integrated into the training routine to prevent injuries in the sense of adjunct compensatory training for climbers (ACT; e.g., www.act.clinic) [10]. In addition, circular elbow tape is recommended during climbing or sport-specific training (see also Chap. 22 Taping) [11]. If there is no alleviation of symptoms, surgical therapy should be considered at the earliest after 9–12 months once conservative treatment options have been exhausted. Fortunately, this only happens very rarely. In such cases, open and arthroscopic procedures with the aim of denervation of the respective epicondyle and excision of diseased tissue are available. Postoperatively, the respective arm is immobilized for 2 weeks. In addition, maximum stress should be avoided for 6 months postoperatively [7].

Lateral and Medial Epicondylitis

- **History:** Degenerative processes in the context of overload-related microtrauma and muscular imbalances
- **Onset:** Chronic
- **Clinical:** Pressure pain at the epicondyle, pain radiating into the forearm muscles, reduction in strength, and pain at rest
- **Diagnostic:** Medical history, clinical finding, sonography, MRI
- **Therapy:** Mostly conservative
- **Outcome:** Often long-lasting complaints

8.3 Tendinopathies of the Biceps, Triceps, and Brachialis Muscles

Although much rarer than epicondylitis humeri, tendinopathies of the biceps, triceps, and brachialis muscles can also be triggers for pain in the elbow region. As with epicondylitis, the cause of the complaints is a repetitive overuse of the muscle attachments with consecutive chronic microtraumas and muscular imbalances that ultimately lead to tendinosis. Clinically, there is an anterior elbow pain in the area of the distal biceps and brachialis muscle tendon or a pain above the dorsal elbow in the area of the tendon insertion of the triceps muscle at the olecranon. As in epicondylitis, clinical examination and sonography are the main diagnostic tools. Further diagnostics by means of X-ray and MRI should only be carried out in the case of persistent complaints.

Tendinopathies around the elbow are also common in other sports, and their therapy is not significantly different for climbers. The therapy is conservative, whereby the reduction of stress as well as prevention by training the antagonist muscle groups has the highest priority. Eccentric exercises of the affected muscles are also an important part of the therapy. Similar to the muscular imbalances of the forearm flexors and extensors for epicondylitis, an imbalance between the pro- and supinator muscles of the forearm plays an essential role in the development of tendinopathies of the biceps and brachialis tendon. The biceps brachii muscle is the strongest supinator (outward rotation) in the elbow joint; the brachialis muscle is the strongest flexor in the elbow joint. The resulting torque is rather unfavorable when climbing, as the forearm is usually turned inward when gripping. Consequently, the sole pull of the biceps muscle on a horizontal hold would cause the hand to turn outward, and the hold could no longer be held. This overload can only be prevented by strongly trained pronator muscles (pronator teres, pronator quadratus, brachioradialis muscles) and flexor muscles (e.g., brachialis muscle). In addition, training pull-ups with complete extension in the elbow joint promote the development of a tendinopathy of the triceps tendon, which should therefore be avoided by means of a certain minimum residual flexion in the elbow joint [10, 11].

Tendinopathies of the Biceps, Triceps, and Brachialis Muscles

- **History:** Degenerative processes in the context of overload-related microtrauma and muscular imbalances
- **Onset:** Chronic
- **Clinical:** Pain around the elbow during load as well as local pressure-induced pain

- **Diagnostic:** Medical history, clinical finding, sonography, if applicable X-ray and MRI
- **Therapy:** Conservative
- **Outcome:** Often tedious complaints

8.4 Functional Compartment Syndrome of the Forearm Muscles

Load-dependent pain in the area of the forearm muscles during climbing is common. In particular, the flexor muscles in the deep compartment of the forearm are exposed to great stress. Their effectiveness is usually the limiting factor for climbing performance, especially on longer climbing routes. However, if the complaints are very intense and persist disproportionately strongly for 1 or 2 days after climbing, one must consider the rare so-called "functional compartment syndrome" of the forearm muscles [4]. Primarily, climbers describe a long-lasting feeling of increased pressure in the forearm flexor muscles. The cause is a stress-induced, excessive swelling of the musculature, especially of the forearm flexors within the muscle fascia, which encloses the so-called compartment. The increased tissue pressure leads to a consecutive impairment of the microcirculation with disrupted outflow of blood and lymph and, in case of a progressive, fulminant course, to disturbances to the neuromuscular functions. In extreme cases, irreversible damage (tissue necrosis, etc.) can occur [4].

This is often preceded by a viral infection, such as glandular fever (infectious mononucleosis) caused by the Epstein-Barr virus (kissing disease), although the exact connection is still unclear [12].

Clinically, during and after exertion while climbing, strong pain in the area of the forearm musculature can be found, going beyond the familiar feeling of "thick forearms" and lasting longer. The pain is often characterized as "dull," "as if the skin on the forearm was too tight."

In advanced stages, paresthesia and circulatory disorders may occur. However, patients typically have no complaints without a corresponding prior exposure.

This rare pathology is difficult to diagnose. In addition to the medical history and the clinical examination, the diagnosis is confirmed by means of compartment pressure measurement under sport-specific stress. For this purpose, special probes are used, which are inserted into the respective compartments via cannulas (Fig. 8.2). In a study of ten competition climbers, reference values for physiological pressures could be determined by means of dynamic compartment measurements. At rest, the pressure should be below 15 mmHg; with sport-specific stress, the pressure can rise to values of up to approx. 30 mmHg. Within a recovery period of 15 min after exertion, pressures up to 15 mmHg are physiological. However, if pressures of 15–30 mmHg occur during the recovery period, there is a high risk of stress-induced functional forearm compartment syndrome. Moreover, pressures of over 30 mmHg confirm the diagnosis. In most cases, conservative therapy is sufficient (e.g., physical rest, stretching of the musculature, ice pads, and loosening massages). Nonsteroidal anti-inflammatory drugs (e.g., ibuprofen) can also have a decongestant effect. Despite these measures, if further deterioration of the findings occurs without further load with progressively increasing compartment pressures, the only remaining treatment option to avoid irreversible damage is a surgical fasciotomy of the forearm (Fig. 8.3). Fortunately, there are only few such climbing-related cases known in literature. Even in these dramatic cases, a complete recovery of the sport-specific performance capacity was achieved in the course of the treatment [4].

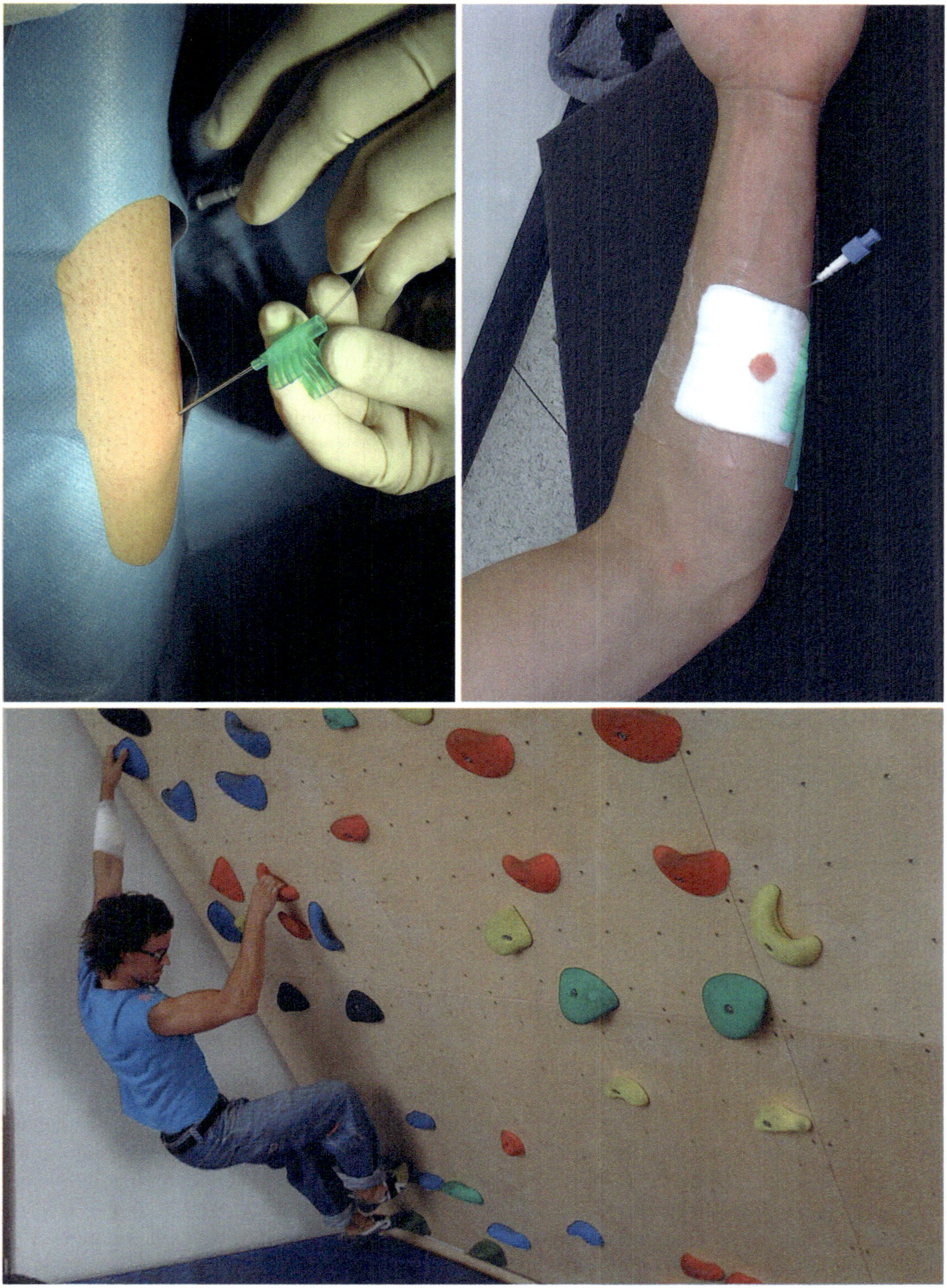

Fig. 8.2 Compartment pressure measurement

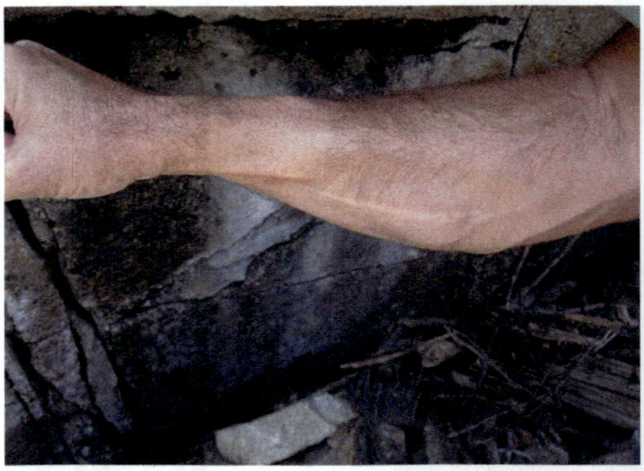

Fig. 8.3 Condition after fasciotomy of the forearm (compartment release) in a climber due to a rare but severe case of functional compartment syndrome

Functional Compartment Syndrome of the Forearm Muscles

- **History:** Swelling of the forearm musculature due to overstrain with consecutively restricted blood and lymph circulation
- **Onset:** Acute to chronic
- **Clinical:** Pain in the muscles of the forearm during and after exercise, possibly swelling
- **Diagnostic:** Medical history, clinical finding, sonography, compartment pressure measurement if necessary
- **Therapy**: Mostly conservative, fasciotomy (compartment release) only necessary in the most severe cases!
- **Outcome:** Good

8.5 Elbow Dislocation (with Vascular Injury)

Severe traumatic injuries, such as dislocations, sometimes even with vascular injuries around the elbow and the forearm, have been rare. Only a few case reports due to climbing accidents are described in literature [13, 14]. However, in the subdiscipline of bouldering, there has been an increasing incidence of elbow dislocations in recent years. This is most likely due to the low fall height with a backward fall vector onto a stretched or slightly flexed forearm. The diagnosis can usually be made clinically due to the malalignment and the often pain-induced complete loss of mobility in the elbow joint. X-ray imaging is required in order to exclude bony injuries.

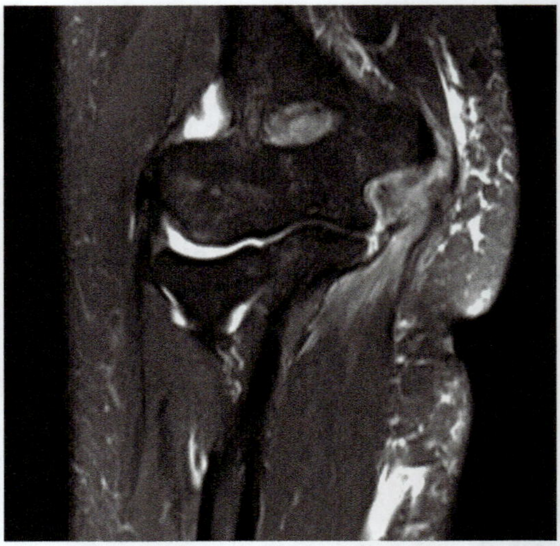

Fig. 8.4 MRI after elbow dislocation with rupture of the ulnar collateral ligament

The reduction should be performed under analog sedation or, if necessary, under anesthesia. The latter has the advantage that the ligament stability of the elbow joint can be tested. Subsequently, immobilization and additional diagnostics are performed by means of MRI (Fig. 8.4) and, in the case of concomitant bony injuries, by means of an additional CT scan to evaluate the extent of the pathology.

Surgery is the standard therapy in case of a tendency toward dislocation at flexion over 30° in the elbow joint or an instability in several examination areas (varus and valgus instability) due to multiple ligament injuries as well as dislocated fractures. The ruptured collateral ligaments are reattached with suture anchors (sometimes additional braced with FiberTape®), and bony injuries are addressed according to the guidelines of orthopedics and trauma surgery. Subsequently, the follow-up treatment consists of an early functional therapy using a movable elbow orthosis and treatment protocol, which is tailored to the extent of the injury.

Good functional results were achieved in climbers with a return to climbing after an average of 6 months [15]. Furthermore, rare second-

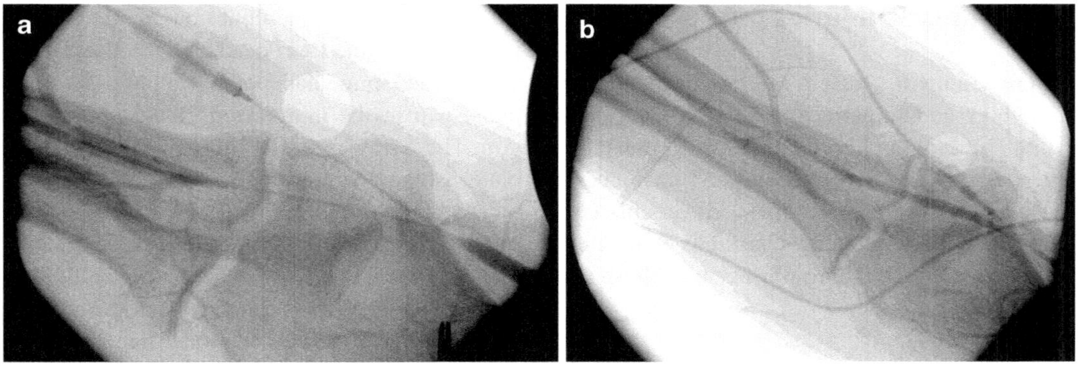

Fig. 8.5 Subtraction angiography depicting a lesion of the brachial artery following an elbow dislocation before (**a**) and after (**b**) patch plasty

ary pathologies, such as vascular lesions, must be excluded or correctly treated within the diagnostic process. The following case study from our own patient population should provide increased awareness of this issue:

8.5.1 Elbow Dislocation with Injury to the Brachial Artery

A 24-year-old, otherwise healthy male climber had suffered a posterior, closed elbow dislocation during a fall while bouldering which was promptly reduced in analog sedation. Subsequently, the peripheral circulation, motor function, and sensitivity were without pathological findings.

The plain X-rays of the elbow showed a non-displaced fracture of the radial head. Afterward, the patient was discharged after immobilization of the respective arm with a plaster splint. In the further course of diagnostics, an MRI was performed, and the patient visited our clinic 2 weeks after the accident. During this consultation, peripheral circulation, motor function, and sensitivity were still intact. However, an ulnar-sided, strand-like induration with local pressure pain, which was initially interpreted as a hematoma after a brachialis muscle tear, was described in the antecubital fossa.

The MRI showed injuries of the medial and lateral collateral ligaments as well as a partial rupture of the brachialis tendon with perifocal hematoma. Due to additional clinical instability, we decided on a surgical therapy. During the operation 5 days later, a long-distance stenosis of the brachial artery in the antecubital fossa and a thrombosis of the radial artery with just sufficient blood flow through collateral vessels were observed as a correlation to the previously palpated induration as a result of an injury to the brachial artery during the previous elbow dislocation (Fig. 8.5). Intraoperatively, an angiography and a subsequent thrombectomy and a venous patch plasty were performed. The ruptured ligaments were reattached, and all pathologies could be addressed. During follow-up 1 year after surgery, the patient was symptom-free even during intensive physical exertion [13].

Elbow Dislocation (with Vascular Injury)
- **History:** Traumatic
- **Onset:** Acute
- **Clinical:** Pain, misalignment, and impaired function
- **Diagnostic:** Medical history, clinical finding, (duplex-) sonography, MRI, and (MRI-) angiography if necessary
- **Therapy**: Reduction in analgosedation or anesthesia, surgery dependent on the extent of the pathology (e.g., instability, bony injuries)
- **Outcome:** Good

8.6 Muscle Injuries

Severe, traumatic muscle injuries in the area of the elbow and forearm as a result of accidents while climbing and bouldering are fortunately a rarity. Nevertheless, these must be excluded in the diagnostic process to avoid irreversible consequences. The following case study is intended to sensitize the attending physician to those rare but severe injuries:

8.7 Rupture of the Short Head of the Biceps Brachii Muscle Belly

A 42-year-old, otherwise healthy male climber fell while climbing in a bouldering gym.

He was climbing around a volume when his foot slipped. Consequently, his right upper arm hit the edge of the volume while the biceps muscle was fully tensed, a blunt trauma that resulted in a rupture of the short head biceps muscle belly. This caused immediate pain, swelling, and a palpable gap in the midportion of the upper right arm. The X-ray and MRI images showed a complete, isolated rupture of the short head of the biceps brachii muscle but no fracture (Fig. 8.6). In accordance with recent literature, we restored the anatomical integrity of the muscle belly itself through an open surgical approach with direct end-to-end modified Mason-Allen stitches, 10 days after the trauma, to prevent the ruptured muscle belly from further retraction, which would later hinder anatomical reduction (Fig. 8.7) [16]. As such, our patient was able to regain full strength and functionality. Six months after surgery, he had already returned to a climbing level close to his pre-injury level. Furthermore, the clinical investigation showed a full range of motion without functional complaints or pain [14]. In general, the majority of muscle injuries

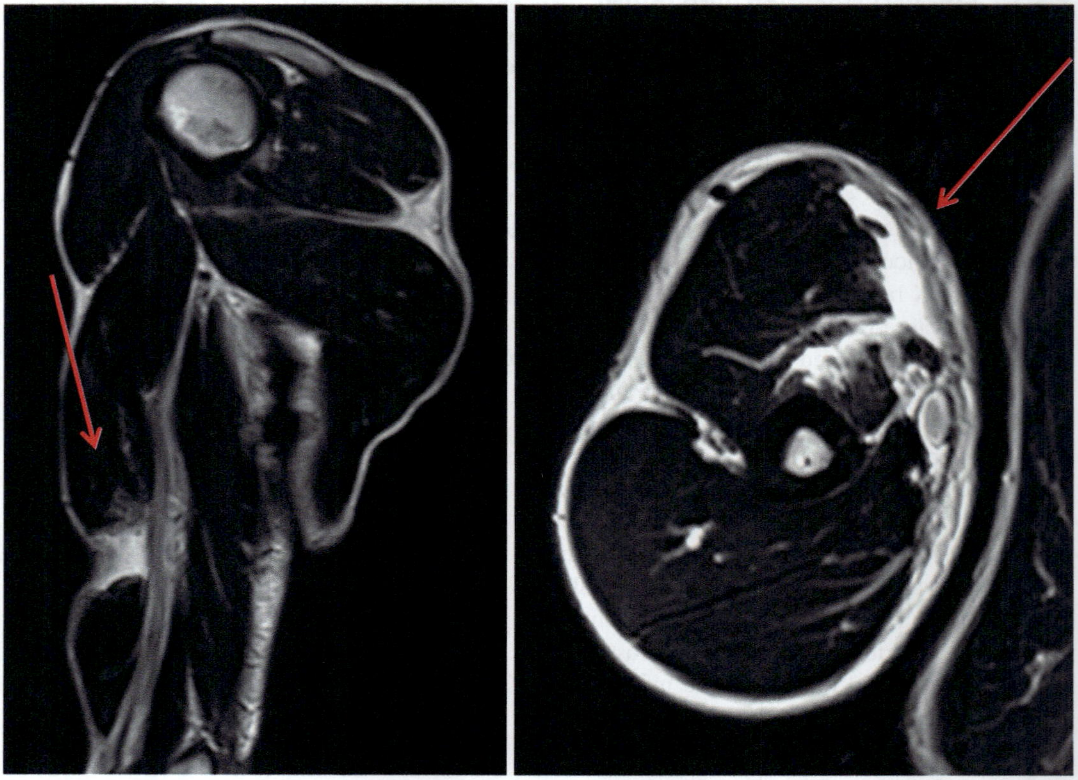

Fig. 8.6 MRI of an upper arm depicting a rupture of the short head of the biceps brachii muscle belly

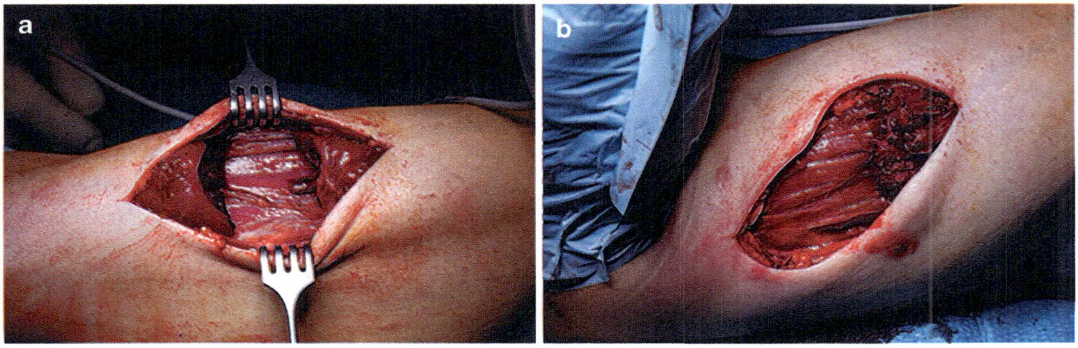

Fig. 8.7 Intraoperative clinical findings of a rupture of the short head of the biceps brachii muscle belly before (**a**) and after (**b**) direct end-to-end suture

can be managed conservatively with measures such as rest, nonsteroidal anti-inflammatory drugs, and physical therapy. There are only a few indications for surgery: a large intramuscular hematoma, tears of more than half of the muscle belly (Grade 3b-4), and the occurrence of a compartment syndrome, which is an absolute emergency [17].

Muscle Injuries
- **History:** Traumatic
- **Onset:** Acute
- **Clinical:** Pain, limitation of strength and function, in some cases palpable gap
- **Diagnostic:** Medical history, clinical finding, sonography, MRI if necessary
- **Therapy**: Conservative for partial muscle ruptures and strains, surgical for subtotal and complete muscle ruptures
- **Outcome:** Good

8.8 Distal Biceps Tendon Rupture

Ruptures of the distal biceps tendon account for only 3% of all tendon injuries of the biceps brachii muscle. However, due to the resulting loss of strength, especially in climbers, it requires a precise diagnosis and therapy. In most cases, an avulsion of the tendon attachment at the tuberosity radii can be found, which occurs mainly in men between 30 and 60 years of age as a consequence of chronic, degenerative changes. Additional risk factors are regular nicotine consumption, excessive strength training, and abuse of anabolic steroids.

Most often, the injury mechanism is described as a sudden force on the flexed and supinated forearm by pulling on an undercling into a "lock-off" position followed by a clicking sound [11].

This is followed by pain in the antecubital fossa and an immediate loss of strength, especially for supination but also in flexion of the elbow joint. The diagnosis can usually be made clinically. In addition to the previously mentioned complaints, the distal biceps tendon can no longer be palpated (hook test), and a so-called "reversed Popeye sign" (the biceps brachii muscle belly retracts proximally) can be seen in the case of a complete rupture. The diagnosis is confirmed by sonography and X-ray. In case of ambiguous findings, such as partial ruptures or bony lesions, MRI imaging is recommended (Fig. 8.8). Since conservative therapy can reduce the force for supination movements by up to 80%, prompt surgical therapy is the gold standard, especially for athletes (even partial ruptures of more than 50%). Thereby, the torn tendon is reattached to its anatomical insertion point on the radius using suture anchors. An early reattachment is essential to prevent tendon retraction and the development of muscle atrophy and fibrosis [18].

Subsequently, a partial immobilization of the elbow joint is necessary for 6 weeks, which is gradually reduced under physical therapy

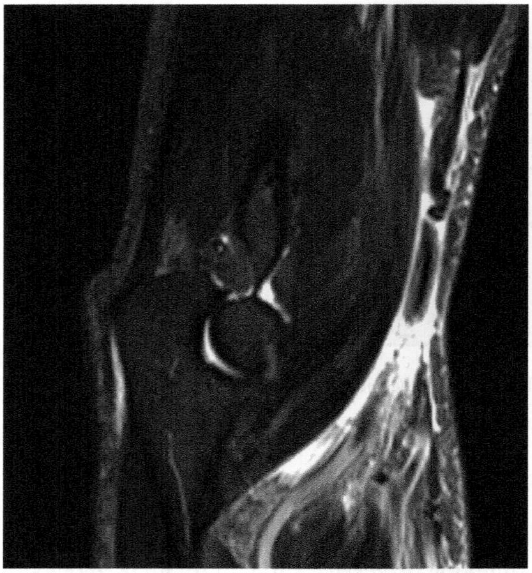

Fig. 8.8 MRI of an elbow depicting a distal biceps brachii tendon rupture

guidance (initial plaster splint, followed by elbow joint orthosis with reduced range of motion) [18]. Regarding the postoperative outcome, a recent review by Thomas and Lawton shows excellent results in athletes with a complete recovery of athletic performance and a full range of motion without functional complaints or pain [19].

Distal Biceps Tendon Rupture

- **History:** Traumatic (mostly in case of preexisting degenerative changes)
- **Onset:** Acute
- **Clinical:** Pain in the elbow and lack of strength for flexion and especially supination in the elbow joint
- **Diagnostic:** Medical history, clinical finding, sonography, X-ray, MRI if necessary
- **Therapy**: Surgical
- **Outcome:** Good

8.9 Conclusion

Most of the pathologies around the elbow are caused by overstrain (e.g., medial and lateral epicondylitis humeri, tendinopathies of the biceps, triceps, and brachialis muscle). Their therapy is almost always conservative. Since training errors are often the trigger for the development of the complaints, prophylaxis through adjunct compensatory training plays an essential role. Traumatic injuries such as distal biceps tendon ruptures as well as rare pathologies such as functional compartment syndrome of the forearm or traumatic vascular and muscle injuries must also be taken into account as a differential diagnosis.

References

1. Schöffl V, Simon M, Lutter C. Finger- und Schulterverletzungen im Klettersport. Orthopade. 2019;48(12):1005–12.
2. Grønhaug G. Self-reported chronic injuries in climbing: who gets injured when? BMJ Open Sport Exerc Med. 2018;4(1):e000406.
3. Schöffl V, Popp D, Küpper T, Schöffl I. Injury trends in rock climbers: evaluation of a case series of 911 injuries between 2009 and 2012. Wilderness Environ Med. 2015;26(1):62–7.
4. Schoeffl V, Klee S, Strecker W. Evaluation of physiological standard pressures of the forearm flexor muscles during sport specific ergometry in sport climbers. Br J Sports Med. 2004;38(4):422–5.
5. Stelzle FD, Gaulrapp H, Pförringer W. Verletzungen und Überlastungssyndrome beim Klettern an künstlichen Kletteranlagen. Sportverletzung-Sportschaden. 2000;14(4):128–33.
6. Rohrbough JT, Mudge KM, Schilling RC. Overuse injuries in the elite rock climber. Med Sci Sports Exerc. 2000;32(8):1369–72.
7. Altintas B, Greiner S. Epicondylitis humeri radialis: konservativ—operativ. Orthopade. 2016;45:870–7.
8. Schöffl V, Willauschus W, Sauer F, Küpper T, Schöffl I, Lutter C, et al. Autologous conditioned plasma versus placebo injection therapy in lateral epicondylitis of the elbow: a double blind, randomized study. Sportverletz Sportschaden. 2017;31(1):31–6.
9. Hastie G, Soufi M, Wilson J, Roy B. Platelet rich plasma injections for lateral epicondylitis of the elbow reduce the need for surgical intervention. J Orthop. 2018;15(1):239–41.
10. Schöffl V, Matros P, Korb L. ACT (Adjunct Compensatory Training). Schöffl V; 2019.
11. Schöffl VR, Hochholzer T, Lightner S. One move too many... Sharp End Publishing; 2016. 176–178 p.
12. Raza H, Mahapatra A. Acute compartment syndrome in orthopedics: causes, diagnosis, and management. Advances in Orthopedics. 2015;2015:543412.
13. Lutter C, Pfefferkorn R, Schoeffl V. Arterial damages in acute elbow dislocations: which diagnostic tests are required? BMJ Case Rep. 2016;2016:bcr2016216336.

14. Simon M, Lutter C, Schöffl V. Rupture of the short head of the biceps brachii muscle belly caused by a rock climbing accident. Wilderness Environ Med. 2020;31(3):327–31.
15. Sandmann GH, Siebenlist S, Lenich A, Neumaier M, Ahrens P, Kirchhoff C, et al. Traumatic elbow dislocations in bouldering. Unfallchirurg. 2014;117(3):274–80.
16. Mueller-Wohlfahrt H-W, Haensel L, Mithoefer K, Ekstrand J, English B, McNally S, et al. Terminology and classification of muscle injuries in sport: the Munich consensus statement. Br J Sports Med [Internet]. 2013;47(6):342–50. http://bjsm.bmj.com/lookup/doi/10.1136/bjsports-2012-091448.
17. Maffulli N, Oliva F, Frizziero A, Nanni G, Barazzuol M, Via AG, et al. ISMuLT Guidelines for muscle injuries. Muscles Ligaments Tendons J. 2013;3(4):241–9.
18. Völk C, Siebenlist S, Kirchhoff C, Biberthaler P, Buchholz A. Rupture of the distal biceps tendon. Unfallchirurg. 2019;122(10):799–811.
19. Thomas JR, Lawton JN. Biceps and triceps ruptures in athletes. Hand Clin. 2017;33(1):35–46.

14. [illegible]
15. [illegible]
16. [illegible]

9 Shoulder Injuries

M. Simon and Volker Schöffl

Christoph Lutter climbing in Mill Creek Canyon, Moab, Utah, USA, Photo Michael Simon

M. Simon (✉)
Department of Orthopedic and Trauma Surgery, Center of Sportsmedicine, Klinikum Bamberg, Bamberg, Germany
e-mail: info@michaelsimon.org

V. Schöffl
Department of Orthopedic and Trauma Surgery, Center of Sportsmedicine, Klinikum Bamberg, Bamberg, Germany

School of Clinical and Applied Sciences, Leeds Becket University, Leeds, UK

Section of Wilderness Medicine, Department of Emergency Medicine, University of Colorado School of Medicine, Denver, CO, USA

Department of Orthopedic and Trauma Surgery, Friedrich-Alexander Universität Erlangen-Nürnberg, Erlangen, Germany

V. Schöffl et al. (eds.), *Climbing Medicine*, https://doi.org/10.1007/978-3-030-72184-8_9

Table 9.1 Current distribution of shoulder injuries in rock climbing (Schöffl et al. [2])

Shoulder injuries 2017–2018 (*n* = 154)	*n*	%	Shoulder injuries 2009–2012 (*n* = 157)	*n*	%
SLAP lesion	37	29.8	SLAP lesion	51	32.5
Impingement syndrome	34	27.4	Impingement syndrome	40	25.5
Shoulder dislocation, Bankart lesion	22	17.7	Shoulder strain	17	10.8
Shoulder strain	16	12.9	Shoulder dislocation, Bankart lesion	16	10.2
Rotator cuff tear	12	9.7	Supraspinatus tendonitis	7	4.5
Acromioclavicular joint injury	12	9.7	Functional instability	7	4.5
Tendinopathy of the long head of the biceps tendon	6	4.8	Tendinopathy of the long head of the biceps tendon	5	3.2
Functional instability	5	4.0	Rupture of the long head of the biceps tendon	5	3.2
Pulley lesion	5	4.0	Rotator cuff tear	5	3.2
Rupture of the long head of the biceps tendon	2	1.6	Acromioclavicular joint injury	3	1.9
Others	2	1.6	Pulley lesion	1	0.6
Supraspinatus tendonitis	1	0.8			

9.1 Overview

Shoulder injuries are the second most common pathologies in rock climbing after finger injuries, whereby numbers have increased in the last couple of years [1].

Within our own patient cohort, we found an increase of the incidence of shoulder injuries from 5% (1998–2001) to 20.2% from 2017 to 2018 [2]. SLAP lesions (superior labrum anterior to posterior) play an essential role as they are the fifth most common diagnosis (5.8%) out of all climbing-related pathologies [2]. Table 9.1 shows the current distribution of shoulder injuries among rock climbers from our own patient cohort. In the following sections, we discuss the specifics of the essential climbing-related shoulder injuries as well as their diagnostics and therapy.

9.2 Impingement Syndrome

Impingement symptoms are the second most common shoulder pathology in rock climbers [3]. Regarding etiology, we have to distinguish between the primary and secondary forms. Primary impingement syndrome is caused by structural changes that mechanically narrow the subacromial space. These changes include bony narrowing from the cranial side (outlet impingement), which is the case in acromial Types II–III according to Bigliani and Morrison, structural variations of the coracoid, and bony malposition after a fracture of the greater tubercle. An increase in the volume of the subacromial tissue (e.g., calcific tendonitis, subacromial bursitis) may also lead to a primary impingement (non-outlet impingement). Secondary impingement is due to a functional disturbance of the centering of the humeral head triggered by muscular imbalance. This leads to an abnormal displacement of the center of rotation in elevation and results in soft tissue entrapment [4, 5]. Another relatively rare form of functional impingement syndrome in climbers is coracoid impingement, which is caused by a hypertrophied subscapularis tendon and muscle (Fig. 9.1) following intense sport-specific training [6]. Finally, internal impingement can also be the reason for impingement symptoms in athletes. In this case, the mechanism is a repetitive squeezing of the supra- and infraspinatus tendons against the glenoid in extreme external rotation and simultaneous abduction (posterosuperior) or rather internal rotation and adduction (anterosuperior) [7, 8]. Additionally, symptoms can be aggravated by

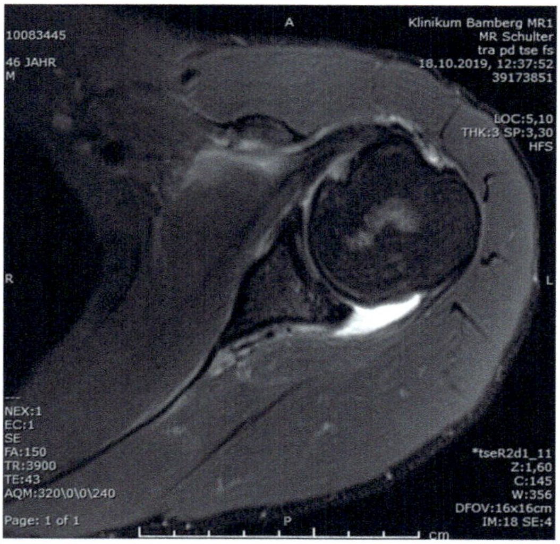

Fig. 9.1 MRI of a shoulder with coracoid impingement

Fig. 9.2 Postural deformity, the so-called "climber's back"

degenerative alterations of the acromioclavicular joint as well as a postural deformity, the so-called "climber's back" (Fig. 9.2) (see also Chap. 12 Spine). Shortened pectoral muscles and pronounced rear shoulder girdle muscles result in an increased thoracic kyphosis, leading to an amplified internal rotation of the shoulders, which ultimately enhances preexisting impingement symptoms [9].

Clinically, climbers complain of persistent pain without a preceding trauma, especially when elevating the arm (in particular between 70° and 120°; painful arc), during forced movement above the head, and also during rest when lying on the affected side [4, 5].

The diagnosis is made mainly by a thorough clinical examination (sensitivity up to 90%; e.g., Hawkins test, Neer's sign, Jobe test, painful arc), ultrasound, and plain X-rays. In case of persistent complaints, an MRI is mandatory to evaluate the condition of the rotator cuff, the bursa, and the labrum-ligament complex. Aspects such as potential tears of the rotator cuff, the degree of atrophy (according to Thomazeau), and fatty degeneration (according to Goutallier) of the muscle bellies of the rotator cuff can be assessed [5]. In general, literature recommends conservative therapy (physical therapy and nonsteroidal anti-inflammatory drugs). In case of sustained complaints, subacromial infiltrations with local anesthetics and corticosteroids (3–4 weeks after onset of symptoms, at the earliest, and no more often than two to three times, no infiltration into the tendon) under ongoing physical therapy are recommended. Corticosteroid injections may be restricted by the World Anti-Doping Agency (WADA) and may need specific documentation (a definitive statement is not possible here, as the regulations are constantly changed and revised.) In the case of an accompanying calcific tendinitis, shockwave therapy shows good results. Overall, more than 70% of all impingement symptoms can be treated successfully through conservative therapy. If conservative therapy fails for 3 months or longer or in the presence of other comorbidities (e.g., rotator cuff tear, SLAP lesion, etc.), surgical therapy is indicated. Hereby, an arthroscopic, subacromial decompression is performed. Besides a bursectomy, this also includes the removal of the anterior and lateral portions of the undersurface of the acromion (5–8 mm). If necessary, the removal of inferior acromial osteophytes and the lateral end of the clavicle without a total resection of the acromioclavicular joint can be performed as part of the same arthroscopic surgery (coplaning) [5].

Impingement Syndrome

- **History:** Congenital deformity, muscular imbalance
- **Onset:** Chronic
- **Clinical:** Pain without trauma, especially on forced movement above the head and when lying on the affected side
- **Diagnostic:** Medical history, clinical finding, sonography, if necessary MRI
- **Therapy:** Mostly conservative
- **Outcome:** Good

9.3 SLAP Lesions

SLAP lesions (superior labrum anterior to posterior) are the most common shoulder injuries in rock climbers (29.8%) [2]. "Shoulder moves" are particularly stressful for the insertion of the long biceps tendon (Schöffl et al. 2011). During these moves, a subsequent internal rotation and anterior tilt of the shoulder under load leads to increased tensile strain on the pulley system and the labro-bicipital complex (LBC). Nevertheless, SLAP injuries are usually caused by repetitive microtrauma rather than one single incident. Thus, it is not surprising that most of the affected climbers are older than 35 years with many years of climbing experience [11, 12]. Patients presenting with acute injuries usually describe a fall onto an outstretched arm or a sudden, unexpected pull on the arm (e.g., a climber trying to hold on while the feet are slipping).

Clinically, climbers complain of a sudden or gradual deterioration of shoulder function and concomitant pain. In the majority of cases, the dominant arm is concerned. The term "dead arm syndrome" specifies the inability to execute sport-specific movements at pre-injury velocity [13]. Other symptoms include sensations of intermittent clicking or popping during the cocking phase and anterior shoulder pain [14]. A high number of clinical tests (e.g., O'Brien test, crank test, dynamic labral shear test) have been described for the detection of SLAP lesions, but no gold standard with regard to sensitivity or specificity has been established yet. Thus, two climbing-specific SLAP provocation tests (Fig. 9.3) are recommended, as they simulate the sport-specific strain on the labro-bicipital complex. To confirm the diagnosis, magnetic resonance arthrography (MRA) is mandatory [15] (Fig. 9.4). According to current literature, SLAP lesions with structural damage, which is more than a fraying (Grade II or higher, according to Snyder) to the biceps tendon anchor or the labrum, should be treated surgically in athletes with a primary tenodesis of the long biceps tendon. SLAP I lesions can be treated conservatively. Our standard surgical approach in overhead sports athletes is a tenodesis of the long biceps tendon at the top of the bicipital groove with a composite tenodesis screw [16, 17]. Against the background of repetitive microtrauma, a solid tenodesis of the long head of the biceps tendon and thus the deactivation of the pain generator in the area of the LBC are reasonable [10, 18]. In

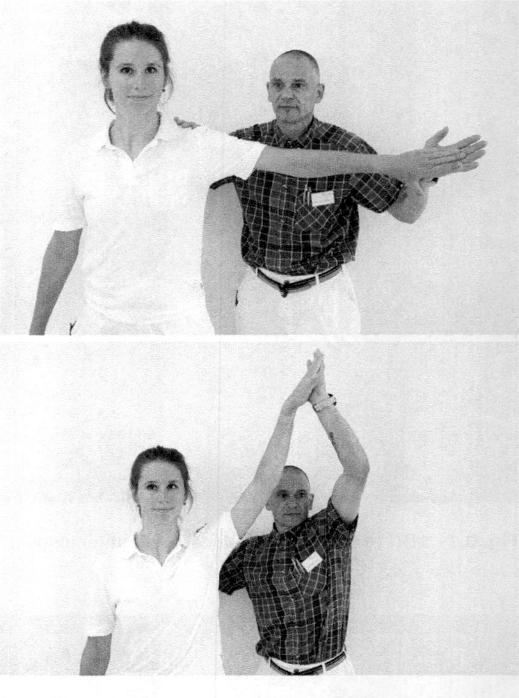

Fig. 9.3 Climbing-specific SLAP tests (Photo: Michael Simon)

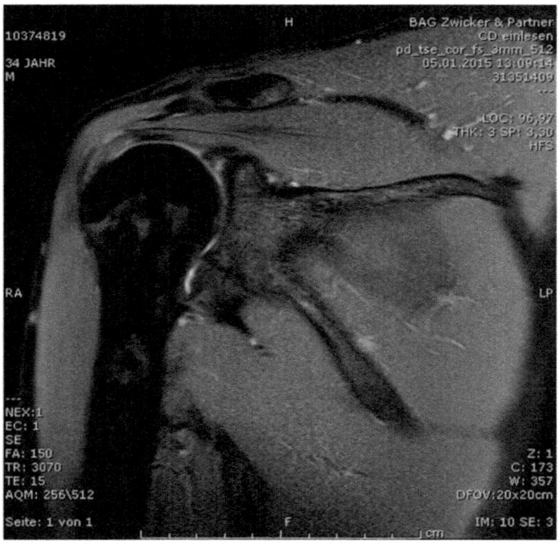

Fig. 9.4 MRI of a shoulder with a SLAP II lesion (Snyder classification)

addition to the tenodesis of the long head of the biceps tendon, the superior labrum is refixed. Although some studies have shown a good outcome in 75–97% of nonathletes treated with an orthotopic refixation of the long biceps tendon to the LBC, other studies focusing on the outcome in overhead athletes found only 22–75% satisfactory results. Thus, for overhead athletes, a primary tenodesis of the long head of the biceps tendon is the therapy of choice [18, 19]. Nevertheless MRI evaluation on asymptomatic climbers recently showed a a high prevalence of labral and other pathologies, thus the clinical presentation is most important to finally decide who needs a surgical revision [20].

This is also highlighted by the findings of a follow-up examination of 30 climbers, who were treated with a primary tenodesis of the long head of the biceps tendon and a refixation of the superior labrum, if necessary. In this cohort, we found good results with an average Constant-Murley score of 89.6 points. All patients had started climbing again after 12 months following surgery, and 67% had already regained their pre-injury climbing level [16].

SLAP Lesions

- **History:** Repetitive microtrauma, acute trauma
- **Onset:** Chronic and acute
- **Clinical:** Sudden or gradual deterioration of shoulder function and concomitant pain
- **Diagnostic:** Medical history, clinical finding, MRA
- **Therapy:** Surgical therapy in case of SLAP lesions Grade II (Snyder) or higher
- **Outcome:** Good

9.4 Pathologies of the Rotator Cuff

A growing number of aging athletes is still climbing at a high difficulty level [12]. As such, the number of degenerative pathologies, such as tears of the rotator cuff (consisting of subscapularis, supraspinatus, infraspinatus, and teres minor muscles), is also increasing. Additionally, injuries to the rotator cuff can also be caused by acute trauma (e.g., shoulder dislocation, a direct fall on the arm in abduction, or a sudden unexpected pull on the arm). The most common type of injury is a rupture of the tendon of the supraspinatus muscle. Clinically, climbers complain of a sudden or gradual onset of pain during rest but mainly during abduction over more than 90° in combination with loss of strength [21]. Clinical symptoms often depend on the location and size of the respective tear. Frequently, there is an inverse correlation between tear size and pain intensity. In practice, partial thickness rotator cuff tears on the bursal surface can be extremely painful, whereas massive tears are mainly associated with a loss of strength and a feeling of instability. The diagnosis is made by a thorough medical history, a clinical examination, ultrasound, and MRI.

Isometric function tests enable the physician to examine the different muscles of the rotator cuff regarding strength and pain (e.g., *supraspinatus muscle*—Jobe test, starter test, external rotation lag sign, drop arm sign; *subscapularis muscle*—lift-off test, internal rotation lag sign, belly press test). The condition of the rotator cuff, the bursa, and the labrum-ligament complex can be investigated with sonography and MRI. The tear size as well as the degree of atrophy (according to Thomazeau) and fatty degeneration (according to Goutallier) of the muscle bellies of the rotator cuff can be assessed [22]

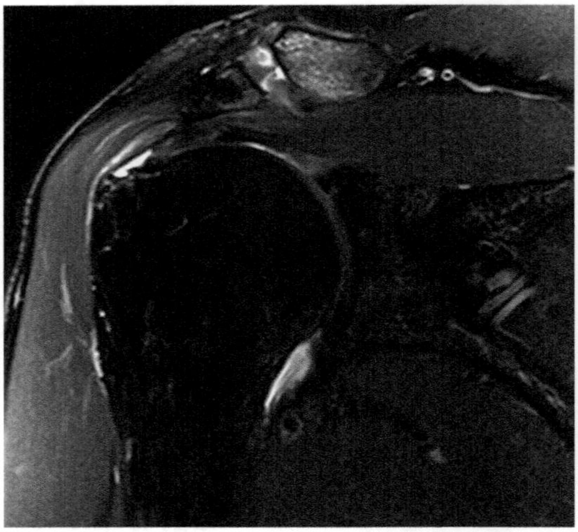

Fig. 9.5 MRI of a shoulder with a complete rotator cuff tear (m. supraspinatus)

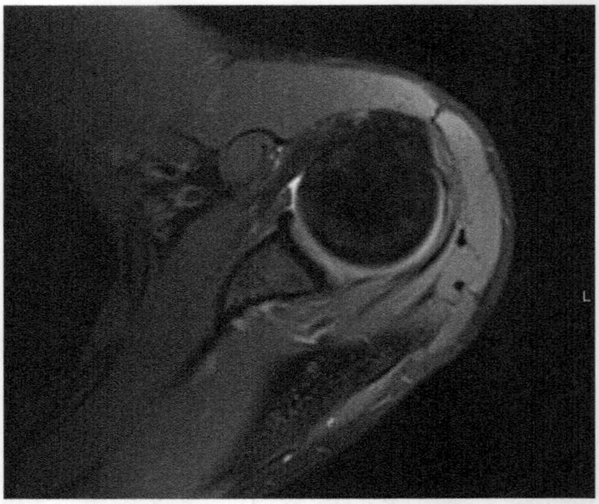

Fig. 9.6 MRI of a shoulder with a Bankart lesion following a traumatic shoulder dislocation

(Fig. 9.5). While small partial thickness tears can be treated conservatively, full thickness tears and severe partial thickness tears in athletes should be treated surgically [11]. According to current literature, arthroscopic surgical techniques using composite suture anchors are the standard procedure. A retrospective study by Simon et al. analyzed the general (Constant-Murley score) and sport-specific (change in International Climbing and Mountaineering Federation [UIAA] grade) outcome after the surgical repair of rotator cuff injuries in 12 rock climbers and found a good functional outcome. The mean Constant-Murley score was 92 points. Furthermore, all patients had restarted climbing within 12 months after surgery [23].

Pathologies of the Rotator Cuff

- **History:** Chronic degenerative changes, acute trauma
- **Onset:** Chronic and acute
- **Clinical:** Pain and loss of strength, especially at abduction over 90°
- **Diagnostic:** Medical history, clinical finding, sonography, MRI
- **Therapy:** Surgical therapy in case of massive full thickness tears or severe partial tears
- **Outcome:** Good

9.5 Shoulder Dislocation

The shoulder joint has the biggest range of motion of all human joints, making it particularly vulnerable to dislocation. Approximately 75% of all primary shoulder dislocations happen during sports, with the shoulder joint in an abduction and externally rotated position.

Unidirectional anterior dislocations make up the majority of cases (95%). As a result of the direction of the dislocation, the particular trauma mechanism, and the age of the patient, injuries of the labrum-ligament complex, the rotator cuff, and bony injuries of the humeral head (Hill-Sachs lesion) or the glenoid (Bankart fracture) may occur. In most cases, these injuries lead to instability with an increased risk of repeated shoulder dislocations, especially during overhead activities. Following a shoulder dislocation, patients complain of load-dependent pain and a pronounced feeling of instability. The diagnosis is made by a thorough medical history and a clinical examination of both shoulders (apprehension test, relocation test, load and shift test, sulcus sign), including neurological tests to exclude injuries to the brachial plexus. Furthermore, an MRI (potentially with an intra-articular contrast agent—MRA) is indispensable [24] (Fig. 9.6). Besides traumatic lesions of the labrum due to traumatic shoulder dislocations (Polar Type I), we also occasionally find pathologies in climbers on the basis of chronic microtrauma (Polar Type III), which lead to functional instability. These cases are accompanied by laxity of the joint capsule and a dysfunctional locomotion pattern, both leading to recurrent shoulder subluxations, whereby structural pathologies of the labrum-ligament complex are not usually found [11].

The climbers' complaints are often associated with specific climbing-related movements, such as passively hanging from one arm without muscular stabilization. These so-called dead hangs can cause mechanoreceptor damage and laxity of the joint capsule. Besides a mechanic insufficiency, a neuromuscular insufficiency of the joint stabilization may also occur [11, 25, 26].

In most cases, the standard therapy for structural lesions of the labrum-ligament complex or fractures of the glenoid in athletes following a shoulder dislocation is an arthroscopic surgery [11]. In this way, the development of a chronic instability can be avoided.

In a recent study, ten climbers from our own patient collective were treated with an arthroscopic refixation of the labrum-ligament complex following a traumatic shoulder dislocation. In the follow-up, we found a very favorable functional and sport-specific outcome. All climbers achieved their pre-injury performance level, no recurrent dislocation had occurred, and the mean Constant-Murley score was 97 points [27].

The therapy for climbers with a functional instability of the shoulder joint (Polar Type III) is always conservative by means of stabilization training and a modification in climbing technique [11, 25, 26].

Shoulder Dislocation

- **History:** Trauma, functional
- **Onset:** Acute, acquired
- **Clinical:** Pain and instability
- **Diagnostic:** Medical history, clinical finding, sonography, MRI (potentially with an intra-articular contrast agent—MRA)
- **Therapy:** Surgical therapy in case of structural pathologies of the labrum-ligament complex
- **Outcome:** Good

9.6 Rupture of the Long Head of the Biceps Tendon

Since climbing can often be performed at a high to very high level into old age, degenerative ruptures of the long head of the biceps tendon (LHBT) and its guidance through the so-called pulley are not uncommon due to the biomechanical load. There is a clear increase in the incidence from the age of 45 onward [12]. Patients usually report preexisting symptoms and often minor trauma, which ultimately leads to a final rupture of the LHBT. A sudden event with stabbing, whiplike pain is indicated in the medical history, after which the impressive "Popeye sign," a full distalization of the muscle belly, is clinically evident (Fig. 9.7). There is usually a tenderness of the intertubercular sulcus, but the loss of strength is minimal. The diagnosis is carried out by sonography and MRI, in addition to the clinical examination [28].

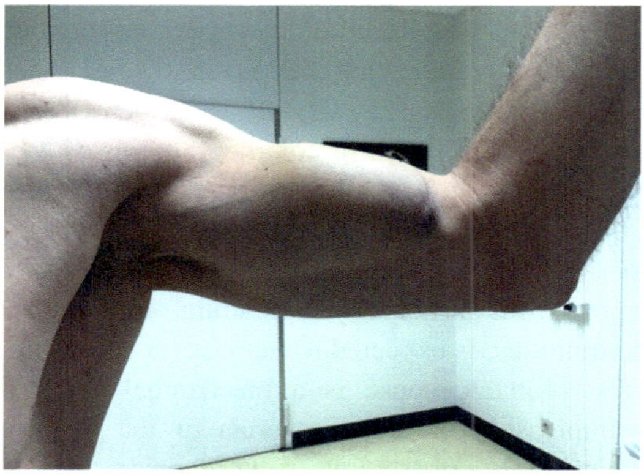

Fig. 9.7 "Popeye sign" following a rupture of the distal biceps tendon

A conservative therapy of LHBT ruptures results in a force reduction of 8–21% for flexion and supination in the elbow and a persistent cosmetic deformity ("Popeye sign"). After surgical refixation, this strength deficit is only 0–10%. Since climbers usually do not accept a strength deficit, surgical therapy (tenodesis) is recommended [11].

All athletes treated in this way regained their previous sport-specific climbing level. In a retrospective study, an average Constant-Murley score of 97.8 points and a self-evaluated biceps function of 97.3% were determined [29].

Rupture of the Long Head of Biceps Tendon
- **History:** Chronic, degenerative changes
- **Onset:** Acute to chronic
- **Clinical:** Pain, Popeye sign, only minimal loss of strength
- **Diagnostic:** Medical history, clinical finding, sonography, MRI
- **Therapy:** Surgical
- **Outcome:** Good

9.7 Thrombosis of the Subclavian Vein (Paget-Schroetter Syndrome)

Deep vein thrombosis of the upper extremities (Paget-Schroetter syndrome) is rare overall (2 per 100,000 patients per year) but, nevertheless, represents one of the most common entities of vascular pathologies in athletes. Especially because of the growing popularity of climbing and bouldering, this diagnosis, which occurs primarily in sports with particularly high strain on the upper extremities, is expected to increase in the future. The probable cause is a mechanical external compression of the subclavian or the axillary vein in its course between the clavicle and the first rib through the musculoligamentous structures (e.g., scalenus anterior, muscle, subclavius muscle, costoclavicular ligament). Compression appears especially during abduction or particularly strenuous arm movements, which often occur during climbing.

Repetitive training protocols might induce intimal lesions and thus cause fibrosis. The resulting increased scar tissue leads to a progressive stenosis. Under load, this can no longer be compensated by collaterals, and even a small thrombus can cause the lumen to become completely blocked. Consequently, the restricted blood flow can no longer be compensated, at which point symptoms appear [30, 31]. Patients report nonspecific shoulder pain and feelings of tension in the shoulder and neck area. In addition, there is often a feeling of pressure in the entire arm, a change of the skin color in the sense of a livid discoloration up to the formation of ecchymoses, and a tendency of swelling during or after climbing. In the case of a subacute course, the development of collateral circulation with increased venous drawing, especially following exercise, can be found in the shoulder area (Fig. 9.8). Since the initial symptoms are often unspecific, it is important for the treating physician to be familiar with the pathology of Paget-Schroetter syndrome in order to make the correct diagnosis. This is the

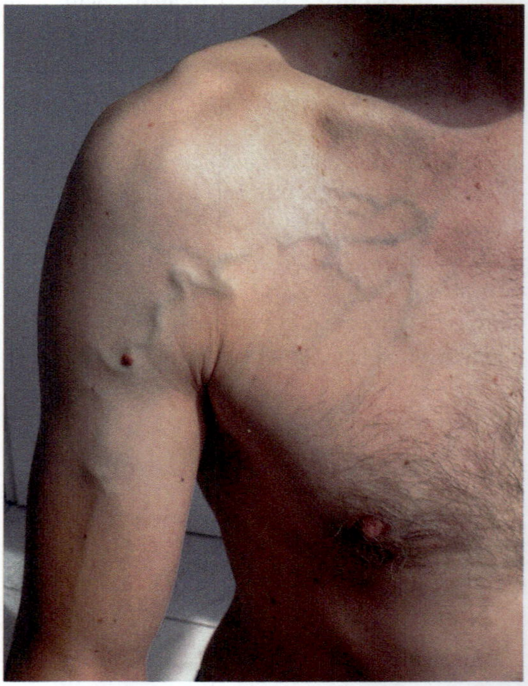

Fig. 9.8 Collateral circulation due to a thrombosis of the subclavian vein (Paget-Schroetter syndrome)

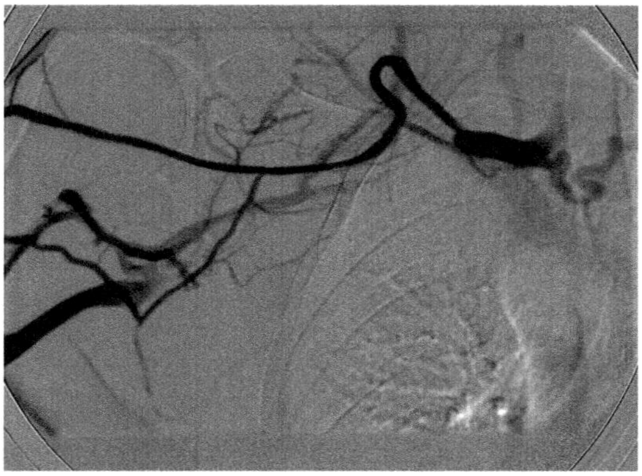

Fig. 9.9 Phlebography of a shoulder showing a thrombosis of the subclavian vein (Paget-Schroetter syndrome)

only way to prevent serious potential complications, such as pulmonary embolism. In addition to the clinical examination, color-coded Doppler sonography is the most important diagnostic tool. Phlebography (Fig. 9.9) (or MRI angiography) is only necessary in rare cases and if the findings are unclear. In addition, CT and MRI may provide important information about additional causes of compression (e.g., cervical rib). Furthermore, thrombophilia diagnosis should be carried out over the course of the disease. Therapy is determined by the extent of the findings and aims to achieve freedom from symptoms as quickly as possible and reduce the risk of developing a post-thrombotic syndrome.

In addition to conservative therapy using oral anticoagulation, surgical options for recanalization and even addressing anatomical variants (e.g., resection of a cervical rib) are available [30, 31].

Thrombosis of the Subclavian Vein (Paget-Schroetter Syndrome)

- **History:** Functional or anatomical compression of subclavian and axillary vein
- **Onset:** Acute to chronic
- **Clinical:** Pain, swelling, and cyanosis of the affected arm
- **Diagnostic:** Medical history, clinical finding, color-coded Doppler sonography, phlebography, or MRI angiography if applicable
- **Therapy:** Conservatively (oral anticoagulation), possibly surgically depending on the extent of the findings
- **Outcome:** Often long-lasting complaints

9.8 Conclusion

Shoulder injuries are the second most common pathologies in rock climbing after finger injuries. SLAP lesions (superior labrum anterior to posterior) play an especially essential role.

Furthermore, the frequency of pathologies of the rotator cuff, the long head of the biceps tendon, impingement symptoms, and injuries due to a shoulder dislocation has increased recently. The basic pillars of diagnosis are a thorough clinical investigation, sonography, and MRI. If surgical therapy is necessary, recent studies have shown very promising postoperative results, enabling rock climbers to resume their pre-injury climbing abilities.

Furthermore, targeted shoulder stability training (e.g., www.act.clinic) as well as avoiding passive dead hangs in the shoulder joint can help to prevent shoulder injuries [26]. Also closed chain eccentric training has recently proofed to help in prevention [32].

References

1. Jones G, Schöffl V, Johnson MI. Incidence, diagnosis, and management of injury in sport climbing and bouldering: a critical review. Curr Sports Med Rep. 2018;17(11):396–401.
2. Schöffl V, Simon M, Lutter C. Finger- und Schulterverletzungen im Klettersport. Orthopade. 2019;48(12):1005–12. https://doi.org/10.1007/s00132-019-03825-3.
3. Schöffl V, Popp D, Küpper T, Schöffl I. Injury trends in rock climbers: evaluation of a case series of 911 injuries between 2009 and 2012. Wilderness Environ Med. 2015;26(1):62–7.
4. Habermeyer, P. Schulterchirurgie. Urban und Fischer. 4th ed. Elsevier; 2010.
5. Garving C, Jakob S, Bauer I, Nadjar R, Brunner UH. Impingement syndrome of the shoulder. Dtsch Arztebl Int. 2017;114(45):765–76.

6. Schöffl V, Schneider H, Küpper T. Coracoid impingement syndrome due to intensive rock climbing training. Wilderness Environ Med. 2011;22(2):126–9.
7. Kirchhoff C, Imhoff AB. Posterosuperior and anterosuperior impingement of the shoulder in overhead athletes-evolving concepts. Int Orthop. 2010;34(7):1049–58.
8. Corpus KT, Camp CL, Dines DM, Altchek DW, Dines JS. Evaluation and treatment of internal impingement of the shoulder in overhead athletes. World J Orthop. 2016;7(12):776–84.
9. Förster R, Penka G, Bösl T, Schöffl VR. Climber's back—form and mobility of the thoracolumbar spine leading to postural adaptations in male high ability rock climbers. Int J Sports Med. 2009;30(1):53–9.
10. Schöffl V, Popp D, Dickschass J, Küpper T. Superior labral anterior-posterior lesions in rock climbers-primary double tenodesis? Clin J Sport Med. 2011;21(3):261–3.
11. Schöffl V, Popp D. Die Sportlerschulter. Grimm C, ME, editors. Stuttgart: Schattauer; 2016. p. 231–234.
12. Lutter C, Hotfiel T, Tischer T, Lenz R, Schöffl V. Evaluation of rock climbing related injuries in older athletes. Wilderness Environ Med. 2019;30(4):362–8.
13. Bedi A, Allen AA. Superior labral lesions anterior to posterior-evaluation and arthroscopic management. Clin Sports Med. 2008;27(4):607–30.
14. Popp D. Superior labral anterior posterior lesions of the shoulder: current diagnostic and therapeutic standards. World J Orthop. 2015;6(9):660. http://www.wjgnet.com/2218-5836/full/v6/i9/660.htm.
15. Popp D, Schöffl V. Superior labral anterior posterior lesions of the shoulder: current diagnostic and therapeutic standards. World J Orthop. 2015;6(9):660–71.
16. Ott B, Popp D, Schöffl I, Küpper T, Hennig F, Schöffl V. Primary long biceps tenodesis as treatment for high grade SLAP lesions: clinical evaluation in overhead sport athletes. Eur J Sport Med. 2017.
17. Popp D, Schöffl V. Shoulder SLAP and biceps tendon repair. Minerva Ortop e Traumatol. 2013;64(3):247–63.
18. Boileau P, Parratte S, Chuinard C, Roussanne Y, Shia D, Bicknell R. Arthroscopic treatment of isolated type II SLAP lesions. Am J Sports Med. 2009;37(5):929–36.
19. Buzzacott P, Schöffl I, Chimiak J, Schöffl V. Rock climbing injuries treated in US Emergency Departments, 2008–2016. Wilderness Environ Med. 2019;30(2):121–8.
20. Cooper JD, Seiter MN, Ruzbarsky JJ, Poulton R, Dornan GJ, Fitzcharles EK, Ho CP, Hackett TR. Shoulder pathology on magnetic resonance in asymptomatic elite-level rock climbers. Orthop J Sports Med. 2022;10(2):23259671211073137. https://doi.org/10.1177/23259671211073137.
21. Witt M. Rotatoren-manschettenruptur. Trauma und Berufskrankheit. 2006.
22. Scheibel M, Habermeyer P. Aktuelle klinische Untersuchung der SchulterCurrent procedures for clinical evaluation of the shoulder. Orthopäde. 2005;34(3):267–83.
23. Simon M, Popp D, Lutter C, Schöffl V. Functional and sports-specific outcome after surgical repair of rotator cuff tears in rock climbers. Wilderness Environ Med. 2017;28(4):342–7.
24. Martetschläger F, Imhoff AB. Schulterluxation des Sportlers. Aktuelle Therapiekonzepte. Orthopäde. 2014;43(3):236–43.
25. Schöffl VR, Hochholzer T, Lightner S. One move too many... Sharp End Publishing; 2016. p. 176–178.
26. Schöffl V, Matros P, Korb L. ACT (Adjunct Compensatory Training). Tradition, UK. 2020.
27. Simon M, Geffel L, Lutter C, Schöffl V. Functional and sport-specific outcome after arthroscopic Bankart repair in rock climbers. In: 34 Jahreskongress der GOTS. 2019.
28. Mendel T, Großstück R, Hofmann GO. Ruptur der langen Bizepssehne und operative Therapiestrategien. Aktueller Überblick. Trauma und Berufskrankheit. 2005.
29. Schöffl VR, Harrer J, Küpper T. Biceps tendon ruptures in rock climbers. Clin J Sport Med. 2006;16(5):426–7.
30. Lutter C, Monasterio E, Schöffl V. Rock climbing-related subclavian vein thrombosis. BMJ Case Rep. 2015;2015:bcr2015212021.
31. Bürger T, Stegemann E, Baumbach H, Weiske N. V.-subclavia-Thrombose bei Thoracic-inlet-Syndrom [Thrombosis of the subclavian vein in thoracic inlet syndrome]. Gefässchirurgie. 2016;21:558–66.
32. Kozin S, Cretu M, Kozina Z, Chernozub A, Ryepko O, Shepelenko T, Sobko I, Oleksiuk M. Application of closed kinematic chain exercises with eccentric and strength exercises for the shoulder injuries prevention in student rock climbers: a randomized controlled trial. Acta Bioeng Biomech. 2021;23(2):159–68.

Part III

Orthopedics of the Lower Extremity

Foot and Ankle

10

Volker Schöffl and M. Simon

Volker Schöffl „La Barre Fixe (assis)", Fontainebleau, France, Photo Enrico Haase

V. Schöffl (✉)
Department of Orthopedic and Trauma Surgery, Center of Sportsmedicine, Klinikum Bamberg, Bamberg, Germany

School of Clinical and Applied Sciences, Leeds Becket University, Leeds, UK

Section of Wilderness Medicine, Department of Emergency Medicine, University of Colorado School of Medicine, Denver, CO, USA

Department of Orthopedic and Trauma Surgery, Friedrich-Alexander Universität Erlangen-Nürnberg, Erlangen, Germany

M. Simon
Department of Orthopedic and Trauma Surgery, Center of Sportsmedicine, Klinikum Bamberg, Bamberg, Germany
e-mail: info@michaelsimon.org

10.1 Introduction

Analysis of injuries in climbers show that most injuries caused by falls are inflicted onto the feet. These are often standard foot injuries, such as strains and sprains, ankle fractures, and talus or calcaneus fractures. In contrast to injuries onto the upper extremities, about which many studies have been conducted, only few studies exist concerning the lower extremities, especially the feet [1–8]. The tight fit and build of modern climbing shoes can lead to specific injuries and overstrain [1, 9]. Furthermore, the overall number of lower limb injuries in climbers is rising [10, 11]. In

V. Schöffl et al. (eds.), *Climbing Medicine*, https://doi.org/10.1007/978-3-030-72184-8_10

summary of the present literature on the incidence of climbing injuries, it can be stated that injuries to the lower extremities are mostly caused by acute trauma, whereas injuries to the upper extremities are mostly chronic and overstrain injuries [1, 12–14]. These fall-related injuries happen either as a "rock hit trauma" when a climber falls while lead climbing and is pulled against the wall by the rope or as a "ground fall trauma" [1, 13, 15]. Both of these result in either ankle sprains and strains or fractures, mostly of the talus, calcaneus, or the ankle joint [1, 13, 15]. Chronic injuries are mostly due to the tight fit of the climbing shoes as well as their cut, build, and biomechanical design [3, 7, 8, 16]. Bowie et al. [15] as well as others [13, 14, 17, 18] report that roughly 50% of climbing injuries concern the lower extremities. Neuhof et al. [19] also found an even injury distribution between the upper (42.6%) and lower extremities (41.3%). Killian et al. [8] found a significant correlation between the incidence of ankle sprains and bouldering as well as between ankle sprains and sport climbing. The following chapter will present these acute injuries as well as the chronic, overstrain conditions.

10.2 Modern Climbing Shoes

Until about 40 years ago, most alpine climbing and rock climbing were done in heavy mountain boots. An extra pair of socks was often worn, and the foot was held in a leather cast that protected it but also took away any sensitivity and the need for strength in the toes. Leather boots were replaced by the legendary E.B. in the 1970s, and a new generation of climbing shoes with sticky rubber soles was just around the corner. Since those early days, climbing shoes have become specialized according to the type of rock climbing one wants to do. Attributes such as downturn, the concave shape that places pressure on the toes, and asymmetry, concentrating the pressure on the big toe, are basic elements of modern climbing shoes [8, 9, 20] (Fig. 10.1). These special climbing shoes help the climber to stand on both slabs and edges, using friction with straight toes and bent toes with precision and proper contact, respectively [7–9, 21, 22]. The majority of foot injuries in climbing result from wearing climbing shoes which are unnaturally shaped or too small in size [7, 8, 21, 22]. The shoe size reduction forces the foot to conform to the shoe and changes the biomechanical position of the foot within the shoe. The foot shortens through supination and contraction of the digits [1, 7, 8, 22]. In front pointing, the proximal and the distal interphalangeal joints are flexed, and the metatarsophalangeal joints are overextended (crimping toes) (Figs. 10.2 and 10.3). Lateral X-rays of feet inside climbing shoes show that the foot weight distribution onto the first and fifth metatarsal head and the heel no longer corresponds to the normal distribution. The foot is front pointing onto the distal toe phalanges [7, 8] (Fig. 10.4). Moreover, the plantar flexion of the metatarsal heads results in a tightening of the plantar fascia [6–8, 22]. High-ability climbers experience more foot deformities and injuries compared to climbers of lower ability due to the common practice of wearing climbing shoes sized smaller than normal streetwear shoes [5, 7, 8]. In 1999, we reported an average shoe size difference between normal shoes and climbing shoes of 2.3 ± 0.73 continental sizes in high-ability rock climbers [mean climbing level 9.7 (UIAA MedCom metric scale)], whereas Killian et al. [8] reported a shoe size difference of one to two continental sizes (mean 1.7) in average climbers [mean climbing level 7.0 (UIAA MedCom metric scale)]. McHenry et al. [5] found ill-fitting and excessively tight footwear in

Fig. 10.1 Modern climbing shoe with "downturn" and a "slingshot"

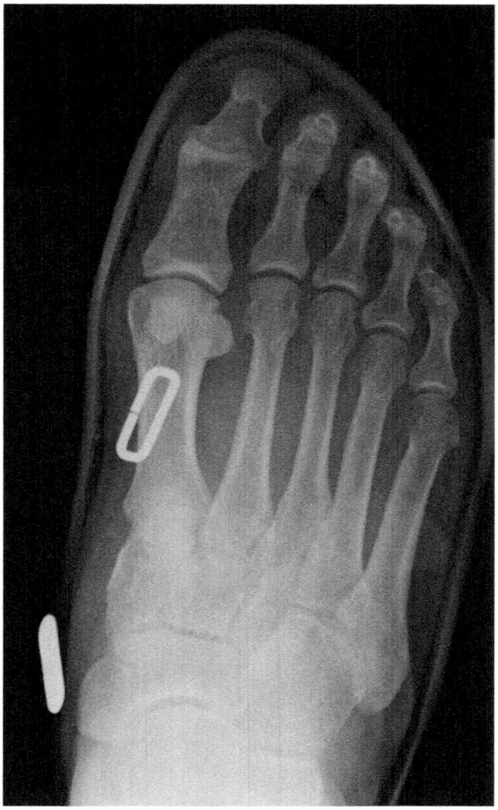

Fig. 10.2 Radiograph of a foot within a climbing shoe in a standing position

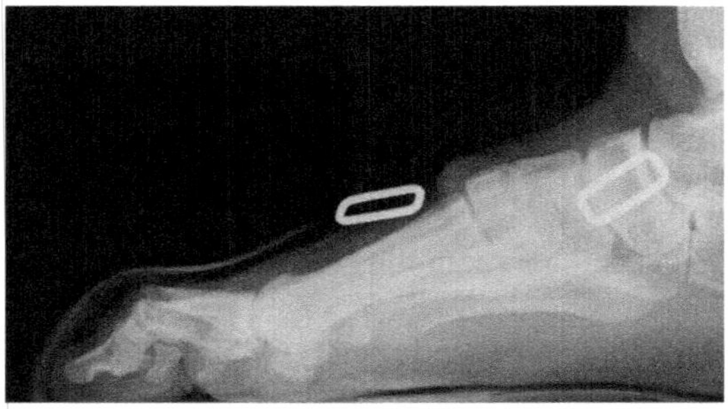

Fig. 10.3 Radiograph of a foot within a climbing shoe in a lateral view. Note the crimping position of the toes and that the forefoot is not resting on the first and fifth metatarsal but on the crimping toes

55 out of 56 rock climbers with a mean difference of climbing shoes to normal shoes of 3.88 UK sizes. Foot pain during activity was also commonplace in 91% of the climbers as well as a mean shoe size reduction of almost four UK sizes between the climbers' street shoe size and that of their climbing footwear using a calibrated foot/shoe ruler. Furthermore, an unfortunate association between higher climbing specific abilities and the urge for a tighter shoe fit ($p < 0.001$) was reported [5].

For further analysis, acute injuries, mostly from falls and overuse (overstrain) injuries, due to the tight-fitting climbing shoes must be distinguished.

10.3 Chronic Foot Conditions From Climbing Shoes

Modern and "aggressive" (high downturn, very pointy last) climbing shoes are not comfortable, and various studies showed that 80–96% of all climbers report feeling pain when wearing their climbing shoes [1, 5, 7, 8, 22]; simultaneously, studies show that these climbers are bearing the discomfort and pain to achieve a better climbing ability [1, 5, 7–9, 22].

Kilian et al. [8] surveyed 100 climbers through a questionnaire on foot conditions. Seventy-three were male, 27 were female, and the mean climbing level was 5.10 US (7.0 UIAA [26]). Eighty-one percent of the climber's complained of discomfort from the climbing shoes. Pain in the toes or hallux valgus was the most frequent complaint. The authors found a significant correlation between the climbing ability and extent of shoe size reduction; however, they found no correlation between the extent of shoe size reduction or skill level and the incidence of pain or discomfort. Sixty-five percent of the surveyed climbers experienced either tingling or numbness, a condition also frequently mentioned by Peters [22] and Largiadèr and Oelz [25]. These neurological complaints and neuromas are known to be greatly exacerbated through constricting shoes [27]. Van der Putten et al. [7] also frequently found mycosis, splinter hemorrhage, nail fractures, hallux valgus, bunions, and callosities. Nevertheless, as the focus of this publication was on the biomechanical development of a climbing shoe, exact numbers of the distribution of these conditions were not given [1].

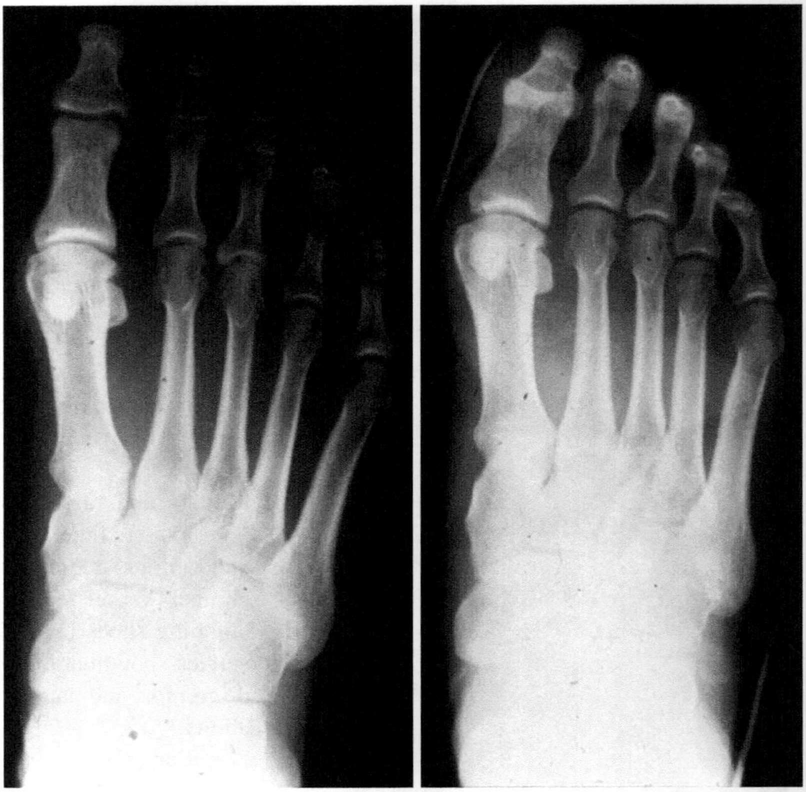

Fig. 10.4 Radiograph of feet with and without climbing shoe in a standing position. Note how the climbing shoes presses the large toe into a hallux valgus position

The shoe size reduction forces the foot to conform to the shoe and changes the biomechanical position of the foot within the shoe. The foot shortens through supination and contraction of the digits [7, 8, 22]. In front pointing, the proximal and especially the distal interphalangeal joints are flexed, and the metatarsophalangeal joints are overextended (crimping toes). This causes a deviation of the distal end of the metatarsals [9]. Repetition of the biomechanical nonphysiologic position can lead to a hallux valgus deformity [1, 5, 7, 22]. Also, the plantar fascia is under excessive tension [1, 5, 7, 22]. Schöffl et al., Killian et al., and McHenry et al. used radiographic analysis of feet within climbing shoes to show that these kind of shoes press the foot into a hallux valgus-like position [1, 5, 8], with an increase of the hallux valgus angle of about 7° [1] (Fig. 10.4). Fifty-three percent of sport climbers showed a unilateral hallux valgus deformity, and 20% showed a bilateral hallux valgus deformity in climbers who had been climbing for more than 12 years at a mean climbing level of UIAA metric 9–66 [4]. This rate of a hallux valgus is much higher than in an age-matched comparison group of non-climbers, in which a prevalence of 4.5% could be found [1]. Killian et al. [8] found an average incidence of a hallux valgus of 34% in climbers of all difficulty levels but in 53% of the group of climbers in the 5.12 range (UIAA metric 8.3–9.0).

As the climbing difficulty level increases, the demand that climbing shoes fit tightly and put extreme pressure onto the toes rises because the footholds become smaller [1]. Thus, high-level climbers are more prone to overstrain injuries of the feet [1, 5, 7–9]. In our own study, we evaluated 30 male high-level climbers with a mean climbing training time of 12.3 h per week and a mean climbing level of UIAA metric 9.66 [4]. The respective shoe seize difference was 2.3 sizes (European size). All climbers showed callosities and pressure marks on the dorsal aspect of the toes. The incidence was the highest onto the large toe (100%) and gradually decreased from toe to toe in the direction of the small toe (8%) [4] (Fig. 10.5).

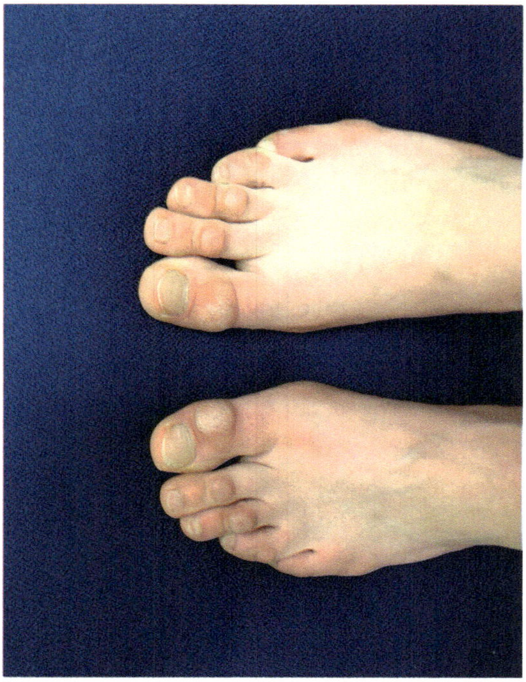

Fig. 10.5 Typical callosity in a 15-year-old female climber

Table 10.1 shows the distribution of chronic overstrain pathologies in this study.

A correlation of chronic deformities of the feet and climbing training intensity and climbing level is most likely, especially if the results of our study [4] are compared to that of Largiadèr and Oelz [25]. In the latter study, the authors only found 28% chronic deformities of the toes and 21% of the feet in climbers who had significantly fewer training hours and a distinctly lower climbing level than the athletes from our study [1].

Most of the previously mentioned chronic pathologies can be diagnosed through a thorough clinical examination and radiography. An MRI is only necessary in rare cases.

Depending on the extent of complaints and the deformity, surgical therapy may be necessary (corrective osteotomy of the first metatarsal) in addition to conservative treatment (e.g., adapted footwear, hallux night storage track) [23]. Moreover, in recent years, we have seen an increasing number of climbers with a hallux rigidus condition, even cases requiring a surgical therapy (cheilectomy or fusion) [20]. Rock climbing is still possible after this procedure; nevertheless, the risk of osteoarthritis in the adjunct joints increases [1] (Fig. 10.6). Furthermore, preexisting enchondroma or bone

Table 10.1 Foot deformities in high level climbers ($n = 30$) [4]

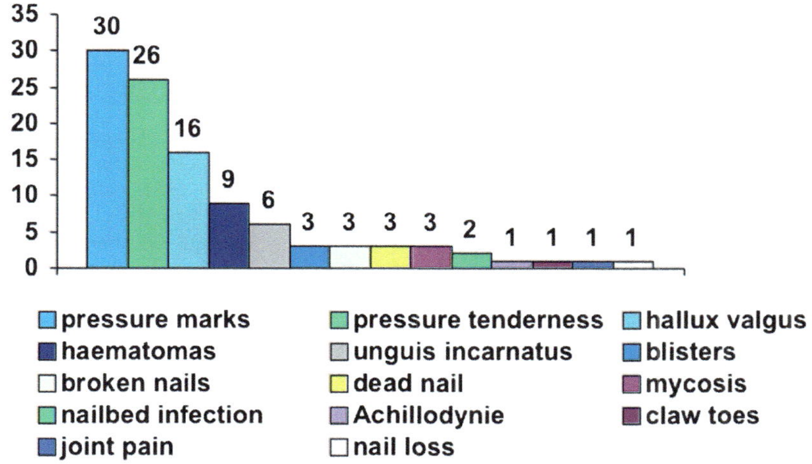

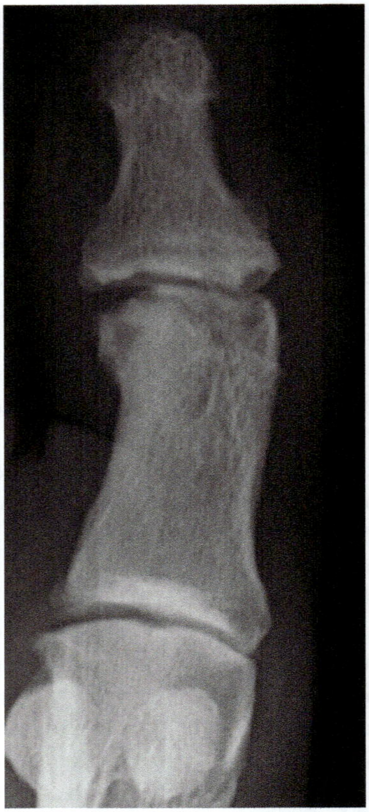

Fig. 10.6 Osteoarthritis auf the large toes base and middle joint in a 35-year-old climber

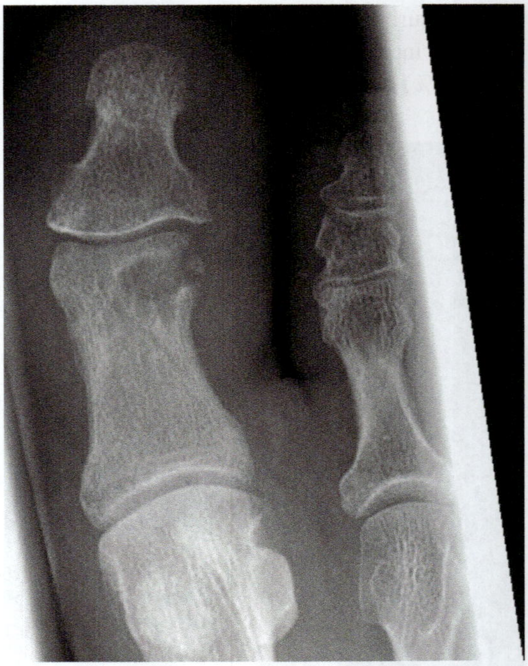

Fig. 10.7 Symptomatic enchondroma of the large toe in a female climber

cysts can lead to increased pain in a tight climbing shoe [1, 5, 7, 8, 18] (Fig. 10.7).

In summary, most climbers are willing to endure some pain in their climbing shoes and accept possible long-term consequences. Climbing ambition and the fun in doing hard moves seem to be too important to the athletes [1, 9] (Fig. 10.8).

Chronic Feet Deformities in Climbers

- **History:** Tight-fitting climbing shoes with toes in a crimping position
- **Onset:** Chronic
- **Clinical:** Pain, callosity, bunions, subungual hematoma, hallux valgus, hallux rigidus
- **Diagnostic:** Clinical finding, radiography
- **Therapy:** Mostly conservative
- **Outcome:** Depending on the shoe size used

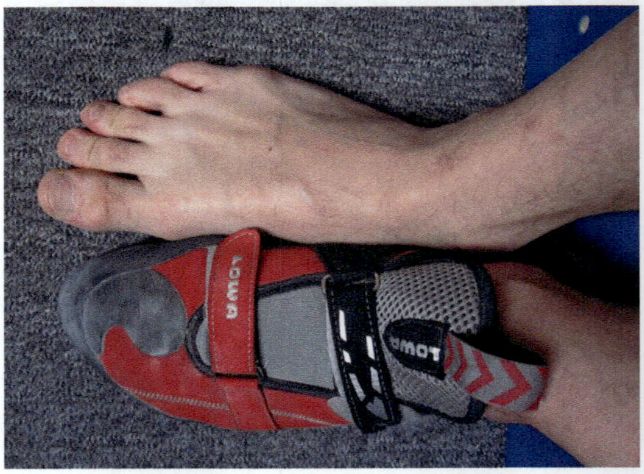

Fig. 10.8 Comparison of a foot with and without climbing shoe

10.4 Traumatic Feet Injuries

In addition to the chronic overstrain injuries caused by tight-fitting climbing shoes, acute injuries, mostly from falls, also need to be evaluated.

Falls are the most common cause for injuries to the lower extremities in rock climbing [15, 19, 28–30]. Thereby, a distinction needs to made between two different types of falls: a wall-collision fall and a ground fall [20]. In a wall-collision fall, the climber hits the wall in an approximately horizontal plane, whereas during a ground fall, the climber experiences the impact in a mainly vertical plane [20].

In a wall-collision fall, the climber takes a fall while lead climbing. When falling, the tensioning of the rope pulls the climber either to or against the rock face [1, 5, 8, 18, 31]. The force of the impact against the rock face varies in accordance with fall height, belay technique, and rope stretch [20] (Fig. 10.9). Contrary to how it may appear, short falls with a high impact factor (generally not far from the ground or the belay) can be the most dangerous. In those cases, the climber is actually pulled back onto the rock, and the angle at which he or she extends the foot (or arm) upon impact creates a lot of stress to this particular location. Additionally, assisted-braking belay devices such as the Gri-Gri® (Petzl, France) and Sirius® (TRE, Germany) can increase the fall factor [20]. Typical injuries are contusions or compound fractures [20].

Fig. 10.9 A long fall with slack rope, reducing the impact

Fig. 10.10 Ground fall trauma

The other pathomechanism, the so-called ground fall, is even more common, especially in bouldering (Fig. 10.10). In a ground fall, the impact of the foot onto the ground causes the trauma. A ground fall from a height of as little as 3 meters can already cause serious fractures of the calcaneus, talus, and the ankle joint (Fig. 10.11) [20]. Other ground fall injuries include sprains and ligamental injuries [13, 14, 17, 20, 32, 33]. These injuries are increasing with the growing popularity of bouldering as a sub-discipline in climbing [10, 34], but they are also fairly common with sport climbers who fall before reaching the first bolt or who are dropped by their belayer [20]. A typical injury mechanism in indoor climbing, especially bouldering, has been a fall onto the mats with the foot stuck in the intermediate section between two mats [28, 29, 35–38]. This has happened quite often, partly because the first indoor

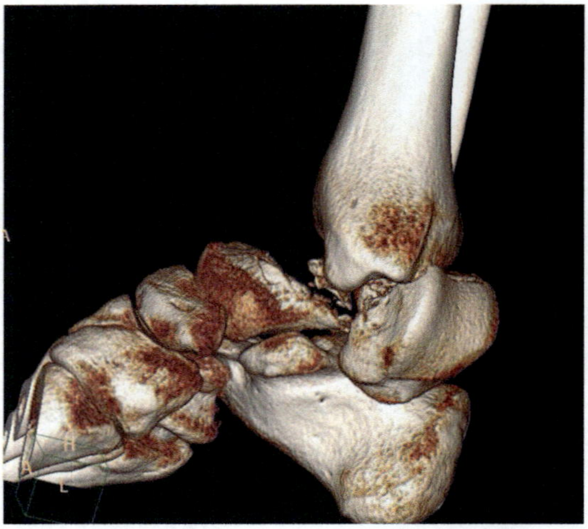

Fig. 10.11 Talus fracture after a climbing fall

Fig. 10.12 Modern indoor bouldering gym with perfect mat covering and no intersections between mats. (Phtoto: Enrico Haase)

bouldering walls in the late 1980s were small places with old mattresses as protection. Since then, this issue has been addressed by the Medical Commission of the UIAA [38] (International Mountaineering Association), and the International Federation of Sport Climbing (IFSC) has implemented a rule that these intersections must be closed in competitions [35]. Furthermore, this also has become a standard in modern bouldering gyms (Fig. 10.12). In addition to mats without gaps, indoor bouldering gyms implement further safety measures such as innovative wall design to decrease the injury rate. Despite being quite rare, falls onto other boulderers must also be considered regarding injury prevention [9]. For example, in some gyms,

Fig. 10.13 "Crashpad belay" in outdoor bouldering. (Photo: Enrico Haase)

kids run around without parental supervision, during which severe injuries can happen. While kids play, they do not care if someone is climbing above them and could fall onto them. Close supervision and the education of and by the parents are essential. An indoor climbing wall should not be seen as a children's playground, especially since most gyms provide designated children's areas [1, 5, 8, 9, 12, 13, 16, 18, 31, 39].

Another risk while bouldering results from items such as chalk bags or water bottles lying underneath the boulders; for example, one climber tore their cruciate ligament at a World Cup after falling onto their own chalk bag.

For bouldering outdoors, which has become very popular in an increasing community, climbers use "crash pads" (portable multiple foam layer mats) as protection and "spotting" to position falling climbers onto these mats. Falls next to or in between the mats also happen outdoors (Fig. 10.13). Most injuries in outdoor bouldering are sprains and strains onto the ankles, as well as talus, calcaneus, or ankle fractures [9, 12, 16, 32, 33, 40, 41]. In rope climbing, ground falls are comparably rare and usually caused by a fall before clipping the first bolt or belay failure [1, 9, 12, 16, 18, 31].

Many studies have already investigated the injury incidence to the lower leg and have reported inhomogeneous injury incidences. Neuhof et al. [19] evaluated 1962 sport climbers and their injuries over a 5-year period and found a total of 699 injuries. Out of these, 345 (41.3%) were to a lower extremity; 29.2% were on the feet. Most of the fractures (48.6%) and most of all contusions

(42.5%) were to the feet, while most ligamental injuries and tendon injuries were to the fingers. Ankle sprains are frequent in boulderers and sport climbers with a significant correlation [8] to these climbing activities. Killian et al. [8] reported an incidence of 0.23 for ankle sprains and 0.03 for foot fractures in rock climbers. Nelsen et al. [13] found that ankle injuries accounted for 19.2% of all climbing injuries; Buzzacott et al. [14] found in a consecutive study that the number increased to 27%. Backe et al. [17] even reported that out of all acute climbing injuries, 50% involved the foot, toe, and ankle. The most common type was a ligament injury, followed by contusions and lacerations. Gerdes et al. [42] also reported that sprains and strains were most common in the fingers (31.9%) and the ankle (23%). Fractures were most common to the foot (24.6%) and the ankle (22.1%). Especially in older athletes, even short falls can lead to foot and ankle fractures caused by the increased brittleness of the bone that occurs with aging [43, 44]. In these cases, even minor falls may result in a serious injury [1, 12, 16, 18, 40] [9, 45].

Typical traumatic feet injuries in rock climbers are [13, 19, 20, 28, 36, 40, 42, 46]: (1) contusion, (2) calcaneus fracture, (3) talus fracture, (4) ankle fracture, and (5) ankle sprain with lateral ligament injury. All of these injuries show an acute onset of symptoms (e.g., pain, swelling, impairment of mobility), and the diagnosis is made by a thorough clinical examination and radiography. In case of an unclear finding, a CT or MRI can be necessary to detect the complete extend of the injury.

The resulting typical injuries should undergo treatment according to standard orthopedic trauma guidelines. In case of displaced fractures, this usually means operative reconstruction (open reduction and internal fixation).

In summary, the literature reports that most injuries in rock climbing are either chronic or occur because of a hard move, leading to pathologies of the upper extremities. Nevertheless, the most serious fractures and sprains are to the feet and ankle joint [1, 10, 14, 16, 33, 39, 40, 42]. As previously discussed, the therapy follows the standard recommendations of the trauma and orthopedic surgical societies.

In analyzing these injuries, one other concomitant effect must be considered. The tight-fitting and non-cushioned climbing shoes described earlier increase the impact and severity of falls because of the change of the foot's biomechanical position within the shoe. The feet are not resting on the first and the fifth metatarsal head but on the toes in a crimp position [4] (Fig. 10.3). This leads to a different, nonphysiologic impact in a fall. Thus, strains and sprains happen more frequently in comparison to falls when normal shoes or mountain boots are worn [1, 5, 7–9, 17, 18].

Traumatic Foot Injuries

- **History:** Ground fall or rock-collision trauma
- **Onset:** Acute
- **Clinical:** Pain, swelling, hematoma, deformity
- **Diagnostic:** Clinical finding, radiography, optional CT/MRI
- **Therapy:** In accordance with orthopedic trauma guidelines
- **Outcome:** Depending on the diagnosis

10.5 Peroneal Tendon Dislocations

Acute peroneal tendon dislocations are rare medical conditions. Although this condition is most often seen in skiers [47–50], rock climbers can also be affected by peroneal tendon and tibialis posterior tendon dislocations [1]. Evaluating 908 climbing injuries over 4 years (2009–2012), we diagnosed three climbers with this pathology [2] (Figs. 10.14 and 10.15). All patients reported a snap of the peroneal retinaculum while having the foot in a plantar flexion with maximum inversion and applying high tension onto the large toe while climbing. These acute peroneal tendon dislocations can be interpreted similarly to pulley injuries in the fingers [51]. With a high-strength impact onto the superior peroneal retinaculum, which serves an analogous function as the pulley system of the fingers, the retinaculum, also enhanced by the calcaneofibular ligament, fails, and the peroneal tendons dislocate.

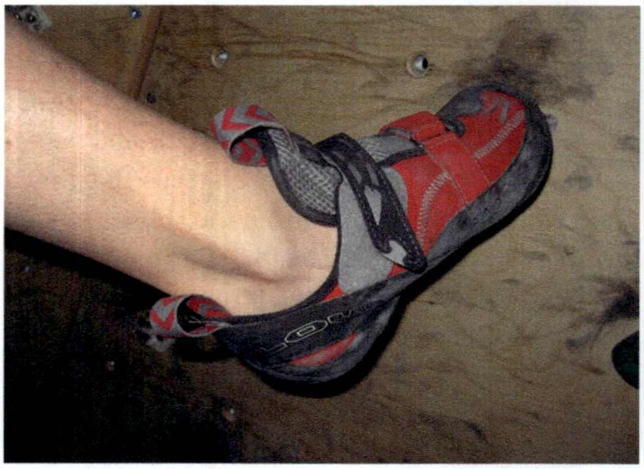

Fig. 10.14 High stress onto the peroneal tendons standing on tiptoe with inward rotation of the ankle

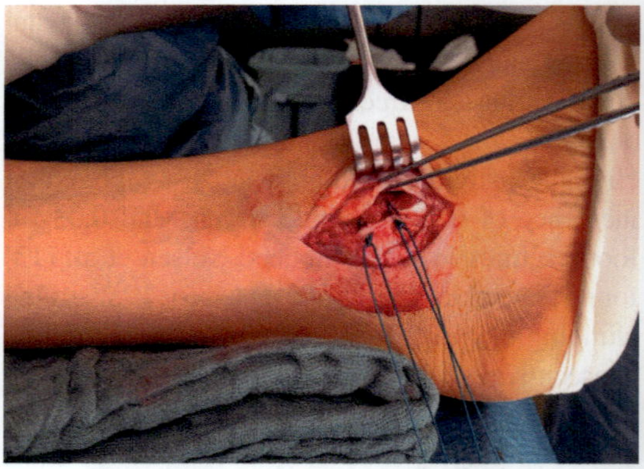

Fig. 10.16 Surgical reconstruction of the peroneal retinaculum with a periosteal flap

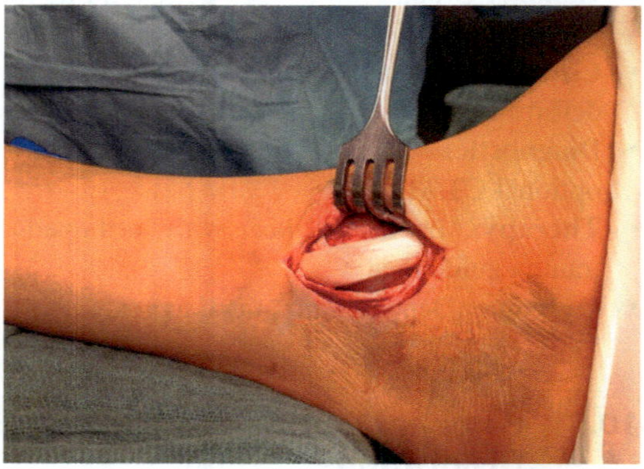

Fig. 10.15 Peroneal tendon dislocation in a climber

In all three cases, the peroneal tendon groove of the fibula was of normal shape; therefore, there was no anatomical predisposition.

Clinically, the patients complain of an acute onset of symptoms. In addition to pain and swelling on the lateral aspect of the upper ankle joint and an impairment of mobility, the peroneal tendon might be visible and palpably dislocated over the lateral malleolus.

The diagnosis is made by a thorough clinical examination and ultrasound, which allows a dynamic imaging of the dislocated tendon. In case of an unclear finding, an MRI can be necessary to detect the complete pathology and determine the shape of the peroneal groove.

In general, literature recommends a nonsurgical approach if the dislocation is treated immediately and the tendons stay in their normal position within a cast [47, 48].

Nevertheless, in active patients, especially in "foot-dependent" sports, surgical repair is recommended [47, 48]. Surgical techniques are mostly described for chronic subluxation and include a peroneal groove deepening, a tenoplasty, or a bone block transfer [47–50, 52]. A direct repair of the ruptured structures is only possible in acute cases, sometimes in combination with a periosteal flap [47, 52]. We performed a surgical repair in all of the previously mentioned cases (Fig. 10.16). In two cases, a direct repair was possible, and in one case, an additional periosteal flap was necessary. After surgery, the area was immobilized for 6 weeks with a cast. Climbing activity was allowed 8 weeks after injury. All athletes regained their full climbing ability with no reoccurrence or consisting problems. Overall, we recommend a surgical repair of dislocations of peroneal as well as tibialis posterior tendons in climbers.

Peroneal Tendons Dislocation

- **History:** High stress onto the foot in inward rotation
- **Onset:** Acute

- **Clinical:** Pain, visible and palpable dislocated tendon
- **Diagnostic:** Clinical finding, ultrasound, MRI optional (mostly to determine the shape of the peroneal groove
- **Therapy:** Surgical
- **Outcome:** Good

10.6 Conclusion

Considering the high incidence of feet pathologies in rock climbers, it is astonishing that so far, very little scientific research has been carried out on this topic. With numbers of up to 90% of climbers with chronic foot conditions, more research needs to be done in this respect (1). To reduce these complaints, climbers need to be advised not to wear their shoes too tight and to have additional, loose-fitting training shoes [20]. The climbing shoe industry should also be involved. New fitting strategies with less stretchy outer materials and biomechanically adjusted constructions are promising approaches in this area [1]. New shoes should have an inner lining to reduce bunions and callosities while still guaranteeing a good perception of the rock. While high-level climbing athletes may require a tight-fitting shoe, the vast majority of recreational climbers need the opposite: a comfortably fitting shoe which supports the foot and does not hurt [1].

References

1. Schöffl V, Küpper T. Feet injuries in rock climbers. World J Orthop. 2013;4(4):218–28.
2. Heid A, Popp D, Schöffl V. Traumatic peroneal tendon dislocations in rock climbers "the climbers pulley lesion of the foot". Med Sport. 2013;17(4):188–92.
3. Morrison AB, Schöffl VR. Physiological responses to rock climbing in young climbers. Br J Sports Med. 2007;41(12):852–61; discussion 61.
4. Schöffl V, Winkelmann HP. [Foot deformations in sport climbers] Fußdeformitäten bei Sportkletterern. D Z Sportmed. 1999;50:73–6.
5. McHenry RD, Arnold GP, Wang W, Abboud RJ. Footwear in rock climbing: current practice. Foot (Edinb). 2015;25(3):152–8.
6. Peters P. Orthopedic problems in sport climbing. Wilderness Environ Med. 2001;12(2):100–10.
7. van der Putten EP, Snijder CJ. Shoe design for prevention of injuries in sport climbing. Appl Ergon. 2001;32(4):379–87.
8. Killian RB, Nishimoto GS, Page JC. Foot and ankle injuries related to rock climbing. The role of footwear. J Am Podiatr Med Assoc. 1998;88(8):365–74.
9. Lutter C, Schöffl V. Trendsportart Bouldern—Verletzungen und chronische Schäden der unteren Extremität. In: Faulhaber M, Schobensberger W, Schobensberger B, Sumann G, Domej W, editors. Jahrbuch 2015—Österreichische Gesellschaft für Alpin- und Höhenmedizin. Innsbruck: Österreichische Gesellschaft für Alpin- und Höhenmedizin; 2015. p. 43–56.
10. Lutter C, Tischer T, Hotfiel T, Frank L, Enz A, Simon M, et al. Current trends in sport climbing injuries after the inclusion into the Olympic program. Analysis of 633 injuries within the years 2017/18. 2020.
11. Lutter C, Tischer T, Cooper C, Frank L, Hotfiel T, Lenz R, et al. Mechanisms of acute knee injuries in bouldering and rock climbing athletes. Am J Sports Med. 2020;48(3):730–8.
12. Schöffl V, Morrison A, Schöffl I, Küpper T. The epidemiology of injury in mountaineering, rock and ice climbing. Med Sport Sci. 2012;58:17–43.
13. Nelson NG, McKenzie LB. Rock climbing injuries treated in emergency departments in the U.S., 1990-2007. Am J Prev Med. 2009;37(3):195–200.
14. Buzzacott P, Schöffl I, Chimiak J, Schöffl V. Rock climbing injuries treated in US Emergency Departments, 2008-2016. Wilderness Environ Med. 2019;
15. Bowie WS, Hunt TK, Allen HA Jr. Rock-climbing injuries in Yosemite National Park. West J Med. 1988;149(2):172–7.
16. Neuhof A, Hennig FF, Schöffl I, Schöffl V. Injury risk evaluation in sport climbing. Int J Sports Med. 2011;32(10):794–800.
17. Backe S, Ericson L, Janson S, Timpka T. Rock climbing injury rates and associated risk factors in a general climbing population. Scand J Med Sci Sports. 2009;19(6):850–6.
18. Schöffl V, Popp D, Küpper T, Schöffl I. Injury trends in rock climbers: evaluation of a case series of 911 injuries between 2009 and 2012. Wilderness Environ Med. 2015;26(1):62–7.
19. Neuhof A, Hennig FF, I. S, V. S. Injury risk evaluation in sport climbing. Int J Sports Med. 2011;32(10)
20. Hochholzer T, Schöffl V. One move too many. 2nd ed. Ebenhausen: Lochner Verlag; 2006.
21. Schöffl V, Winkelmann HP. Foot deformations in Sport climbers. Ö J Sportmed. 1999;3:58.
22. Peters P. Nerve compression syndromes in sport climbers. Int J Sports Med. 2001;22(8):611–7.
23. Santrock RD, Smith B. Hallux valgus deformity and treatment: a three-dimensional approach: modified technique for lapidus procedure. Foot Ankle Clin. 2018;23(2):281–95.

24. Schöffl V, Morrison A, Hefti U, Ullrich S, Küpper T. The UIAA Medical Commission injury classification for mountaineering and climbing sports. Wilderness Environ Med. 2011;22(1):46–51.
25. Largiader U, Oelz O. An analysis of overstrain injuries in rock climbing. Schweiz Z Sportmed. 1993;41(3):107–14.
26. Schöffl V, Morrison AB, Hefti U, Schwarz U, Küpper T. The UIAA Medical Commission Injury Classification for Mountaineering and Climbing Sports. Wilderness Environ Med. 2011;22(1):46–51.
27. Bouche RT, Sullivan K, Ichikawa DJ. Athletic injuries. In: Oloff LM, editor. Musculoskeletal disorders of the lower extremities. 1st. 1st ed. Philadelphia: WB Saunders; 1994. p. 34.
28. Schöffl V, Morrison A, Schöffl I, Küpper T. Epidemiology of injury in mountaineering, rock and ice climbing. In: Caine D, Heggie T, editors. Medicine and sport science—epidemiology of injury in adventure and extreme sports. Vol. 58: Karger; 2012. p. 17–43.
29. Schöffl V, Morrison AB, Schwarz U, Schöffl I, Küpper T. Evaluation of injury and fatality risk in rock and ice climbing. Sport Med. 2010;40(8):657–79.
30. Mosimann U. Notfälle beim Eisklettern. Ther Ber. 2006;4:70–3.
31. Schöffl VR, Hoffmann G, Küpper T. Acute injury risk and severity in indoor climbing-a prospective analysis of 515,337 indoor climbing wall visits in 5 years. Wilderness Environ Med. 2013;24(3):187–94.
32. Jones G, Schöffl V, Johnson MI. Incidence, diagnosis, and management of injury in sport climbing and bouldering: a critical review. Curr Sports Med Rep. 2018;17(11):396–401.
33. Jones G, Asghar A, Llewellyn DJ. The epidemiology of rock-climbing injuries. Br J Sports Med. 2008;42(9):773–8.
34. Lutter C, Tischer T, Schöffl V. Olympic competition climbing: the beginning of a new era. British Journal Sportsmed (in press). 2020;
35. Schöffl V, Schöffl I. Competition climbing. Sport ortho Sport trauma. 2012;28:22–8.
36. Schöffl V, Küpper T. Injuries at the 2005 World Championships in Rock Climbing. Wilderness Environ Med. 2006;17:187–90.
37. Schöffl V, Winkelmann HP. [Accident statistics at "indoor climbing walls"] Unfallstatistik an "Indoor-Kletteranlagen". Sportverletz Sportschaden. 1999;13(1):14–6.
38. Schöffl V, Schlegel C. Statement Competition Climbing: www.uiaa.ch; 2000. www.uiaa.ch.
39. Schöffl VR, Küpper T. Injuries at the 2005 World Championships in Rock Climbing. Wilderness Environ Med. 2006;17(3):187–90.
40. Josephsen G, Shinneman S, Tamayo-Sarver J, Josephsen K, Boulware D, Hunt M, et al. Injuries in bouldering: a prospective study. Wilderness Environ Med. 2007;18(4):271–80.
41. Jones G, Johnson M. A critical review of the incidence and risk factors for finger injuries in rock climbing. Curr Sports Med Rep. 2016;15(6):400–9.
42. Gerdes EM, Hafner JW, Aldag JC. Injury patterns and safety practices of rock climbers. J Trauma. 2006;61(6):1517–25.
43. Hartsock LA, Feagin JA Jr, Ogilvie BC. Climbing and the older athlete. Clin Sports Med. 1991;10(2):257–67.
44. Lutter C, Hotfiel T, Tischer T, Lenz R, Schöffl V. Evaluation of rock climbing related injuries in older athletes. Wilderness Environ Med. 2019;30(4):362–8.
45. Schöffl V, Lutter C. The "Newbie" syndrome. Wilderness Environ Med. 2017;28(4):377–80.
46. Jones G, Asghar A, Llewellyn DJ. The epidemiology of rock climbing injuries. Br J Sports Med. 2007;42:773–8.
47. Arrowsmith SR, Fleming LL, Allman FL. Traumatic dislocations of the peroneal tendons. Am J Sports Med. 1983;11(3):142–6.
48. Brage ME, Hansen ST Jr. Traumatic subluxation/dislocation of the peroneal tendons. Foot Ankle. 1992;13(7):423–31.
49. Mendicino RW, Orsini RC, Whitman SE, Catanzariti AR. Fibular groove deepening for recurrent peroneal subluxation. J Foot Ankle Surg. 2001;40(4):252–63.
50. Kollias SL, Ferkel RD. Fibular grooving for recurrent peroneal tendon subluxation. Am J Sports Med. 1997;25(3):329–35.
51. Schöffl V, Hochholzer T, Winkelmann HP, Strecker W. Pulley injuries in rock climbers. Wilderness Environ Med. 2003;14(2):94–100.
52. Tan V, Lin SS, Okereke E. Superior peroneal retinaculoplasty: a surgical technique for peroneal subluxation. Clin Orthop Relat Res. 2003;410:320–5.

Hip and Knee Injuries

11

Christoph Lutter and Volker Schöffl

Volker Schöffl in „Bamboozallad", Thakekh, Laos, Photo Michael Simon

C. Lutter (✉)
Department of Orthopedic Surgery, University Hospital Rostock, Rostock, Germany

V. Schöffl
Department of Orthopedic and Trauma Surgery, Center of Sportsmedicine, Bamberg, Germany

School of Clinical and Applied Sciences, Leeds Becket University, Leeds, UK

Section of Wilderness Medicine, Department of Emergency Medicine, University of Colorado School of Medicine, Denver, CO, USA

Department of Orthopedic and Trauma Surgery, Friedrich-Alexander Universität Erlangen-Nürnberg, Erlangen, Germany

V. Schöffl et al. (eds.), *Climbing Medicine*, https://doi.org/10.1007/978-3-030-72184-8_11

11.1 Overview (Statistics)

Injuries and overuse damage to the hip, thigh, and knee joint have become more frequent in climbing and bouldering in recent years, with a significant increase of acute knee injuries [1, 2]. Changes within the sport and current trends in wall construction and route setting are considered to be the main cause for this change [1, 3]. Thereby, complex, three-dimensional wall structures and climbing routes often require unconventional physical effort of the athletes; gymnastic elements and new techniques are increasingly used. A comparative analysis of more than 600 climbing athletes recently showed an impressive increase of knee injuries between the survey periods of 1998–2001 and 2017–2018 [1, 2]. Thereby, acute knee injuries currently represent the fourth most common injury region in the sport, accounting for 7.1% of climbing-related injuries. Overuse injuries or other symptoms of the hip/groin (0.8%) or thigh (0.9%) are diagnosed less frequently [1] and are not among the ten most frequently injured body regions in climbing. However, a stringent diagnostic and therapeutic procedure is also of utmost importance for these complaints.

11.2 Diagnosis

The diagnostic procedure for complaints in the hip, thigh, or knee initially follows the usual criteria and should be structured as follows:

1. Anamnesis: How long have the complaints existed? After which strain or movement did the pain start? Can the complaints be triggered (e.g., specific body positions)? Do the complaints radiate? Have the pain characteristics changed? Which treatment has already been administered?
2. Clinical examination: Inspection of abnormalities (swelling, limping, etc.), palpation of the corresponding region (pain exacerbation by palpation, etc.), and joint-specific tests (e.g., FADIR test of the hip, meniscus tests, etc.). This should always be accompanied by a comparative examination of the opposite side.
3. Imaging procedures: Ultrasound, X-ray, CT, and MRI, depending on the suspected diagnosis.

A detailed anamnesis and a detailed clinical examination are essential for the clarification of hip, thigh, and knee problems in climbers and usually allow a correct diagnosis. Imaging procedures can be selected based on the anamnesis and clinical examination. MRI is the preferable imaging technique for more precise clarification of the symptoms.

11.3 Femoroacetabular Impingement

Femoroacetabular impingement (FAI) is a common cause of hip pain in climbers and must be considered and recognized as it represents a major contributing factor to arthritis of the hip [4]. The underlying pathomechanism is usually a subtle deformation of the femoral neck (Fig. 11.1), whereby repeated abnormal bone contact (impingement) between the femoral neck and acetabulum occurs, particularly in cases of

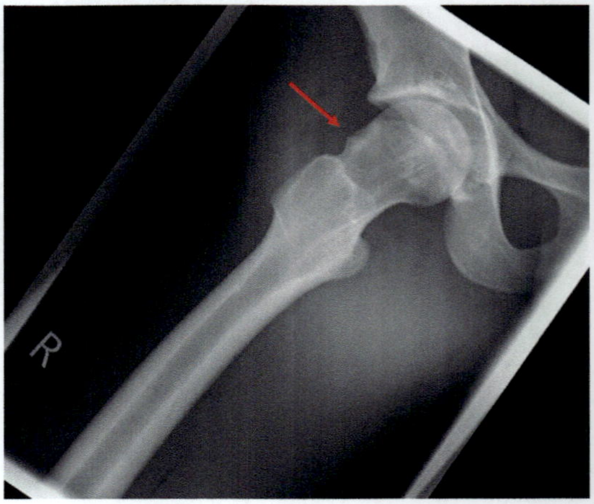

Fig. 11.1 X-ray (lateral view) of a climbers right hip joint with radiological evidence of femoroacetabular impingement (red arrow marks the so-called "bump" at the femoral neck)

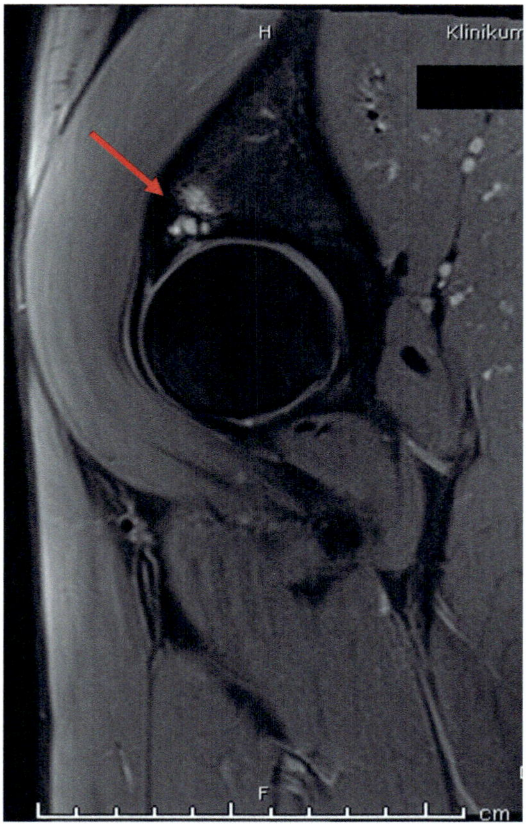

Fig. 11.2 MRI of a labral lesion with acetabular roof cysts (see arrow) in femoroacetabular impingement

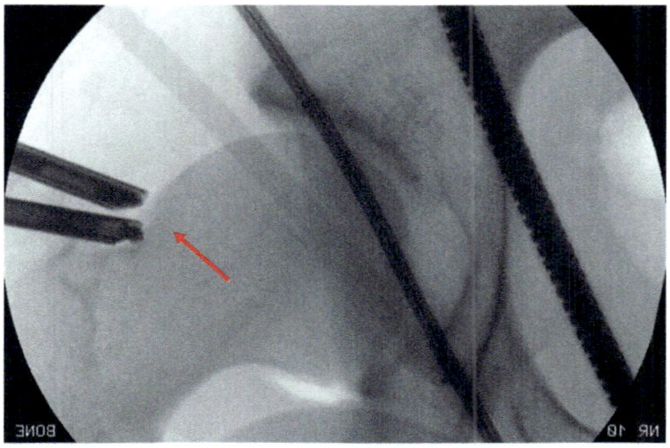

Fig. 11.3 Intraoperative fluoroscopy during arthroscopic femoral neck osteoplasty (see arrow) in femoroacetabular impingement

strong flexion of the hip joint (e.g., use of the "high step" position), which can cause joint damage, pain, radiological changes, and arthritis [5] (Figs. 11.1 and 11.2). Two major types of FAI are classified. "Cam" impingement is a bony prominence (so-called "bump") of the femoral head and neck transition (Fig. 11.1). During hip flexion and internal rotation, the bump is pressed against the acetabular rim, which can cause cartilage damage and labral tears (Fig. 11.2) [4, 5]. "Pincer" impingement is a protruding acetabulum. Flexion of the hip leads to contact between the femoral head and the acetabulum, which can also cause labral injuries and cartilage damage. Cam impingement is more common in younger, athletic men and is associated with reactive remodeling reactions of growth plates in children/adolescents, who are very active in sports. Pincer impingement is usually more likely to be detected in middle-aged women. Mixed images are common [4].

Clinically, patients usually report pain during sports, especially when competing at high speed with maximum flexion and/or internal rotation of the hip (e.g., "drop knee" position). After a physical examination with appropriate stress tests (including the so-called FADIR test = flexion, adduction, and internal rotation of the hip), the radiological clarification (radiographs of the pelvis and the hip) and MR arthrography (Figs. 11.1 and 11.2) are performed. Depending on the findings, an arthroscopy of the affected hip joint should be considered (partial resection of the femoral bump, reconstruction of the labrum or arthroscopic cartilage therapy) (Fig. 11.3).

In the clarification of groin and hip pain, differential diagnoses must be considered both in the patient's medical history and during the clinical examination. As shown in Fig. 11.4, acute injuries around the origin of the hip flexors/knee extensors can also trigger groin pain. The pictured athlete experienced severe pain in his groin while stepping on a high foot hold in an unfavorable body position. Magnetic resonance imaging confirmed a traumatic rectus femoris muscle tear with detachment of the pelvis (Fig. 11.4).

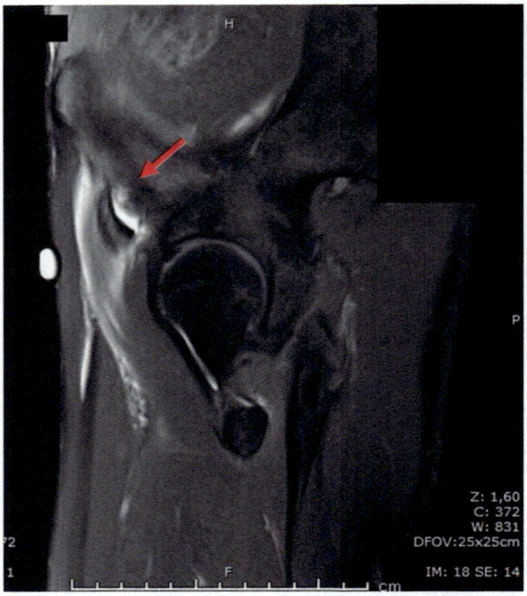

Fig. 11.4 MRI of an avulsion lesion of the rectus femoris muscle (see arrow)

- **Etiology:** Anatomical variants of the femoral head/neck transition and/or the acetabulum.
- **Onset:** Gradual, progressive, and load-dependent complaints, especially in cases of strong flexion/internal rotation of the hip joint.
- **Clinic:** Dull and deep hip/groin pain.
- **Diagnosis:** Clinical, radiographic, and MR arthrography.
- **Therapy:** Depending on the labrum and cartilage, arthroscopy of the hip joint with femoral neck shaping, labral reconstruction, cartilage therapy, acetabular roof plastic, etc.
- **Result:** Often asymptomatic after surgical correction. Subsequent progress of the joint destruction cannot always be safely avoided.

Fig. 11.5 Heel hook position of the right lower extremity during bouldering with external rotation and flexion in the hip and knee joint

11.4 Heel Hook Injuries

An injury mechanism that has become increasingly common in recent years is the heel hook-related injury, which is attributable almost exclusively to bouldering [6] (Fig. 11.5). Modern bouldering gyms or competition facilities offer complex wall structures, which require the use of the legs and feet in a much more complex motion sequence than formerly known and performed [7]. The increasing popularity of outdoor bouldering also contributes to the accumulation of heel hook-related injuries (Fig. 11.5).

To stabilize the body, especially in bouldering, heel or toe hooks are often used. In steep wall sections, the heels or toes are thereby used to transfer the tensile load of the ischiocrural muscles to the corresponding wall structures to stabilize the body position (Fig. 11.5). Additionally, these movements often require an external rotation of the lower leg with maximum external rotation of the knee. This position can lead to the maximum load on collateral ligaments, lateral meniscus, posterior cruciate ligament, popliteus tendon, posterior joint capsule, iliotibial tractus, and ischiocrural muscles. In the case of heel hook

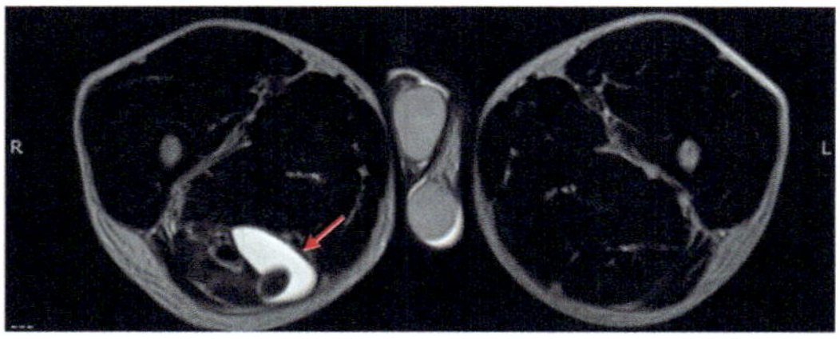

Fig. 11.6 MRI of an avulsion injury with hematoma formation in the ischiocrural muscles of the right lower extremity after acute trauma during heel hook (see arrow)

injuries, athletes often report a loud snapping sound in the event of the accident, which is most likely explained by a phenomenon in which the extended iliotibial tractus snaps over the lateral condyle, similar to a maximum pivot shift. The injuries can be divided into two main locations: injuries near the knee and near the pelvis. Injuries near the pelvis usually involve a tendon avulsion (partial or total) of the biceps femoris or the rectus femoris insertion (Fig. 11.6). Stretching techniques and sufficient warm-up might have a preventive effect. Clinical examination usually includes stress tests and palpation. In addition, sonography often provides indications of peritendinous or intramuscular hematomas around the ischiocrural muscles. Reliable visualization of tendon avulsions or muscle tears is also possible in some cases. To exclude any bony lesions, it is recommended to perform an X-ray diagnosis. For detailed evaluation of the muscle-tendon injuries, MRI is usually performed (Fig. 11.6). Depending on the extent of the muscle-tendon injury, conservative therapy is performed by resting and avoiding the corresponding load on the affected muscles. Higher graded lesions usually require a surgical procedure (Fig. 11.7). Ehiogu et al. recently evaluated current rehabilitation concepts for climbers after acute hamstring injuries [8]. The authors thereby highlighted the fact that rehabilitation should address the loading capacity of injured tissues early in the treatment regime using primarily concentric muscle-strengthening strategies.

Heel hook injuries close to the knee joint have already been described in other sports (e.g., martial arts), in which the opponent is often attacked

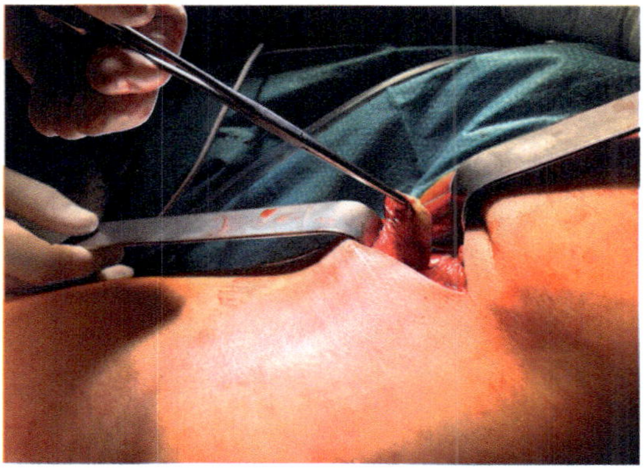

Fig. 11.7 Intraoperative findings after avulsion injury of the ischiocrural musculature. The figure shows the tendon stump sheared off from the pelvis before refixation

with the legs. A higher injury risk seems obvious due to the extreme forces and the fast movements in these sports; in climbing, however, heel hooks are often performed statically. Regarding the underlying pathologies of heel hook injuries around the knee, please refer to Sect. 11.5 (knee injuries). The selective use of imaging techniques is crucial in the supervision of climbing athletes. If an athlete reports a loud snapping noise described during a heel hook without corresponding persistent symptoms, imaging can be omitted with reference to the "snapping" of the iliotibial tractus. However, a clinical follow-up investigation is recommended. If the patient reports symptoms, an MRI of the knee joint and the distal thigh is recommended in addition to regular X-rays.

A sufficient warm-up before climbing/bouldering and regular stretching exercises are recommended to avoid heel hook injuries.

- **Etiology:** Instant pain during the heel hook movement, especially in bouldering. A painless loud snapping noise around the knee can occur during the heel hook.
- **Onset:** Acute onset of pain, usually on the dorsal thigh. Alternatively, snapping phenomenon around the knee, often painless.

- **Clinic:** Variable, depending on localization and underlying pathology.
- **Diagnosis:** Often by anamnesis (snapping phenomenon); in cases of complaints—ultrasound, X-ray, and MRI.
- **Therapy:** Depending on the underlying pathology.
- **Result:** Depends on pathology. Often complete healing, surgical therapy necessary occasionally.

11.5 Knee Injuries

While knee injuries are generally among the most common sports injuries, few injuries of this type have been diagnosed [9, 10].

However, the current worldwide indoor boulder "boom" is causing a rapid increase in the absolute number of injuries and especially acute injuries (see Chap. 2, Injury Statistics). Among these acute injuries, a recently growing number of knee injuries has been noticed. In this context, Josephsen et al. reported a higher prevalence of knee injuries in indoor bouldering compared to outdoor bouldering [11]. The increase in acute injuries around the knee is attributed to the design of modern climbing and bouldering facilities as well as the changed training methods in modern bouldering.

A comparative analysis of the current injury statistics (database of the authors' patient collective) showed a doubling of knee injuries between 2015 and 2018 compared to the years 1998–2001. While 6.7% of the climbing-related injuries in the collective were in the knee joint area, other authors report percentages of up to 10% among the sport-specific injuries [12].

The underlying mechanisms of injury in bouldering and lead climbing have long been unclear and have recently been identified and published for the first time. The following four trauma mechanisms were identified: (1) high step (Fig. 11.8), (2) drop knee (Fig. 11.9), (3) heel hook (Fig. 11.9), and (4) falls on the ground (Fig. 11.10) [13].

Fig. 11.8 High step position of the right lower extremity during bouldering with maximum flexion in the knee joint

Fig. 11.9 Drop knee position of the left lower extremity during bouldering with flexion in the knee joint and strong internal rotation in the hip joint

Meniscal lesions, strains of the iliotibial tractus, and cruciate ligament injuries are the most common diagnoses. Individual mechanisms of injury and the resulting injuries were reported as follows (Table 11.1):

Fig. 11.10 Uncontrolled fall on the mat in bouldering

Table 11.1 Distribution of knee injuries [13]

Injury type	% in own patients
Medial meniscal tear	28.6
Iliotibial band sprain*	19.5
ACL tear +	9.1
LCL tear	7.8
Lateral meniscal tear	5.2
Patellar dislocation	3.9
Minor joint effusion	3.9
Cartilage injury	3.9
Others	3.9
ACL rupture	2.6
Partial ACL tear	2.6
Medial meniscal tear w. MCL tear	2.6
MCL tear	2.6
Popliteal muscle tear	2.6
Joint fracture	1.3

- *High step*: Mainly medial meniscus lesions
- *Drop knee*: Mainly medial/lateral meniscus lesions and medial collateral ligament injuries (Fig. 11.11)
- *Heel hook*: Mainly strains of the iliotibial tractus, the ischiocrural muscles, and lateral meniscus lesions (Fig. 11.12)
- *Falls on the ground*: Mainly ACL injuries

In terms of diagnostics and therapy, the supervision of climbing athletes does not differ significantly from that of other athletes. Results from the authors' patients showed an overall fast *return to sport* in climbing athletes after a knee injury.

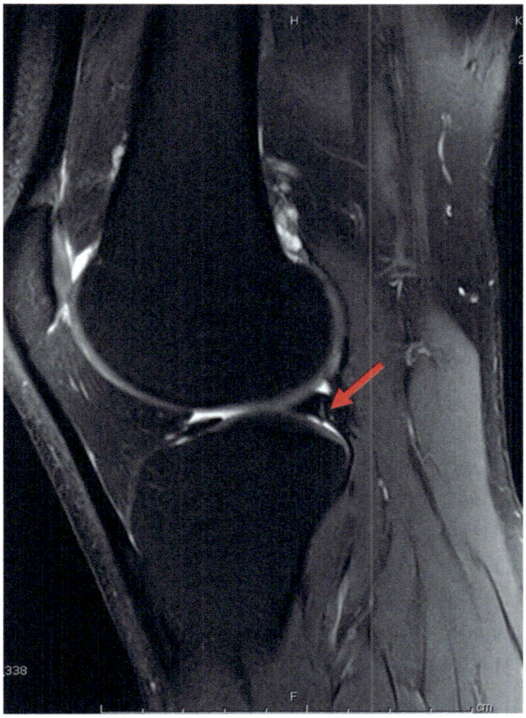

Fig. 11.11 MRI of a lateral meniscus posterior horn lesion after a heel hook injury in bouldering (see arrow)

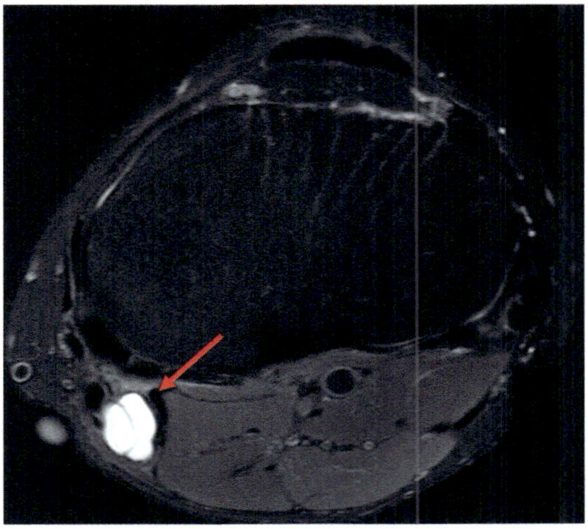

Fig. 11.12 MRI of a ganglion cyst at the ischiocrural muscles near the knee (see arrow)

Sport-specific stabilization exercises are recommended to prevent injuries from falls.

- **Etiology:** Four different trauma mechanisms: *high step, drop knee, heel hook,* and *fall on the ground*
- **Onset:** Acute injuries
- **Clinic:** Depending on injury
- **Diagnosis:** Clinical investigation, X-rays, and MRI
- **Therapy:** Often conservative therapy, surgical procedure required in approximately one-third of all injuries
- **Result:** Depending on injury; mostly good outcome

11.6 Conclusion

In summary, complaints in the hip/groin, thigh, and knee form a large group of clinical patterns in climbers. Sufficient knowledge of the underlying pathologies as well as mechanisms of injuries allows a quick and correct diagnosis. A conservative therapeutic approach is possible in many cases; acute injuries of the knee or specific anatomical variation around the hip sometimes require surgical therapy. Healing processes are sometimes prolonged, and adjustment of training after returning to sport is recommended.

References

1. Lutter C, Tischer T, Hotfiel T, Frank L, Enz A, Simon M, et al. Current trends in sport climbing injuries after the inclusion into the Olympic program. Analysis of 633 injuries within the years 2017/18. Muscles Ligaments Tendons J. 2020;10(2):201–10.
2. Schöffl V, Hochholzer T, Winkelmann HP, Strecker W. Pulley injuries in rock climbers. Wilderness Environ Med. 2003;14(2):94–100.
3. Schöffl V, Lutter C. The "newbie" syndrome. Wilderness Environ Med. 2017;28(4):377–80.
4. Lung R, O'Brien J, Grebenyuk J, Forster BB, De Vera M, Kopec J, et al. The prevalence of radiographic femoroacetabular impingement in younger individuals undergoing total hip replacement for osteoarthritis. Clin Rheumatol. 2012;31(8):1239–42.
5. Pun S, Kumar D, Lane NE. Femoroacetabular impingement. Arthritis & Rheumatol (Hoboken, NJ). 2015;67(1):17–27.
6. Schöffl V, Lutter C, Popp D. The heel hook—a climbing-specific technique to injure the leg. Wilderness Environ Med. 2016;27(2):294–301.
7. Lutter C, El-Sheikh Y, Schöffl I, Schöffl V. Sport climbing: medical considerations for this new Olympic discipline. Br J Sports Med. 2017;51(1):2–3.
8. Ehiogu UD, Stephens G, Jones G, Schöffl V. Acute hamstring muscle tears in climbers-current rehabilitation concepts. Wilderness Environ Med. 2020;31(4):441–53.
9. Buzzacott P, Schöffl I, Chimiak J, Schöffl V. Rock climbing injuries treated in US emergency departments, 2008-2016. Wilderness Environ Med. 2019;30(2):121–8.
10. Schöffl V, Popp D, Küpper T, Schöffl I. Injury trends in rock climbers: evaluation of a case series of 911 injuries between 2009 and 2012. Wilderness Environ Med. 2015;26(1):62–7.
11. Josephsen G, Shinneman S, Tamayo-Sarver J, Josephsen K, Boulware D, Hunt M, et al. Injuries in bouldering: a prospective study. Wilderness Environ Med. 2007;18:271–80.
12. Asakawa D, Sakamoto M. Retrospective survey of sport climbing injuries and self-care in the Gunma prefecture. J Phys Ther Sci. 2019;31:332–5.
13. Lutter C, Tischer T, Cooper C, Frank L, Hotfiel T, Lenz R, et al. Mechanisms of acute knee injuries in bouldering and rock climbing athletes. Am J Sports Med. 2020;48(3):730–8.

Part IV

Orthopedics of the Trunk

The Spine

12

J. Liße and Volker Schöffl

Volker Schöffl, Le Daume Joanne, Fontainebleau, France, Photo Kilian Reil

J. Liße (✉)
Department of Orthopedic and Trauma Surgery, Klinikum Bamberg, Bamberg, Germany

V. Schöffl
Department of Orthopedic and Trauma Surgery, Center of Sportsmedicine, Klinikum Bamberg, Bamberg, Germany

School of Clinical and Applied Sciences, Leeds Becket University, Leeds, UK

Section of Wilderness Medicine, Department of Emergency Medicine, University of Colorado School of Medicine, Denver, CO, USA

Department of Orthopedic and Trauma Surgery, Friedrich-Alexander Universität Erlangen-Nürnberg, Erlangen, Germany

V. Schöffl et al. (eds.), *Climbing Medicine*, https://doi.org/10.1007/978-3-030-72184-8_12

12.1 Traumatic Spine Injuries

The formerly more frequent and dreaded ground fall has become less frequent, thanks to new belay devices, belaying trainings, and increasingly better protection of climbing areas. In boulder gyms, fractures of the spine are very rare due to the safety mats, and in climbing gyms, fall protection floors are increasingly used and belaying trainings are frequently given. This is shown by the accident statistics of alpine and climbing clubs [6–8]. Nevertheless, ground falls leading to serious injuries still occur occasionally. Falls into the climbing rope with wall collision seem to have increased as a consequence of the development of semiautomatic belaying devices. However, these falls usually result in injuries to the lower extremities. An exception is the "head over heels" fall, during which the climber turns and hits the wall with his skull or back. Particularly, severe hyperextension injuries of the spine can also result from a simple fall into the rope with high impact, if the climber falls unexpectedly, has had an unfavorable body position during the fall, or does not build up enough tension in the back during the fall.

12.1.1 Back Strain

Not only rope falls or sudden (rotational) movements but also a decrease in the strength of the muscles in a tense movement can lead to strains of the back, just as in other parts of the body. Mostly, these strains occur in eccentric movement. These injuries are small tears in muscles or ligaments that are rarely detectable in radiological imaging. Sometimes, sonography or MRI scans show increased fluid accumulation. Such strains can cause considerable pain when the damaged muscle or ligament is activated and, thus, impair performance for several weeks, although they are considered harmless. In this case, the main aim is to rule out the presence of more serious injuries. Therapeutically, relief can be achieved immediately after the accident by local cooling and in later phases through heat therapy and massages (e.g., through fascial rolls) as well as functional physiotherapy. Ointments that promote blood circulation also have a positive effect. To prevent strains of the back, sufficient warming-up should be ensured before starting the high-performance part of the sport.

12.1.2 Fractures of the Spinous or Transverse Process

Isolated spinous and transverse process fractures are a rarity. There are reports of rarely seen clay shoveler's fractures through sudden flexion force on the back and neck muscles [9]. Still, the majority of these fractures occur in direct impact trauma (ground fall or wall collision) or due to hyperflexion of the spine. More often, they occur in combination with other spinal column injuries. Therefore, if a spinal or transverse process fracture occurs in isolation, the presence of a second, previously overlooked spinal column injury must always be considered and searched for. In the case of frustrating conventional X-ray diagnostics, this can also be carried out with cross-sectional imaging. In the final analysis, MRI should be administered to exclude a ligamental injury.

The isolated spinal or transverse process fracture is a domain of conservative therapy. The leading symptom is pain, accompanied by swelling and, sometimes, hematoma. Immobilization is hardly possible because the pain is triggered by activation of the inserting muscles and their traction or by movement of the attached ribs. Therefore, sufficient pain therapy is absolutely necessary.

12.1.3 Hyperextension Fractures and Injuries

Hyperextension injuries are rare but highly severe injuries of the spine. They occur either in the thoracolumbar transition or in the cervical spine. Such injuries can occur when a climber falls into the rope at an unexpected moment, from an unfavorable body position, or when the body tension is low. In these cases, a tight pull of the rope by the belayer results in a strong pull on the belay loop and, thus, a quasi-fixation of the pelvis and the lower lumbar spine. The upper body bends above the fixation, which leads to a frontal tearing and a rear compression of the spine. The injury can be purely ligamentous and transdiscal or completely or partially osseous. It is often accompanied by neurological deficits and usually associated with paraplegia. These unstable injuries require surgical treatment by ventral or dorsal stabilization. If there is also a sensorimotor deficit, this injury is an emergency, requiring immediate surgery.

As a purely ligamentous injury, hyperextension injuries can easily be overlooked in conventional X-rays. If the clinical symptoms and the mechanism of the accident show evidence for the suspicion of such an injury, cross-sectional imaging is always indicated.

Climbers with preexisting conditions such as ankylosing spondylitis or DISH (diffuse idiopathic skeletal hyperostosis) have an increased risk of suffering such an injury during a rope fall due to the reduced flexibility of the spine.

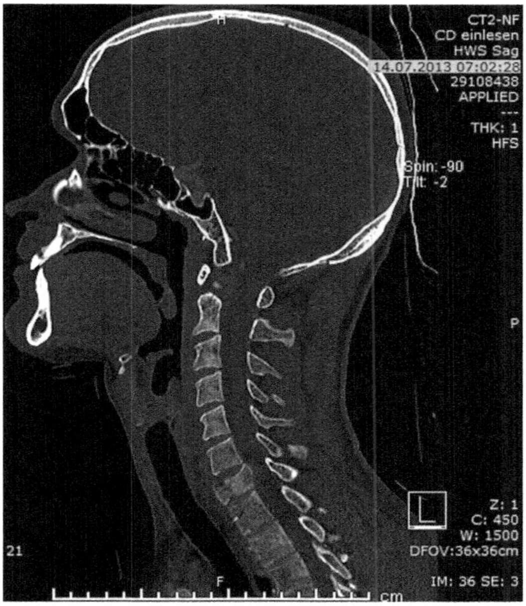

Fig. 12.1 Sagittal CT scan of a C7 burst fracture in a 30-year-old male climber before surgery

12.1.4 Compression and Burst Fractures

Compression and burst fractures occur as a result of climbing accidents with a ground fall or wall collision. Biomechanically, a distinction can be made between direct impact (floor or wall) and spinal compression or shearing. Torsion fractures, as they occur in high-speed traumas (e.g., traffic accidents), are an absolute exception. Clinically, patients with compression and burst fractures present with pain, swelling, hematoma, and, in severe cases, neurological deficits. Conventional X-ray diagnostics are usually followed by cross-sectional imaging (CT/MRI). For falls from a height of more than 3 m, cross-sectional imaging is always indicated (Fig. 12.1).

Conservative treatment can be administered for pure compression fractures without dislocation of the posterior vertebral body edge with a height reduction of less than 50% of the vertebral body and segmental kyphosis of less than 20%. These are considered to be stable fractures. X-ray monitoring after 4–8 days and adequate pain therapy should be performed [10].

Compression fractures with posterior vertebral body edge dislocation, more than 50% height reduction of the vertebral body, or more than 20% segmental kyphosis are surgically addressed (Fig. 12.2). The same applies to burst fractures and all fractures with additional sensorimotor deficits. In the case of incomplete or complete paraplegia, this is a surgical emergency [10].

In the case of an isolated injury to a vertebral body, dorsal stabilization with a small-segment fixation is indicated. In this case, stabilization is performed in the vertebral body above and below the injured vertebra. If fractures occur in the thoracolumbar transition with distraction or torsion, two segments are used for stabilization (Figs. 12.3 and 12.4). If there is also a destruction of the intervertebral disc, ventral stabilization with the insertion of a bone graft or a vertebral body replacement is required.

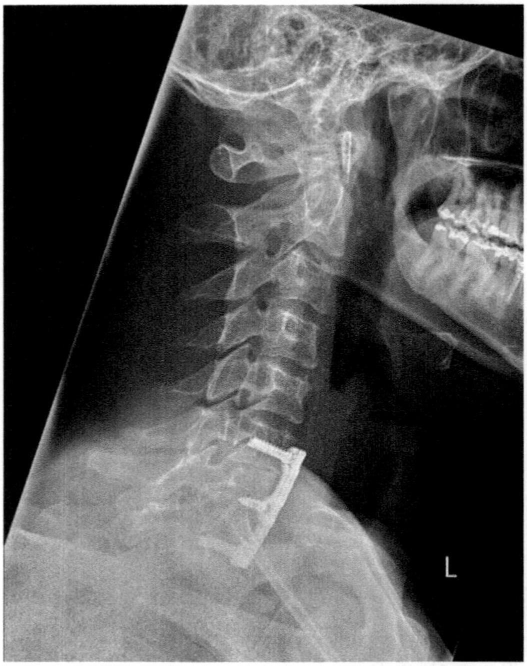

Fig. 12.2 Radiographic image postoperative of the same patient after surgery with a ventral plate stabilization C6–Th1

Traumatic Spine Injuries

- **History:** Ground fall or wall collision, rarely rope fall or twisting motion
- **Onset:** Acute
- **Clinical:** Pain, swelling, hematoma, deformity, crepitations, sensorimotor deficit
- **Diagnostic:** Clinical finding, radiography, X-ray functional images, CT/MRI
- **Therapy:** In accordance with orthopedic trauma guidelines
- **Outcome:** Depending on the diagnosis, often bad with a primary sensorimotor deficiency

12.2 Chronic and Degenerative Back Diseases

The entire range of chronic and degenerative spinal diseases can be found among sport climbers as in the normal population—from easily treatable muscular tension, lumbalgia, and sciatica to conditions which are more difficult to treat, such as herniated discs and spondylolisthesis.

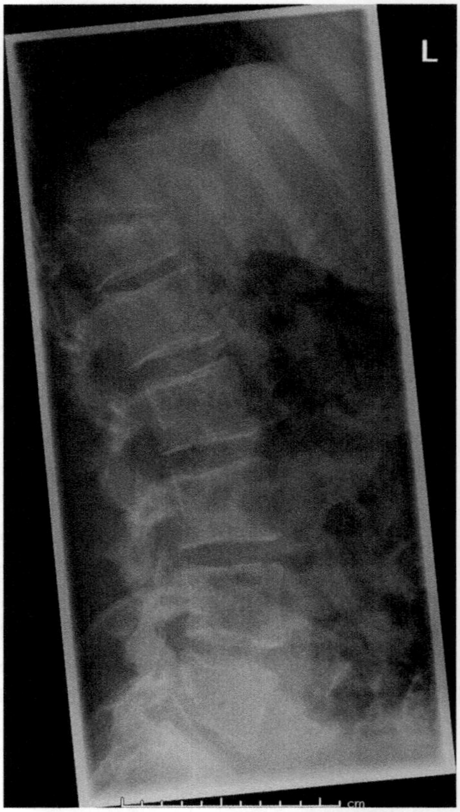

Fig. 12.3 Radiographic image of a Th12 compression fracture in a 26-year-old male climber due to ground fall, before surgery

Even though there are still too few comparative and reliable studies, it can be stated that sport climbing in general is a good training for the back [2, 4, 11]. The preventive value of sport climbing is shown in studies [12], as well as the fact that sport climbing has already been incorporated in rehabilitation and physiotherapy for a long time. Nevertheless, climbing without additional stretching and balancing exercises is a one-sided load that can manifest bad posture and associated suffering.

Special forms of chronic back problems in climbers are the "climber's back" [3] and the "belayer's neck" [5]. These are false postures developed by chronic overuse, uniform stress, and the biomechanical requirements of the climbing sport, which can lead to severe pain and degenerative changes in the spine.

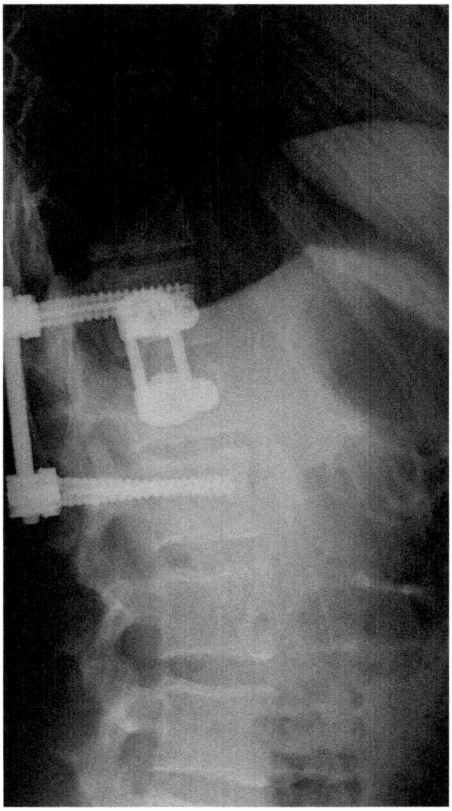

Fig. 12.4 Radiographic image of the same patient after surgery with short-segment posterior fixation Th11–L1 and accomplishment with anterior lumbar corpectomy via a mini-open, extreme lateral, transpsoas approach

It is not yet known how much damage is caused to the spine by the repetitive falls sport climbers take. It seems possible that the chronic microtrauma these falls create for the intervertebral discs could lead to early degeneration. This would be hard to prove, because most climbers who suffer from nucleus pulposus prolapse (a herniated disc) don't report it happening in a fall but when executing some sort of twisting motion.

12.2.1 Myalgia and Myogelosis

While myalgia simply describes the symptom of muscle pain which can occur in conjunction with many various diseases, myogelosis are palpable knots in muscles, which persist even when the muscle returns to relaxation (e.g., in narcosis). This is the main difference to a simple muscular tension. The symptoms of myogelosis are palpable knots within the muscle, a tenderness to pressure, and muscle tightness. Histological findings showed atrophic muscle fibers and a thinning of the endomysium in these muscular knots, which is known to be a consequence of malfunctional metabolism through poor blood flow. Often through misuse or overuse of the back, the muscles become strained, a process which repeats itself in a vicious cycle. The muscles seize up from overuse and then, due to the stress of being flexed, receive poor blood flow. Lacking good nutrition, the muscle has to work a bit harder for any motion, increasing the tension and further decreasing the blood flow. This condition is especially common in the neck and shoulder area but also known for the lumbar spine. Until the muscles relax and receive blood flow and therefore a chance to recover, they will be stiff. A muscle that is too tight does not receive the appropriate amount of blood flow, which, in turn, allows the acids that are created during exercise to build up in the muscle body. Myogelosis can be painful and inhibit a climber's ability. If the condition is not treated as soon as possible, it can scar the muscle permanently. Through heating, massaging, and stretching the muscle, the excessive tightness can often be worked out and the blood flow can be stimulated. To prevent myogelosis from occurring during periods of intense training, it is important that climbers warm up correctly and then go through a "cooldown" period at the end of each workout [13].

12.2.2 Lumbago

The term lumbago subsumes a broad range of painful symptoms, leading to reduction in the mobility of the lumbar vertebrae, or lower spine. Most cases are harmless and easy to treat, but some rare back diseases can appear initially with lower back pain. Climbers experience lower back pain and reduction of mobility due to overuse and false postures or excessive training. This can also be the initial symptom of a "climber's back" [3].

For diagnosis, a clinical examination by an experienced physician might be enough, and radiography or cross-sectional imaging may not be required. Pathologies such as ankylosing spondylitis, Scheuermann's disease, disc herniations, kissing spine, spinal stenosis, scoliosis, and spondylolisthesis/spondylolysis must be ruled out when symptoms persist and present therapy-refractory. Therapy consists of mobility supporting physical therapy, heat therapy, and analgetic therapy.

12.2.3 Blocking of Facet Joints and Rip Joints

Many climbers report a blocking of joints, which frequently occurs and has a high tendency of reoccurrence. These blockings must be distinguished from joint dislocations [5]. To clarify this, a blocked joint is not actually a dislocated joint; this is only seen in trauma. Additionally, neither the small intervertebral joints nor the rib joints physically dislocate; the blockages are minor "dislocations" or shifts, which cannot be shown in cross-sectional imaging like CT or MRI scans. Increased stress to the chest spine, false postures, or sudden torsional movements can lead to stress onto the small intervertebral joints (facet joints). High stress to these facet joints leads to minor shifts, which stress their joint capsule and ligaments. Nevertheless, there is constant activation of these joints, leading to pain, discomfort, muscular tension, and sometimes even a decreased range of motion in parts of the spine. These symptoms might result in inflammation. The upper to middle thoracal vertebrae are affected the most. During climbing, the rhomboid muscles work hard to stabilize the scapula. If these muscles get weak and are no longer capable of stabilizing the scapula, it leads to the phenomenon of the "chicken winging" or "scapular winging" experienced by climbers during hard climbing sections. With the major and minor rhomboid muscle finding their insertions on the lower cervical and upper thoracic spine, this could also lead to the blocking of joints [5].

The increased kyphosis of a "climber's back" might predispose a climber to frequent blockages of the intervertebral joints, resulting in a decreased range of motion in certain parts of the spine. A proper clinical examination can differentiate between blockages, nerve compressions, and herniations. If a herniation or nerve compression has been excluded, therapy can be initiated. In early stages, a simple chiropractic maneuver can provide complete pain relief. Some patients report giving themselves the treatment, e.g., with two tennis balls or Blackroll® balls. If the blockage is already accompanied by muscular tension or inflammation, massages, heat therapy, and physiotherapy can help to ease the muscular tension. Often, spontaneous relief of the blockage occurs when the muscle tension is gone.

A similar pathology applies for the adjunct rib joints, meaning the proximal attachments of the ribs to the vertebra. A minor change of the positioning of the vertebrae together with the increased tension of the muscles leads to similar blockages of the rib joints. These blockings can be even more painful while breathing, as these joints move with every inhalation and exhalation. If these problems persist for a long time, they will lead to even more muscular tension and relieving postures.

12.2.4 Herniated Disc or Nucleus Pulposus Prolapsus

A herniated disc is more common in the lower back than in the mid back or the neck. The overall incidence is high. Intervertebral disc surgery is the most frequently performed spine surgery. Between each vertebra, there are shock absorbers that allow the spine to move but stay connected. These shock absorbers, the discs, are made up of a nucleus of thick liquid surrounded by an annular fibrous ring. A nucleus pulposus prolapsus occurs when pressure on the disc creates a bulge in the fibrous ring. This bulge can then put pressure on the nerve that branches out of the spine or on the spine itself. As previously mentioned, it is difficult to distinguish whether there has been a predisposition and a damage to the spine before

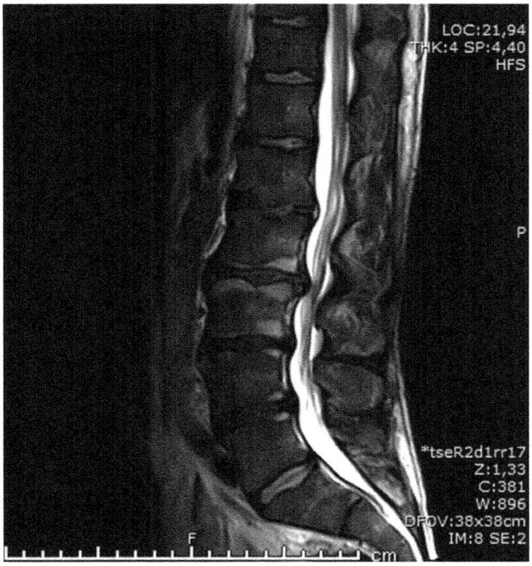

Fig. 12.5 Sagittal MRI T2 TSE scan of a 42-year-old male climber showing a L2–L3 disc herniation with a sequestrum

or if a certain activity (such as climbing) caused the prolapse. It is still not known how much damage is caused to the spine by the repetitive falls sport climbers take. It seems possible that the chronic microtrauma these falls cause to the intervertebral discs could lead to early degeneration. This would be hard to prove, because most climbers who suffer from nucleus pulposus prolapse (a herniated disc) don't report it happening in a fall but when doing some sort of twisting motion [14]. The classification differs between protrusion, prolapse, and a free sequestered disc. This results in typical symptoms causing pain, numbness, or weakness in according muscles. The key to a correct diagnosis is to relate the clinical findings (diminished reflexes, motor deficiency and lack/loss of sensibility) to the cross-sectional imaging findings (Fig. 12.5). If there is no congruence between clinical symptoms and MRI imaging, other pathologies (such as shingles neuralgia, multiple sclerosis, or peripheral nerve lesions) have to be considered. It is important to remember that many disc protrusions and even prolapses are completely asymptomatic. We have seen a lot of disc prolapses and protrusions as incidental findings on MRI scans in patients who came complaining of other problems. Most cases (up to 90% of the lumbar disc herniations) respond well to conservative treatment. Analgetic therapy, physical therapy, and nerve root and epidural infiltration (steroids and local anesthetic) alleviate the patients' suffering. In severe cases with paralysis or persistent pain, surgery is inevitable.

12.2.5 Spondylolisthesis/Spondylolysis

There is a rare condition known as spondylolisthesis, or "sliding vertebra," which can be due to a birth defect, such as a dysplastic type, or mechanical overuse from athletics (e.g., in wrestlers, gymnasts, contortionists, football players, and climbers) [10]. Sports with causative mechanisms, such as lumbar hyperextension and rotation found in rowing, artistic gymnastics, and throwing sports, record high spondylolysis prevalence rates—16.88%, 16.96%, and 26.67%, respectively, compared to the general population's 3.0–8.02% [15]. Climbing also involves these causative mechanisms, and reports suggest that climbing may initiate the development of spondylolysis, but its prevalence has not yet been investigated [5]. Spondylolisthesis describes the condition of a ventral or dorsal shift of two vertebral bodies against each other, whereas spondylolysis describes the interruption of the pars interarticularis of the vertebral body. There is a genetic predisposition to dysplastic type of spondylolisthesis, which can also occur at the end of multi-segmental vertebral body fusions (after surgery) or as a consequence of Scheuermann's disease. Patients suffer from lumbar pain during movement, especially reclination; there may also be radiculopathy. Diagnosis is made by functional X-ray in extension and flexion to detect segmental hypermobility (Fig. 12.6). Cross-sectional imaging like CT and MRI can display the interruption of the pars interarticularis of the vertebral body (Fig. 12.7). In severe cases, surgery as a small-segment instrumentation is necessary. For climbers diagnosed with spondylolisthesis graded Meyerding I and II but not suffering from

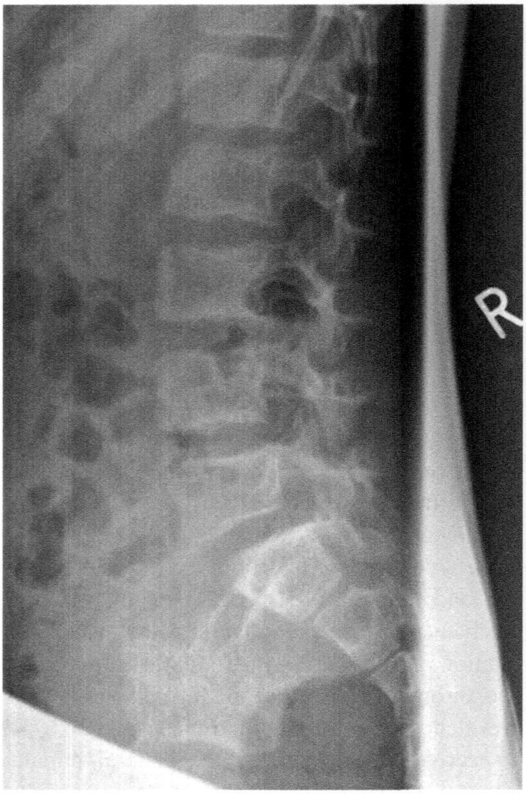

Fig. 12.6 Radiographic image of a young male climber (9 years old) with a congenital spondylolisthesis/spondylolysis L5/S1 graded Meyerding II

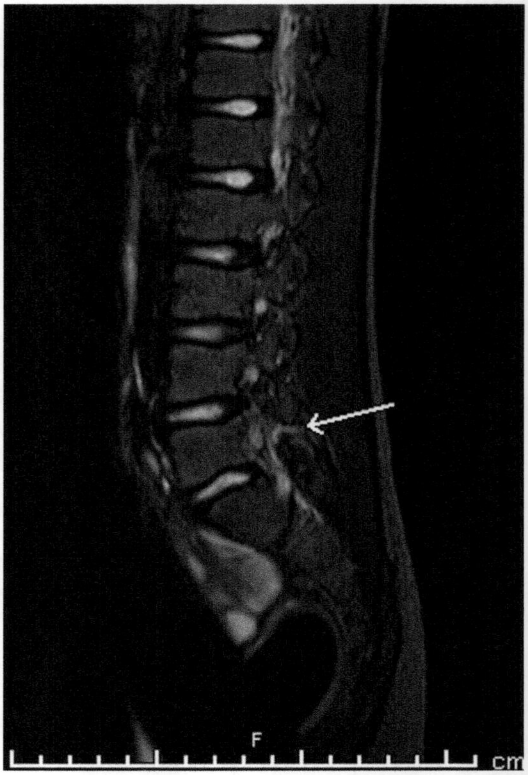

Fig. 12.7 MRI STIR sequence scans of the same boy show the spondylolysis in the pars interarticularis L5/S1 graded Meyerding I

any symptoms, we recommend X-ray controls in an annual interval. For these climbers, it is not necessary to stop climbing.

12.2.6 Thoracic Outlet Syndrome

Excessive tension of a shortened pectoralis minor muscle or scalene muscle can lead to an entrapment of the nerves and blood vessel which run into the arm. Though rare, thoracic outlet syndrome is a nerve or blood vessel compression syndrome that affects the shoulder and arm. Overdevelopment and tightening of the small chest muscles and the scalene muscles (in the neck) can put pressure on the brachial plexus or/and subclavian artery [16]. Swelling of muscles after stress, overstretching, or anatomic abnormalities might be the reasons for this syndrome; for example, people with an extra rib in their upper body can also suffer from this syndrome. If the artery is compressed, paleness and cold skin are symptoms due to reduced blood flow. Compression of the nerve might lead to diffused pain in the arms, numbness, or a heavy sensation of the hands and lower arms. The condition is most noticeable when the arms are raised above the head. The symptoms vary according to whether the muscles are placing pressure on the nerves or the blood vessels. The syndrome can be diagnosed with the Adson test in a clinical examination or provocation duplex sonography.

Intensive physical therapy such as stretching the pectoralis and scalene muscles and the thoracic spine can often reduce the pain. Surgery is only necessary in very severe cases.

Thoracic Outlet Syndrome

- **History:** Congenital (additional rib), overuse, muscular disbalance
- **Onset:** Chronic
- **Clinical:** Pain in the arms, numbness
- **Diagnostic:** Clinical findings, Adson test, provocation duplex sonography
- **Therapy:** Conservative—stretching and balancing exercises, surgery in severe cases only
- **Outcome:** Good

12.2.7 Climber's Back

"Climber's back" is characterized by anterior positioned shoulders and a "humpback" spine. The presentation of this postural adaptation is a combination of hyperkyphosis of the thoracal spine and a hyperlordosis of the lumbar spine, as well as a recurvated body hiatus [5, 17], which is similar to "swimmer's back" [18, 19]. The origin of these spinal postural dysfunctions is probably a functional adaptation of the form of the spine designed to meet the sport-specific biomechanical stresses of the sport [20, 21]. It is thought that shortened pectoralis muscles are one of the main factors leading to this spinal postural dysfunction [17, 22]. Our studies showed [3] that the climbing ability level was strongly correlated to the postural adaptations found frequently in elite male climbers who climb routes graded UIAA 10 or above. Given the greater differences of both the kyphosis and lordosis angles between a sport climbers group and a recreational climbers group (control group) in our study, it is suggested that an even greater difference between performance-oriented climbers and a normal population is likely. We only examined male climbers in our study. In female climbers, there seem to be different postural adaptions to climbing; this might be due to their different muscle morphology as well as their different climbing technique [22].

Although sport climbing has been used in physical therapy for back disorders and rehabilitation as well as physiotherapy for scoliosis [23, 24], excessive training and/or the neglect of compensatory exercises may lead to the development of "climber's back." The functional adaptation of the thoracolumbar spine to specific sporting exposure is not a performance precursory variance of the norm but has a latent disease potential, especially during the growth period [19, 25, 26]. The biomechanical adaptations to intensive climbing that can result in a "climber's back" may cause an aberrant erect posture but may be an essential adaptation to facilitate elite climbing kinetics. Nevertheless, we think that this adaptation to the specific postural requirements of higher-grade climbing is not advantageous or a necessary prerequisite of elite climbing kinetics; we feel that it is a precursor to the associated health conditions brought on by the development of thoracic outlet/inlet syndrome or early joint disorders due to the increased stress onto the pars interarticularis of the spine [21]. Therefore, it is not only to be seen as a physiological adaptation of the specific postural requirements of intensive climbing but also as a pathological finding, which demands compensatory training and exercise, especially in adolescent athletes (Fig. 12.8). A difference in the length of the legs (a trait most people unknowingly have) and congenital problems (like scoliosis) can contribute to false posture. Climbers developing thoracal hyperkyphosis, lumbar hyperlordosis, and anterior positioned shoulders as well as shortened pectoralis muscles might experience back or shoulder pain, reduced mobility, and sometimes even radiculopathy. In clinical examination, an appearance of the characteristically posture with thoracal hyperkyphosis and lumbar hyperlordosis should lead to examination of shortening of pectoralis muscles (e.g., using protocol given by Janda). Preventative stretching exercises (for pectoralis and iliopsoas muscles) should be recommended to the athletes. Here, we refer to our ACT (Adjunct Compensatory Training for Rock Climbers, www.act.clinic), which provides exercises for stretching and balancing the spine and core musculature.

Fig. 12.8 Typical climber posture (climber's back): overextension of the vertebral spine and a concave chest. An elite climber displaying the hyperkyphosis and hyperlordosis characteristic of a "climber's back" (reprinted with permission)

12.2.8 "Belayer's Neck" and Cervical Spine Syndrome

Cervical spine syndrome or cervical pain syndrome is a well-known and widespread set of symptoms. The "belayer's neck" is a postural adaption to the constant overextension the cervical spine experiences in hours of belaying and climbing. This could be worsened by the increased thoracic kyphosis from a "climber's back," leading to a compensatory overextension of the cervical spine. In combination with shortened pectoralis muscles, this brings high stress onto the cervical spine. This can become a problem while belaying long and steep routes and spending a lot of time belaying and climbing. Due to the increased overextension of the cervical spine, high pressure is applied to the small intervertebral joints. This irritates the joint cartilage and joint capsule. As a protective mechanism, the paravertebral muscles raise their tension to protect the spine, resulting in cervical spine syndrome-like symptoms. Even though many climbers and belayers suffer from these symptoms, there is almost no scientific work on the consequences and circumstances of "belayer's neck." The range of symptoms includes neck and back pain, headaches, or even paresthesia and hypesthesia (prickling sensation and numbness), as well as impaired vision and ringing in the ears. A proper clinical examination (e.g., Spurling test) by a physician can exclude more serious pathologies, such as herniated discs or nerve root compression and can lead to diagnosis. In doubt, MRI can provide diagnosis, whereas conventional radiography is obsolete. The therapeutic regimen covers analgetic, massage, and physical therapy as well as local infiltration and muscle relaxant medication.

For prevention purposes, prism glasses are available for purchase, e.g., CU belay glasses®, LACD belay glasses®, or Y&Y glasses®, which can be highly recommended for people with cervical spine problems (Fig. 12.9). Through mirroring prisms, the belayer can have an image of the climber without keeping his head in an upright and overextended position (Fig. 12.10). Using prism glasses requires initial training and requires some time for adjustment, which is why we recommend starting to belay with prism glasses in top rope climbing.

Climber's Back and Belayer's Neck

- **History:** Overuse, bad postures, and functional adaptation to sport-specific stress
- **Onset:** Chronic
- **Clinical:** Lumbar or cervical spine pain, typical humpback, shortened pectoralis, and iliopsoas muscle
- **Diagnostic:** Clinical findings, typical humpback with hyperkyphosis of the thoracal and a hyperlordosis of the lumbar spine
- **Therapy:** Conservative—stretching and balancing exercises, prism glasses (belayer's neck)
- **Outcome:** Good, although can become chronic with degenerative injuries

Fig. 12.9 Female climber belaying with CU belay glasses®. (Reprinted with permission)

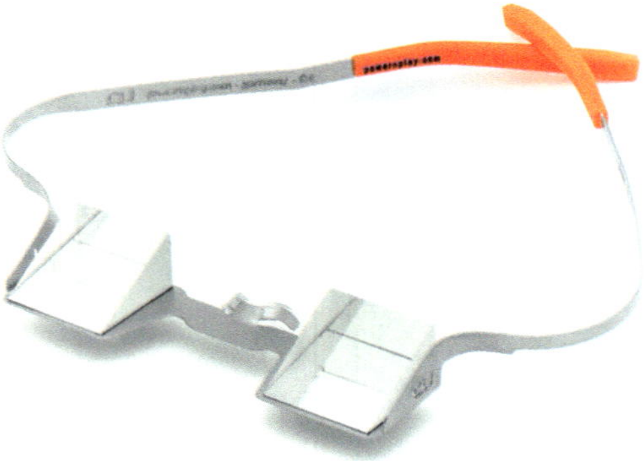

Fig. 12.10 Prism glasses CU belay glasses® (reprinted with permission)

References

1. Schöffl V. Akute und chronische Schäden durch Sportklettern. In: Berghold F, et al., editors. Alpin- und Höhenmedizin. Berlin, Heidelberg: Springer; 2015.
2. Muehlbauer T, Stuerchler M, Granacher U. Effects of climbing on core strength and mobility in adults. Int J Sports Med. 2012;33(6):445–51.
3. Förster R, Penka G, Bösl T, Schöffl VR. Climber's back—form and mobility of the thoracolumbar spine leading to postural adaptations in male high ability rock climbers. Int J Sports Med. 2009;30(1):53–9.
4. Hawrylak A, Chromik K, Ratajczak B, Barczyk-Pawelec K, Demczuk-Włodarczyk E. Spinal range of motion and plantar pressure in sport climbers. Acta Bioeng Biomech. 2017;19(2):169–73.
5. Schöffl VR, Hochholzer T, Lightner SJ. One move too many: how to understand the injuries and overuse. 2016.
6. Rauch S, Wallner B, Ströhle M, Dal Cappello T, Maeder MB. Climbing accidents—prospective data analysis from the international alpine trauma registry and systematic review of the literature. Int J Environ Res Public Health. 2020;17(1):203.
7. Nelson NG, McKenzie LB. Rock Climbing Injuries Treated in Emergency Departments in the U.S., 1990-2007. Am J Prev Med. 2009;37(3):195–200.
8. German alpine club (DAV) accidents statistic [Internet]. Munich; 2019. https://www.alpenverein.de/Bergsport/Sicherheit/Unfallstatistik/.
9. Kaloostian PE, Kim JE, Calabresi PA, Bydon A, Witham T. Clay-shoveler's fracture during indoor rock climbing. Orthopedics. 2013;36(3):e381–3.
10. Ruchholtz S, Wirtz DC. Orthopedics and trauma surgery. 3rd ed; 2019.
11. Heitkamp HC, Wörner C, Horstmann T. Sport climbing with adolescents: effect on spine stabilising muscle strength. Sportverletz Sportschaden. 2005;19(1):28–32.
12. Schinhan M, Neubauer B, Pieber K, Gruber M, Kainberger F, Castellucci C, et al. Climbing has a positive impact on low back pain: a prospective randomized controlled trial. Clin J Sport Med. 2016;26(3):199–205.
13. Olsen O, Sjøhaug M, Van Beekvelt M, Mork PJ. The effect of warm-up and cool-down exercise on delayed onset muscle soreness in the quadriceps muscle: a randomized controlled trial. J Hum Kinet. 2012;35:59–68.
14. Schäfer J, Gaulrapp H, Pförringer W. Verletzungen und Überlastungssyndrome beim extremen Sportklettern. Sport Sport. 1998;12(1):21–5.
15. Soler T, Calderón C. The prevalence of spondylolysis in the Spanish elite athlete. Am J Sports Med. 2000;28(1):57–62.
16. Nichols AW. The thoracic outlet syndrome in athletes. J Am Board Fam Pract. 1996;9(5):346–55.
17. Schlageter M. Funktionelle Störungen des Bewegungsapparates bei Kletterern. 2004.
18. Dalichau S, Scheele K. Der Einfluss sportmechanischer Anforderungen im Leistungstennis auf das thorakolumbale Wirbelsäulenprofil. Sport Sport. 2002;16(2):64–9.
19. Wilson FD, Lindseth RE. The adolescent "swimmer's back". Am J Sports Med. 1982;10(3):174–6.

20. Schöffl VR, Hochholzer T, Imhoff AB, Schöffl I. Radiographic adaptations to the stress of high-level rock climbing in junior athletes: a 5-year longitudinal study of the German Junior National Team and a group of recreational climbers. Am J Sports Med. 2007;35(1):86–92.
21. Neugebauer H. Radsport und Rundrücken. Sportärztliche Praxis; 1961.
22. Rolland L, Allet L, Linhart C, Ziltener J-L. Females "Climber's back". 2014.
23. Heitkamp HC, Mayer F. Effekte eines Klettertrainings im Ver- gleich zu isokinetischem Krafttraining auf die wirbelsäulenstabilisierende. Muskulatur. 1999;
24. Heitkamp HC, Fichter C. Klettern und paravertebrale Muskulatur bei jugendlichen Skoliotikern. 2005.
25. Ashton-Miller JA. Thoracic hyperkyphosis in the young athlete: a review of the biomechanical issues. Curr Sports Med Rep. 2004;3(1):47–52.
26. Wojtys EM, Ashton-Miller JA, Huston LJ, Moga PJ. The association between athletic training time and the sagittal curvature of the immature spine. Am J Sports Med. 2000;28(4):490–8.

Part V
Climbing and Age

Long-Term Effects of Intensive Rock Climbing to the Hand and Fingers

13

Thomas Hochholzer and Volker Schöffl

Jens Lisse is climbing "Götterdämmerung", Frankenjura, Germany.
Picture by Michael Simon

T. Hochholzer (✉)
Private Clinic Hochrum, Innsbruck, Austria

V. Schöffl
Department of Orthopedic and Trauma Surgery, Center of Sportsmedicine, Klinikum Bamberg, Bamberg, Germany

School of Clinical and Applied Sciences, Leeds Becket University, Leeds, UK

Section of Wilderness Medicine, Department of Emergency Medicine, University of Colorado School of Medicine, Denver, CO, USA

Department of Orthopedic and Trauma Surgery, Friedrich-Alexander Universität Erlangen-Nürnberg, Erlangen, Germany

V. Schöffl et al. (eds.), *Climbing Medicine*, https://doi.org/10.1007/978-3-030-72184-8_13

13.1 Introduction

The connection between rock climbing and degenerative arthritis of the finger joint has been debated and investigated by various authors [1–4]. Our analysis includes an overview of data and patients spanning approximately 30 years [5]. Some have ended their sporting careers over the years; others—some of them over 60—are still more or less active in their sport. The risk of polyarthritis of the finger joints can be well assessed on the basis of these cases. There is certainly a correlation between the number of years of climbing, the level of climbing, and a higher incidence of osteoarthrotic changes in the finger joints [1, 2, 5–10]. In addition to radiographic examinations, however, the symptoms and clinical signs are also decisive for a further prognosis for the climber.

13.1.1 Clinical Findings

Pain in the area of the middle and, to a lesser extent, in the end joints of the fingers is a common problem faced by sports climbers. These complaints mostly occur after intensive training or climbing periods. Climbers report load-dependent pain but also pain at rest, sometimes with a dull sensation of swelling [1, 11]. At times, the finger middle joints (PIP) can no longer be fully extended and show an extension deficit of 5–15°. Climbers also often describe a loss of fine motor skills [12–14]. In the morning, an unpleasant stiffness of the joints becomes noticeable, which usually improves after a few movement exercises. At the end joints, an adaptive thickening of the joints often becomes apparent, sometimes with small "nodules" as is also known from the osteoarthritis of the DIP joint (Heberden's osteoarthritis). Nevertheless, the underlying pathology of Heberden's osteoarthritis is different. During the examination, in addition to a slightly spindle-shaped thickening of the middle joints, elastic joint effusions and slight instabilities of the radial and ulnar ligaments at the PIP joints can sometimes be found. The frequently found slight instability is usually caused by the joint effusions.

Pain in the dorsal part of the PIP joint requires special attention. Pressure pain combined with painful movement restrictions and a snap phenomenon at the beginning of flexion is sometimes experienced by older climbers.

Noticeable radial or ulnar instabilities—as known in severe osteoarthritis of the DIP joint—are extremely rare. The metacarpophalangeal (MCP) joints of the fingers are clinically and radiologically rarely affected. Only a radiographic examination can confirm the diagnosis of arthrosis of the PIP or DIP joints and thus differentiate between inflammation of the capsule or joint effusions.

13.2 Adaptive Osseous Reactions

In the analysis of radiographic images of sport climbers, it is important to separate bony adaptive reactions from degenerative changes in the joints [15] (Table 13.1). This distinction is not always easy and requires precise radiological diagnostics. Schöffl et al. [15, 16] used the following protocol to distinguish between adaptive osseous reactions and osteoarthrotic findings (Table 13.1). In contrast to the scale of Kellgren and Lawrence [17], subchondral sclerosis was added to the group of physiologic, adaptive stress reactions and was not defined as osteoarthritis [8, 15]. Furthermore, osteoarthrosis was diagnosed if a Kellgren and Lawrence grade ≥ 2 was present [2, 6, 8, 15]. Prior studies showed that reactions such as cortical hypertrophy and subchondral sclerosis in the fingers could be adaptive signs of high stress in climbing and could not be attributed to pathologic osteoarthritic changes [14, 18, 19]. The visual guidelines given in the *Atlas of Standard Radiographs* [20] were used for comparison with normal radiograph results [2, 6, 8, 15].

One of the most frequent adaptive reactions—which can often be seen clinically—is an enlargement of the joint outlines (Fig. 13.1).

In about 45% of climbers, this broadening of the joint base and surrounding connective tissue

Table 13.1 Protocol for radiographic evaluation

Protocol for radiographic evaluation (Schöffl et al. [8, 9, 15])	
Finding	Definition
Cortical hypertrophy	Visible cortical hypertrophy at the middle point of the phalanx; length in AP view or biconvex cortical hypertrophy in lateral view
Subchondral sclerosis, increased thickness of epiphysis	Stress reaction in contradiction to Kellgren-Lawrence scale
Calcification of insertion of flexor digitorum superficialis or flexor digitorum profundus tendon	Visible calcification
Broadened proximal interphalangeal joint base	Increased radius of lateral or medial base of phalanx
Broadened distal interphalangeal joint base	Increased radius of lateral or medial base of phalanx
Early osteoarthritic reaction	Kellgren and Lawrence grade ≥ 2 (note: subchondral sclerosis defined as adaptive stress reaction) or epiphyseal fracture

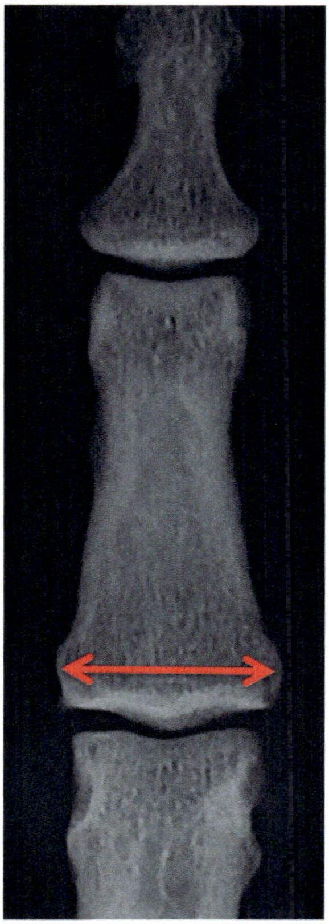

Fig. 13.1 A 48-year-old climber: climbing time over 30 years. Widening of the joint bases at the DIP and PIP joint (red arrow). Also, small mini-osteophyte at the PIP joint

is already visible after 5–10 years [14, 16]. The widening of the joint surfaces should not be confused with osteophytic structures, which show a different radiographic picture. The intensity of the osseous reactions correlates with the number of active climbing years and intensity of the sport; they become increasingly pronounced with increasing duration and intensity [5, 8]. However, the first signs of adaptations can already be observed after about 3 years of intensive climbing. A study of athletes of the German National Climbing Team from 1999 to 2004 [15] showed interesting findings. Young athletes on the German Junior Climbing Team (GJNT), starting at an average age of 14 years, were compared with a control group over 5 years and again after 11 years. Stress reactions and adaptation symptoms of the skeleton were already found in about 30% at the beginning of the longitudinal study [15]. In both the GJNT and the group of recreational climbers (RC), one case with an epiphyseal growth plate fracture was detected in each group. These were classified as early osteoarthrotic findings, nevertheless both had healed without any further damage in later controls [16]. Thus, they do not stand out as osteoarthrotic cases in the longitudinal analysis (Table 13.2). Slightly increased subchondral sclerosis of the cortical bone close to the joint or cortical thickening of the middle phalanges are seen frequently (Fig. 13.2). In addition, ossification of the tendon insertions (profundus and superficialis tendon) occur frequently in long-term climbers (Fig. 13.3).

Table 13.2 Radiographic changes of the German Junior National Team (GJNT), recreational climbers (RC), and the non-climbing control group (CG) (Schöffl et al. [15])

	GJNT	RC	CG
Number	19 (m = 15, f = 4)	18 (m = 16, f = 2)	12 (m = 8, f = 2)
Stress reactions	9 (47%)	5 (28%)	0
Cortical hypertrophy	5 (26%)	2 (11%)	0
Subchondral sclerosis, increased thickness of epiphysis	9 (47%)	1 (6%)	0
Calcification of insertion of flex. sup. or flex. prof. tendon	0	0	0
Broadened joint base PIP	8 (42%)	5 (28%)	0
Broadened joint base DIP	3 (16%)	0	0
Early osteoarthrotic reactions	1 (5%)	1 (6%)	0
Osteophytes PIP	0	0	0
Osteophytes DIP	0	0	0
Decreased jointspace	0	0	0
Subchondral cysts	0	0	0
			0
Epiphyseal fractures (stress fractures)	1	1	0
Calculation			
Barnett-Nordin index: (cortical thickness of the middle phalanx/ total osseous thickness of the middle phalanx (cortical and cancellous bone))	0.49 ± 0.05 (mean ± sd)	0.49 ± 0.07 (mean ± sd)	0.48 ± 0.08 (mean ± sd)

First number = n; second number = %; flex sup = M. flexor digitorum superficialis; flex prof = M. flexor digitorum profundus; *PIP* proximal interphalangeal joint, *DIP* distal interphalangeal joint

13.3 Long-Term Damage

Whether rock climbing leads to degenerative arthritis of the finger joint has been debated and already investigated by various authors [1–5]. Radiographic adaptations and changes in long-term adult climbers are a well-known fact, and a positive correlation to their years of climbing has been shown [1, 2, 5–8, 21–23]. Many authors described radiographic changes of finger joints of long-term climbers such as osteophytes, subchondral sclerosis, and joint space narrowing [4]. Nevertheless, none of the previously mentioned authors could provide clear evidence of an increased rate of degenerative arthritis compared to an age-matched group of non-climbers [1–4]. Most of this research concerns adult long-term climbers (>5 years), whereas data on young climbers or the long-term effects of intensive climbing since childhood are still sparse [24].

For example, Rohrbough et al. [2] could not prove a significantly higher rate of osteoarthrosis among climbers than non-climbers. These findings contradict a recent study by Allenspach et al. [6]. They compared 31 male sport climbers to 67 non-climbers and found a significantly higher risk of osteoarthritis in the climbers ($p < 0.001$) [6]. While another study by Sylvester et al. [3] could not prove a higher risk of osteoarthritis in climbers' hands, it showed that climbers' fingers have an increased diameter with additional bone deposited subperiosteally. Significant predictors include the highest level achieved in bouldering and sport climbing [3]. Bollen et al. [1] and Hochholzer and Schöffl [5, 8, 21] found that 28–39% of climbers with more than 15 years of climbing experience showed osteoarthritic findings on radiographs. However, these were mostly minor degenerative changes such as so-called mini-osteophytes in the PIP or MCP joints without any clinical relevance. Clear osteoarthroses could only be found in less than 10% of the climbers, whereas the climbers with condition after growth plate joint injuries stood out again (7%) (see Sect. 6.3.9). In contradiction to these radiographic findings, the respective climbers are sometimes surprisingly free of complaints [5] (Table 13.3).

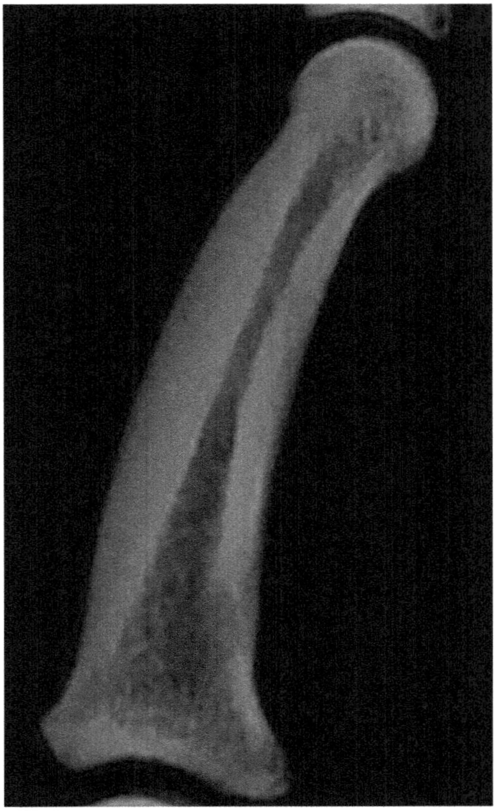

Fig. 13.2 Lateral image of the ring finger of a male sport climber (age, 24 years; climbing time, 6 years). Distinct convex hypertrophy of the cortex leads to the narrowing of the medullary canal as an example of an adaptive reaction of the bone to the load during climbing

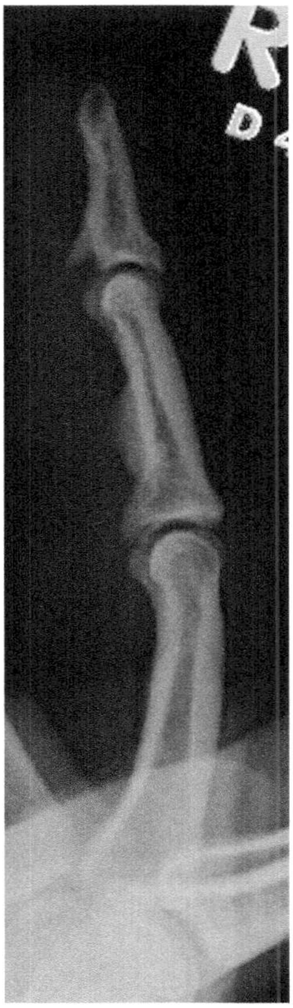

Fig. 13.3 Calcification of the tendon insertions of the superficialis, profundus, and extensor tendons in a climber after about 20 years of climbing

Table 13.3 Radiographic findings in 165 high-level climbers (second number in bracket in percent). The high number of osteoarthritic signs in climbers 5–10 years is due to epiphyseal fractures

Climbing years number	5–10 (*n* = 65)	>10 (*n* = 48)	>15 (*n* = 29)	>20 (*n* = 13)
No reaction	14 (22)	4 (8)	0 (0)	0 (0)
Stress reactions	51 (13)	44 (91)	29 (100)	13 (10)
Cortical hypertrophy	40 (37)	34 (70)	21 (72)	12 (92)
Subchondral sclerosis	28 (43)	27 (56)	23 (79)	12 (92)
Calcification of flexor tendons	12 (18)	16 (33)	8 (27)	7 (53)
Broadened joint base PIP	29 (45)	32 (66)	19 (65)	7 (53)
Broadened joint base DIP	37 (57)	37 (77)	26 (89)	8 (61)
Osteoarthrotic signs	11 (17)	20 (41)	18 (62)	4 (30)
Osteophytes PIP	3 (5)	11 (22)	17 (58)	4 (30)
Osteophytes DIP	0 (0)	4 (8)	4 (13)	4 (30)
Joint space narrowing	2 (3)	1 (2)	0 (0)	2 (15)
Incongruent joint space	6 (9)	1 (2)	0 (0)	3 (23)
Subchondral cystes	2 (3)	1 (2)	0 (0)	0 (0)

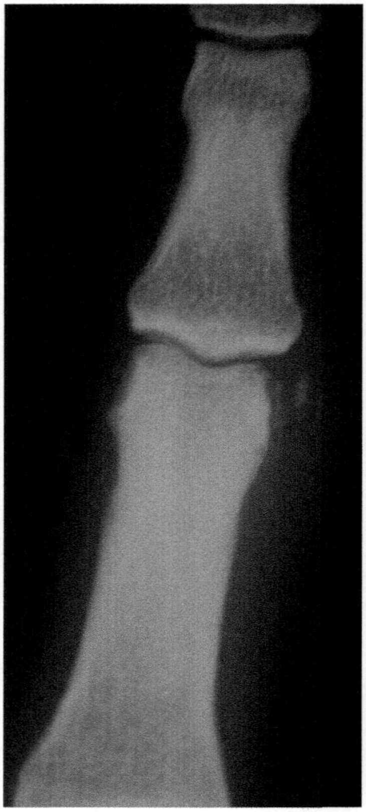

Fig. 13.4 Periarticular calcifications of the capsule and ligaments of a 35-year-old climber

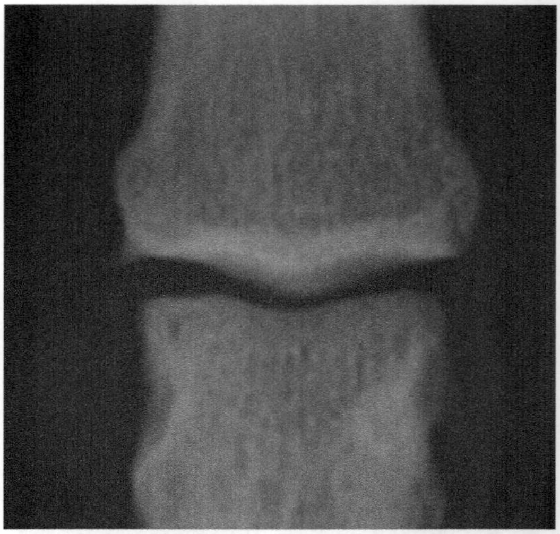

Fig. 13.5 An a.p. X-ray of the ring finger of a male climber of the PIP joint (age, 32 years; climbing time, 12 years). Small osteophytes at the radial side of the PIP joint

Periarticular calcifications are relatively common, especially in the area of the ulnar and radial collateral ligament of the PIP joint. However, they rarely correlate with true joint space narrowing or other degenerative changes. They are usually due to poorly healed or unhealed injuries to the collateral ligaments. More often, however, the cause seems to be microinjuries of the capsule and ligaments in the PIP joint. When a "one finger pocket" is used in climbing, considerable torsional stress occurs, which stretches the capsular ligament apparatus in the PIP joints and causes stress (Fig. 13.4).

An increasing number of small osteophytes have been found on the MCP, PIP, and DIP joints in athletes who have been climbing for more than 10 years. These bone spurs are clearly visible in the a.p. image, mostly without changes in the lateral image. Clinically, these mini-osteophyte formations did not show any symptoms or movement restrictions (Fig. 13.5).

However, it is extremely important to not only take a.p. X-rays but also lateral X-ray images. This is where the most serious degenerative changes become apparent in long-term climbers (>10 years). In 2010, we were able to examine a total of 13 climbers who showed pronounced dorsal and, less frequently, palmar osteophyte formation in the PIP and DIP joints. These mostly dorsal osteophyte (bone spur) formations are also called "extensor hood syndrome" [22] (see Chap. 6) (Table 13.4). Some of these climbers have impressive ossifications in the area of the capsule and extensor tendon apparatus. Clinically, a frequent additional observation to pressure pain was a slipping phenomenon in the area of the dorsal extensor apparatus at the PIP joint. This phenomenon could be regularly triggered by movement. The fact that the dorsal extensor hood and capsule are irritated in this case can be easily seen on the X-ray image. All of these climbers had a high climbing level and had already been climbing between 5 and 30 years (average 17.8 years); the average age was 32.5 years.

In addition to the dorsal and volar osteophytes, however, changes in the joint space were also evi-

Table 13.4 Description of climbers with extensor hood syndrome (Schöffl et al. [22])

No.	Side	Finger	Joint	Age	Climbing level	Climbing years	Kellgren score
1	Right	Middle	DIP	31	9.5	22	3
2	Right	Middle	PIP/DIP	44	10.3	30	4
3	Both	Middle	PIP	39	10.3	22	3 and 3
4	Both	Middle	PIP	42	10.3	21	3 and 4
5	Both	Middle	PIP	25	11.0	6	2 and 3
6	Both	Ring	PIP	37	11.3	17	2 and 2
7	Both	Middle	PIP	20	10.7	8	2 and 3
8	Right	Middle	DIP	33	10.0	23	3
9	Left	Ring	PIP	28	9.7	14	3
10	Right	Middle	DIP	43	10.3	18	3
11	Both	Middle	PIP	25	9.7	20	3 and 3
12	Both	Middle	PIP/DIP	55	9.0	30	3 and 4
13	Right	Ring	PIP	17	10.3	5	2

Climbing level UIAA metric, *DIP* distal interphalangeal joint; *PIP* proximal interphalangeal joint

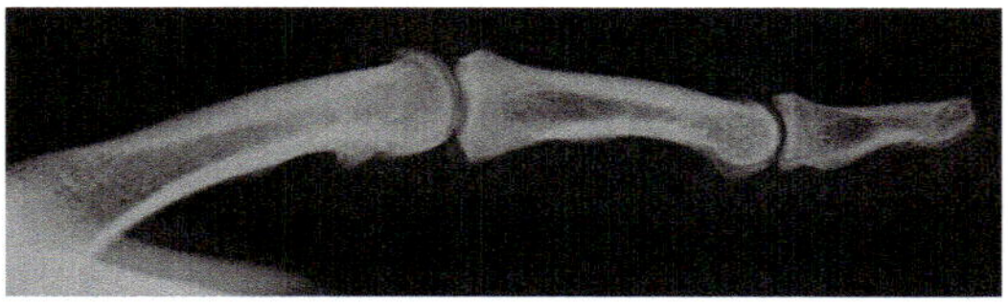

Fig. 13.6 Lateral X-ray of a long-time climber with dorsal and palmar osteophytes

dent, less with clear narrowing than more frequently with irregularities in the joint surfaces. After classification using the Kellgren-Lawrence classification, most of the changes in the score ranged between 2 and 4 (Fig. 13.6).

In contrast, fewer clinical complaints were reported at the DIP joint, despite the considerable formation of osteophytes. In all cases, there was a limitation of extension and flexion of up to 10°. These thickenings and swellings tend to be more painful for younger climbers, but the older climbers have probably learned to adapt the load to the pain (Figs. 13.7 and 13.8).

The importance of lateral X-ray views are also supported by a recent study by Allenspach et al. [6]. They investigated the anterior-posterior and lateral radiographic views of the fingers in a group of 31 high-level Swiss sport climbers. The mean years of climbing was 20 years, and the mean of the highest reached redpoint sport climbing level was 8b (UIAA 10) [6]. Almost all of the climbers had only little or no symptoms in the finger joints, but up to 84% showed osteophytes at the PIP-joints and up to 68% in the DIP-joints. According to the Kellgren and Lawrence score, six climbers (19%) had signs of an osteoarthritis (significant), whereas the age-matched non-climbing group had practically no signs of radio-

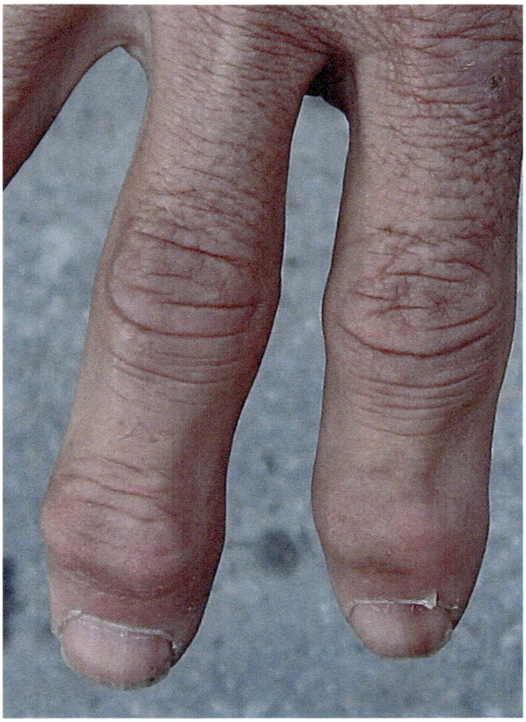

Fig. 13.7 Widening of the DIP joints of a 55-year-old climber. Clinical limitation of movement, minor complaints

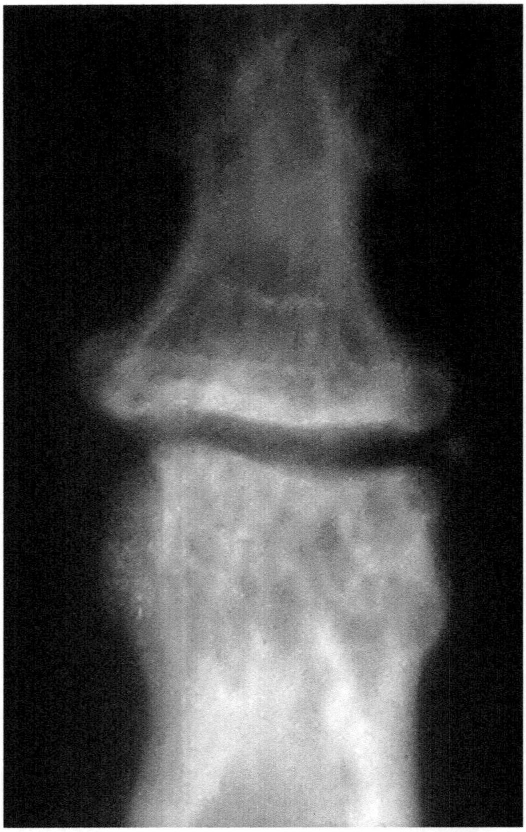

Fig. 13.8 X-ray of the DIP of the same climber: significant widening of the joint surfaces and small periarticular calcification

graphic changes in the finger joints. They observed that these bone spurs were best detected in the lateral radiographs [6]. These bone spurs can also be detected through ultrasound examination [25].

In comparison to known degenerative diseases of the hand (Bouchard's and Heberden's osteoarthrosis), the absence of any inflammatory, destructive, or unstable joint changes is noticeable in the climber patients. The visible radiographic changes differ fundamentally from the images of known finger polyarthritis. There is also no climber known to us who has had to be fitted with a finger prosthesis.

In a recent study of the 11-year follow-up of our once young high-level climbers from the German Junior National Team, we found that one quarter of climbers who performed at a high level in their youth showed a "mild" form of osteoarthritis (Kellgren II) at an age of less than 30 years [16]. During follow-up, 40% of the subjects of the GJNT and 23% of the RC group complained of swollen finger joints in the morning. Nevertheless, in all cases except one, the swelling was reported without any accompanied pain. Training intensity in 1999 ($p < 0.05$) and body weight in 1999 were significant ($p < 0.05$) for the development of radiographic stress reactions in 2011. Significant statistical influences for the development of early-onset osteoarthritis could be found for the overall sum of training years ($p = 0.024$), use of campus board training in 1999 ($p = 0.033$), and climbing level ($p = 0.030$).

13.4 Exemplary Cases

Three individual, exemplary cases show how the finger joints of long-time sport climbers react to the strain.

Climber A, born in 1951, started climbing at the age of 14 and climbed at a high level for decades (UIAA 10th grade (higher 5.13 YDS)). He was one of the pioneers for the introduction of the UIAA 7th grade (5.10 YDS) of difficulty in the alpine area. Between 1984 and 1987, the climber underwent surgeries on both hands for carpal tunnel syndrome and ulnar tunnel syndrome. For many years, the climber suffered from painful swelling of the fingers and a 15–20° extension deficit of all of the middle finger joints. The climber's semi-professional climbing career ended in 1994, when he was still suffering from significant pain and movement restrictions of the finger middle and end joints. In 1998, a radiographic examination was carried out on both hands. The finger joints were largely free to move and no longer showed any extension deficit; the climber no longer suffered from pain at rest or during exertion for recreational climbing tours in the lower difficulty range. Overall, only minor "mini-osteophytes" were found, particularly in the middle joints, and no destruction or narrowing of the joint space was found (Figs. 13.9, 13.10, and 13.11).

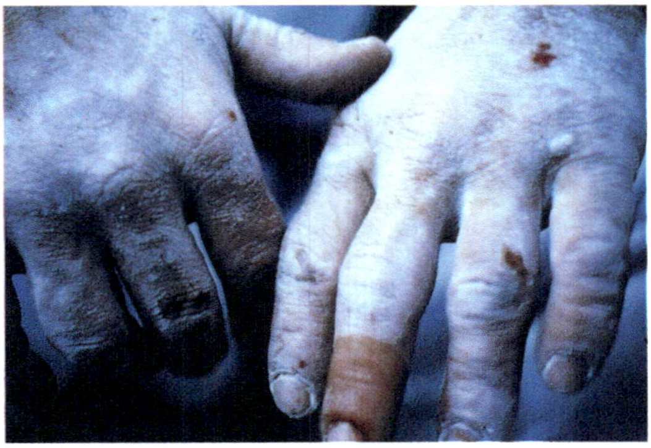

Fig. 13.9 Massively swollen middle finger joints at the time of active climbing

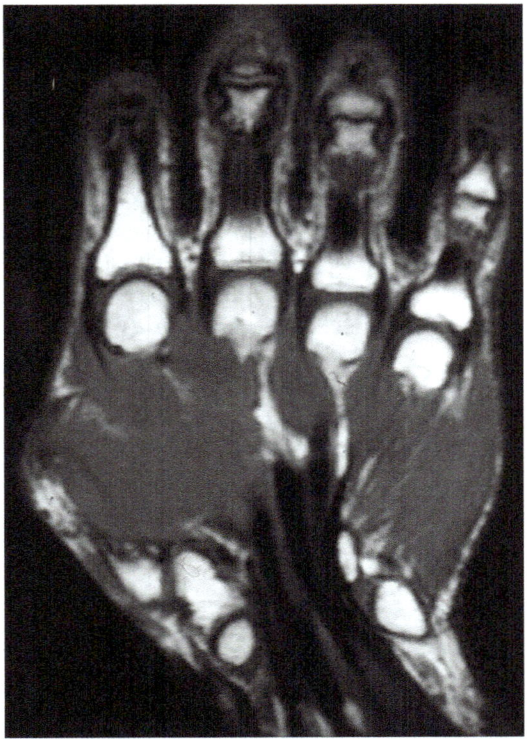

Fig. 13.10 MRI of the hands. The extension deficit in the PIP joints is clearly visible, as they are only partially depicted. Clearly hypertrophic small hand muscles, flexor tendons, and collateral ligaments. In the PIP joint, also minor effusion is evident. The MRI was performed in 1988

Climber S, born in 1965, started climbing at the age of 15. Soon afterwards, the level of difficulty of climbing increased to that of the top of the world with participation and achievements in World Championships and international competitions. Climber S established sensational first ascents on all continents. There were no major injuries to the hand or fingers and small wounds had always healed well. After 30 years of climbing, the X-rays of the hands are completely unremarkable (Fig. 13.12).

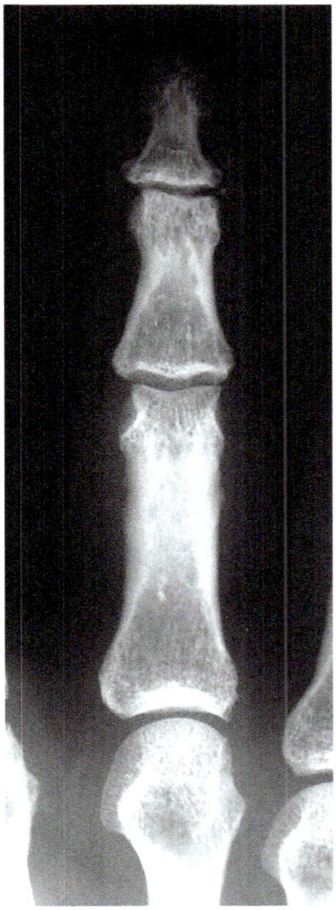

Fig. 13.11 X-ray image showing small capsule calcification in the DIP joint on the ring finger with significant widening of the end joint and less in the middle joint. Small, pointed osseous extension at the base joint. The joint spaces are wide and intact. X-rays were taken in 1998

Climber K, born in 1954, started climbing at the age of 14. For decades, he climbed sport climbing routes and alpine routes up to the lower 10th UIAA grade (up to 5.13c) of difficulty and established many classic first ascents on all continents. In the course of his career, he sustained numerous injuries to fingers and hands, often including lengthy finger swellings, many of which probably did not heal properly. Climber K

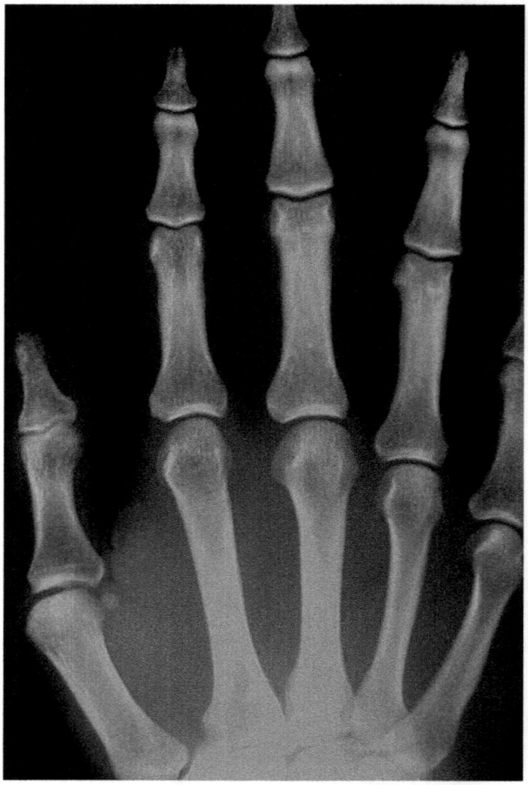

Fig. 13.12 Hand a.p. at the age of 42 years. No degenerative changes. Slightly increased sclerosis of the joint surfaces with joint base widening

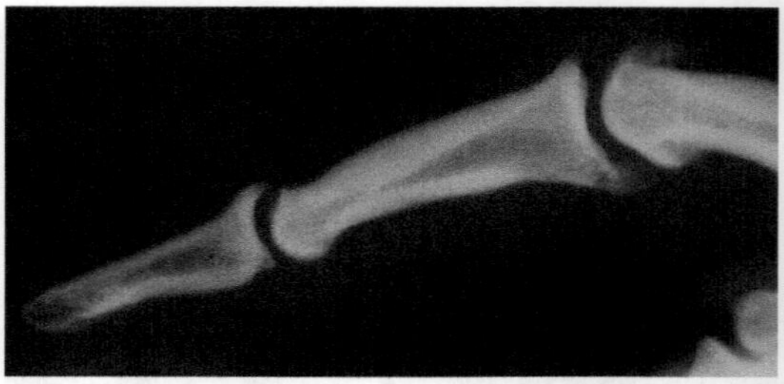

Fig. 13.13 PIP joint of a male climber (age, 54 years; climbing time, more than 30 years). The joint space is well preserved, even if it appears somewhat asymmetrical in the DIP joint and in the palmar part of the PIP joint. Noticeable large traction osteophytes at the PIP joint, which often irritate the extensor tendons and the capsule

trained with one-finger pockets and hanging exercises on small edges for many years (Fig. 13.13).

13.5 Causes and Therapy

Many studies have shown that intensive and long-lasting sporting activity lead to reactions of the soft tissues and bone [26]. Leal et al. [27] were the first to report radiographic abnormalities in the fingers and hands of rock climbers. Heuck et al. [18] and Hochholzer et al. [19] demonstrated radiographic adaptations to the high stress of climbing using radiographic and MRI analyses. They found an adaptive hypertrophy of the joint capsule in the PIP and DIP joints, thickening of the collateral ligaments, cortical hypertrophy, and a hypertrophy of up to 50% of the flexor tendons themselves. They also showed that certain reactions, such as cortical hypertrophy, could be adaptive signs of the high impact to the fingers in climbing and could not be attributed to pathologic osteoarthritic changes. Similar findings of connective tissue adaptations were lately also reported by Schreiber et al. [14]. These reactions are probably due to subperiosteal bone formation. Hahn et al. [28] found an average 6% increase in cortical thickness and a 20% decrease in cancellous bone width in high-level sport climbers compared to a control group. Our study on young climbers of the German national climbing team shows how fast these reactions occur. After only 3 years of intensive climbing training, we found first significant cortical thickening and widening of the joint surfaces [15].

No major development of osteoarthritis is to be expected in sports with cyclical movements [29]. On the other hand, axial loads and high-pressure peaks in the joints under unfavorable joint positions, such as those frequently encountered in climbing, are very stressful for the joints [26, 30, 31]. Osteoarthrotic conditions in long-term climbers are well-known, and several studies have shown that osteoarthrotic changes in the hands of rock climbers increase with the number of years and the level of climbing [1–3, 6, 8, 9, 14, 28].

The development of finger polyarthrosis in sport climbers is certainly multifactorial and depends on many different exogenous and individual genetic factors. The genetically determined connective tissue structure and possible hypermobility influence the stress on joints and thus the development of osteoarthritis.

In the field of sports, the most common cause for the development of degenerative diseases is probably overloading with microinjuries. Epiphyseal fractures in young climbers can lead to an early onset of osteoarthritis [32–34].

The crimping finger position with hyperflexion in the middle joint and a final extension in the DIP joint probably leads to microinjuries in the sensitive cartilage area [18]. There seems to be evidence that climbers show abnormal radiographic findings that have been underestimated or unhealed overload reactions with joint swelling. There may also be reason to discuss that some forms of training, such as unusually intensive use of the smallest crimps and the campus board, are more likely to cause degenerative changes. Our recent longitudinal study on the onset of osteoarthritis in young rock climbers found significant statistical influences on the development of early-onset osteoarthritis from the overall sum of training years ($p = 0.024$), use of campus board training in 1999 ($p = 0.033$), and the climbing level ($p = 0.030$) [16]. One quarter of climbers who performed at a high level in their youth showed a "mild" form of osteoarthritis (Kellgren II), at ages of under 30. Following the training regimes of the climbers for more than 10 years showed that intensive finger training (e.g., campus board training) can lead to early-onset osteoarthritis of the hand.

Therefore, the most sensible step to avoid degenerative changes of the finger joints seems to be the optimization of the load during training. This includes the avoidance of overstrain reactions with swelling of the finger joints, sensible training schedules with conscious recovery phases, and the complete healing of injuries and overstrain reactions. Stressful forms of training involving with hanging exercises on very narrow crimps or intensive training on campus boards seem to increase the overload reactions of the finger joints.

After intensive training, it is certainly advisable to "post-treat" the finger joints, for example, with hand putty, small bags filled with sand, or acupressure rings. Concentric movements of the joints without force peaks promote blood circulation and regeneration.

If complaints occur, the first measure should be a rest from climbing, must be taken until the complaints have subsided. The joint effusions of the PIP joints should always be taken and treated first. In addition to ice treatments of all kinds, topical anti-inflammatory ointments, short-term oral NSAID administration, and minimal injections of a corticoid with a corresponding break of 2–3 weeks can result in improvement in therapy-resistant cases. Chondroitin and glucosamine seem to modulate the symptoms in the long term. In the case of symptomatic complaints of the extensor hood, a single targeted injection at the point of pain in the PIP joint with a highly diluted corticosteroid can often bring long-term improvement. In few cases, surgical removal of the traction osteophytes was necessary, yielding mostly good results. Radial shockwave therapy and hand baths with anti-inflammatory sulfur solutions, as used in rheumatoid arthritis patients, also help.

In the overview of the X-rays, especially among climbers who have been practicing this sport for a long time (longer than 10 years), it is astonishing that there appear to be relatively few severe degenerative changes. Only a few climbers were found to have slightly incongruent joint spaces at the PIP and DIP joints, without the joint space being massively narrowed. It is interesting to note that the clinical changes with movement restrictions often regressed after months to years, and the radiologically minor changes were clinically insignificant.

Another interesting aspect is that some of the long-time high-level climbers display no degenerative changes at all. However, the extent to which a very early start with climbing in childhood or adolescence influences the onset of early finger arthrosis remains to be seen.

13.6 Summary

In summary, it can be said that many years of climbing in the high difficulty range leads to increased degenerative changes in the middle and end joints of the fingers. This can be seen radiologically in about 30% of patients who have climbed for more than 15 years. Clinically, even very significant changes are surprisingly rarely accompanied by serious impairments. Individual factors, intensity of climbing, number of climbing years, injuries, and the type of training methods seem to be decisive for the development of finger joint arthrosis.

References

1. Bollen SR, Wright V. Radiographic changes in the hands of rock climbers. Br J Sports Med. 1994;28(3):185–6.
2. Rohrbough JT, Mudge MK, Schilling RC, Jansen C. Radiographic osteoarthritis in the hands of rock climbers. Am J Orthop. 1998;27(11):734–8.
3. Sylvester AD, Christensen AM, Kramer PA. Factors influencing osteological changes in the hands and fingers of rock climbers. J Anat. 2006;209(5):597–609.
4. Schöffl V, Schweizer A. Sportclimbing related injuries and overuse syndromes. In: Seifert L, Wolf P, Schweizer A, editors. IRCRA handbook of climbing & mountaineering from science to performance. Research in sport and exercise science. London, New York: Routledge; 2016. p. 59–75.
5. Hochholzer T, Schöffl V. [Osteoarthrosis in finger-joints of rockclimbers] Degenerative Veränderungen der Fingergelenke bei Sportkletterern. D Z Sportmed. 2009;60(6):145–9.
6. Allenspach P, Saupe N, Rufibach K, Schweizer A. Radiological changes and signs of osteoarthritis in the fingers of male performance sport climbers. J Sports Med Phys Fitness. 2011;51(3):497–505.
7. Schöffl V, Popp D, Küpper T, Schöffl I. Injury distribution in rock climbers—a prospective evaluation of 911 injuries between 2009-2012. Wilderness Environ Med. 2015;26(1):62–7.
8. Schöffl VR, Hochholzer T, Imhoff AB, Schoffl I. Radiographic adaptations to the stress of high-level rock climbing in junior athletes: a 5-year longitudinal study of the German junior national team and a group of recreational climbers. Am J Sports Med. 2007;35(1):86–92.
9. Schöffl VR, Schöffl I. Finger pain in rock climbers: reaching the right differential diagnosis and therapy. J Sports Med Phys Fitness. 2007;47(1):70–8.
10. Morrison AB, Schöffl VR. Review of the physiological responses to rock climbing in young climbers. Br J Sports Med 2007;43(3).
11. Della Santa DR, Kunz A. [Stress syndrome of the fingers related to rock climbing]. Schweiz Z Sportmed 1990;38(1):5–9.
12. Burtscher M, Jenny E. Häufigste trainingsbedingte Beschwerden und Verletzungen bei Sportkletterern. Prakt Sport Trauma Sportmed. 1987;2:15–21.
13. Krause R, Hochholzer T. Verletzungen von Sportkletterern in der Sportambulanz. Sport Sports. 1995;9:30–3.
14. Schreiber T, Allenspach P, Seifert B, Schweizer A. Connective tissue adaptations in the fingers of performance sport climbers. Eur J Sport Sci. 2015;15(8):696–702.
15. Schöffl V, Hochholzer T, Imhoff A. Radiographic changes in the hands and fingers of young, high-level climbers. Am J Sports Med. 2004;32(7):1688–94.
16. Schöffl VR, Hoffmann PM, Imhoff A, Kupper T, Schöffl I, Hochholzer T, et al. Long-term radiographic adaptations to stress of high-level and recreational rock climbing in former adolescent athletes: an 11-year prospective longitudinal study. Orthop J Sports Med. 2018;6(9):2325967118792847.
17. Kellgren JH, Lawrence JS. Radiological assessment of osteo-arthrosis. Ann Rheum Dis. 1957;16:494–502.
18. Heuck A, Hochholzer T, Keinath C. Die MRT von Hand und Handgelenk bei Sportkletterern: Darstellung von Verletzungen und Überlastungsfolgen. Radiologe. 1992;32:248–54.
19. Hochholzer T, Heuk A, Hawe W, Keinath C, Bernett P. Verletzungen und Überlastungssyndrome bei Sportkletterern im Fingerbereich. Prakt Sport Trauma Sportmed. 1993;2:57–67.
20. Kellgren JH, J.S. L. Atlas of standard radiographs: the epidemiology of chronic rheumatism. Oxford: Blackwell Scientific Publications; 1963. p. 1–18.
21. Hochholzer T, Straub G, Schöffl V, Bayer T. Knöcherne Stressreaktionen der Hand bei jugendlichen Sportkletterern. Alp Med Rundbrief. 2013;48:20–5.
22. Schöffl V, Hochholzer T, Schöffl I. Extensor hood syndrome—osteophytic irritation of digital extensor tendons in rock climbers. Wilderness Environ Med. 2011;21(3):253–6.
23. Schlegel C, Buechler U, Kriemler S. Finger injuries of young elite rock climbers. Schweizerische Zeitschrift für "Sportmedizin und Sporttraumatologie". 2012;50(1):7–10.
24. Schöffl V, Lutter C, Wollings K, Schöffl I. Kids in adventure and extreme sports: too risky? Rock climbing. Res Sports Med. 2018. (In press).
25. Pastor T, Fröhlich S, Spörri J, Schreiber T, Schweizer A. Cartilage abnormalities and osteophytes in the fingers of elite sport climbers: an ultrasonography-based cross-sectional study. Eur J Sport Sci. 2020;20(2):269–76.
26. Warden SJ, Fuchs RK, Castillo AB, Turner CH. Does exercise during growth influence osteoporotic fracture risk later in life? J Musculoskelet Neuronal Interact. 2005;5(4):344–6.
27. Leal C, Rane A, Herrero R. Soziologie, Trainingszeit und Fingerverletzungen beim Sportklettern. Prakt Sport Trauma Sportmed. 1987;2:44–6.

28. Hahn F, Erschbaumer M, Allenspach P, Rufibach K, Schweizer A. Physiological bone responses in the fingers after more than 10 years of high-level sport climbing: analysis of cortical parameters. Wilderness Environ Med. 2012;23(1):31–6.
29. Leadbetter W, Schneider M. Cycling injury. In: Krausz J, van der Reis Krausz V, editors. Bicycling book. Boston: Dial Press; 1982.
30. Mote CD. The forces of skiing and their implications to injury. Int J Sport Biomech. 1987;3:309–25.
31. Spitzenpfeil P, Mester J. Vibrationsbelastungen beim alpinen Skilauf. Sport Ortho Sport Trauma. 1997;13:209–12.
32. Bartschi N, Scheibler A, Schweizer A. Symptomatic epiphyseal sprains and stress fractures of the finger phalanges in adolescent sport climbers. Hand Surg Rehabil. 2019;38(4):251–6.
33. Schöffl I, Schöffl V. Epiphyseal stress fractures in the fingers of adolescents: biomechanics, pathomechanism, and risk factors. Eur J Sports Med. 2017.
34. Hochholzer T, Schöffl VR. Epiphyseal fractures of the finger middle joints in young sport climbers. Wilderness Environ Med. 2005;16(3):139–42.

Pediatric Aspects in Young Rock Climbers

14

Isabelle Schöffl and Volker Schöffl

Luc Schöffl in Halfway Log Dump, Ontario, Canada, Photo Michael Simon

I. Schöffl (✉)
Department of Pediatric Cardiology, Friedrich-Alexander Universität Erlangen-Nürnberg, Erlangen, Bayern, Germany

V. Schöffl
School of Clinical and Applied Sciences, Leeds Becket University, Leeds, UK

Department of Orthopedic and Trauma Surgery, Center of Sportsmedicine, Klinikum Bamberg, Bamberg, Germany

Section of Wilderness Medicine, Department of Emergency Medicine, University of Colorado School of Medicine, Denver, CO, USA

Department of Orthopedic and Trauma Surgery, Friedrich-Alexander Universität Erlangen-Nürnberg, Erlangen, Germany

V. Schöffl et al. (eds.), *Climbing Medicine*, https://doi.org/10.1007/978-3-030-72184-8_14

14.1 Significance of Climbing for the Early Child Development

Exercise plays a central role for the physical and mental development of children. Climbing represents an exceptionally adequate exercise modality as climbing promotes several developments at once.

First of all, climbing combines endurance with power [1]. On the one hand, climbing therefore influences the maximal functional capacity in children (VO_2peak), a parameter, which has proven to be the best predictor for overall mortality and morbidity [2]. On the other hand, the fact that climbing is a sport in which the whole body is concerned, no muscle groups are at a disadvantage. However, muscular disbalances need to be kept in mind (Fig. 14.1).

In addition to its power and endurance characteristics, climbing demands a high amount of technical skills, balance, and coordination [3]. Especially, the motoric and cognitive promotion represents a distinctiveness which adds many benefits to a child's development. Interestingly, children know naturally how to move in the vertical environment. A lot of technical skills, like keeping one's central weight below the holding hand or how to turn one's body as to minimize the weight having to be supported, have to be laboriously learned by grown-ups, whereas children move correctly right from the start without ever being instructed. The movement engrams being learned during climbing can then improve locomotion in daily life. In a study investigating the effects of a climbing intervention in children with infantile cerebral palsy was able to show that children improved their general functional capacity within 3 weeks [4].

Finally the mental and psychological aspects of climbing need to be discussed. Climbing demands a high amount of responsibility. The ability to judge real dangers and differentiate objective danger from irrational fear from falls is important and needs to be learned by the athlete [5] (Fig. 14.2). Whereas climbing has been deemed safe in several studies [6–8], certain precautions are necessary, and children need to be made aware of the objective dangers of the sport. Route-reading skills and a high demand on concentration [9] are a further advantage for the mental development of children.

Overall, climbing needs to adapt to the circumstances and respective capabilities of each child. The physical-mental development of small children often develops in large steps rather than

Fig. 14.1 Climbing advances the whole musculature

Fig. 14.2 Full concentration and responsibility already at an early age

in a continuous pattern. Physical, mental, and educational changes have a greater impact on the behavior and competence of children than usually observed in adults. In a sport, where a high amount of individual responsibility, self-assessment, and motoric capabilities is important, special attention must be paid to the development of each individual.

14.2 Specifics of Climbing with Children

The most important thing when climbing with children is safety. When climbing, the athlete places his/her life in the hands of another person, namely, the one holding the rope at the other end. This responsibility represents a special challenge for children but also teaches them a better comprehension of reality. Climbing, thus is the exact contrary of computer games, where actions performed in the game world lead to no comprehensible consequences for the child [10]. In order to learn from the climbing experience, children need to be able to assess themselves correctly and deduce the correct decisions. They need to learn to judge risks and act accordingly. This responsibility needs to be introduced slowly and according to their respective skills which need to be reevaluated constantly. Even skills the child had seemingly already acquired can change over the course of development and deteriorate. As an example, most children can understand that they shouldn't be playing underneath bouldering adults around the age of 6 years. However, when surrounded by pears and in a stressful or exciting situation, the importance of such rules is easily forgotten. However, if the child is slowly introduced to personal responsibility, a heightened self-esteem can be the result (Fig. 14.3).

The age, at which children can start belaying others has declined with the use of semi-automatic belaying devices. Under surveillance of adults with an added belay device, an 8-year-old child can, when adequately developed, belay another child of comparable weight in "top-rope." Lead-belaying of a weight-comparable peer can be performed, dependent on mental development, at the age of 12 [11]. Again, constant reevaluation of the skills of the concerned child is essential, always bearing in mind that regression of concentration can occur at any time.

Fig. 14.3 Even children can take responsibility, but it is essential to follow safety guidelines. Responsible "spotting" and crash pads are necessary

With respect to the need for movement, we need to differentiate between children (up to 14 years of age) and adolescents (14–18 years of age). Especially up to the 12th birthday, body perception and movement knowledge should be in the center of the physical performance. Sometimes, it may be necessary to limit the movement compulsion in order to prevent one-sided strain or overuse syndromes. However, in most cases, children rarely perform exercises that cause pain or overuse. Physically motivated adolescents are different. Early disbalances and muscular contractions need to be observed early on and treated (s. below). It is difficult to give overall recommendations for certain age-groups as age does not always conform with development. Even though a team may consist of the same age-groups, their individual developments could differ over a wide range. This is even more pronounced in mixed teams as girls tend to be ahead of boys in their mental and physical development by one and a half to 2 years. However, even within the same gender, there are substantial differences: a 15-year-old boy can still be on the developmental stage of puberty with open growth-plates and a body of a child with subsequent coordination deficits as a consequence of "unharmonic growth." Another adolescent of this age may be already bodily mature and experience none of these difficulties. The maturity can be determined using the skeletal age (X-ray of the hand for determining biological age). If there is concern regarding developmental delay, endocrinological workup may be necessary. As restrictive energy intake is a cause for developmental delay, a proper workup with respect to the diet and mental status of the athlete is essential. A good solution for following the development is keeping growth, weight, and BMI curves of each athlete over time.

14.3 Training with Children and Adolescents

Main principal when training with children and adolescents is a high fun aspect and keeping the training age-appropriate. There are no concerns about cardiopulmonary overuse to date [12]. Even though, the cardiopulmonary function of each child/adolescent needs to be checked at regular intervals (s. Chap. 19, Sportsmedical Supervision of Climbers), an optimal climbing-specific training sets the necessary foundations at the right moments. A sport-specific power-training should be postponed until the skeletal growth is completed and climbing-specific coordination is already well developed. Sport-specific power training with high intensities, like campus boarding, should be avoided when working with children or adolescents [6]. As a consequence of a very high strength to body-weight ratio, such training is unnecessary and highly questionable for the athlete's health. A study investigating the correlation between high-intensity campus boarding and the development of early arthrosis in the finger joints was able to prove movement constraints in finger extension [13]. Often, adolescents try to mimic the methods applied by high-level adult athletes, not knowing about the possible consequences. Even though high-level athletes may be able to tolerate such training modalities, they could be detrimental to the undeveloped skeletal system of adolescents. During childhood and adolescence, the main focus in training should therefore lie on learning motoric skills and improving movement coordination. When setting routes, the endurance aspect should be more pronounced than the ability to hold on to extremely small holds. The sensitive areas are the growth plates, the insertions of the tendons on the bone, and the relatively soft joint cartilage. Therefore, the most common injuries are epiphyseal fractures and disturbances [14], especially at or around of the time of the pubertal growth spurt [15]. This injury is explained in detail in the chapter regarding finger injuries and was lately found more often in adolescents training for speed climbing [16]. In rare cases, we have observed osteonecrosis of other bones, such as the capitulum humeri (M. Panner) (Fig. 14.4.) or the

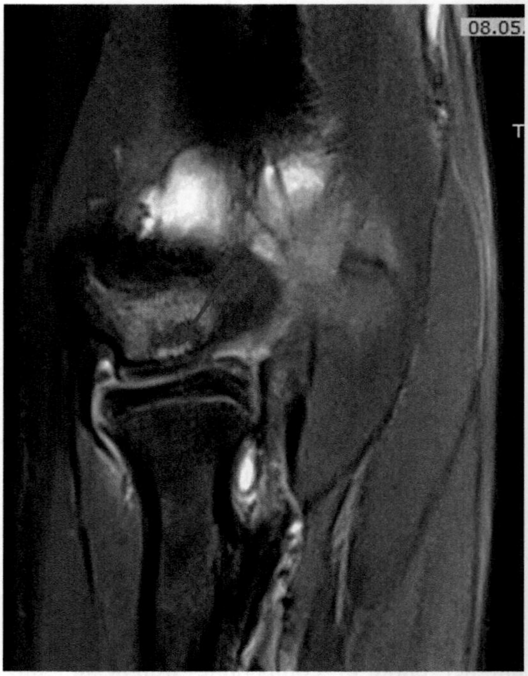

Fig. 14.4 Osteonecrosis of the capitulum humeri—M. Panner in a 12-year-old climber

metacarpalia. In these cases, it is difficult to be certain whether climbing itself was responsible for these rare cases or whether climbing only helped demask the pathology, and the cause is probably multifactorial. Special attention needs to be paid on the proper development of the spine in order to detect early rounding in the sense of a climbers back [17] or scoliosis. Functional "scapula alatae" can be observed regularly and most commonly correct themselves, whereas special caution is needed with respect to gliding vertebrae.

Another point that needs to be discussed when working with children in climbing is the proper choice of shoes. Even though adults have the tendency to choose especially small and tight shoes, this does not work with children. Climbing shoes should never be too small or too tight as they could restrict bone growth (Figs. 14.5 and 14.6).

In addition to that, a climbing backpack has to be adapted to the child's weight and height. Even though the school bag of many children may be exceptionally heavy and large, this does not mean they can carry the same amount in an alpine expedition. The weight of a climbing backpack should not surpass 10% of the child's bodyweight so as not to hinder the growing skeletal system [18] (Fig. 14.7).

14.4 Conclusion

Climbing represents a sport which is beneficial for the health and development of children and adolescents (Fig. 14.8). Besides training, the skeletal musculature also has a cardiopulmonary effect, though to a lesser extent. The intensity and volume of the training should be adapted to the biological age of each athlete. High-intensity training modalities with added weights should be avoided before the end of puberty.

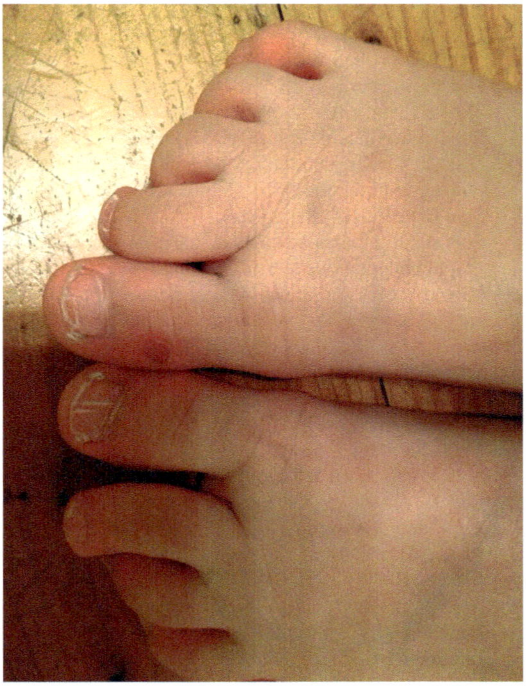

Fig. 14.5 Climbing shoes in children should never be too tight. Pressure marks as seen in this 6-year-old must be avoided

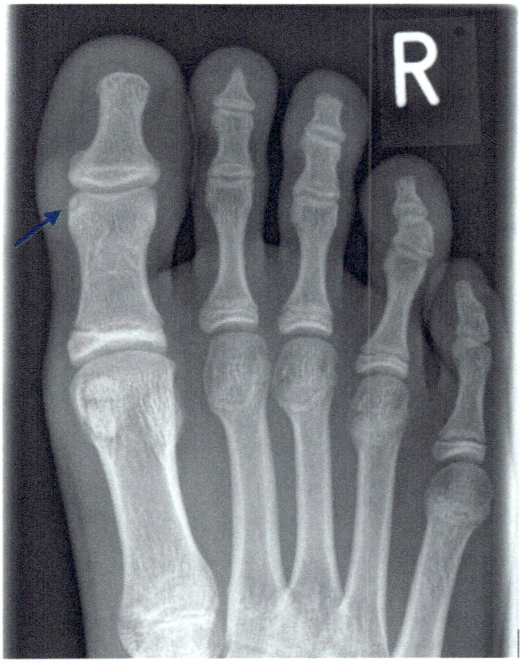

Fig. 14.6 Stress fracture—osteonecrosis of the bunion of a 16-year-old climber

Fig. 14.7 Children have no subjective risk assessment—it is therefore the responsibility of the person in charge to assure the safety in case of a fall. What may look spectacular in this picture is in reality sufficiently assured with mats and spotters

Fig. 14.8 Children enjoy bouldering. (Photo: Kilian Reil)

References

1. Cutts A, Bollen SR. Grip strength and endurance in rock climbers. Proc Inst Mech Eng H. 1993;207(2):87–92.
2. Laukkanen JA, Zaccardi F, Khan H, Kurl S, Jae SY, Rauramaa R. Long-term change in cardiorespiratory fitness and all-cause mortality: a population-based follow-up study. Mayo Clin Proc. 2016;91(9):1183–8.
3. Orth D, Davids K, Seifert L. Coordination in climbing: effect of skill, practice and constraints manipulation. Sports Med. 2016;46(2):255–68.
4. Schram Christensen M, Jensen T, Voigt CB, Nielsen JB, Lorentzen J. To be active through indoor-climbing: an exploratory feasibility study in a group of children with cerebral palsy and typically developing children. BMC Neurol. 2017;17(1):112.
5. Giles LV, Rhodes EC, Taunton JE. The physiology of rock climbing. Sports Med. 2006;36(6):529–45.
6. Schöffl V, Lutter C, Woollings K, Schöffl I. Pediatric and adolescent injury in rock climbing. Res Sports Med. 2018;26(sup1):91–113.
7. Woollings KY, McKay CD, Kang J, Meeuwisse WH, Emery CA. Incidence, mechanism and risk factors for injury in youth rock climbers. Br J Sports Med. 2015;49(1):44–50.
8. Woollings KY, McKay CD, Emery CA. Risk factors for injury in sport climbing and bouldering: a systematic review of the literature. Br J Sports Med. 2015;49(17):1094–9.
9. Bourdin C, Teasdale N, Nougier V. Attentional demands and the organization of reaching movements in rock climbing. Res Q Exerc Sport. 1998;69(4):406–10.
10. Schöffl I, Schöffl V. Kann das noch gesund sein. In: Alpenverein D, editor. Berg 2018. München: Deutscher Alpenverein; 2017. p. 107–13.
11. Hochholzer T, Schöffl V. Soweit die Hände greifen. 4te ed. Ebenhausen: Lochner Verlag; 2014.
12. Schöffl I, Wüstenfeld J, Jones G, Dittrich S, Lutter C, Schöffl V. Cardiopulmonary function of adolescent elite sport climbers. 2020. (In press).
13. Schöffl VR, Hoffmann PM, Imhoff A, Kupper T, Schöffl I, Hochholzer T, et al. Long-term radiographic adaptations to stress of high-level and recreational rock climbing in former adolescent athletes: an 11-year prospective longitudinal study. Orthop J Sports Med. 2018;6(9):2325967118792847.
14. Schöffl V, Popp D, Küpper T, Schöffl I. Injury distribution in rock climbers—a prospective evaluation of 911 injuries between 2009-2012. Wilderness Environ Med. 2015;26(1):62–7.
15. Schöffl I, Schöffl V. Epiphyseal stress fractures in the fingers of adolescents: biomechanics, pathomechanism, and risk factors. Eur J Sports Med. 2017;3(1). (In press).
16. Meyers RN, Howell DR, Provance AJ. The association of finger growth plate injury history and speed climbing in youth competition climbers. Wilderness Environ Med. 2020;31(4):394–9.
17. Förster R, Penka G, Bosl T, Schöffl VR. Climber's back—form and mobility of the thoracolumbar spine leading to postural adaptations in male high ability rock climbers. Int J Sports Med. 2009;30(1):53–9.
18. Küpper T. Non-traumatic aspects of sport climbing. Wien Med Wochenschr. 2005;155(7–8):163–70.

Climbing in Older Athletes

15

Christoph Lutter and Volker Schöffl

Volker Schöffl in Mill Creek Canyon, Moab, Utah, USA, Photo Michael Simon

C. Lutter (✉)
Department of Orthopedic Surgery, University Hospital Rostock, Rostock, Germany

V. Schöffl
Department of Orthopedic and Trauma Surgery, Center of Sportsmedicine, Klinikum Bamberg, Bamberg, Germany

School of Clinical and Applied Sciences, Leeds Becket University, Leeds, UK

Section of Wilderness Medicine, Department of Emergency Medicine, University of Colorado School of Medicine, Denver, CO, USA

Department of Orthopedic and Trauma Surgery, Friedrich-Alexander Universität Erlangen-Nürnberg, Erlangen, Germany

V. Schöffl et al. (eds.), *Climbing Medicine*, https://doi.org/10.1007/978-3-030-72184-8_15

15.1 Introduction

Physical activity has been established to be one of the most influential contributing factors to remaining healthy in later stages of life. While many population structures suffer from obesity, immobilization, and improper nutritional behavior, many older adults are very active in sports, some even in the higher age ranges of the population. Rock climbing is one of the sports that are well known to engage athletes on a life-long basis. The sport has proven to have health-promoting effects and an overall low injury hazard in the past [1–4]. However, modern climbing and training methods have changed, and a wider spectrum of climbing-related injuries can now be detected [5]. Several studies on climbing injuries have been conducted in the past, focusing on injury grading and incidences as well as on diagnostic and treatment strategies [6]. However, injury surveillance studies for rock climbing injuries in older athletes are still rare [7]. Furthermore, the rapid increase of athletes in today's generation of climbers aged between 20 and 30 will likely lead to a wave of elder athletes in the future.

Available studies underlined the fact that overall climbing injury incidence is minor and that most injuries are of a minor grade [4, 6]. Acute injuries in climbing are related to a fall onto the lower leg or from performing a hard move and injuring the upper extremities [3, 8]. Beyond this, overuse injuries can result from permanent undue stress and overload, mainly to the upper extremities. Most climbing-related scientific studies tend to focus on sport-specific injuries such as flexor tendon pulley injuries, since these injuries occur the most frequently [6, 9]. Expecting a rising number of very active, older rock climbing athletes due to the increase in popularity of climbing and bouldering, it is important to further analyze this patient collective [8].

15.2 Acute and Overuse Injuries

A recently published study investigated 198 patients ≥35 years of age (150 [76%] male, 48 [24%] female) with a total number of 275 independent injuries caused by rock climbing or bouldering [10]. Among the 275 injuries, 187 (68%) were overuse injuries and 88 (32%) were acute injuries. The percentage of bouldering in the climbing routine and training was stated by the athletes as 44.5 ± 6.6 (0–100). None of the athletes in the age group >65 years of age reported partaking bouldering on a regular basis. Representing 15% of all injuries, a finger pulley laceration (single or multiple pulley rupture) was the most frequent diagnosis, followed by finger joint capsulitis (12%), subacromial impingement (10%), and finger tenosynovitis (7%) (Table 15.1). With a ratio of three to one, a significantly higher rate of injured male climbers was found compared to injured female climbers. Climbing levels and climbing experience were both significantly higher in males. Thereby, sex did not influence the severity of injury. These findings are in accor-

Table 15.1 Ten most frequent injuries in the older rock climbing athlete [10]

Injuries	All (n = 275)	Age 35–49 (n = 217)	Age 50–64 (n = 54)	Age ≥ 65 (n = 4)
Pulley injury	41 (15%)	33 (15%)	8 (15%)	None
Capsulitis (finger)	34 (12%)	27 (12%)	6 (11%)	1 (25%)
Impingement (shoulder)	28 (10%)	19 (8%)	8 (15%)	1 (25%)
Tenosynovitis	20 (7%)	19 (8%)	1 (2%)	None
SLAP tear	19 (6%)	17 (7%)	2 (4%)	None
Epicondylitis	15 (6%)	13 (5%)	2 (4%)	None
Osteoarthritis (finger)	12 (4%)	8 (4%)	4 (7%)	None
Dupuytren disease	9 (3%)	6 (3%)	3 (6%)	None
Strain finger joint capsule	9 (3%)	8 (4%)	1 (2%)	None
Strain finger flexor tendon	8 (3%)	7 (3%)	1 (2%)	None

SLAP superior labral tear from anterior to posterior

dance with most other available data [3, 4, 9]. The average UIAA injury score was overall low (2.0 ± 0.3 [1–4]). None of the athletes suffered a UIAA grade 5 injury, and none of them died (UIAA 6). Climbing level (UIAA level, $p < 0.001$) and climbing experience (climbing years, $p < 0.001$) were both significantly higher in men than in women. Against the findings of two previously published studies, we could not confirm a correlation between the climbing difficulty (UIAA score) and the development of overuse injuries [3, 4]. This finding is remarkable, as one would expect overuse injuries to be most frequent in very old athletes with decades of climbing history. No significant association between the climber's UIAA climbing level, climbing hours/week, years of climbing, height, weight, and sex and the development of an acute or overuse injury was found. The UIAA injury score was also not significantly influenced by any of these individual parameters.

Athletes affected by acute injuries did not have significantly higher UIAA scores than those affected by overuse injuries. As reported in investigations of younger rock climbing athletes, the vast majority of injuries in the study (90%) affected the upper extremities [2, 6, 9]. In a prospective analysis of more than half a million entries to an indoor climbing and bouldering gym, the authors found an injury rate of 0.02/1000 h, of which only 15 were of UIAA injury scale grade 2, 13 UIAA 3, and 2 UIAA 4 [11]. No higher graded injuries (UIAA injury scale) were reported, and no fatalities occurred in the study. The severity of injuries in older athletes is also mostly low, with a UIAA score of 2 (defined as "moderate severe injury,...") in 93% of all cases [12]. Only six acute cases were classified as UIAA grade 3 injuries, and one polytraumatic patient was classified as a UIAA grade 4 injury. This overall minor grade of injury severity is mostly in accordance with the previously published data. It is notable that there is a slight increase of more severe injuries graded UIAA 3 and higher compared to prior studies on younger athletes [3, 4]. Contrary to expectations that more severe injuries would be more probable in younger, more aggressive, and occasionally reckless athletes, the study might confirm the recent trend in which we have seen an overall shift into more serious injuries over the last 5 years [5]. This finding is most likely caused by indoor bouldering and the rising number of people being enthusiastic about this sport.

15.3 Degenerative Conditions

Among overuse injuries, degenerative conditions were defined as non-reversible injuries that had worsened over time. Subgroup analysis showed a weak correlation of the climber's age with the development of a degenerative condition (Tables 15.1 and 15.2). Among the degenerative condi-

Table 15.2 Most frequent degenerative conditions in the older rock climbing athlete [10]

Injuries	All ($n = 88$ degenerative injuries)	Age 35–49 ($n = 63$ degenerative injuries)	Age 50–64 ($n = 24$ degenerative injuries)	Age ≥ 65 ($n = 1$ degenerative injury)
Impingement (shoulder)	28 (32%)	19 (29%)	8 (33%)	1
SLAP tear	19 (22%)	17 (27%)	2 (8%)	None
Osteoarthritis (finger)	12 (14%)	8 (13%)	4 (17%)	None
Dupuytren disease	9 (10%)	6 (10%)	3 (13%)	None
Biceps tendon rupture (LBT)	6 (7%)	4 (9%)	2 (8%)	None
Osteoarthritis elbow	6 (7%)	3 (5%)	3 (13%)	None
Osteoarthritis foot	4 (5%)	4 (6%)	None	None
Other	4 (5%)	2 (3%)	2 (8%)	None

SLAP superior labral tear from anterior to posterior; *LBT* long biceps tendon

tions, subacromial impingement syndrome was the most frequent diagnosis (32%), followed by chronic SLAP-lesions (superior labral tear from anterior to posterior, 22%), osteoarthritis of the fingers (14%), and Dupuytren's contracture (10%) (Table 15.2, Fig. 15.1). The subacromial impingement syndrome (Table 15.1, 10% of all patients) was found to occur twice as frequently compared to younger athletes in existing literature [4]. This is most likely influenced by the overall higher prevalence of impingement syndrome in older patients [13]. However, osteoarthritis of the finger has previously described as representing only 3% of all finger injuries in younger rock climbers [4] (Figs. 15.1 and 15.2). While this condition is the seventh leading cause of complaints in older rock climbers overall, it represents the third leading diagnosis among the degenerative conditions (Table 15.2). Our findings are in agreement with a recently published study on long-term radiographic adaptations of the fingers to the stress of long-time high-level and recreational rock climbing [14].

15.4 Conclusions

While the injury distribution in older athletes is overall comparable to that of younger athletes, degenerative conditions such as impingement syndrome of the shoulder or osteoarthritis of the fingers are common injuries in older athletes. We therefore recommend careful, sport-specific medical supervision and evidence-based education under consideration of age-specific factors to prevent injuries among all age groups and to decrease the number of degenerative injuries in older rock climbers.

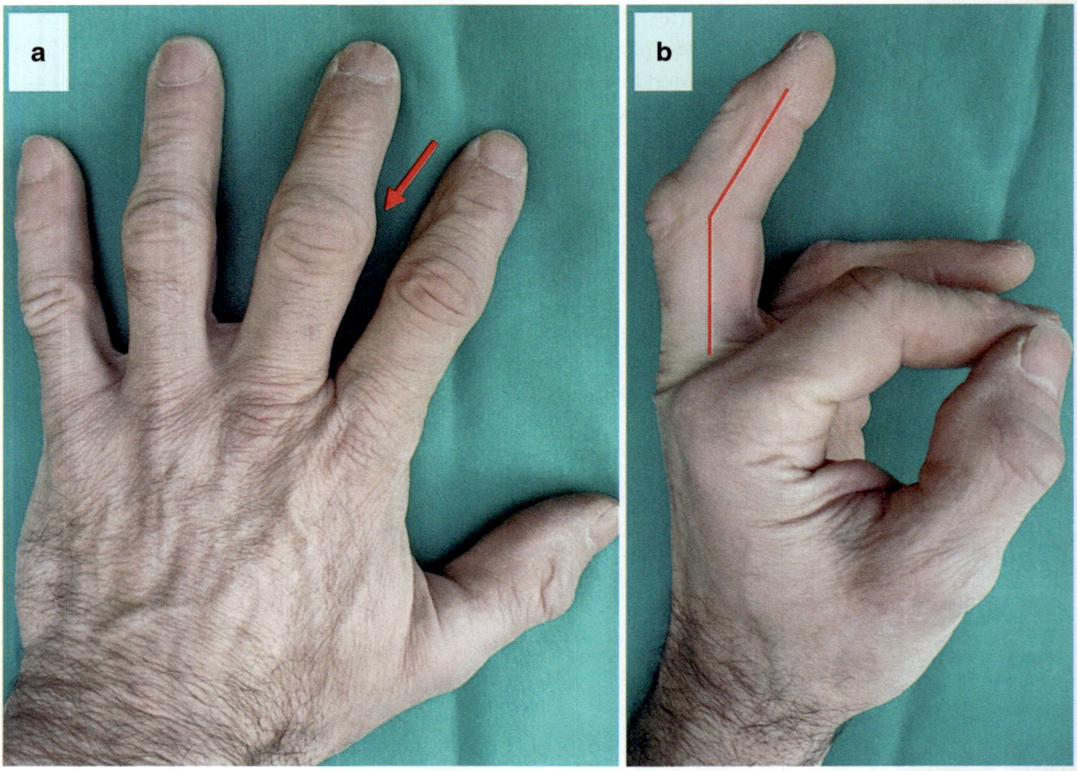

Fig. 15.1 (**a**) Left hand of a long-time senior rock climber (experience, approximately 40 years; UIAA level, X) with distinct distension of the PIP (red arrow) and DIP joints. (**b**) In addition, there is a clear extension deficit of the fingers, here approximately 40° extension deficit of the Dig. III

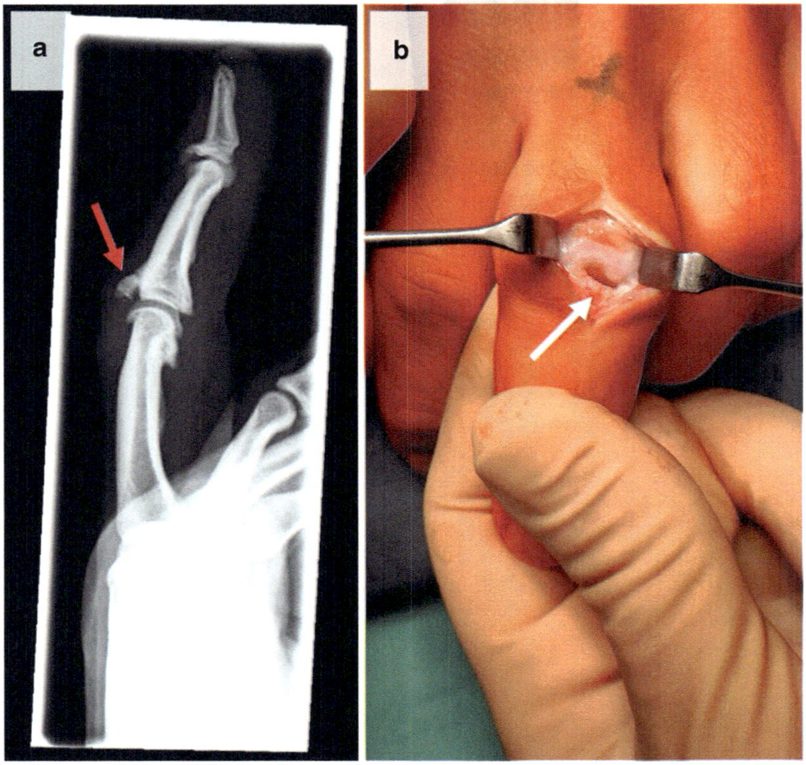

Fig. 15.2 (**a**) X-ray of a 38-year-old male rock climber with severe bone spurs (osteophytes) at the dorsal aspect of the PIP joint (see arrow). A similar finding can be noticed in the DIP joint area. (**b**) Surgical treatment with removal of bone spurs around the PIP joint. The arrow marks the site of the resected osteophyte

References

1. Dores H, de Araujo GP, Monge J, Costa R, Tata L, Malhotra A, et al. Subclinical coronary artery disease in veteran athletes: is a new preparticipation methodology required? Br J Sports Med. 2018;54(6):349–53.
2. Neuhof A, Hennig FF, Schöffl I, Schöffl V. Injury risk evaluation in sport climbing. Int J Sports Med. 2011;32(10):794–800.
3. Schöffl V, Hochholzer T, Winkelmann HP, Strecker W. Pulley injuries in rock climbers. Wilderness Environ Med. 2003;14(2):94–100.
4. Schöffl V, Popp D, Küpper T, Schöffl I. Injury trends in rock climbers: evaluation of a case series of 911 injuries between 2009 and 2012. Wilderness Environ Med. 2015;26(1):62–7.
5. Schöffl V, Lutter C. The "newbie" syndrome. Wilderness Environ Med 2017;28(4):377–380.
6. Woollings KY, McKay CD, Emery CA. Risk factors for injury in sport climbing and bouldering: a systematic review of the literature. Br J Sports Med. 2015;49(17):1094–9.
7. van Mechelen W, Hlobil H, Kemper HC. Incidence, severity, aetiology and prevention of sports injuries. A review of concepts. Sports Med. 1992;14(2):82–99.
8. Lutter C, El-Sheikh Y, Schöffl I, Schöffl V. Sport climbing: medical considerations for this new Olympic discipline. Br J Sports Med. 2017;51(1):2–3.
9. Nelson NG, McKenzie LB. Rock climbing injuries treated in emergency departments in the U.S., 1990-2007. Am J Prev Med. 2009;37(3):195–200.
10. Lutter C, Hotfiel T, Tischer T, Lenz R, Schöffl V. Evaluation of rock climbing related injuries in older athletes. Wilderness Environ Med. 2019;30(4):362–8.
11. Schöffl VR, Hoffmann G, Küpper T. Acute injury risk and severity in indoor climbing-a prospective analysis of 515,337 indoor climbing wall visits in 5 years. Wilderness Environ Med. 2013;24(3):187–94.
12. Schöffl V, Morrison A, Hefti U, Ullrich S, Küpper T. The UIAA medical commission injury classification for mountaineering and climbing sports. Wilderness Environ Med. 2011;22(1):46–51.
13. Garving C, Jakob S, Bauer I, Nadjar R, Brunner UH. Impingement syndrome of the shoulder. Deutsches Arzteblatt Int. 2017;114(45):765–76.
14. Schoffl VR, Hoffmann PM, Imhoff A, Kupper T, Schoffl I, Hochholzer T, et al. Long-term radiographic adaptations to stress of high-level and recreational rock climbing in former adolescent athletes: an 11-year prospective longitudinal study. Orthop J Sports Med. 2018;6(9):2325967118792847.

Part VI

General Medical Considerations

16 Anorexia Athletica and Relative Energy Deficiency

Isabelle Schöffl and Volker Schöffl

Michael Simon in Kalymnos, Photo Archive Michael Simon

I. Schöffl (✉)
Department of Pediatric Cardiology, Friedrich-Alexander-Universität Erlangen-Nürnberg, Erlangen, Bayern, Germany

V. Schöffl
School of Clinical and Applied Sciences, Leeds Becket University, Leeds, UK

Department of Orthopedic and Trauma Surgery, Center of Sportsmedicine, Klinikum Bamberg, Bamberg, Germany

Section of Wilderness Medicine, Department of Emergency Medicine, University of Colorado School of Medicine, Denver, CO, USA

Department of Orthopedic and Trauma Surgery, Friedrich-Alexander Universität Erlangen-Nürnberg, Erlangen, Germany

V. Schöffl et al. (eds.), *Climbing Medicine*, https://doi.org/10.1007/978-3-030-72184-8_16

16.1 Introduction

Several sporting disciplines have a recognized benefit from low body weight, namely, sports with weight categories like martial arts and sports in which the body weight needs to be moved against gravity like gymnastics and of course sport climbing. As a consequence, some climbers control their weight through at least one unhealthy way [1]. This is called relative energy deficiency in sports or RED-S and can have serious long-term complications: dental caries, osteoporosis, renal insufficiency, anemia, and also psychiatric conditions. Even though RED-S, formerly known as anorexia athletica, is not recognized as a true disorder, it can lead to psychiatric disorders like anorexia nervosa or depression. Some of the athletes are so convinced of the dependence of their weight on their performance, and this wrong belief is so firmly ingrained that they suffer their whole life from the mental consequences.

So far there are no restrictions on athletes diagnosed with RED-S by the IFSC, as long as it is medically justifiable, especially as restricting the athlete could lead to a serious aggravation of the athlete's situation. Instead of receiving help from his or her federation, the athlete could become isolated increasing their risk of serious disorders following in the wake of RED-S. However, a regular monitoring of BMI is being conducted during world cup competitions. If concerns are raised, the IFSC can oblige the athlete to get a medical check-up [2]. An early detection of possible warning signs and support for the athlete must be a high priority in order to avoid long-term damage.

16.2 RED-S (Relative Energy Deficiency in Sport)

Relative energy deficiency in sport (RED-S) describes a physiology as a consequence of relative energy deficiency and includes limitations to the metabolism, the menstrual cycle, the bone health, the immune system, the protein synthesis, and the cardiopulmonary function [3]. In theory, every athlete is at an increased risk of RED-S. However, some sports present a higher risk factor. These are sports with an aspect of aesthetics and appearance, weight-class sports, endurance sports [3], and sports in which the own body weight needs to be moved against gravity (climbing, gymnastics). In 2014, the IOC published a "consensus statement" in which all the signs and risks of RED-S were explained [4]. One year late, a diagnostic tool was developed, the RED-S CAT [3].

16.3 RED-S in Climbing

An optimal balance between low body weight, or a low body fat percentage, respectively, with still a high amount of strength plays an essential role in climbing [5]. This is a consequence of the fact that the own body weight needs to be propelled against gravity often with only the fingers doing the whole work. The endurance aspect in climbing is rather low compared to endurance sports but probably comparable to team sports or martial arts [6, 7]. Training, competition, and even climbing outdoors most likely resemble interval training [8]. Restricting the diet and reducing especially the amount of incorporated fatty acids in order to keep the body fat low is therefore a regular occurrence in climbers as the need for a sufficient amount of nutrients is not felt as much as in endurance sports. Another option is using an endurance sport for losing weight and body fat, a method employed by many climbers excessively.

16.3.1 Energy and Nutrient Supply in Sport Climbers

Male elite climbers have a lower BMI than elite athletes from other sports, even when compared to long-distance runners [1, 9, 10]. The mean energy supply in climbers is comparable to elite athletes from other sports in which the body weight needs to be low (gymnastics, dancing) or sports with weight classes (wrestling, martial arts) and is decidedly lower than in endurance athletes [1]. However, endurance athletes also have a much higher energy demand as a consequence of their specific training regimes, and therefore a direct comparison is not feasible. In addition to that, it is not only essential how much food is consumed but its components. A study on climbers nutritional intake was able to show that a small part comes from visible fat or meat (12% of the climbers vs. 60% general pediatric population [1]). This is less than in any other sport discipline [1]. Especially the female elite climbers tend to restrict their calories with a very low protein content leading to very low protein and caloric levels in their blood [1]. Furthermore, the composition of the food is mainly inappropriate. The combination of little food intake with high training volumes leads to malnutrition on the one hand and sudden ravenousness which can only be satisfied with fast food low on nutrients [1]. This results in a shortage of micronutrients.

16.3.2 Correlations Between Climbing Performance and BMI/Body Fat

The influence of a low body weight on the climbing performance needs to be questioned critically. Several studies have shown that good climbers tend to be exceptionally small and thin [1, 5, 11, 12]. However, it is not possible to draw the conclusion that there exists a causality between body weight and climbing level. It also needs to be stressed that especially the mental belief that a low body weight can influence the climber's performance is so deeply embedded in some climbers that it becomes a self-fulfilling prophecy. Even though this may seem harmless at first, the energy and nutrient restrictions following such beliefs can cause long-term impairment of health and performance and may therefore limit the performance of the athlete in the long term.

16.3.3 Eating Disorders in Climbers

The belief that a low body weight represents the key to success in climbing is widely spread among climbers. However, are this belief and the actions taken in order to reach certain weight goals already considered a disorder?

The full scale of anorexia nervosa or bulimia nervosa is a recognized psychiatric disorder with clear diagnostic criteria. Climbers have many characteristics in common with people who have a tendency toward eating disorder. They are perfectionists, reliant, and ready to work hard, have a high need for approval, can tolerate pain and discomfort well, and plan their eating manners according to the demands of their sport [1]. On the other hand, there are some characteristics that do not fit: low self-esteem coupled with being discontent with one own performance; a distinct distorted perception of the own body coupled with a fear of gaining weight; a rigid eating and training regimen, which is kept up during season breaks. These characteristics lead to athletes who can have successful careers in spite of their disorders and never acknowledge their disease [1, 13]. In some cases, athletes have even taken laxatives and diuretics in order to lose weight [13].

RED-S is observed more frequently in women than in men and occurs mainly between the ages of 15–25. However, we've had male athletes who were affected severely.

The consequences of restricting one's diet can be severe: amenorrhea, stress fractures, osteoporosis, changes in metabolism, kidney and liver conditions, depression, and other psychiatric diseases. In many cases, the self-restricting diet leads to permanent damage as has been reported by some athletes openly. Angela Eiter, former world champion in lead climbing reported openly about her ongoing problems with the topic in a European Newspaper (Neue Zürcher Zeitung). After having been asked to lose some weight in order to be more performant in climbing, she was more successful, and this has impacted on her entire eating habits to this day.

Since a low body weight is considered normal in climbing, a pronounced vigilance is necessary from coaches, physicians, and peers in order to detect early warning signs.

16.4 Definition RED-S CAT and Its Implementation

The RED-S CAT is a clinical assessment tool for the evaluation of athletes/active individuals suspected of having relative energy deficiency [3]. It is meant for use by medical professionals in the clinical evaluation and management of athletes with this syndrome [3]. The goal is to screen suspected athletes according to this tool and then place them in one of the categories. It is based on the IOC Consensus Statement on RED-S [4].

16.4.1 Red Light

In the red light, we find athletes with one of the following diagnosed disorders:

- Anorexia nervosa and other eating disorders
- Other serious medical (psychological and physiological) conditions related to low energy availability
- Use of extreme weight loss techniques leading to dehydration-induced hemodynamic instability and other life-threatening conditions
- Severe ECG abnormalities (i.e., bradycardia)

Athletes in the red light need medical help, and due to the seriousness of the presentation, sport participation represents a serious threat. It may also distract the athlete from devoting the attention needed for treatment and recovery. Athletes in this category should be forbidden to participate in competitions or serious training [3].

16.4.2 Yellow Light

Here, we find athletes with a moderate risk which can be identified with the following characteristics:

- Prolonged abnormally low body fat (measured using DXA or anthropometric measurements)
- Substantial weight loss of 5–10% in 1 month
- Attenuation of expected growth and development in adolescents
- Low energy intake over a longer time period
- Abnormal menstrual cycle with amenorrhea over more than 3 months
- No menarche at 15 years of age
- Low mineral bone density (DXA measurements)
- Stress fractures
- Psychological complications from low energy intake
- Disordered eating behavior negatively affecting other team members
- Lack of progress in treatment and/or noncompliance

Athletes in the yellow light are allowed to participate in competitions if medically supervised and treated. Reevaluation of the athlete's risk assessment should occur at regular intervals of 1–3 months depending on the clinical scenario to assess compliance and to detect changes in clinical status [3].

16.4.3 Green Light

Athletes in the green light only have a low risk. They can be classified according to the following criteria:

- Appropriate physique that is managed without unhealthy diet/exercise strategies
- Healthy eating habits
- Healthy endocrine system
- Healthy bone mineral density
- Healthy musculoskeletal system

There are no restrictions for these athletes [3].

16.4.4 Treatment

Athlete in the red light should therefore not participate in competitions or training. They need to sign a written contract in which the therapy goals are spelled out so that they know when a return to sport is possible. Athletes in the yellow light can participate in competitions but should be monitored closely and follow an exact treatment plan. Athlete in the green light can continue as before [3].

This represents also the approach already practiced by the IFSC. In order to monitor the developments in the IFSC athletes, BMI measurements for screening during world cup competitions have become routine. If an abnormality during these routine screenings is observed, the national federation is informed and is asked to take appropriate steps in order to categorize the athlete and then provide appropriate treatment plans if indicated. It is also up to the national federation if the athlete needs to leave the competition circuit or not. If the athlete is below a minimum BMI the athlete is only granted an international permit after the federation's physician has provided a written statement allowing the athlete to keep competing [2]. However, this concept is in perpetual evolvement.

16.5 Conclusion

The health of each athlete must be the priority for each physician working with climbers. If an eating disorder or relative energy deficiency is suspected, this needs to be addressed promptly. There can be no beating about the bush since the consequences can be permanent. Especially, adolescents can experience worsening of a little energy restriction into severe psychiatric disorders from which they may never recover. The RED-S CAT represents a useful tool for detecting and judging the extent of an eating disorder in athletes and provides guidelines for helping these athletes.

References

1. Zapf J, Dingerkurs M, Fichtl B, Wielgross C. Ernährung und essverhalten beim sportklettern. Sport Sport. 2002;18:97–103.
2. IFSC MC. International Federation of Sport Climbing Turin: International Federation of Sport Climbing; 2014. http://www.ifsc-climbing.org/.
3. Mountjoy M, Sundgot-Borgen J, Burke L, Carter S, Constantini N, Lebrun C, et al. The IOC relative energy deficiency in sport clinical assessment tool (RED-S CAT). Br J Sports Med. 2015;49(21):1354.
4. Mountjoy M, Sundgot-Borgen J, Burke L, Carter S, Constantini N, Lebrun C, et al. The IOC consensus statement: beyond the Female Athlete Triad—Relative Energy Deficiency in Sport (RED-S). Br J Sports Med. 2014;48(7):491–7.
5. Watts PB, Martin DT, Durtschi S. Anthropometric profiles of elite male and female competitive sport rock climbers. J Sports Sci. 1993;11(2):113–7.
6. Schöffl I, Wüstenfeld J, Jones G, Dittrich S, Lutter C, Schöffl V. Athlete's heart in elite sport Climbers? cardiac adaptations determined using ECG and echocardiography data. Wilderness Environ Med. 2020;31(4):418–42.
7. Schöffl I, Wüstenfeld J, Jones G, Dittrich S, Lutter C, Schöffl V. Cardiopulmonary function of adolescent elite sport climbers. 2020. (In press).
8. Mermier CM, Robergs RA, McMinn SM, Heyward VH. Energy expenditure and physiological responses during indoor rock climbing. Br J Sports Med. 1997;31(3):224–8.
9. Kemmler W, Roloff I, Baumann H, Schöffl V, Weineck J, Kalender W, et al. Effect of exercise, body composition, and nutritional intake on bone parameters in male elite rock climbers. Int J Sports Med. 2006;27(8):653–9.
10. Morrison AB, Schöffl VR. Review of the physiological responses to rock climbing in young climbers. Br J Sports Med. 2007;
11. Schöffl I, Schöffl V, Dotsch J, Dorr HG, Jungert J. Correlations between high level sport-climbing and the development of adolescents. Pediatr Exerc Sci. 2011;23(4):477–86.
12. Smith EJ, Storey R, Ranchordas MK. Nutritional considerations for bouldering. Int J Sport Nutr Exerc Metab. 2017;27(4):314–24.
13. Schöffl I, Schöffl V. Kann das noch gesund sein. In: Alpenverein D, editor. Berg 2018. München: Deutscher Alpenverein; 2017. p. 107–13.

17 Sport Climbing with Pre-existing Medical Conditions

T. Küpper and A. Morrison

Christoph Lutter bouldering at Charmonix, France, Photo Michael Simon

T. Küpper (✉)
Institute of Occupational, Social and Environmental Medicine, RWTH Aachen Technical University, Aachen, Germany
e-mail: tkuepper@ukaachen.de

A. Morrison
Royal Free London NHS Foundation Trust, London, UK

Pre-existing medical conditions may—but not always—set limitations on physical exercise and sporting participation. Some adjustments in frequency, intensity, time, and type (FITT) may be required, including to sport climbing. The

V. Schöffl et al. (eds.), *Climbing Medicine*, https://doi.org/10.1007/978-3-030-72184-8_17

most important message first—it is very rare that patients must stop climbing completely!

There is a broad consensus that regular physical activity and exercise are an essential part of a healthy lifestyle with significant health benefits and are regularly prescribed for the prevention and treatment of many diseases [1]. For example, aerobic exercise and aerobic with resistance exercise can significantly improve the aerobic capacity of chronic heart failure patients, whereas resistance exercise alone cannot [2]. Other meta-analyses confirm that both aerobic and dynamic resistance training can reduce hypertension [1, 3], and small studies show climbing can reduce backache [4], even depression [5].

Limb (1995) reported no cardiac events at UK indoor climbing walls with over 1.7 million users. All exercise adapts the heart accordingly. A trained sport climber's heart is characterized by a submaximal cardiac output, high to maximal heart rate, and moderate to marked increase in peripheral resistance [6]. This is explored later, but sport climbing is essentially a mix of aerobic plus resistance exercise combined with significant hand-gripping, where performance limiting or terminating "climbing fatigue" is not systemic but generally localized to the forearm flexor muscles responsible for gripping. It's unlike standard cycling or running treadmill exercise testing where performance is predominately aerobic and terminated by breathlessness and fatigue of the exercising large muscle groups. Both standard ergometric exercise tests can also prescribe a steady exercise state for the heart rate (HR), which is impossible in climbing. Climbing's unique physiological response to each route combines submaximal aerobic and anaerobic pathways, and a disproportionate rise in HR and blood pressure (BP), especially when climbing nearer the maximal grades of climber [6].

There is no standardized method to test climbing exercise. However, Sheel (2004) compared cycling ergometry VO_2 to the VO_2 when climbing on grades 2 and 3 below the maximum grade of the climber [7, 8]. Sheel found the respective climbing grade'sVO_2 to be 45 and 51% of the cycling VO_2 max.

In assessing over one million standard ergometry exercise tests, there were no fatal or nonfatal complications in the athletes, and in the remaining 712,285 predominantly with coronary heart disease, there were 17 deaths and 96 life-threatening complications, most frequently regarding ventricular fibrillation [9]. Gibbons et al. conducted 71,914 maximal exercise tests with uniform conditions in a single medical facility in a population with low prevalence of known heart disease from 1971 to 1987 [10]. No complications occurred from 1977 in 45,000 maximal tests. The overall cardiac complication rate in women and men is 0.8 complications per 10,000 tests with 95% confidence intervals of 0.3–1.9 complications per 10,000 tests. Standardized maximal exercise testing was found to be safer than previously published, but there still remains a very minor risk to be addressed. Fletcher et al. clearly defines various parameters for ergometry testing in those with cardiovascular disease, with epidemiological data demonstrating the necessity of regular exercise in these patients, and the resulting positive health benefits [11].

Climbers are generally dedicated to the sport, be it recreational or competitive. They may have long walks with rucksacks to outdoor crags and/or regularly participate in other exercise which may include cycling to a climbing gym, strength training, or other sports participation. However, whenever there are pre-existing medical conditions, the climber must seek the advice of a doctor who is well informed about the respective diagnosis in sport medicine and who is preferably also experienced in climbing. Sport climbing's physiology is often misunderstood or dismissed as there's limited data published with respect to exercise prescription. The patient should not be unnecessarily limited in their regular activities as the benefits of regular exercise and sports are significant and almost exclusively outweigh the risks.

There must always be regard for any disease process itself, and each climber presents with a different FITT training background. They may require a baseline or maximal level of fitness to be established and/or comprehensive history reported.

In seeking the best possible medical advice, the following questions must be answered:

1. Is there any risk that the disease may be worsened by sport climbing?
2. Alternatively, does the disease limit the ability to climb?
3. Does the disease cause any risk while climbing?
   (a) When leading?
   (b) When belaying?
4. Are there any additional risk/s from environmental factors which may worsen the disease?

The following are examples of replies to the four questions raised above.

1. Diabetics should monitor their blood sugar during and after sport as exercise stresses glucose metabolism.
2. Limited mobility in the case of rheumatic or orthopedic diseases may cause limitations when climbing.
3. Seizures may be a risk when leading or belaying but not when seconding or top rope climbing;
4. If a diabetic has peripheral angiopathy or a climber suffers from Raynaud's disease or other vascular dysfunction, cold weather may cause significant problems and may increase the risk of frostbite.

For further clarification, the following questions should also be asked:

1. Which—if any—may be a limiting factor to climb?
2. Is the disease in an acute phase, or has it been stabilized?

Based on these answers and—if necessary—additional clinical tests, the individual recommendation should be given. Again, in most cases, the patient will be able to continue climbing if he/she accepts some rules. Principally, any acute or non-stabilized situation should be stabilized by adequate therapeutic procedures before any assessment of fitness to climb is answered. It should be noted that there are nearly no evidence-based recommendations and data specifically about climbing with pre-existing diseases [6], but there are for related sports. Therefore, the bulk of the recommendations, also those from international panels—e.g., the UIAA Medical Commission (Union Internationale des Associations d'Alpinisme)—are based on a consensus of international experts who are climbers themselves and who transfer sports medical knowledge from other disciplines to sport climbing [12–16]. It should be noted that most of these recommendations focus on "classic" alpine climbing and not sport climbing.

The following recommendations for sport climbing with pre-existing medical conditions will consider each condition in turn. However, some physiological understanding of the sport will first be discussed to avoid further repetition in the respective diagnoses that follow.

17.1 Physiology of Sport Climbing Seen with the Eyes of Exercise Physiology

Climbing is often said to be "strenuous" without any understanding or qualification of what this means. How fit must a person with pre-existing disease be to climb without taking inadequate risk for himself or fellow climbers? Answering this requires an understanding of baseline physiological parameters induced by sport climbing. Sport climbing is an intermittent activity where each unique route has a graded difficulty level once climbed, and this is typically followed by a short rest or belaying one's partner. It involves complex static isometric and dynamic moves.

Higher grades become incrementally more strenuous, as are routes with increasing incremental overhangs angles or "roof" routes. The lowest grade is equivalent to a scramble walk. Trained climbers can climb more routes of higher grades in a single session because of their enhanced training status; and if more experienced, they have additionally honed efficient climbing techniques to reduce strenuous loading (i.e., not over gripping and thus delaying the onset on climbing-specific fatigue, better footwork, and/or core strength). It is well established that climbing increases heart rate (HR) disproportionately relative to the "exercise" of climbing and can fluctuate between 129 and 180 bpm during a climb, with peaks greatest on increasing inclines on overhanging routes [6].

Accordingly, a trained sport climber's heart is characterized by a *submaximal cardiac output, high to maximal heart rate, and moderate to marked increase in peripheral resistance* [6]. As the sport is not predominately aerobic, but also involving resistance training, heart rate and blood pressure will be disproportionately high during the climb itself, with irregular breathing, which return to normative values soon after the climb [6]. A typical climb lasts 2–7 min, with blood lactates remaining elevated over 20 min post-climb [17]. Sport climbing does not require a high level of cardiovascular fitness with an adult VO_{2peak} during the climb estimated to be 52–55 mL/kg^{-1}/min^{-1} for maximum oxygen uptake, averaging 20–25 mL/kg^{-1}/min^{-1}. The latter is equivalent to the stress tolerance in an untrained population [17]. The VO_{2max} results for sedentary people undergoing isometric exercises are <45 mL/kg^{-1}/min^{-1}, while elite cyclists have >68 mL/kg^{-1}/min^{-1}. Therefore, measurements of the maximal oxygen uptake (VO_{2max}) are not representative for the real strain of sport climbing [18].

In a study investigating whether there was any advantage in an active pause when sport climbing, the probands who climbed a 20 m route with a crux of VIII+ (UIAA scale) showed a mean heart rate (HR) of 148 bpm (+/−16), a mean oxygen uptake (VO_2) of 1660 mL/min (+/−340), and a relative oxygen uptake of 24.7 mL/kg^{-1}/min^{-1} (+/−5.3) [18]. Short peak loads require 31.9 mL/kg^{-1}/min^{-1} (+/−5.3). An active pause had no advantage over a passive one. However, it must be criticized that lactate was not investigated although it is well known that its increase requires a significantly longer recovery. With a passive pause, lactate is still increased for 30 min post-climb before getting back to normal levels or continued increasing for 20 min after an active pause [18].

Given this physiological understanding, it can be concluded that sport climbing makes no significant demands on general endurance nor does it require a high cardiovascular fitness for heart patients; the data suggests sport climbing exercise corresponds to loads which are recommended as secondary prevention in heart cases.

As expected, climbing routes closer the individual's grade limit will cause a significantly higher peaks in HR and higher VO_2 [6]. However, when climbing moderate routes, HR is at 67%, and VO_2 is 45% of the individual's maximum VO_2 for cycling ergometry [8]. This again indicates that standard exercise ergometry tests should not be used for assessing climbing fitness because the (largely) linear correlation between HR and VO_2 does not exist during climbing. Such correlation will be found in aerobic endurance disciplines where the body's large muscle groups become fatigued and the subject breathless, unlike climbing where performance limiting fatigue is generally localized to hand gripping function. In consequence, the measurement of the ischemic threshold of a patient with coronary heart disease may be used, but the VO_2 of a pulmonary patient is irrelevant if this patient climbs.

Some authors also mention the psychological factors that may additionally impact on the disproportionality between HR and VO_2 when climbing, e.g., stress and fear [8, 19, 20]. Williams et al. examined the heart's response to climbing outdoors with experienced and novice climbers using ECG and placebo or beta blockade to limit HR peaks [20]. Williams concluded, "*…this popular sport represents more an anxiety-type of psychological stress than a physical stress and as such is likely to increase moral fibre rather than muscle fibre*." However, the testing day was wet, route slippery and insecure, especially for novices. Experienced outdoor climbers will generally deal with inclement weather and bold movements much better than novices. Beta blockade does not improve climbing performance, though incorrectly thought to be important to climbing performance among some climbers in the 1970s. Although studies are few, experienced climbers generally manage the "fear" of a bold move much better than novices, irrespective of weather, and have lower HRs; and more significantly, experienced climbers also don't overgrip holds which will raise HR [6]. Perhaps it is not surprising that climbing training is relevant for 59% of the performance only [21].

Interestingly, the anaerobic strain is much higher when comparing indoor climbing to outdoor. While the HR, VO_2, and lactate at rest were identical, the maximal lactate was significantly higher (10.4 mmol/L) for indoor climbers compared to climbing natural rocks of similar difficulty [22]. All studies demonstrate that lactate is also not a valid parameter to measure fitness of climbers as there are no steady states in exercise to be achieved while climbing, and each graded route differs in its "exercise" intensity. This may be different when considering local lactate, especially if the arms would be measured, but this testing is not realistic for normal climbing.

Although there are several limitations when applying results from traditional exercise testing protocols (cycling, threadmill) to sport climbing, some important conclusions may be drawn for climbers with pre-existing conditions [19, 23]. With an increase of 54–85% of HR_{max} and a VO_2 which corresponds to easy jogging, climbing may be recommended as discipline for secondary prevention, e.g., for diabetes or coronary heart disease.

17.2 Acute Infections

The most frequently encountered and probable cause acutely affecting the heart is acute febrile infection. This is assumed to be an important factor for sudden and severe complications or even deaths in sports via myocarditis as third frequent cause of death [24], and arrhythmia [25–28]. Although there are no valid data available, obviously different viruses behave differently, and those that are "critical" are Parvovirus B19, herpes viruses, and entero- and adenoviruses [29–31]. When outdoor climbing, it should be taken into account that in several regions a myocarditis may be a complication of a borreliosis [30].

Independent from the sporting discipline, there is international consensus in sport medicine and occupational medicine that demanding activities should not be performed at least for twice the number of days that the patient had fever. Non-demanding activities like compensatory training is possible if the patient feels well enough to do so.

17.3 Sport Climbing with Diabetes Mellitus

Summary

- **Limiting factor:** Endurance, not maximal force, climbing grade difficulty level, characteristics of the route, altitude, etc.
- **Additional risk factors:** Cold, heat, minor trauma, (shoes too tight!)
- **Common mistake:** Blood sugar measured before and while climbing and also some hours after exercise

There are no studies available concerning sport climbing and diabetes. However, there is international consensus in sport medicine that regular sport is a good secondary prophylaxis for diabetics and especially for type 2 diabetes; this is supported with good data. The main effect is the build-up of metabolic highly active muscle tissue.

If there is no significant neuropathy, the maximal force and short-time-endurance of diabetics is not reduced, and, vice versa, side glucose metabolism will not be influenced significantly by the activity. Therefore, the risk for hypoglycemia is reduced compared to endurance sports, e.g., mountaineering. Hence, such complications are extremely rare in climbing gyms (Küpper, unpublished data). However, consequent monitoring of the blood sugar should be integral part of any sportive activity of diabetics with another measurement some hours after climbing included. Often patients will forget the latter when they climb in the evening, and the test should be performed then after midnight. Overall, blood sugar management is much easier now with modern equipment which is small and lightweight, some of which offer automatic and/or continuous measurement. So far, no contraindications were found for such devices, but the user should always keep up-to-date with the respective specifics of the different systems.

Basically there are continuous glucose monitoring systems (CGM) and flash glucose monitoring systems (FGM). CGMs measure the sugar in the liquid of the subcutaneous tissue more or less continuously, at least in short intervals. For FCMs, the sensor is located also in the subcutaneous tissue in short intervals, but the results will be stored on a microchip and then read by a special device or via smartphone. Data from both systems may be stored and evaluated digitally and used later for advice by the sports medicine specialist or the diabetologist. For both systems, a very small platinum electrode must be placed into the subcutaneous fat. It doesn't disturb the patient, but dependent on the system, it must be replaced every few weeks. Independent unsystematic observations from a limited number of cases have shown that the systems are well tolerated by the climbers. The location of the sensor at the dorsal upper arm generally does not constrict. This would be the case only for a very limited number of moves, e.g., back and footing, where the arm is used backward.

CGM and FGM must be differentiated from noninvasive continuous measurement (NCM). The latter is of advantage for persons with phobia for needles. No device must penetrate the skin. However, this system has a significant and unratable disadvantage. NCM is based on the reflexion of laser light, which points at the skin's sweat. Here, several substances are solved (glucose, protein, water, uric acid, etc.), and their collective measurement allows a conclusion of blood sugar concentration. For people at rest, the results are of acceptable validity. However, during sports, the composition of sweat differs significantly from those at rest. Another problem is that in people who are used to sweating regularly (sports, sauna, etc.), sweat is produced more effectively than in other situations which may mean their sweat contains much more water and less of other (valuable) substances, especially salt. Research demonstrated that this effect cannot be taken into account by available systems thus far. Seen from the eyes of sport medicine, such systems cannot be generally recommended in contrast to CGM and FGM systems.

The necessary validity and exactness of measurement has been included into international standards. Since 2016, only those devices meeting the testing criteria set out in ISO 15197 can be sold. With an accepted tolerance of +/−15%, CGM and FGM systems fulfill this standard, but NCM show an error of measurement of +/−30% or more which is not acceptable according to ISO 15197. The exactness results of watches or wristband fail to meet these criteria. These systems will give an alarm to the user when the limits should be exceeded, but this is not reliable enough because of data variance. However, these systems may be used as backup for others. The exactness of measurement must be taken into account also when the "traditional" test with strips should be used. Such tests are linked chemical-optic tests and are therefore temperature dependent. When used below +14 °C, the reported blood sugar concentration is false-low (below 0 °C any measurement is impossible). When the temperature should be higher than +35 °C, the result will be a false high. Therefore, these devices should be carried under warmth and shade of jackets on cold days, and similarly in the shade during hot days. As the measurement of these test substance devices are UV light sensitive, ensure the container with the testing strips is opened for as short a time as possible and in the shade.

Checkup for Climbing Diabetics

- **History:** Experience in climbing, training status, nutrition, blood sugar management, hypoglycemic phases in the past (especially when correlated to sports), medication, knowledge of individual symptoms for hypoglycemia
- **Actual status:** Vessels (peripheral angiopathy, especially of the fingers or feet?), neurology (peripheral neuropathy?), ocular fundus
- **Laboratory:** Creatinine, urine status, HbA1c
- **Technical investigations:** ECG at rest, ergometry
- **In the case of individual indication:** Stress echocardiography, 24 h ECG, ultrasound of the kidneys, etc.

A precondition for diabetics to undertake safe sporting participation is the continued knowledge about the disease, good compliance to improve diabetes control, and the absence of any secondary organ damage which may cause limitations. The latter would necessitate special advice with a careful risk and benefit assessment included.

Independent from the discipline, fellow climbers should be informed of the diabetic's diagnosis and first aid involving the following options should they arise. They should also know where the diabetic carries some carbohydrates (normally readily accessible in the top compartment of the rucksack) [16]. Any emergency kit should include some carbohydrates. On the strength of past experiences using cereal bars as the carbohydrate, often recommended especially in Anglo-American literature, this is not the best choice. They are bone-dry and make the athlete thirsty. Much better are carbohydrate gels which are state of the art for diving diabetics. They can be consumed directly from the packet, and thirsty persons tolerate them much better. To carry along glucagon or insulin in the emergency bag is a topic of discussion. If climbers do carry these, they must be confident in using them correctly.

The limited temperature tolerance of diabetic drugs should also be taken into account as they are quite resistant against heat but should never be allowed to freeze [32]. Only about 15% of the normal population knows the local emergency number. Therefore, this should be included in the emergency kit as the number can vary. For example, in most countries worldwide, 112 is the emergency calling number, but its 911 in North America. In some African and Asian countries, there are other local numbers.

With respect to tactics for managing blood sugar levels, the following points should be noted. After a strenuous day, the blood sugar level may be a little higher than normal in the following morning, and this is fine. When significant endurance exercise is anticipated in a given day, the amount of carbohydrates consumed at breakfast should be increased by about 30%. In hot climates, the hydration status should be monitored carefully. Dehydration is more dangerous for diabetics than a mild hyperhydration, and therefore intake should be liberal [16]. Special attention should be given to the treatment and prevention of injuries, also to minor wounds [16].

Even with good compliance, diabetics will develop some minor microangiopathy and neuropathy in their daily routine which is irrelevant. In cold weather, this increases the risk of frostbite as the diabetic person will not feel the effects of cold as much as a healthy person due to the neuropathy; and if frostbite has already occurred, the healing will be more inferior for the diabetic due to the microangiopathy. Protection from the cold should be actioned early by wearing non-constrictive shoes and gloves and, if necessary, fingerless gloves for climbing. Perfectly fitted shoes and regular care for the nails and feet are a must for diabetics [16]. Never wear shoes sized too small. Comfort is the measure of all things! As soon as pressure marks occur from ill-fitting footwear, climbing must be stopped immediately, and it should be considered whether to buy larger or better fitting shoes. If wounds should occur at the hands or feet, however minor, these should be treated carefully. This includes perfect disinfection, covering, and no climbing until fully healed.

17.4 Sport Climbing with Cardiovascular Disease

For most cardiovascular diseases, there is no doubt that exercise is an important factor for the prescription of secondary prevention [2, 3, 33–35]. Therefore, it must be strongly warned to advise a climber to stop climbing without significant arguments or not to continue climbing with specific advice by a specialist.

17.4.1 Hypertension

One may call hypertension a pandemic of modern civilization, and, of course, it also affects climbers. Some sport medicine specialists unfamiliar with climbing physiology give the undifferentiated advice to "stop climbing" when presented with climber with hypertension. It is conspicuous that the former often "never in their life have climbed higher than a bar stool" (cit. Ulf Gieseler,

Heidelberg/Germany). They should consider the following argument. If assessed from an exercise physiology viewpoint, a hypertensive person, who is treated perfectly, who does not suffer from secondary damages of hypertension, and who shows a normal blood pressure regulation at rest and during exercise, should be viewed as a healthy person able to climb [6]. Meta-analyses confirm that both aerobic and dynamic resistance training can also reduce hypertension, among many other health benefits to be derived from exercise, and are endorsed by multiple international hypertensive consensus recommendations [1, 3].

Summary

- **Limiting factors:** Maximal force, especially when combined with a whole-body load and forced respiration during expulsion (and therefore indirectly the difficulty of the climb), may cause risk but not the length of the climb, endurance exercise, altitude, climate, etc.
- **Additional risk factor:** Competition situation, "pushing the limits"
- **Common mistakes:** Compensation of a lack of technique with a more of power, climbing of significant overhanging routes only, not adequate pauses between the climbs

Also for hypertension, it should be noted that data are insufficient. However, some observations indicate strongly that well stabilized hypertensive patients without secondary organ damages show a normal exercise-adequate increase and decrease of their blood pressure [6, 36]. The specific problem is that the method of Riva-Rocci cannot be used during climbing, and for technical reasons, the earliest measurement will only be possible at least 1 min after the climb when blood pressure has already decreased. A study by the authors with continuous blood pressure measurement during climbing has been approved by the ethical commission but has not yet commenced.

There is one study which tried to evaluate whether the climbing of a hypertensive patient might cause a risk for the myocardium. It is unfortunate that the study design was weak, but no effect was found in 19 participants [37]. However, although data are scarce, some conclusions may be drawn to advise hypertensive climbers. Experienced climbers showed a systolic blood pressure increase of about 13 mmHg, while novices showed even more [38]. From this, it may be concluded that good climbing technique is always advantageous to develop and learn, especially for hypertensive climbers. Near the individual limit, blood pressure will increase disproportionately during the climb and reduce post-climb [6, 36].

Checkup for a Hypertensive Climber

- **History:** Compliance? Check blood pressure recordings! Experience in climbing? Exercise-related symptoms? Secondary organ damage?
- **Actual status:** Blood pressure, body weight, heart, lungs, vessels, ocular fundus
- **Laboratory:** Creatinine, urine status
- **Technical investigations:** ECG at rest, ergometry, 24 h blood pressure
- **In special situations/**individual indication: Echocardiography (stress), thorax X-ray, ultrasound of the kidneys, etc.

There is no climbing limitation when blood pressure has been stabilized within a physiological range, the diurnal rhythm is intact, blood pressure response is adequate to the respective exercise, and there are no secondary damages of organs (these might define a limitation) [6].

When the systolic blood pressure is higher than 160 mmHg, or diastolic above 100 mmHg, the situation should be handled with care. The climber may continue climbing but at a lower level until an adequate therapy has been established. Overhanging routes should be strictly avoided for this period. However, if the pressure should be higher than 200 mmHg systolic and above 120 mmHg diastolic, any sport should be stopped until sufficient therapy has lowered such pressure into an acceptable range. Then, secondary damage of organs should be excluded before the continuation of climbing may be discussed [6]. If such damage should be identified, it should be considered whether they cause a limitation for climbing, or vice versa, whether climbing may worsen the damage. In most cases, it will be possible to continue climbing at a "level of joy" without competition and "pushing the limits." Routes with significant overhangs should be avoided.

Although there are no data so far, a careful view on the characteristics of the respective climb should be recommended. Maximal loads near or at the individual limit and therefore with and forced respiration during expulsion should be avoided. This would be especially the case in hard overhangs or roofs. Counterforce techniques ("Piazen") would have a similar effect, although less pronounced. In contrast, slabs and walls would be beneficial.

For optimal therapy, there are international recommendations which will be regularly updated. Because maximal aerobic endurance exercise and maximal heart rate (HR) do not really play a significant role in sport climbing, the decrease of HR_{max} and VO_{2max} by beta-blockers can be neglected. However, a careful follow-up of any update of the World Anti-Doping Agency (WADA) regulations should be provided whether the ban of such drugs—actually limited to some disciplines, e.g., all shooting—may be expanded to climbing. Here beta-blockers do not decrease performance but increase it probably by a decrease of nervousness and anxiety. Then, the hypertensive climber with beta-blockers for therapeutic use needs a TUE certificate ("therapeutic use exception") if a competition should be planned, the same for diuretic drugs if they are part of the therapy. All other antihypertensive drugs are so far not critically concerning anti-doping rules.

17.4.2 Coronary Heart Disease (CHD)

There is an increase of cardiac emergencies with increasing age, also in active or sporting people independent from the discipline, although less than in sedentary persons (survey in [6]). CHD is one of the most important underlying diseases. It is less known that the positive predictive value of a single ergometry of an asymptomatic person is 55–65% only and increases only after several ergometries [39–41]. Gibbons et al. found the overall cardiac complication rate for cycling ergometry in healthy adults to be 0.8 complications per 10,000 tests, though later the complications were nil for many years [10]. Scherer and Kaltenbach used ergometry testing in 712,285 patients with predominantly CHD, and there were 17 deaths and 96 life-threatening complications, most frequently regarding ventricular fibrillation [9]. The safety of various ergometry testing for this population has been refined and aids personalized exercise prescription and significantly improves the health of such patients [11].

Based on this data, most authors recommend to check persons who start with any sport at the age of about 40+, but a general screening of the symptom-free population doesn't make any sense [41]. Consequently, this is also not included in most national or international guidelines.

Summary

- **Limiting factor:** Maximal workload (myocardial) but not maximal workload of peripheral muscles like arms, hands, or legs. Length of the climb, especially when it is a hard one. Altitude (hypoxia), climate, and others
- **Additional risk factors:** Competition, "pushing the limits"
- **Common mistakes:** Inadequate technique will be compensated for by more muscular strength or an abrupt increase of climbing difficulty—both causing forced respiration during expulsion.

However, the above recommendations need a modifying statement to calm these down. Thousands of elderly climbers, including those with CHD, visit climbing gyms regularly, and cardiac emergencies in sport climbing are extremely rare events. Limb reported no cardiac events at UK indoor climbing walls with over 1.7 million users [42]. It should be reminded that the general exercise tolerance does not reflect the physiological situation when climbing. However, this is fortunately not exactly the problem for the climbing CHD patient. Such patients don't need data concerning their maximum workload; they need their so-called ischemic threshold. This is the pulse rate at the workload when first symptoms of CHD should occur or first typical changes of the ECG curve.

Checkup for CHD Patients Who Climb

- **History:** Angina pectoris symptoms in the past? General and specific experience in sports? Medication?
- **Actual status:** Heart, lungs
- **Laboratory:** If applicable, parameters of blood clotting if the patient should take anticoagulants
- **Technical investigations:** ECG, ergometry with analysis of the "ischemic threshold," echocardiography
- **With individual indication only:** 24 h ECG, stress echocardiography, X-ray thorax, myocardial radio imaging, etc.

Data concerning the necessary exercise tolerance in sport climbing are scarce; most of them originate from very small collectives. Watts et al. recommend from their data encouraging amateur climbers to tolerate a threshold of at least 150/min [18]. Own data indicate a threshold of 120–130/min. (Küpper, unpublished data). All recommendations include the precondition that exercise should not be performed at levels higher than 75–80% of the ischemic threshold [43, 44], which includes a safety margin.

When the climber wants to keep safe, he should respect the following principles. Any symptoms should be stabilized and adequately treated, and exercise tolerance should be sufficient. To monitor the exercise, a pulse watch may be used with its alarm threshold set at 85% of the ischemic threshold. Note: Watches with a fingertip sensor and which measure only if the person puts the finger on this sensor are not acceptable. In general, climbing for fun should be the goal and not "pushing the limits" or competitions. Forced respiration during expulsion should be avoided whenever possible, and slabs or walls should be favored over roofs or overhangs. CHD patients should climb only if they feel completely well and must accept their individual limit of exercise tolerance. This is based on observations which indicate that pulse frequency is correlated with the difficulty of the respective climb [17]. HR is clearly lower on easier overhangs than on more difficult slabs and crimpy holds.

According to Gieseler (personal communication), a generally low risk can be assumed when left ventricular function is normal, ergometry fulfills age-adjusted normal values, exercise does not induce ischemia or complex arrhythmia, and relevant stenoses does not exist, but also after a non-complicated myocardial infarction when after 2–3 months climbing will be tried carefully and with slow increase of load.

Be careful if the HR during climbing is often at, or higher than, the ischemic threshold. If exercise-induced arrhythmia should occur, or in cold climate when cold-induced angina is known, the patient should stay for some minutes in moderate cold before he enters the cold zone. Here any exercise should be increased step by step (as the high breathing volume of cold air is the trigger of symptoms!).

No climbing when angina pectoris symptoms should occur at low exercise. When angina is not stable, in the case of severe ventricular dysfunction (ejection fraction <40%), pathological decrease of blood pressure of more than 15 mmHg during exercise, ST segment depression of more than 2 mm while exercising, resuscitation after myocardial infarction with complex arrhythmia less than 6 months ago.

17.4.3 Arrhythmia

Again it must be stated that there are no specific studies for climbers. However, because emergencies caused by arrhythmia are extremely rare in climbing gyms, it can be concluded that the risk should be quite low. A "healthy worker effect" may give additional risk reduction.

Summary

- **Limiting factor:** Maximal load (HR), but not or to a very limited extent maximum muscular force, length of the climb, endurance, altitude, climate, etc.
- **Additional risk factor:** Combination of some specific types of arrhythmia, e.g., in the case of arrhythmia absoluta and a thrombus in the atrium, a sudden increase of HR may mobilize the thrombus and cause a stroke.

- **Common mistake:** Ignorance and missing the diagnosis of occasional "hops" (citation of a high-risk person who later was vitally endangered by Torsade-des-Pointes arrhythmia!)

In sports medicine, the most important question is whether the respective arrhythmia may become life-threatening for the patient (which would mean "no sports until the problem has been solved"). If this is excluded, the next question is whether the arrhythmia may worsen during exercise. Then the limit should be defined as described for CHD (see above) and which should not be exceeded by the patient during sports. Again the pulse watches as described above with alarm set at 85% of the defined threshold should be used.

Checkup for a Climber with Arrhythmia

- **History:** Cardial history (symptoms, collapse, etc.) general and specific sport experience, medication, etc.
- **Actual status:** Symptoms when working or exercising in daily routine?
- **Laboratory:** Normally not necessary, probably electrolytes
- **Technical investigations:** ECG, ergometry, 24 h ECG, echocardiography (especially to detect atrial thrombus in the case of arrhythmia absoluta)
- **Individual indication only:** Stress echocardiography, X-ray thorax, etc.

Concerning exercise tolerance the recommended minimum values were already discussed in the chapter CHD (see above). In analogy one may call it "arrhythmogenic threshold." Before any advice, the etiology of the respective arrhythmia must be clear because arrhythmia is a symptom and not a diagnosis. It is possible that the main limitation of the person is defined by the underlying disease and not by the risk caused by the arrhythmia.

A low risk may be assumed for climbers when there are some ventricular extrasystoles only which do not cause hemodynamic consequences, and when there are some supraventricular extrasystoles without tendency for tachycardia or in the case of stable and hemodynamically non-relevant arrhythmia absoluta without thrombus in the atrium [6].

On the other hand, a high risk should be assumed if a pre-excitation syndrome was diagnosed, in the case of intermittent tachycardia independent from its cause and in the case of hemodynamically or complex arrhythmias. Of course, the question whether the patient is fit to climb again may be answered when the therapy has been proven to be sufficient [6].

Having a pacemaker, implanted defibrillator, or an event recorder, is not a general contraindication against climbing or sport in general. Here, the clinical situation is guiding. However, it must be considered that a harness or the belt of a rucksack may apply pressure on the system, and this must be avoided [6]. With modern sport climbing harnesses, the former is not a problem anymore. However, about the latter, any patient should be informed. It is a misbelief that anticoagulant therapy is an absolute contraindication against climbing. Of course any bleeding of minor wounds will be prolonged, or hematoma will be larger. The danger for the patient, however, is intracranial bleeding. Therefore, the patients should wear a helmet without any exception, also indoors in the gym.

17.4.4 Cardiac Insufficiency and Valvular defects

Summary

- **Limiting factor:** Workload, not short maximal muscular forces
- **Additional risk factors:** Climbing at moderate altitude
- **Common mistakes:** Ignorance and overconfidence. Climbing with cardiac insufficiency means climbing for fun, not for "pushing the limits" or competition!

Again the advising physician must live with missing data and studies. However, there is consensus that the climber should strictly avoid massive increases of blood pressure especially on longer athletic routes, routes near or beyond the

Table 17.1 NYHA classification [46], the most commonly used classification of heart failure in Germany since the guidelines were published in 2005 [50]

Class	Definition
1	Heart disease with no limitation to any physical activity. Normal workload and daily routine does not cause undue exhaustion, arrhythmia, shortness of breath, or angina pectoris
2	Heart disease with a mild decrease of the exercise tolerance. No symptoms at rest. Normal daily activities and workload may cause exhaustion, arrhythmia, shortness of breath, or angina pectoris
3	Significant heart disease with marked decrease of exercise tolerance. No symptoms at rest, but easy exercise causes exhaustion, arrhythmia, shortness of breath, or angina pectoris
4	Heart disease with symptoms at rest and during all physical activity, bed confinement

climber's abilities, or the use of excessive force when climbing technique is lacking. Beyond that any advice orientates mainly on the cardiac exercise capabilities. This will be usually classified according to the New York Heart Association (NYHA) since 1928 (Table 17.1), although it has been modified several times [45, 46]. Again it must be noted that cardiac insufficiency is a symptom and not a diagnosis. The latter must be clear because it may cause the main limitation to the climber and not the insufficiency.

With Stage 1 high end climbing or competition climbing is limited significantly, but motivated recreational climbers will be able to continue their sport without major problems. Some of them may also join recreational competitions. A check by a cardiologist which includes an echocardiography should be performed annually [6].

In Stage 2 competitions and athletic hard routes should be avoided. The limitation is too significant, and an acute problem at such high loads cannot be excluded for sure. Climbing for fun at walls or slabs should be possible without major problems in most cases [6].

In Stage 3 climbing is more or less dominated by social factors. Some easy short and sloped routes may be possible. Belaying is also possible; leading may be critical. Any kind of "pushing the limits" or competition is strictly forbidden.

Stage 4 limits the patient to look and meet others when they are climbing, often with supplemental oxygen. The patient will not be able to climb anymore. This is not a question for sport medicine anymore [6].

Valvular defects are similar to cardiac insufficiency when seen from the viewpoint of exercise physiology. The advice will be therefore guided by the recommendations for cardiac insufficiency [6]. In the case of valvular transplant, the usual infection prophylaxis or consequent therapy of any infection is a must.

17.5 Pulmonary Diseases

Summary

- **Limiting factors:** Whole body load or maximal force, depending on age and type of pulmonary disease. Avoid strictly forced respiration during expulsion when bullae are known or a pneumothorax has happened in the patient's history! In the case of COPD or silicosis, breathing minute volume or VO_2 may define the limitation.
- **Additional risk factors:** Exercise with active infections in the respiratory system, ignorance of the problem or overconfidence

Again data are scarce. However, there are some studies concerning relevant parameters. In their already referenced papers, the groups of Watts and those of Mermier draw different conclusions from each other regarding the breathing volume and VO_2 of motivated amateur climbers (which means difficulties up to UIAA grade VIII) and increases about three- to fourfold compared to resting values [18, 19]. In competition climbers, Booth et al. (1999) found an six- to eightfold increase of the respective values. These studies give at least some evidence about the minimal requirements of the lung differentiated according to the sportive intention of the climber.

Checkup for a Climber with Pulmonary Disease

- **History:** Exact diagnosis of the disease, allergies with symptoms of the respiratory system, exercise-dependent symptoms, general and specific sport experience, environmental factors (cold or exercise-induced asthma)
- **Actual status:** Actual exercise tolerance
- **Labor:** Red blood cell count, blood gases at rest and during exercise
- **Technical investigations:** Spiroergometry or, alternatively, ergometry with blood gas analysis at every loading increment, X-ray of the lungs
  With individual indication only: CT thorax, exclusion of exercise-induced asthma (spirometry before and 5–10 min after maximal exercise)

A low risk may be assumed and sport climbing therefore continued when the symptoms have been therapeutically stabilized and the minimal criteria as described above are fulfilled [6].

A high risk with the consequence that climbing should be stopped at least for a while until the situation has been stabilized, e.g., by drugs. This would be the case if the maximum breathing volume or VO_2 should be significantly lower than mentioned above. At least a modification of climbing toward less strenuous climbs will be necessary. Beyond this there is a significantly increased risk in the case of large bullae, recurring spontaneous pneumothorax in the patient's history, any acute decompensated or otherwise unstable situation, not or insufficiently treated exercise-induced asthma, cold-induced asthma without special advise how to handle that in the field, relevant hypoxemia when exercising, and severe pulmonary arterial hypertension (systolic >40 mmHg).

Extreme forced respiration during expulsion should be principally avoided, although there are only a few information about incidences and which are all from other sports [47, 48]. Climbing tactics should be adjusted to the sometimes significantly changing status of the patient. For asthma patients or those with COPD, this means that regular peak flow measurements should be performed and documented. With this simple self-test, the patient can monitor and be forewarned of any early worsening in asthma symptoms and adjust medication accordingly before the symptoms become much worse.

17.6 Renal Insufficiency

Summary

- **Limiting factor:** Fluid and electrolyte balance, but not style or intensity of climbing
- **Additional risk factors:** Hot climate/dehydration, catabolic metabolism from training too hard
- **Common mistakes:** Ignorance of fluid or electrolyte balance, climbing during acute infections of the urinary system

Is renal function worsened by climbing? Direct effects are not expected, but indirect ones may occur. The etiology of the underlying condition should be diagnosed and—if possible—treated effectively. In the case of autoimmune diseases, although fortunately rare, an immuno-depressive therapy may be necessary creating limitations for several sports, and because additional infections may further harm the kidneys. The advice to the climber should be to comply with the severity of the disease.

The German classification system for renal insufficiency (Table 17.2) shows two subgroups for Class 3, and neither has any relevance on the advice given to the climber. In stadium 1 and 2, the patient will normally not have any problems when climbing. In stadium 3, the monitoring of fluid balance will be more and more important; in stadium 4, it is a must. If an arteriovenous shunt for dialysis is inserted, it must be protected against any trauma, also minor ones (cover it with soft material). Competition climbing is principally possible in most cases of stadium 1–3 under optimal nephrological coaching. Ideally, the nephrologist is also a climber or has at least some personal experience in climbing.

Table 17.2 Differentiation of chronic renal insufficiency according to ICD-10

ICD-10 code	Description
N18.1	Chronic renal insufficiency, Stadium 1—defect with normal or increased glomerular filtration rate (GFR 90 mL/min or more)
N18.2	Chronic renal insufficiency, Stadium 2—defect with mild decreased glomerular filtration rate (GFR 60–90 mL/min)
N18.3	Chronic renal insufficiency, Stadium 3—defect with moderate decreased glomerular filtration rate (GFR 30–60 mL/min)
N18.4	Chronic renal insufficiency, Stadium 4—defect with significant decreased glomerular filtration rate (GFR 15–30 mL/min)
N18.5	Chronic renal insufficiency, Stadium 5—chronic uremia, terminal renal insufficiency

After a kidney transplantation, the transplant is not well protected somewhere deep in the body but quite superficially situated, and therefore there is a significant risk for injury. Here, the construction and the fitting of the climbing harness are of special importance. Most sit harnesses are constructed in a manner which applies pressure on the region where the transplant is located, especially in the case of a fall [49]. The advice here is to select a harness that fits perfectly on the lower part of the hip that avoids any lifting that might press on the transplant in the event of a fall. The fit should be checked regularly when climbing. If this is observed, there are no other medical contraindications to climb for those with kidney transplant [49].

17.7 Seizures

On the topic "seizures and sport climbing," there are no systematic studies available. However, there is one international group of experts who has been working intensively about the topic "seizures and mountaineering" [13]. The following statements are mainly based on their work but adapted slightly for sport climbing.

Summary

- **Limiting factors:** Risk caused by acute seizures. During leading, risk of uncontrolled falls. During belaying, complete failure of the safety chain with the possibility of fatal falls of the leading climber
- **Additional risk factors:** Change of plasma concentrations of anticonvulsive drugs in very hot weather (=dehydration), mistakes in drug intake, and others
- **Common mistakes:** Underestimating of the potential danger when there were no seizures for a longer time

Check of Climbing Persons with Seizures

- **History:** Frequency of attacks? When was the last one? Is there an aura? Medication/compliance?
- **Actual status:** Nothing special
- **Laboratory:** Normally not necessary, if indicated drug monitoring
- **Technical investigations:** Normally not necessary

With this group of diagnoses, the tactics of climbing is of extraordinary importance. When bouldering near the ground, there should be a height which allows the drop and with a spotter behind the climber—this is safe in nearly all situations. Take care that there are no big boulders nearby where the climber might land after a fall! When climbing on "French routes" where there is a bolt every 1 or 2 m, it is also safe for the patient to lead. Realistically, these are not real leads but more of a "self-established top-rope" as two bolts can be clipped from the same position and are in front of the climber's body. In the case of a seizure during the lead, the belayer should let the climber down with controlled speed, while—if possible—another climber cares and protects the head of the victim.

The advice differs once the distance between bolts becomes longer. Here grounders (falls where the climber lands on the ground instead of in the rope/harness) are a special danger for persons with seizures. When sport climbing outdoors on rocks, the situation again is completely different. Here, the distances between the bolts are usually longer, and falls may be long. In such situations, leading is not recommended for persons with seizures. If abseiling should be necessary, it is possible for these patients; however, they should be belayed with another rope or by a companion who may put load on the rope to block the brake in the case of a problem of the patient.

The role of the belayer if he should suffer from a seizure is completely different as this may cause a complete failure of the safety chain. If seizure patients belay others, they should at least use fully automatic belaying systems (e.g., grigri).

The important question is how long an increased risk should be assumed and when limitations may be eased. In contrast to urban traffic, this has not been regulated by legislation. However, traffic regulations may usefully serve as a reference model since there are also dangerous situations that may arise if a person should become incapacitated when driving. Driving regulations differentiate several situations but also take into account that probably a car may put several persons at risk which is not the case in climbing. Therefore, a seizure-free period of at least 1 year may be used as benchmark to resume leading. In some cases, it may be possible to reduce this to 6 months when a specialist advises the patient and no seizures have occurred without drugs. For alpine climbing, the situation should be handled more strictly. Here, at least 2 years without seizures should be recommended.

It must be pointed out that these are consensus recommendations of specialists from 35 countries (UIAA Medical Commission) which were put together in all conscience from non-systematic studies, observations from practice and case reports. They are not evidence-based. It may be necessary to modify them if new studies should be available. The most recent version will be always found on the website of UIAA MedCom at www.theuiaa.org/medical_advise.html.

17.8 Conclusion

Most pre-existing diseases do not mean "stop climbing!" However, a detailed and specific assessment of the patient is a must, requiring a specific competence. It is advantageous if the advising physician is also a climber as many may not be aware of climbing's unique physiology and thus unnecessarily advise terminating all climbing participation, when all that may be needed is a slight modification to climbing participation. In this way, it is possible to retain the ongoing physical and mental health benefits of sport climbing for the patient, which also improves the symptoms and outcome of the disease itself.

References

1. Liguori G. In: (ACSM) ACoSM, editor. ACSM's guidelines for exercise testing and prescription. 11th ed. Wolters Kluwer Health, Inc.; 2021.
2. Zhang Y, Xu L, Yao Y, Guo X, Sun Y, Zhang J, et al. Effect of short-term exercise intervention on cardiovascular functions and quality of life of chronic heart failure patients: a meta-analysis. J Exerc Sci Fit. 2016;14(2):67–75.
3. Pescatello LS, MacDonald HV, Lamberti L, Johnson BT. Exercise for hypertension: a prescription update integrating existing recommendations with emerging research. Curr Hypertens Rep. 2015;17(11):87.
4. Engbert K, Weber M. The effects of therapeutic climbing in patients with chronic low back pain: a randomized controlled study. Spine. 2011;36(11):842–9.
5. Karg N, Dorscht L, Kornhuber J, Luttenberger K. Bouldering psychotherapy is more effective in the treatment of depression than physical exercise alone: results of a multicentre randomised controlled intervention study. BMC Psychiatry. 2020;20(1):116.
6. Küpper T, Morrison A, Gieseler U, Schöffl V. sport climbing with pre-existing cardio-pulmonary medical conditions. Int J Sports Med. 2009;30:395–402.
7. Sheel AW, Seddon N, Knight A, McKenzie DC, DE RW. Physiological responses to indoor rock-climbing and their relationship to maximal cycle ergometry. Med Sci Sports Exerc. 2003;35(7):1225–31.
8. Sheel AW. Physiology of sport rock climbing. Br J Sports Med. 2004;38(3):355–9.
9. Scherer D, Kaltenbach M. [Frequency of life-threatening complications associated with exercise testing (author's transl)]. Dtsch Med Wochenschr 1979;104(33):1161–1165.
10. Gibbons L, Blair SN, Kohl HW, Cooper K. The safety of maximal exercise testing. Circulation. 1989;80(4):846–52.
11. Fletcher GF, Ades PA, Kligfield P, Arena R, Balady GJ, Bittner VA, et al. Exercise standards for testing and training: a scientific statement from the American Heart Association. Circulation. 2013;128(8):873–934.
12. Milledge J, Küpper T. Consensus Statement of the UIAA Medical Commission Vol. 13: People with pre-existing conditions going to the mountains. Bern/Switzerland: The Medical Commission of the Union Internationale des Associations d'Alpinisme (UIAA); 2008. www.theuiaa.org/medical_advice.html.

13. Angelini C, Giardini G. Consensus Statement of the UIAA Medical Commission Vol. 16: Travel to altitude with neurological disorders. Bern/Switzerland: The Medical Commission of the Union Internationale des Associations d'Alpinisme (UIAA); 2009.
14. Donegani E, Hillebrandt D, Windsor J, Küpper T, Gieseler U, Rodway GW. Consensus statement of the UIAA Medical Commission Vol. 21: People with pre-existing cardiovascular conditions going to the mountains. Bern/Switzerland: Union International des Associations d'Alpinisme, Medical Commission; 2012. www.theuiaa.org/medical_advice.html.
15. Donegani E, Hillebrandt D, Windsor J, Gieseler U, Rodway GW, Schöffl V, et al. People with pre-existing cardiovascular conditions going to the mountains consensus statement of the medical commission of the Union Internationale des Associations d'Alpinisme (UIAA MedCom). Travel Med Inf Dis (TMAID). 2014;12(3):237–52.
16. Hillebrandt D, Gurtoo A, Küpper T, Richards P, Schöffl V, Shah P, et al. UIAA medical commission recommendations for mountaineers, hillwalkers, trekkers, and rock and ice climbers with diabetes. High Alt Med Biol. 2018.
17. Watts PB. Physiology of difficult rock climbing. Eur J Appl Physiol. 2004;91(4):361–72.
18. Watts PB, Daggett M, Gallagher P, Wilkins B. Metabolic response during sport rock climbing and the effects of active versus passive recovery. Int J Sports Med. 2000;21(3):185–90.
19. Mermier CM, Robergs RA, McMinn SM, Heyward VH. Energy expenditure and physiological responses during indoor rock climbing. Br J Sports Med. 1997;31(3):224–8.
20. Williams ES, Taggart P, Carruthers M. Rock climbing: observations on heart rate and plasma catecholamine concentrations and the influence of oxprenolol. Br J Sports Med. 1978;12(3):125–8.
21. Mermier CM, Janot JM, Parker DL, Swan JG. Physiological and anthropometric determinants of sport climbing performance. Br J Sports Med. 2000;34(5):359–65; discussion 66
22. Booth J, Marino F, Hill C, Gwinn T. Energy cost of sport rock climbing in elite performers. Br J Sports Med. 1999;33(1):14–8.
23. Limonta E, Brighenti A, Rampichini S, Ce E, Schena F, Esposito F. Cardiovascular and metabolic responses during indoor climbing and laboratory cycling exercise in advanced and elite climbers. Eur J Appl Physiol. 2018;118(2):371–9.
24. Maron BJ, Levine BD, Washington RL, Baggish AL, Kovacs RJ, Maron MS. Eligibility and disqualification recommendations for competitive athletes with cardiovascular abnormalities: task force 2: preparticipation screening for cardiovascular disease in competitive athletes: a scientific statement from the American Heart Association and American College of Cardiology. J Am Coll Cardiol. 2015;66(21):2356–61.
25. Boden BP, Breit I, Beachler JA, Williams A, Mueller FO. Fatalities in high school and college football players. Am J Sports Med. 2013;41(5):1108–16.
26. Gouttebarge V, Ooms W, Tummers T, Inklaar H. Mortality in international professional football (soccer): a descriptive study. J Sports Med Phys Fitness. 2015;55(11):1376–82.
27. Morentin B, Suarez-Mier MP, Monzo A, Molina P, Lucena JS. Sports-related sudden cardiac death due to myocardial diseases on a population from 1–35 years: a multicentre forensic study in Spain. Forensic Sci Res. 2019;4(3):257–66.
28. Roberts JA. Viral illnesses and sports performance. Sports Med (Auckland, NZ). 1986;3(4):298–303.
29. Breinholt JP, Moulik M, Dreyer WJ, Denfield SW, Kim JJ, Jefferies JL, et al. Viral epidemiologic shift in inflammatory heart disease: the increasing involvement of parvovirus B19 in the myocardium of pediatric cardiac transplant patients. J Heart Lung Transplant. 2010;29(7):739–46.
30. Kindermann I, Ukena C, Mahfoud F, Böhm M, Yilmaz A, Klingel K. Myokarditis-update. Kardiologe. 2016;10(5):311–30.
31. Yilmaz A, Klingel K, Kandolf R, Sechtem U. A geographical mystery: do cardiotropic viruses respect national borders? J Am Coll Cardiol. 2008;52(1):82; author reply-3.
32. Küpper T, Schraut B, Rieke B, Schöffl V, Steffgen J. Drugs and drug administration in extreme climates. J Travel Med. 2006;13(1):35–47.
33. Löllgen H. Bedeutung und Evidenz der körperlichen Aktivität zur Prävention und Therapie von Erkrankungen. Dtsch Med Wochenschr. 2013;138(1):1–8.
34. Löllgen H, Dickhuth HH, Dirschedl P. Vorbeugung durch körperliche Bewegung. Dt Ärztebl. 1998;95(24):B1228–B34.
35. Löllgen H, Gerke RJ, Steinberg T. Der kardiale Zwischenfall beim Sport. Dt Ärztebl. 2006;103(23):B1382–B7.
36. Küpper T. Sportklettern mit Hypertonie. In: Kupper T, Ebel K, Gieseler U, editors. Moderne Berg- und Höhenmedizin. 2nd. ed. Stuttgart: Gentner Verlag; 2011. p. 563–5.
37. Aras D, Akalan C, Koz M, Ileri M. Does long term sport rock climbing training affect on echocardiography and heart rate variability in sedentary adults? A randomized, and controlled study. IJAEP. 2016;5(5):26–35.
38. Ferguson RA, Brown MD. Arterial blood pressure and forearm vascular conductance responses to sustained and rhythmic isometric exercise and arterial occlusion in trained rock climbers and untrained sedentary subjects. Eur J Appl Physiol Occup Physiol. 1997;76(2):174–80.
39. Samek L, Löllgen H. Belastungs-EKG: Ischämiediagnostik. In: Lollgen H, Winter HJ, Erdmann E, editors. Ergometrie—Belastungsuntersuchungen in Klinik und Praxis. Berlin/Heidelberg: Springer Verlag; 1995. p. 111–26.

40. Löllgen H. Kardiopulmonale Funktionsdiagnostik [Cardiopulmonary function diagnosis]. 2. Aufl. ed. Basel: Ciba; 1990.
41. Lollgen H, Winter UJ. Ergometrien in der Prävention—Reihenuntersuchungen. In: Lollgen H, Winter UJ, Erdmann E, editors. Ergometrie—Belastungsuntersuchungen in Klinik und Praxis. Berlin/Heidelberg: Springer Verlag; 1995. p. 319–22.
42. Limb D. Injuries on British climbing walls. Br J Sports Med. 1995;29(3):168–70.
43. Hollmann W, Hettinger T. Sportmedizin, Grundlagen für Arbeit, Training und Präventivmedizin. Stuttgart: Schattauer; 2000.
44. Morgan BJ, Alexander JK, Nicoli SA, Brammell HL. The patient with coronary heart disease at altitude: observations during acute exposure to 3100 meters. J Wilderness Med. 1990;1:147–53.
45. (anonymous). The criteria committee of the New York Heart Association. Diseases of the heart and blood vessels: nomenclature and criteria for diagnosis. In: N.N., editor. The Criteria Committee of the New York Heart Association diseases of the heart and blood vessels: nomenclature and criteria for diagnosis. 6th. ed. Boston, MA: Little Brown; 1928.
46. (anonymous). The Criteria Committee of the New York Heart Association. Nomenclature and Criteria for Diagnosis of Diseases of the Heart and Great Vessels. In: N.N., editor. The Criteria Committee of the New York Heart Association Nomenclature and Criteria for Diagnosis of Diseases of the Heart and Great Vessels. 9th ed. ed. Boston, Mass.: Little, Brown & Co; 1994. p. 253–6.
47. Sadarangani S, Patel DR, Pejka S. Spontaneous pneumomediastinum and epidural pneumatosis in an adolescent precipitated by weight lifting: a case report and review. Phys Sportsmed. 2009;37(4):147–53.
48. Davis PF. Primary spontaneous pneumothorax in a track athlete. Clin J Sport Med. 2002;12(5):318–9.
49. Welch TR. Climbing harness fit in kidney transplant recipients. Wilderness Environ Med. 1999;10(1):3–5.
50. Hoppe UC, Böhm M, Dietz R, Hanrath P, Kroemer HK, Osterspey A, et al. Leitlinien zur Therapie der chronischen Herzinsuffizienz (2005). Z Kardiol. 2005;94:488–509.

Sport Climbing During Pregnancy 18

T. Küpper and A. Morrison

Isabelle Schöffl in the first ascend of "Dreamcatcher",
Na Pha Daeng, Laos, Photo Volker Schöffl

T. Küpper (✉)
Institute of Occupational, Social and Environmental
Medicine, RWTH Aachen Technical University,
Aachen, Germany
e-mail: tkuepper@ukaachen.de

A. Morrison
Royal Free London NHS Foundation Trust,
London, UK

V. Schöffl et al. (eds.), *Climbing Medicine*, https://doi.org/10.1007/978-3-030-72184-8_18

With its low risk of trauma, broad variety of moves, freedom to choose the individual load intensity, and technical difficulty, sport climbing developed into a popular recreational and competition sport [1]. Many years ago, climbing during pregnancy was considered to be a "no go" as data was sparse. The sport was also subjectively perceived as being a high-risk activity, even though significant objective data demonstrated its safety and low risk of injury and trauma compared with other sports (e.g., contact sports). The majority of medical professionals do not have an appreciation or understanding of the risks involved in the sport, and how they can be managed. The aim here is to provide information about sport climbing during pregnancy to care for the mother and fetus without being too restrictive.

It is now broadly encouraged that women with uncomplicated pregnancies should be encouraged to continue their aerobic and strength-conditioning exercises before, during, and after pregnancy, albeit with some modification due to anatomic and physiologic changes, type of exercise, and fetal requirements [2].

There is considerable medical literature about sports and pregnancy—but not including the different disciplines of climbing. However, even with this wealth of data in other sports, the quality of studies is often limited and subject to a more or less significant bias [3]. This includes the effect of physical exercise of moderate to vigorous intensity and frequency in the different trimesters as additional independent variables [4].

Some exercises that are recognized as safe for women with uncomplicated pregnancies who participated in such exercise previously in a meta-analysis include swimming, yoga (modified), pilates (modified), and strength training in consultation with an obstetric care provider [4]. This is also true for moderate strength training which shows some biomechanical similarities to sports climbing [5–9]. There are few maternal medical conditions where aerobic exercise is absolutely contraindicated [2]. However, sports featuring sudden deceleration or risk of collision are to be avoided—e.g., downhill ski racing, ice hockey, equestrian activities, pole vaulting [10].

With respect to sport climbing exercise, it should be differentiated whether the pregnant woman is an experienced climber or not. The former can continue climbing with some minor constrains, and the latter should be advised not to start climbing until the pregnancy has been completed. This strategy is recommended for any sport with complex strong and supple moves, (potential) risk of trauma, requirement for experience in using specialist equipment and techniques for safety (e.g., belaying and abseiling), and increased intra-abdominal pressure or which shows characteristics of competition.

Sports to be avoided for physiological reasons include diving as the fetus is not protected from decompression problems which risks malformation and gas embolism after decompression disease. Sport climbing or training at high altitude also presents physiological challenges, a topic discussed later.

All recommendations provided below are for active and experienced sport climbing women with uncomplicated pregnancies. The pathbreaking principle is the question of how the woman feels. In the case of a high-risk pregnancy, any advice depends significantly on the individual risk and also on the mentality of the woman and her need for security. Such advice would fall beyond the scope of this paper.

18.1 Synopsis

Specific factors of sport climbing with relevance for pregnant women are the following:

- Experienced climber with uncomplicated pregnancy
- Risk of trauma
- Level of exertion
- Abdominal pressure applied by the sit harness vs. whole body harness
- Stage (first, second, or third trimester)
- Shift of the body's center of gravity during pregnancy
- Reduced stability of ligaments and joints

Principally, there is a need to weigh the advantages of continued sport participation during pregnancy against the possible disadvantages or risks. There are clear advantages to any woman who continues to exercise during her pregnancy: increase of psychological and physical well-being, better weight control, a reduction of gestational diabetes by almost 50% and a decrease of the risk of pre-eclampsia by about 40%, a lower incidence of preterm birth and Cesarean birth, and also an improvement of other factors, e.g., quality of sleep and less gestational depression [6, 10–18].

Three meta-analysis found that birth weights between those who exercised during pregnancy, or did not, was minimal to none. However, those who exercised "vigorously" during the third trimester were more likely to deliver infants weighing 200–400 g less than comparable controls, although there was not an increased risk of fetal growth restriction [2]. Vigorous was defined as up to 85% of capacity, though it is still not known what absolute level of intensity and duration, or both, that if exceeded could place the fetus at risk.

Trained women may continue their exercise during pregnancy and can realize significant strength gains [10, 19]. Table 18.1 provides an overview of the relative and absolute contraindications and warning signs which should be considered.

The risk of trauma is very low in sports climbing compared to other sports—e.g., the injury incidence is significantly lower than in contact sports or even school volleyball. Table 18.2 provides more detail. Most of the injuries sustained in sports climbing are of minor severity [20]. Prolonged medical treatment or in-patient treatment is extremely rare in relation to the large number of climbers [1, 21, 22]. The injury incidence in pregnant versus non-pregnant women was marginally higher, 0.28 versus 0.23/10.000 h of climbing, respectively, but all of these injuries were again of very minor severity, and such data are scarce [20]. The annual accident statistics of the German Alpine Club and others show that more than 50% of all severe accidents in sports climbing are caused by significant mistakes and failures during belaying [23]. This reinforces the need that all climbers—pregnant or not—must always have regard for perfect belaying technique.

Table 18.1 Relative and absolute contraindications for sport climbing during pregnancy and warning symptoms (Modified and amended from ACOG Committee Opinion, 2020)

Relative contraindication	Absolute contraindication	Warning symptoms
Severe anemia	Incomplete closure of the cervix	Vaginal bleeding
Extreme underweight (anorexia)	Persistent bleeding in the second or third trimester	Dizziness or sight disorder
Not clarified arrhythmia of the mother	Premature rupture of the membranes	Headache
Chronic bronchitis	Gestational hypertension	Thoracal pain, palpitations, conspicuous fatigue
Stress-induced asthma	Cardial disease with significant hemodynamic effect	Muscular weakness affecting balance
Heavy smoking	Restrictive pulmonary disease	Dyspnea before onset of exercise
Any of the following situations if not stabilized perfectly: Type 1 diabetes, hypertension, seizures, preeclampsia, hyperthyreosis	Multiple high-risk pregnancies or preterm birth	Edemas of the lower leg until deep vein thrombosis has been excluded
Orthopedic limitations	Placenta previa later than week 26	Reduced fetal activity
Couch potato	Premature uterine contractions	Premature uterine contractions or abdominal pain
Intrauterine retardation		Outflow of amniotic liquor
Patent foramen ovale with significant right-left shunt		Sudden onset of edemas of the ankle, face, or hands

Table 18.2 Injury and trauma risk of several sport disciplines. Note: The table does not give any information about the severity of the respective injuries (Modified and amended from [20, 21, 24, 46–48])

Discipline	Accidents per 1000 h of activity	Discipline	Accidents per 1000 h of activity
Rugby, amateurs	283	Soccer National League Germany	3.10
Rugby, professionals	150	Competition climbing	3.10
Ice hockey, professionals	83	Triathlon	2.50
Rugby, youth	57	Boxing	2.00
Handball, women	50	Mountain biking	1.00
Soccer, champions league	31.6	Ski/Snowboard (downhill)	1.00
"Classic" mountaineering, 1980s	37.5	Nordic walking	0.90
Motorbike sport	22.4	"Classic" mountaineering, 2010s	0.56
American football	15.7	Wind surfing	0.41
Sailing	8.8	Sport climbing	0.027–0.079
Polo	7.8	Sport climbing, men	0.19
Kite surfing	7.0	Sport climbing, women	0.23
Volleyball at school	6.7	Sport climbing, pregnant women	0.28
Ice climbing	4.07		

Generally, it is recommended that pregnant women should avoid disciplines with increased risk for trauma. Compared to other sports, the risk of injuries in sports climbing is very low (Table 18.2). Additionally, it should be pointed out that most of the few injuries which occur are minor (e.g., abrasions or wounds), which do not require professional medical care with a score of NACA I or NACA II, and no fatalities [20, 24, 25]. However, there are some considerations which may reduce this risk further. First of all, climbing during pregnancy does not mean "pushing your limits" or competition climbing. During pregnancy, other aspects prevail as there are social factors and activity and exercise in general. Top rope belaying and avoiding the "sharp end" of the rope also increase the safety for the women although the height of a fall is usually low in sports climbing, and the bolts are perfectly fixed. Special care should be given to avoid "grounders" (falls to the ground), a type of accident which is still needlessly common in climbing gyms [23].

In general, the environment of an indoor climbing gym is safer than outdoor climbing where there are objective risks like breaking handholds, rock fall, scree, and others. However, other risks are present at climbing gyms—crowded halls and mass gatherings especially in early evening or at weekends. Here, climbers should take care to be socially distanced to minimize collision accidents when a climber falls from above or nearby [23].

A specific problem for pregnant climbers concerns the risk of trauma by wearing a sport climbing harness. In the first trimester, the fetus is situated completely within the pelvis, and this protects it as a knight's armor would do. Therefore, there is no problem with the fitting of the harness. During the second trimester, the abdomen will enlarge and emerges from the protective pelvis presenting two problems. The first concerns a shift forward of the center of gravity, a change in postural balance which will often go unnoticed and uncompensated for some long time by the women because it develops slowly. For example, falling is a common cause of injury in the general pregnant population as pregnant woman are two to three times more likely to fall compared to non-pregnant women due to changes in center of gravity and joint laxity [10]. The climber's growing bump and larger breasts will push her further away from the wall influencing both grip and footwork. Try climbing with more "open" positions rather than twisting, deep bending, or crunch moves as the pregnancy progresses. The second problem that may arise is that the climbing harness may cause direct pressure on the lower abdomen, especially in the case of a fall, though there is no data. Nevertheless in the third trimester, the woman should consider switching to using a classic whole-body harness where the abdominal strap is located extremely low. Wearing one when abseiling or being lowered can feel awkward and "pushed out" as the woman's center of balance will be moved as the rope is tied in at the chest and not the waist as with a sport climbing harness.

Some women develop edema, including to the feet, especially in the third trimester [10]. This may necessitate the need for new climbing shoes. Generally, comfortable climbing shoes which do not constrict are recommended [26]. Edema may be linked to pre-eclampsia and hypertension which should be investigated; however, there is no published data linking edema to prenatal exercise [10].

The normal course of pregnancy causes significant physical and metabolic load for the body which necessitates regular rests. Resting cardiac output increases by 50% [6, 27], and tidal volume and minute volume also increase significantly [6, 28]. Add to this a steadily increasing rise in the diaphragm's position, and it is not surprising that many pregnant women complain about shortness of breath, especially when exercising [6, 29]. Unlike cycling or running where exercise programs can simply prescribe a steady state heart rate to aim for, this steady state goal has never been applied to sport climbing whether pregnant or not as it is short bursts of intermittent exercise. When feeling breathless or uncomfortable, climbers generally enjoy a break at an early stage of such complaints [26]. For any kind of sports or work, pregnant women should know that any subjective estimation of load by pulse rate is not reliable during pregnancy and that the load is significantly higher than the subjective estimation would be [6, 30].

The increase of body weight and changes on the angles of the joints may cause significant musculoskeletal adjustments and loads [6]. The stress of the knees and the hip joints may increase by 100%. This may worsen preexisting instabilities or defects of these joints. Physiologically, the different hormonal pattern during pregnancy causes an increasing relaxation of the ligaments. Overload problems of the shoulders, ankles, and fingers in pregnant climbers were found [31, 32]. Some research found a general increased risk for injuries of the ligaments around the ankles in women, independent of pregnancy [20], while others did not [31]. Therefore, the influence of relaxed ligaments on the injury risk of pregnant women is somehow unclear. So far there is a lack of studies on sportive pregnant women [6]. Nevertheless, good climbing technique is recommended to avoid damages of ligaments and joints.

Choosing an appropriate climbing route is very important. First of all, to avoid "pushing the limits" means that the climb should be at least one grade easier (better two) than the woman was able to climb before her pregnancy. If she climbs routes which she already knows, this will further reduce the load and therefore the strain on her body. She should also not climb overhangs or roofs to avoid a significant increase of intra-abdominal pressure and favor slabs. High intra-abdominal pressure may decrease stability of the pelvic floor, and then problems like incontinence or prolapse may occur [6].

One of the rare studies concerning pregnant women climbers revealed some very interesting and relevant results, with the following figures reported as an average and the range in brackets [1]. The study group consisted of experienced climbers who had climbed for 11.1 years [2–21] before pregnancy, and 55.6% were normally leading. In the main, they continued climbing for 6.3 months of their pregnancy and were belaying for 5.8 months [2–9]. There was an inverse relationship whereby every 3 months the portion of women leading climbs decreased significantly, while top rope belaying increased. The climbing grade difficulty while leading decreased in the first trimester to 6.3 (4.00–9.33, UIAA decimal scale), to 5.8 (4.33–7.33) in the second trimester, and to 5.1 (4.33–7.33) in the third. The tendency was similar in top roping or seconding: 6.7 (4.33–8.0), 6.3 (4.00–7.66), and 5.6 (4.00–

7.66), respectively. The frequency of climbing also decreased from 32.2 h/month before pregnancy to 20.6 h in the first trimester, 16.3 h in the second, and 6.1 h in the third. These women did not receive any sport-specific advice, and yet the majority instinctively adapted their climbing exercise with their pregnancy to meet the same advice a doctor or midwife who is experienced climber would give. This also indicates that the acceptance of specific advice should be high in contrast to other situations in daily life. Those women who stopped climbing because of the pregnancy did so because of fear or unsureness concerning health of the fetus (33.9%), and in 7.1% they were pressed by their partner to do so. Both these reasons would indicate seeking professional medical advice to allay any concerns.

When the recommendations given above were followed, no negative consequences from the climbing activity on the fetus were found [1]. However, data are sparse. There is only one study available where this was investigated more in detail [33]. The study group consisted of 81% ($N = 32$) who were primipara (singleton pregnancy), and had a mean climbing experience of 9 years [2–24]. About half of them climbed until the third trimester, and 90% of them with reduced intensity. With a mean duration of pregnancy of 39.5 weeks (+/−1.7), a mean birth weight of 3543gr (+/−403) and a size of 50.9 cm (+/−2.1) of exclusively healthy and completely developed babies were born after a normal duration of pregnancy. Two women gave birth at week 36, but there was no connection to their climbing activity.

Pregnant women should be made aware of environmental factors. It does not make any sense to climb in extreme heat [34]. They should also care for good fluid balance to optimize peripheral and placental perfusion.

Since there are sport climbing facilities not only at sea level but also at altitude in alpine villages, some notes are added here concerning altitude sojourns of pregnant women, especially if they do some exercise. Again we are only considering otherwise healthy women without a high-risk pregnancy here. Any recommendations are limited as most studies researched pregnant indigenous populations who live permanently at high altitude (e.g., Tibetans) and whose generations of genetics will increase placental blood and oxygen supply to protect the fetus. Knowledge of pregnant lowlanders with acute exposure at moderate altitude (hours), exercising or not, is scarce (Leal et al., 2008).

We exclude the topic altitude diseases here for two reasons: (1) there are no gender-dependent differences concerning incidence, symptoms, prevention, and therapy of acute mountain sickness (AMS) [35] and (2) because pregnant women should prevent any altitude-related problem by adequate altitude profiles even more than non-pregnant people.

With respect to altitude medications and exotic locations, the use of acetazolamide is contraindicated during the first trimester (risk of teratogenicity) and after 36 weeks of pregnancy (risk of severe neonatal jaundice). Some drugs used for prophylaxis or treatment are also contraindicated during pregnancy: most antimalarials, quinolones, sulfonamide, and others [36].

Exercise presents competition for oxygenated blood supply to go to working muscles and the placenta, so there's a risk of fetal hypoxia or preterm labor [36]. Short stays without exercise, whether a few hours to a few days, are very low risk up to 2500 m, but no data is available. Long stays without exercise above 2500 m for weeks or months presents a higher maternal incidence of hypertension, pre-eclampsia, and placental abruption; and for the fetus there is intra-uterine growth retardation during the third trimester and low birth weight [36]. It is suspected there is a higher incidence of spontaneous abortion in the first trimester, but not proven [36]. The most commonly reported pregnancy complication by visitors to Colorado obstetrical care is preterm labor and bleeding complications [35]. Therefore, if the climber experienced difficulty becoming pregnant, it is recommended that she avoid high altitude.

There is no reason for any concern that the fetus may be in trouble if the woman goes to moderate altitude. For years it is well known that neurohumoral factors protect the fetus at rest and during exercise of the mother [37, 38]. An additional effect is caused by the hyperventilation of the mother which slightly overcompensates the effect of the reduced oxygen partial pressure at least up to 3100 m [39, 40]. Thousands of skiers prove these findings every winter. The fear that fetal growth retardation may occur has been proven for high altitude and prolonged stay only [41]. This can be neglected for sport climbing facilities in the European Alps.

The Medical Commission of the International Mountaineering and Climbing Federation (UIAA) gives conservative recommendations compared to the USA: acclimatized pregnant women may stay at 3000 m without any problem (USA: 4000 m). Significant exercise or workload should not be performed higher than 2500 m [36, 42, 43]. Other authors are even stricter—without supporting this by data—and recommend a limit of 1500–2000 m for significant workload [10, 44]. However, all such recommendations are fulfilled by sport climbing facilities in the European Alps and at most other places worldwide. There is a general consensus that above 2500 m any significant exercise should be started after 3–4 days of acclimatization only to avoid concurrency between the working muscles with the perfusion of the placenta [36, 41, 42, 45]. For the same reason, pregnant women should avoid hard exercise above 3000 m. It should be noted that these UIAA consensus recommendations were given by specialists from 35 countries which were established to their best knowledge because of a lack of good studies.

18.2 Conclusion

Extensive global literature strongly supports women with uncomplicated pregnancies to continue their aerobic and strength-conditioning exercises before, during, and after pregnancy, albeit with some modification due to anatomic and physiologic changes, type of exercise, and to fetal requirements. Experienced and active climbers with uncomplicated pregnancies can continue climbing throughout their pregnancy. Those who haven't climbed previously should wait until after their pregnancy before beginning. Sport climbing overall has a low incidence of injury, with the majority of injuries minor in severity (NACA I or NACA II). The risk of trauma is very low in sports climbing compared to other sports (e.g., contact sports). Though data specifically on pregnant climbers is sparse, most pregnant climbers already instinctively adopt the recommendations presented in this paper, while others unnecessarily stop early as unsure about whether it is safe to continue as there is very limited data to inform this specifically. Sport climbing is also often incorrectly and subjectively perceived as being a high-risk activity by medical professionals who do not understand or have an appreciation of the sport and the ways to manage the risks. This paper presents those risks and how they can be modified. This includes selecting your routes carefully, avoiding climbing roofs and overhangs, not pushing your climbing grade, and considering switching to a whole body harness in the third trimester.

References

1. Drastig J, Küpper T. Ist Sportklettern in der Schwangerschaft sicher? Sport Ortho Trauma. 2012;28:274–81.
2. (anonymous). Physical activity and exercise during pregnancy and the postpartum period: ACOG Committee Opinion, Number 804. Obstet Gynecol. 2020;135(4):e178–e88.
3. Kramer MS, McDonald SW. Aerobic exercise for women during pregnancy. Cochrane Database Syst Rev. 2006;2006(3):CD000180.
4. Poyatos-Leon R, Garcia-Hermoso A, Sanabria-Martinez G, Alvarez-Bueno C, Sanchez-Lopez M, Martinez-Vizcaino V. Effects of exercise during pregnancy on mode of delivery: a meta-analysis. Acta Obstet Gynecol Scand. 2015;94(10) 1039–47.
5. Avery ND, Stocking KD, Tranmer JE, Davies GA, Wolfe LA. Fetal responses to maternal strength conditioning exercises in late gestation. Can J Appl Physiol. 1999;24(4):362–76.
6. Bo K, Artal R, Barakat R, Brown W, Davies GA, Dooley M, et al. Exercise and pregnancy in recreational and elite athletes: 2016 evidence summary from the IOC expert group meeting, Lausanne. Part 1-exercise in women planning pregnancy and those who are pregnant. Br J Sports Med. 2016;50(10):571–89.
7. Clapp JF 3rd. Is exercise during pregnancy related to preterm birth? Clin J Sport Med. 2009;19(3):241–3.

8. Clapp JF 3rd. Does exercise training during pregnancy affect gestational age? Clin J Sport Med. 2009;19(3):241–3.
9. Olson D, Sikka RS, Hayman J, Novak M, Stavig C. Exercise in pregnancy. Curr Sports Med Rep. 2009;8(3):147–53.
10. Bo K, Artal R, Barakat R, Brown W, Dooley M, Evenson KR, et al. Exercise and pregnancy in recreational and elite athletes: 2016 evidence summary from the IOC expert group meeting, Lausanne. Part 2-the effect of exercise on the fetus, labour and birth. Br J Sports Med. 2016;50(21):1297–305.
11. (anonymous). SMA statement the benefits and risks of exercise during pregnancy. Sport Medicine Australia. J Sci Med Sport. 2002;5(1):11–9.
12. Baker JH, Rothenberger SD, Kline CE, Okun ML. Exercise during early pregnancy is associated with greater sleep continuity. Behav Sleep Med. 2018;16(5):482–93.
13. Bo K, Artal R, Barakat R, Brown WJ, Davies GAL, Dooley M, et al. Exercise and pregnancy in recreational and elite athletes: 2016/2017 evidence summary from the IOC expert group meeting, Lausanne. Part 5. Recommendations for health professionals and active women. Br J Sports Med. 2018;52(17):1080–5.
14. Garnaes KK, Morkved S, Salvesen KA, Salvesen O, Moholdt T. Exercise training during pregnancy reduces circulating insulin levels in overweight/obese women postpartum: secondary analysis of a randomised controlled trial (the ETIP trial). BMC Pregnancy Childbirth. 2018;18(1):18.
15. Magro-Malosso ER, Saccone G, Di Tommaso M, Roman A, Berghella V. Exercise during pregnancy and risk of gestational hypertensive disorders: a systematic review and meta-analysis. Acta Obstet Gynecol Scand. 2017;96(8):921–31.
16. Straughen JK, Sealy-Jefferson S, Bazydlo M, Helmkamp L, Misra DP. Physical activity and risk of preterm birth in a cohort of African American Women. Womens Health Issues. 2018;28(6):488–94.
17. Vargas-Terrones M, Barakat R, Santacruz B, Fernandez-Buhigas I, Mottola MF. Physical exercise programme during pregnancy decreases perinatal depression risk: a randomised controlled trial. Br J Sports Med. 2019;53(6):348–53.
18. Yu Y, Xie R, Shen C, Shu L. Effect of exercise during pregnancy to prevent gestational diabetes mellitus: a systematic review and meta-analysis. J Matern Fetal Neonatal Med. 2018;31(12):1632–7.
19. Bell R. The effects of vigorous exercise during pregnancy on birth weight. J Sci Med Sport. 2002;5(1):32–6.
20. Neuhof A, Hennig FF, Schöffl I, Schöffl V. Injury risk evaluation in sport climbing. Int J Sports Med. 2011;32(10):794–800.
21. Schöffl V, Morrison A, Schwarz U, Schöffl I, Küpper T. Evaluation of injury and fatality risk in rock and ice climbing. Med Sport. 2010;40(8):657–79.
22. Schöffl VR, Küpper T. Injuries at the 2005 World Championships in Rock Climbing. Wilderness Environ Med. 2006;17(3):187–90.
23. (anonymous). Kletterhallenunfallstatistik 2018 des Deutschen Alpenverein und KLEVER Kletterhallenverbandes; 2019. https://www.alpenverein.de/bergsport/sicherheit/unfallstatistik/kletterhallen-unfallstatistik-2018_aid:34127.html.
24. Schöffl VR, Hoffmann G, Küpper T. Acute injury risk and severity in indoor climbing-a prospective analysis of 515,337 indoor climbing wall visits in 5 years. Wilderness Environ Med. 2013;24(3):187–94.
25. (anonymous). NACA Score. Ames Research Centre, Moffett Field, California National Advisory Committee for Aeronautics; 1960 (rev. 1980).
26. Morrison A. Sportklettern in der Schwangerschaft. In: Küpper T, Ebel K, Gieseler U, editors. Moderne Berg- und Höhenmedizin. Stuttgart: Gentner Verlag; 2011. p. 571–4.
27. Pivarnik JM. Cardiovascular responses to aerobic exercise during pregnancy and postpartum. Semin Perinatol. 1996;20(4):242–9.
28. Heenan AP, Wolfe LA. Plasma acid-base regulation above and below ventilatory threshold in late gestation. J Appl Physiol (1985). 2000;88(1):149–57.
29. Milne JA, Howie AD, Pack AI. Dyspnoea during normal pregnancy. Br J Obstet Gynaecol. 1978;85(4):260–3.
30. O'Neill ME, Cooper KA, Mills CM, Boyce ES, Hunyor SN. Accuracy of Borg's ratings of perceived exertion in the prediction of heart rates during pregnancy. Br J Sports Med. 1992;26(2):121–4.
31. Drastig J, Küpper T. Gibt es geschlechtsspezifische Verletzungsmuster beim Sportklettern? Dt Ztschr Sportmed. 2014;65(3):66–71.
32. Folkl AK. Characterizing the consequences of chronic climbing-related injury in sport climbers and boulderers. Wilderness Environ Med. 2013;24(2):153–8.
33. Drastig J, Hillebrandt D, Rath W, Küpper T. Pregnant women in sport climbing—is there a higher risk for preterm birth? Z Geburtsh Neonatol. 2016;221(1):25–9.
34. Ravanelli N, Casasola W, English T, Edwards KM, Jay O. Heat stress and fetal risk. Environmental limits for exercise and passive heat stress during pregnancy: a systematic review with best evidence synthesis. Br J Sports Med. 2019;53(13):799–805.
35. Niermeyer S. The pregnant altitude visitor. Adv Exp Med Biol. 1999;474:65–77.
36. Jean D, Leal C, Meijer H. Consensus Statement of the UIAA Medical Commission Vol. 12: Women Going to Altitude: The Medical Commission of the Union Internationale des Associations d'Alpinisme (UIAA); 2008. www.theuiaa.org/medical_advice.html.
37. Parer JT. Effects of hypoxia on the mother and fetus with emphasis on maternal air transport. Am J Obstet Gynecol. 1982;142(8):957–61.

38. Parer JT. Evaluation of the fetus during labor. Curr Probl Pediatr. 1982;12(6):1–58.
39. Moore LG, Jahnigen D, Rounds SS, Reeves JT, Grover RF. Maternal hyperventilation helps preserve arterial oxygenation during high-altitude pregnancy. J Appl Physiol Respir Environ Exerc Physiol. 1982;52(3):690–4.
40. Moore LG, Brodeur P, Chumbe O, D'Brot J, Hofmeister S, Monge C. Maternal hypoxic ventilatory response, ventilation, and infant birth weight at 4,300 m. J Appl Physiol (1985). 1986;60(4):1401–6.
41. Moore LG. Fetal growth restriction and maternal oxygen transport during high altitude pregnancy. High Alt Med Biol. 2003;4(2):141–56.
42. Jean D, Moore LG. Travel to high altitude during pregnancy: frequently asked questions and recommendations for clinicians. High Alt Med Biol. 2012;13(2):73–81.
43. Baumann H, Huch R. [Altitude exposure and staying at high altitude in pregnancy: effects on the mother and fetus]. Zentralbl Gynakol 1986;108(15):889–99.
44. Entin PL, Coffin L. Physiological basis for recommendations regarding exercise during pregnancy at high altitude. High Alt Med Biol. 2004;5(3):321–34.
45. Jean D, Leal C, Kriemler S, Meijer H, Moore LG. Medical recommendations for women going to altitude. High Alt Med Biol. 2005;6(1):22–31.
46. Schöffl V. Unfallrisikoanalyse Klettersport. Erlangen-Nürnberg: Friedrich-Alexander-Universität; 2009.
47. Schöffl V, Schöffl I, Schwarz U, Henning F, Küpper T. Injury-risk evaluation in water ice climbing. Med Sportiva. 2009;13(4):1–9.
48. Schöffl V, Lutter C, Woollings K, Schöffl I. Pediatric and adolescent injury in rock climbing. Res Sport Med. 2018;26(51):91–113.

19 Sports-Medical Supervision of Competition Climbers and Climbing Competitions

Volker Schöffl and Isabelle Schöffl

Keo Schöffl, Fontainebleau, France, Photo Kilian Reil

V. Schöffl (✉)
Department of Orthopedic and Trauma Surgery, Center of Sportsmedicine, Klinikum Bamberg, Bamberg, Germany

School of Clinical and Applied Sciences, Leeds Becket University, Leeds, UK

Section of Wilderness Medicine, Department of Emergency Medicine, University of Colorado School of Medicine, Denver, CO, USA

Department of Orthopedic and Trauma Surgery, Friedrich-Alexander Universität Erlangen-Nürnberg, Erlangen, Germany

I. Schöffl
Department of Pediatric Cardiology, Friedrich-Alexander Universität Erlangen-Nürnberg, Erlangen, Bayern, Germany

V. Schöffl et al. (eds.), *Climbing Medicine*, https://doi.org/10.1007/978-3-030-72184-8_19

19.1 Introduction

In the advent of the inclusion of climbing into the Olympic programs of Tokyo and Paris, the sports-medical supervision of the climbing team has become more. We have already been conducting these exams for the German Alpine club for over 20 years, having performed Olympic-standard examinations long before climbing was on the Olympics list. In other countries, these examinations are newer and need to be organized by the national federations and Olympic centers. As we also take care of regional and state teams, a step concept was developed, in which we start with physical examinations, history, finger ultrasound, rest ECG, and a laboratory test with the lower teams and increase the extent and number of examinations up to the Olympic levels. Adolescent climbers need to be evaluated especially carefully because of their high risk for epiphyseal growth plate fractures [1]. Thus, long-term damage can be avoided and, in an injury, early intervention is possible. The general stages of children's development and maturation are monitored, and early signs of malnutrition must be detected. In the following sections, we discuss the most important components of the team examinations and competition care as well as the most important pathologies.

19.2 Climbing Team Examinations

The yearly team examination is an important tool for prevention and early detection of orthopedic and/or medical conditions. Reversible conditions, such as restricted range of motion in the fingers or growth plate fractures, can be detected early and counteracted. Additionally, the general stages of development of the children and adolescents must be considered. Only then is it possible to detect possible malnutrition, which is a frequent problem in young ambitious climbers. Musculoskeletal disorders such as muscle shortenings of the forearm flexors or a climber's back (humpback) must also be looked for. Highly intensive finger board and campus board training should not be performed in these age groups. While some youth climbing teams may have medical supervision, very few adult climbers do [2].

The international standards for these team examinations are recommended by the International Sports Climbing Federation (IFSC) Medical Commission in 1999 [3] and include:

Sports-medical examination (morphology climbing including weight, height, BMI, plicometry (body fat), flexibility, lung and heart auscultation, skin examination)
Standard Laboratory test (blood cell count, as indicated by the clinical examination)
ECG (12 channels) standard and stress test (either step test or bicycle)
Spiroergometry (bicycle or step test)
Echocardiography (first visit of a new team member, then at least every two years)
Orthopedic examination (joints, posture, muscular dysbalance)
Further examinations (X-ray, MRI, Ultrasound) as medically necessary (by IFSC MedCom, Paris 2009 [3])

Even though the IFSC MedCom recommends yearly examinations, they are not mandatory for World Cup athletes in this form. The representing federations only need to certify that a medical examination has been performed. In Germany, we have been examining our national team athletes according to the standards listed above, every year for the last 20 years. Additionally, all athletes' fingers received an ultrasound examination to check for injuries (e.g., tenosynovitis, pulley injury, ganglion cysts, etc.) in adult athletes and warning signs of growth plate fractures in adolescent athletes [4, 5]. In local, regional, or state teams, a stepwise reduction of these examinations is performed. The extent of the examinations also depends on the financial means of the respective team. Nevertheless, a physical examination, patient history, finger ultrasound, rest ECG, a laboratory test (blood count, urea, creatinine, electrolytes, cholesterol, triglycerides, iron, ferritin, magnesium, and blood clotting parameters, in accordance to the recommendations of the German Olympic Board), and plicometry (using the Jackson-Pollock skin measurements [6]) for body fat detection are always performed. All athletes answer a questionnaire with regard to injuries and training patterns. In addition to the medical and family history, the questionnaire contains questions about sleep disorders, eating behaviors, vaccination status, and general risk factors, e.g., alcohol or nicotine consumption, allergies, medications, and supplements. We always inform the athletes about anti-doping and prohibited substances and how to proceed if medication is necessary [7].

19.2.1 Orthopedic Aspects

We focus specifically on the general and athletic history, physical examination, and finger ultrasound [4, 8]. In prior analyses, we found a history of swollen finger joints in up to 42% and a minor decreased range of motion (extension deficit >5° in the proximal interphalangeal finger joint) in 32% of Youth National Team Members [9]. Fortunately, these problems are not as prevalent anymore; recently, we had only 7.5% [3] climbers (40 climbers of the German national team over 2 years) with mild finger contractures [1].

The ultrasound examination is used to see if pathologies are already detectable and to determine if the growth plates have started to close [1, 4]. Special attention is given to the time frame during which the growth plates start closing, visible through a palmar-sided closure of the plate while still open on the dorsal side. As the growth plate is more prone to injury in this time frame [4], special information is given to the young climbers and their parents to look out for pain at the dorsal aspect of the finger under and after stress, if they are in the growth phase of finger middle phalanx growth plate closure [1].

In our recent analysis of the German national team examinations of 2016/2017, the results of 17 girls and 23 boys were presented [1]. The mean age was 17.5 years (18.3 years for the boys and 17.4 years for the girls), and they had been climbing for 10.9 years (the boys for 11.3 and the girls for 10.3 years). Typical findings were proneness to infections, musculoskeletal injuries, and, in one case, severe headaches. Thirty-six percent of our athletes had to take a break from climbing due to injury or illness. Most of these breaks were due to infections (7 in total), mainly minor respiratory or gastrointestinal, with one severe case of pneumonia during a World Cup in Japan. The rest had to take a break due to finger injuries [2], foot injuries [2], and injuries to the shoulder and biceps [1, 2]. There was also one case of biceps tendonitis and one case of tibialis anterior syndrome. One climber needed arthroscopic surgery for a CAM impingement of the hip.

19.2.2 Pediatric and Medical Aspects

It is also important to consider the fact that in children and adolescents, the team doctor may be the first doctor these young athletes have seen since their early childhood. Therefore, it is important to check their vaccination schedule, as oversea travel may require further immunizations, or their immunizations could be incomplete (Schöffl et al., 2019), which raises the debate as to whether vaccinations should be mandatory for high-level athletes competing in Olympic sports [10]. Certain vaccinations such as Hepatitis B are recommended internationally in all athletes [10] and should be present in all athletes. For a thorough history, the HEADSS (home, education, activities, depression, drugs, suicide, and sex) mnemonic can be used in order to assess the patient's psychological status. In the past, inquiring into these fundamental aspects has allowed us to detect several pathologies. Cases of burnout, anorexia athletica, and bulimia, or, in other words, chronic energy deficiency in sport (RED-S), were especially suspected. In these cases, a psychological or psychiatric evaluation has to be initiated. If anorexia athletica is suspected, an MRI of the head is performed to rule out intracranial tumors, and, if indicated, an extensive pediatric examination with additional hormonal workup is carried out. In addition to the physical examination, a detailed analysis by a child psychiatrist is essential for evaluating the mental health of the athlete. In a study investigating the annual German team examination in the years 2017 and 2018, there were no athletes with a suspected anorexia athletica (Schöffl et al., 2019).

There was one case of a suspected immunodeficiency, which was later discarded, but a high susceptibility to infections during the competition period in all athletes (36% of all athletes had to take a break from training or competitions during the season) (Schöffl et al., 2019). One athlete even contracted severe pneumonia while in Japan for a World Cup and had to be hospitalized on site (Schöffl et al., 2019). Two girls received further workup due to secondary amenorrhea and therapy-resistant headaches, respectively. One athlete presented with an iron deficiency anemia, and one athlete presented with hyperkalemia but showed no symptoms (Schöffl et al., 2019).

Another important function of the yearly team examination is the education of the athletes with regard to doping and permissible and forbidden use of medications. It needs to be stressed that there are "white lists" provided by national anti-doping agencies with detailed explanations of supplements and medications. The athletes need to understand that the use of supplements and medications is their responsibility and that every substance must be checked before being applied internally or externally. Most positive doping results in sports are a consequence of dietary mistakes and are based in the unknowing intake of forbidden substances.

19.2.3 Cardiac Aspects

Even though there is little doubt that physical activity has positive effects on reducing the risk of cardiovascular disease [10], elite athletes train between 10 and 20 h per week at high intensity. These high training loads could potentially lead to adverse adaptations resulting in sudden cardiac death (SCD), or an underlying cardiovascular abnormality could lead to SCD as a consequence of an exercise-induced catecholamine surge acting on an arrhythmogenic substrate [11].

The resting and stress ECG and echocardiography are standard sports-medical examinations required once a year [12]. While these are established standards for Olympic sports, they are also recommended by the IFSC Medical Commission for climbing athletes [3, 13].

Echocardiography is vital for the detection of dangerous conditions such as long QT syndrome, myocardial hypertrophy, serious arrhythmias, myo- or pericarditis, and Brugada syndrome [14]. Even though these conditions are rare, it is necessary to exclude them as possible diagnoses. Adolescent athletes present a greater prevalence of training-related and unrelated ECG changes than non-athletes such as a longer PR interval, greater frequency of sinus bradycardia, first-degree AV block, incomplete right bundle branch block (IRBBB) and voltage criteria for left ventricular hypertrophy, and early repolarization without representing a pathology [14]. Even though its use for detecting athletes at risk of SCD is controversial, the ECG can be used for identifying adaptations such as left ventricular hypertrophy (LVH), the detection of dangerous arrhythmias (WPW and long-QT syndrome) and Brugada syndrome [14].

However, most cardiac structural changes are missed using ECG, and this is where echocardiography becomes essential. Hypertrophic cardiomyopathy is the leading cause of SCD in young athletes, followed by coronary anomalies [15], both of which can be diagnosed using echocardiography. However, HCM can be especially difficult to differentiate from left ventricular hypertrophy (LVH) often seen in athletes, as measurements can overlap. Key parameters for investigating the adolescent heart using echocardiography and ruling out cardiomyopathy are left ventricular end-diastolic diameter (LVEDD), maximal septal thickness (IVSD), left ventricular wall thickness (LVPW), the dimension of the left atrium, and the transmitral E/A ratio [16, 17]. The adaptations of the heart to athletic activity vary according to the type of sport. Dynamic (endurance) sports are characterized by a high amount of movement and only a minimal amount of force, which leads to increases in heart rate and cardiac output. This in turn means that dynamic sports cause a volume load on the heart, which leads to an enlargement of the left ventricular internal diameter and a proportional increase of wall thickness, termed eccentric LVH [18]. Static, or so-called resistance, sports (weightlifting, martial arts, field throwing) involve minimal movement but a high amount of force [19], which leads to an elevation of cardiac output as a consequence of an increase in heart rate and blood pressure. This pressure load on the heart leads to a thickening of the ventricular wall but unchanged internal dimension, termed concentric left ventricular hypertrophy [20]. A study investigating the effects of elite climbing on the heart of athletes from the German Junior National Team showed adaptations more in line with concentric left ventricular hypertrophy, so that a classification as a static sport could be suggested [14]. Interestingly, the dimensions of the left ventricular wall were comparable to those of an age-matched control of athletes from the German Junior National Nordic Skiing Team, suggesting that the adaptations of elite climbing on the athlete's heart are noticeable [14].

These observations were strengthened by the fact that the cardiopulmonary exercise results of these athletes were also comparable to athletes in team and combat sports and inferior to endurance athletes. Cardiopulmonary exercise testing (CPET) is usually implemented to determine endurance performance and cardiopulmonary fitness in elite athletes. The most important parameter studied using CPET is the VO_2 peak. Power athletes have lower VO_2 peak than team and endurance athletes [20, 21]. Thus, evaluating VO_2 peak during a standard cardiopulmonary exercise test could allow for a classification of climbing. In a study which compared elite climbers with elite Nordic skiers using a cycle exercise test, the climbers performed much worse than the Nordic skiers but were comparable to team and combat athletes [22]. Furthermore, CPET is an important instrument for assessing the risk of sudden cardiac death in athletes [23, 24] especially when employed in combination with an ECG. It allows for differentiation between physiologic left ventricular hypertrophy and hypertrophic cardiomyopathy [25, 26], one of the main reasons for SCD in athletes and difficult to diagnose using only echocardiography and ECG. Finally, CPET can be used to individualize training programs for athletes. Aerobic capacity has been shown to be important for climbing-specific recovery [24, 27] (Fig. 19.1).

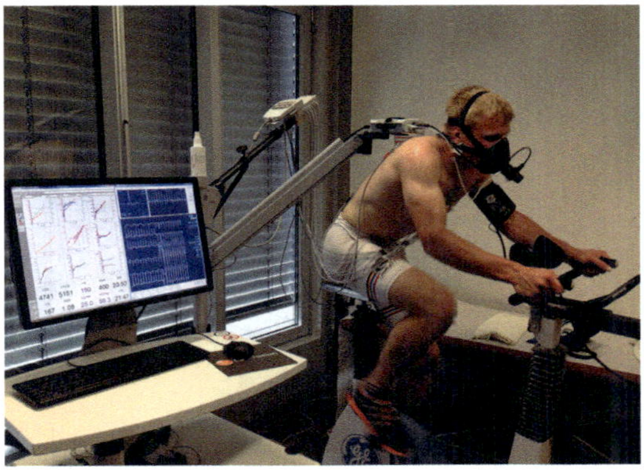

Fig. 19.1 Spiroergometry during the yearly team exam

19.3 Climbing Competition Care

With the growth in the climbing gym business, sport climbing competitions have seen a dramatic increase in popularity. In addition to the large events such as World Cups, National Championships, and Masters (e.g., Adidas Rockstars, Arco Rockmasters), smaller competitions are held on state and regional levels, as well as city championships and fun competitions in almost every commercial climbing gym.

For the larger events, a medical doctor should be serving on site as a competition medical officer; in the smaller events, paramedics normally cover the competitions. The demands and regulations vary between various countries, based on the respective legal situation. For all IFSC events, a medical doctor must be on site from the opening of the isolation until the end of the competition. The following sections we will present some aspects and considerations for the medical officer of a climbing event.

Most injuries which require treatment during a climbing event are minor. In lead climbing, rope burns, skin bruising, and the occasional "wall rash" from rubbing against the rough texture are the most common ailments of competitors [28]. More severely graded trauma, e.g., dislocated ankle fracture based on a ground fall or polytraumatic injuries, are rare [1]; nevertheless, if they occur, they demand goal-oriented action, professional skills, and the necessary equipment (Fig. 19.2). Higher graded injuries are

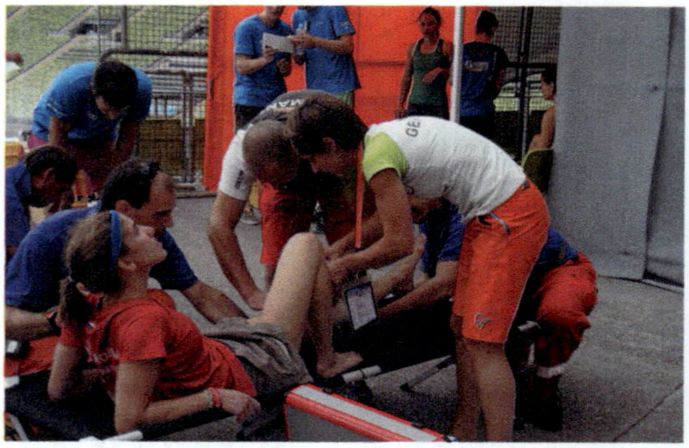

Fig. 19.2 On-site (Boulder World Cup) treatment of a knee injury

Fig. 19.3 Taking care of athletes during the World Championships

more likely in bouldering events than in lead or speed climbing [2, 28]. As in other sport events, the medical officer is faced with a different setting than the normal working environment. In a hospital ER setting and as an emergency physician in pre-hospital care, we always work in teams with nurses, paramedics, and others. If a major incident at a climbing competition happens, the medical officer is, at least initially, alone. This is complicated by the environment, in which speakers are blaring music, access to and from the base of the wall is often blocked by competitors, and the officer is often literally in the spotlight as a crowd of people watch to see what happens. To alleviate some of the stress, there are a few questions medical personnel should have answered before the competition begins [29].

> "Questions to be clarified before the competition begins:
> What are the medical facilities and potential circumstances at the competition site (e.g., medical room, EMT equipment, another emergency physician at site, etc.)?
> How and who do I call to get an ambulance to the site (phone numbers, EMT stations)?
> Can an ambulance or gurney get to the base of the wall?
> What are my communication systems? (radio, mobile phone numbers, speakers, microphone)
> What other people working on site have medical training and can be of help?" [30–32]

If there is another medical service on site (e.g., mountain rescue, paramedics), contact should be established before the beginning of the event. This helps to reduce "friction" in the event of a real emergency. Radio or cell phone contact with these other services should be established, as well as with the head of the jury and the competition's delegate [2]. The medical area at the competition must be defined and marked. Access to the wall and road access must be clarified, as well as knowledge about the hospital environment around the competition site. The role of the medical officer must be clearly defined: is he or she "just" in charge of the competitors or does he or she serve as a doctor for the whole event, including all spectators? This can hardly be guaranteed by one person who, as the competition officer, needs to be always close to the wall, because if things happen, they happen fast. Being in charge of the whole event in the Munich World Cup with three doctors, we have already treated non-climbing conditions such as sunburns, heatstroke, heart attack, and a miscarriage. Nevertheless, the athletes themselves often complain about non-climbing-related conditions, e.g., headaches, a cold, a runny nose, etc. (Figs. 19.3 and 19.4.). With respect to all of these aspects, attention must be paid to the WADA positive and prohibited lists. In general, we only have medication in our kits which is WADA approved. This excludes our emergency kit, which is only used in a real medical emergency. Make sure to note the yearly changes of the WADA lists. In Germany, an annual anti-doping course is mandatory for all national team physicians. We strictly confirm the "no needle policy" of the IOC. It is not the job of a competition's medical officer to inject athletes

Fig. 19.4 Acute treatment of a finger injury

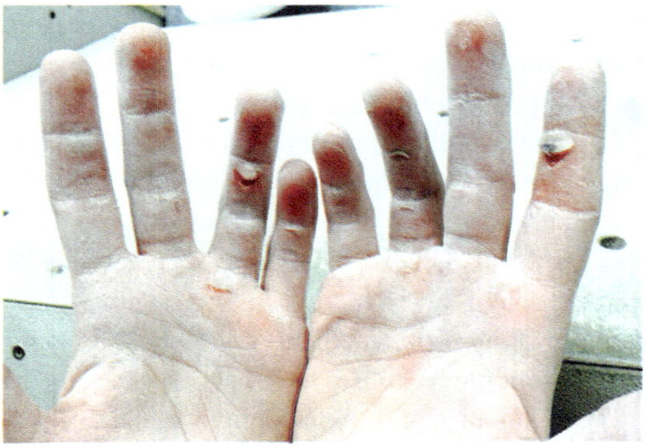

Fig. 19.5 Skin bleeds on the fingers need to be stopped and covered before a climber can continue to compete

with substances so that they are pain-free and can compete! The doping control system per se has nothing to do with the medical officer and is purely handled by an independent third party. At World Cup events, the IFSC Medical Commission has an injury surveillance system that requires any and all injuries to be reported. An IFSC delegate is supposed to collect these reports from the medical officer and submit them to the IFSC for further analysis.

The IFSC has serious regulations about participating in an event with a bleeding wound. This includes "flappers" on the fingers, for which an athlete can be disqualified if the bleeding is not brought under control (Fig. 19.5). Bouldering events, during which knee bars and heel hooks are common, have a lot of serious scratches that are technically "bleeding events" and thus must be attended by a physician. In the last few competitions at which we have worked, our doctors have been quite busy controlling such bleedings. The same move attempted by numerous boulderers can often create the same injury, competitor after competitor, and make for a hectic afternoon for the medical personnel. One method for dealing with bleeding scrapes or cuts is to clean the wound and then swab it with vasoconstrictive nose drops (e.g., oxymetazoline or epinephrine) to stop the bleeding before applying a bandage. Epinephrine seems to work the fastest and is the widely accepted bleeding medication in boxing events. Remember that epinephrine is on the WADA list and is only allowed as a topical medical drug (dental, nasal, or ophthalmological) during competitions and not intramuscular. Therefore, we generally try not to use it at all [2]. Blood stains on holds need to be removed before the next climber can compete to avoid blood-to-blood transmission of HIV or hepatitis [33].

The organizers and jury committee are responsible for the safety of the athletes during an event. However, if the medical officer on site sees a problem (a split between bouldering mats, large runout between bolts, etc.), he or she can and should protest the problem. If this is done with some tact and manners, it is generally accepted and even appreciated [2].

19.4 Conclusion

The annual team examination is an important tool for prevention and early detection of orthopedic and/or medical conditions. It also helps to establish an element of trust between the athlete and the team doctor. Early signs of overstrain injuries, e.g., epiphyseal growth plate injuries, can be detected, and further harm can be prevented. The internal medical and pediatric components should not be underestimated, as recurrent infections or an incomplete state of vaccination is frequent. A cardiologic evaluation is also important to minimize the risk of sudden cardiac death in athletes. While the work on competition site is mainly treating acute injuries and disorders, the yearly exams are important for prevention and securing the athlete's future, both as an athlete and as a healthy person. Our main goal in working with athletes is not the success of the athlete but the prevention of sport-related harm and the preservation of health. We are happy to see our athletes become world champions, but our goal as team doctors is to give them the care that they will continue to receive after completion of their athletic career: a healthy future without long-term medical consequences. Our patients should be able to look back at their time as a professional athlete with delight and look forward to a prosperous future.

References

1. Schöffl I, Lutter C, Schöffl V. What to examine in youth climbing athletes: yearly examination of the german climbing team and its consequences. Sportverl Sportschad. 2019;
2. Schöffl V, Hochholzer T, Lightner S. One move too many. Boulder CO: Sharp End Publishing; 2016.
3. Coscia F, Schöffl V. Final Report IFSC MedCom 2009. Barcelona: Medical Commiccion of IFSC; 2009.
4. Schöffl I, Schöffl V. Epiphyseal stress fractures in the fingers of adolescents: biomechanics, pathomechanism, and risk factors. Eur J Sports Med. 2015;3(1):27–37.
5. Garcia K, Jaramillo D, Rubesova E. Ultrasound evaluation of stress injuries and physiological adaptations in the fingers of adolescent competitive rock climbers. Pediatr Radiol. 2017;48(3):366–73.
6. Jackson AS, Pollock ML. Practical assessment of body composition. Phys Sports Med. 1985;13:76–90.
7. Lutter C, Tischer T, El-Sheikh Y, Schoffl V. Doping in sport climbing: status quo in a new olympic discipline. Curr Sports Med Rep. 2019;18(10):351–2.
8. Schöffl V, Lutter C, Woollings K, Schöffl I. Pediatric and adolescent injury in rock climbing Rock Climbing Research in Sports Medicine. 2018;26(sup 1):91–113.
9. Schöffl V, Hochholzer T, Karrer A, Winter S, Imhoff A. [Finger problems in adolescent top level climbers—a comparison of the German junior national team with recreational climbers] Fingerschäden jugendlicher Leistungskletterer—Vergleichende Analyse der deutschen Jugendnationalmannschaft sowie einer gleichaltrigen Vergleichsgruppe von Freizeitkletterern. D Z Sportmed. 2003;54:317–22.
10. Schöffl V, Morrison A, Kupper T. Risk of transmission of blood borne infections in climbing—consensus statement of UIAA Medcom. Int J Sports Med. 2010;32(3):170–3.
11. Lutter C, Tischer T, El-Sheikh Y, Schöffl V (2019) Doping in Sport Climbing: Status Quo in a New Olympic Discipline Curr Sports Med Rep 2019;18(10):351–2.
12. Ruiz JR, Castro-Pinero J, Artero EG, Ortega FB, Sjostrom M, Suni J, et al. Predictive validity of health-related fitness in youth: a systematic review. Br J Sports Med. 2009;43(12):909–23.
13. Sharma S, Merghani A, Mont L. Exercise and the heart: the good, the bad, and the ugly. Eur Heart J. 2015;36(23):1445–53.
14. Schöffl I, Wüstenfeld J, Jones G, Dittrich S, Lutter C, Schöffl V (2020) Athlete´s heart in elite sport climbers: Cardiac adaptations determined using ECG and Echocardiography data. Wild Env Med 2020; 31(4):418–25.
15. Lammlein KP, Stoddard JM, O'Connor FG. Preparticipation screening of young athletes: identifying cardiovascular disease. Prim Care. 2018;45(1):95–107.
16. De Innocentiis C, Ricci F, Khanji MY, Aung N, Tana C, Verrengia E, et al. Athlete's heart: diagnostic challenges and future perspectives. Sports Med. 2018;48(11):2463–77.
17. Yim ES, Basilico F, Corrado G. Early screening for cardiovascular abnormalities with preparticipation echocardiography: utility of focused physician-operated echocardiography in preparticipation screening of athletes. J Ultrasound Med. 2014;33(2):307–13.
18. Pavlik G, Major Z, Csajagi E, Jeserich M, Kneffel Z. The athlete's heart. Part II: influencing factors on the athlete's heart: types of sports and age (review). Acta Physiol Hung. 2013;100(1):1–27.
19. Galderisi M, Cardim N, D'Andrea A, Bruder O, Cosyns B, Davin L, et al. The multi-modality cardiac imaging approach to the Athlete's heart: an expert consensus of the European Association of Cardiovascular Imaging. Eur Heart J Cardiovasc Imaging. 2015;16(4):353.
20. Barbier J, Ville N, Kervio G, Walther G, Carre F. Sports-specific features of athlete's heart and their relation to echocardiographic parameters. Herz. 2006;31(6):531–43.
21. Fagard R. Athlete's heart. Heart. 2003;89(12):1455–61.
22. Schöffl I, Wüstenfeld J, Jones G, Dittrich S, Lutter C, Schöffl V. Cardiopulmonary function of adolescent elite sport climbers. 2020. (In press).
23. Bassett DR Jr, Howley ET. Limiting factors for maximum oxygen uptake and determinants of endurance performance. Med Sci Sports Exerc. 2000;32(1):70–84.
24. Stephens P. Sudden cardiac death in the young: the value of exercise testing. Cardiol Young. 2017;27(S1):S10–S8.
25. Degens H, Stasiulis A, Skurvydas A, Statkeviciene B, Venckunas T. Physiological comparison between non-athletes, endurance, power and team athletes. Eur J Appl Physiol. 2019;119(6):1377–86.
26. Sharma S, Elliott PM, Whyte G, Mahon N, Virdee MS, Mist B, et al. Utility of metabolic exercise testing in distinguishing hypertrophic cardiomyopathy from

physiologic left ventricular hypertrophy in athletes. J Am Coll Cardiol. 2000;36(3):864–70.
27. Schöffl V, Möckel F, Köstermeyer G, Roloff I, Küpper T. Development of a performance diagnosis of the anaerobic strength endurance of the forearm flexor muscles in sport climbing. Int J Sports Med. 2006;27:205–11.
28. Schöffl V, Küpper T. Injuries at the 2005 World Championships in Rock Climbing. Wilderness Environ Med. 2006;17:187–90.
29. Schöffl V, Burtscher E, Coscia F. Injuries and medical incidences during the IFSC 2012 Climbing World Cup Series. Med Sport. 2013;17(4):168–70.
30. Schöffl V, Lutter C. The "Newbie" syndrome. Wilderness Environ Med. 2017;28(4):377–80.
31. Lutter C, Tischer T, Hotfiel T, Frank L, Enz A, Simon M, et al. Current trends in sport climbing injuries after the inclusion into the Olympic program. Analysis of 633 injuries within the years 2017/18. 2020. (In press).
32. Lutter C, El-Sheikh Y, Schoffl I, Schoffl V. Sport climbing: medical considerations for this new Olympic discipline. Br J Sports Med. 2017;51(1):2–3.
33. Schöffl V, Morrison A, Kupper T. Risk of transmission of blood borne infections in climbing—consensus statement of UIAA Medcom. Int J Sports Med. 2011;32(3):170–3.

Part VII

Rehabilitation and Prevention

20 Climbing Injury Rehabilitation

Uzo Dimma Ehiogu, G. Jones, and M. I. Johnson

Drs. Chris Lutter and Volker Schöffl practicing rehabilitation at "Kesslerloch", Swiss, Photo Isabelle Schöffl

U. D. Ehiogu
Research and Training Department, Birmingham Royal Orthopaedic Hospital, Birmingham, UK

Birmingham Medical School, College of Medical and Dental Sciences, University of Birmingham, Birmingham, UK

School of Clinical and Applied Sciences, Leeds Becket University, Leeds, UK

G. Jones (✉) · M. I. Johnson
School of Clinical and Applied Sciences, Leeds Becket University, Leeds, UK
e-mail: g.j.jones@leedsbeckett.ac.uk; m.johnson@leedsbeckett.ac.uk

V. Schöffl et al. (eds.), *Climbing Medicine*, https://doi.org/10.1007/978-3-030-72184-8_20

20.1 Introduction

Sports climbing imposes high load demands on the musculoskeletal system, and periodically climbers, like all sports participants, will sustain injury. The evidence base in regard of the aetiology and pathology of climbing-related injuries is increasing, but currently there is a paucity of published research on rehabilitation [1]. In recent years, the rise in the popularity of sport climbing means it is increasingly likely that healthcare professionals will be required to diagnose and treat climbing-related injuries. It is therefore vital that those involved in the care of climbers have an understanding of the principles of rehabilitation [2] and an explicit knowledge of climbing-related injuries and their management [3].

20.2 Rehabilitation and Physical Preparation

Rehabilitation is the enhancement of tissue, functional, emotional and psychological function after injury to structure and/or impairment in physiology [4]. It is a planned and systematic approach to the judicious application of physical load for the restoration of function [5]. However, this definition may be too restrictive for the athlete as it does not fully address the high-performance outcomes sought from a rehabilitation programme. The philosophy must include both a rehabilitation and a physical preparation mindset in which the performance outcome is the key driver for all decision-making [6].

Rehabilitation for use in the general population is focused on the restoration of basic function such as activities of daily living. This does not adequately address the high functioning athlete and successful return to sport [7]. Therefore, levels of physical restoration must go beyond traditional rehabilitation paradigms if the athlete is to return to climbing in optimal condition [8]. The principles of rehabilitation and physical preparation are influenced by tissue healing time frames. The early-to-middle stages of recovery are often characterised by low tissue loading, restoration of optimal joint range of motion, soft tissue mobility and the re-establishment of motor control and movement patterns [9]. In contrast, the late phases and return to climbing are the physical preparation stages. Its focus is on the development of high load/force interventions to improve physical qualities, tissue capacity, technical skill and tactical abilities necessary for optimal climbing performance [10].

20.3 Principles of Management

Rehabilitation for the climber regardless of the actual injury should be based upon evidence-informed principles. At the core of any management, intervention is the judicious application of appropriate research, the clinician's experience and shared decision-making with the patient [11]. This awareness of the need for a balanced appraisal of practice evidence is critical for situations where empirical research is limited which is the current situation in climbing-specific rehabilitation [3].

20.3.1 Early Stage

The principles of management can be characterised into anatomically specific interventions and regional and global interventions [12]. Anatomically specific interventions involve management of local impairments and pathology specific to the injury. This is often associated with the early to middle stages of the climber's rehabilitation. For example, interventions that improve the material and morphological properties of isolated muscular, tendon and connective tissues occur at this stage [13].

Injury is often associated with a period of deconditioning and disuse because of pain and immobilisation [7, 14, 15]. It is well documented that lengthy periods of immobilisation can lead to reduced muscle strength and impaired load tolerance of connective and tendon tissues [16, 17]. The restoration of optimal joint and soft tissue mobility is important to establish early during the rehabilitation of the climbing athlete. Appropriate joint and soft tissues mobility is important from a biomechanical perspective. Optimal joint mobility provides a basis for force generation, force transfer and force absorption [18, 19]. Suboptimal mobility prevents the climber from obtaining the most efficient body shape to facilitate optimal biomechanics. The integration of kinetic chain training early in the climber's rehabilitation is of critical importance [20]. Climbing is a whole-body strength and power sport reliant on upper and lower body interactions for optimal performance [21, 22].

The trunk musculature [23, 24] and lower body [18] has been shown to provide a strong base for force production and kinetic energy transfer from the ground to the upper extremity in non-climbing studies. Anecdotally, many climbing coaches will attest to the importance of trunk muscle force production for optimal performance during climbing (see Fig. 20.1). The case for training the hip musculature and lower limb strength is more controversial as this is not as well recognised by climbing coaches for influencing climbing performance. Research from other upper limb-based sports increasingly recognises the importance of developing appropriate strength and power development in the lower limb as a basis for athletic performance [15, 20, 25].

Fig. 20.1 Retraining trunk, shoulder and lower limb integration

Fig. 20.2 Skill-related training at the physical preparation stage

20.3.2 Middle–Late Stage

Regional- and global-specific interventions which increase kinetic chain integration are associated with the middle to latter stages of rehabilitation and physical preparation [26]. At this stage, the management aims are twofold:

1. To increase sport relevant physical capacities
2. To improve and/or maintain climbing skill acquisition and tactical awareness

The improvement in sport relevant physical capacities such as maximal muscle strength, explosive strength (rate of force development) are not sports specific but are sport relevant to climbing [27, 28]. This type of training is best achieved with mainly non-climbing-based activities that can quantifiably overload specific muscle groups without the complication of the climbing wall. It involves the development of generic physical qualities using traditional resistance equipment that form the foundation for sports-specific climbing, skill acquisition and tactical development. It is the expression of these physical qualities in a sports-specific situation which allows for optimal biomechanical movement patterns and climbing performance (see Fig. 20.2). These generic capacities form the basis of a physical preparation programme for full return to unrestricted climbing at a high level of performance [29, 30].

20.4 Other Professionals Allied to Performance

Optimal management of the climbing athlete requires a multi-disciplinary team approach [31]. The rehabilitation professional must understand their scope of practice in terms of knowledge, expertise and their limitations. This may often mean collaborative working, and communication with other professionals allied to the climber as required to the climber is required. The climber must always be at the centre of the decision-making process may be coordinated by a designated performance manager. Therefore, an understanding of when to initiate onward referrals, for example, to coaching staff, dietitians, surgeons, strength and conditioners and podiatry is important for optimal performance management [32].

20.5 Exercise and Mechanical Loading

The management of mechanical load on the human musculoskeletal system is one the fundamental tenants of effective rehabilitation in athletic populations [32, 33]. Mechanical forces applied to biological tissues has been termed mechanotransduction [34]. The application of mechanical load on biological tissues has been shown to facilitate adaptations in the morphological, material and neurological properties of tissues [35]. In general, many of the physical therapies including soft tissue massage, electrotherapeutics, exercise therapy and strength and conditioning modalities share commonalities with the principle of mechanotransduction [36]. The commonality shared by these approaches is the application of load to musculoskeletal tissues facilitates changes at cellular, organ and system level [37]. For example, the use of electrotherapies such as shockwave therapy or pulsed ultrasound in the early phases of rehabilitation involve the application of mechanical load to tissues at the cellular level. There is preliminary evidence to suggest adaptation in tissue morphology at cellular level with various electrotherapies. Connective tissues have been shown to sense mechanical stimuli via signalling molecules in the extracellular matrix (ECM) and cytoskeletal structures of cells [38].

At the cellular level, mechanical forces tension the adherent ECM which then signal mechanical receptors via the intergrin. It has been shown that simulation of the intergrin receptors in cultured cells respond to mechanical load by increasing focal adhesions [39]. This may be an active adaptation to mechanical load promoting cellular growth, survival and morphological changes. For example, when the intergrin is stimulated experimentally with mechanical forces, the cellular mechanisms can be seen to realign along the lines of stress or applied tensional loads [40]. While this theoretical perspective may appear far removed from the climbing athlete in clinical practice, it is noteworthy that this underlies and supports all the higher-level adaptations at tissue, organ and system levels and hence their performance outcome. At the tissue and organ levels, changes in morphology, physiology and material properties of muscular and connective tissues can be achieved by long-term programming of the climbing athlete based upon mechanical loading [41–43]. For example, the cornerstone of physical rehabilitation and physical preparation programmes is the improvement of bio-motor capacities such as strength, endurance, power and rate of force development in muscular and connective tissues [44, 45].

20.6 Blood Flow Restriction

In recent years, adjuncts such as blood flow restriction (BFR) have shown promising results in clinical populations with high-quality studies [40, 46].

Blood flow restriction training has been shown to produce significantly improved gains in muscle strength and hypertrophy in injured load-compromised populations [47]. The underling mechanism of action is not known for sure, but the occlusion of blood proximal to the muscle belly is thought to increase fibre type recruitment, metabolic accumulation, cellular swelling and stimulation of protein synthesis [48]. This modality shows promising results in that high intensity muscular adaptations can be achieved with very modest training loads [49]. For example, the American college of sports medicine (ACSM) recommend load intensities of 70–80% of the one repetition maximum (1RM) to achieve adaptations in muscular strength, power and hypertrophy [27]. However, this type of load is often contraindicated in an injured population because of pain and soft tissue insufficiency [29].

BFR has been consistently shown to elicit hypertrophic adaptations in muscular tissues with training intensities as low as 20% of 1RM [50]. This has several advantages for the climbing athlete during rehabilitation. Firstly, it allows the early therapeutic application of load in contractile tissues that either cross or are closely associated with load-compromised structures such as the radioulnar joint or annular pulley ligaments [51]. For example, after sustaining an annular pulley injury, a climber could potentially begin resisted finger exercises earlier in their rehabilitation. This would minimise the risk of applying excessive mechanical loads through the healing tissues while still maintaining muscle strength in the forearm flexors. Secondly, it theoretically allows the training of non-injured regions to a high level without compromising injured areas [52]. For example, hang board training or other specific training modalities could be utilised to overload the finger and forearm muscles without compromising a load-impaired shoulder joint.

20.7 Electrophysical Therapy and Other Therapeutic Adjuncts

20.7.1 Electrophysical Therapy

Electrophysical therapies have been long established in clinical practice as adjuncts or standalone interventions for the management and/or rehabilitation of injuries and disease. Electrophysical therapies are used for symptomatic relief of pain and discomfort, to reduce muscle spasm, to resolve inflammation and to facilitate tissue healing.

During electrophysical therapy, apparatus (devices and accessories) is used to deliver energy (stimuli) into the body to provoke physiological responses at cellular, tissue, organic and whole-body levels to maintain and optimise health [International Society for Electrophysical Agents in Physical Therapy (ISEAPT) https://www.wcpt.org/iseapt]. Energy is delivered across the intact surface of the skin via conduction, convection, conversion, radiation or evaporation to alter physiological processes in 'target tissue' for therapeutic benefit. Physiological actions include alterations in the structure and/or activity of substrates involved in metabolism and healing (enzymes and proteins) or the modulation of neural activity. Despite much research, there is insufficient robust clinical research evidence from randomised controlled clinical trials to make confident judgements about the effectiveness of most electrophysical therapies.

There are many types of therapeutic device available to practitioners and most have preprogrammed menu-driven settings that manufacturers claim to be beneficial for a bewildering array of medical conditions, often leaving practitioners confused. The principles of different types of electrophysical techniques have largely remained unchanged over time. Different types of electrophysical therapies can be categorised as follows:

Thermal and microthermal

- Superficial heating using contact or radiant techniques
- Pulsed and continuous shortwave and radiofrequency therapies
- Cooling (cryotherapy) using contact and radiant heat

Nonthermal, microthermal and light energies

- Therapeutic ultrasound (US)
- Low level laser therapy (LLLT)
- Ultraviolet therapy

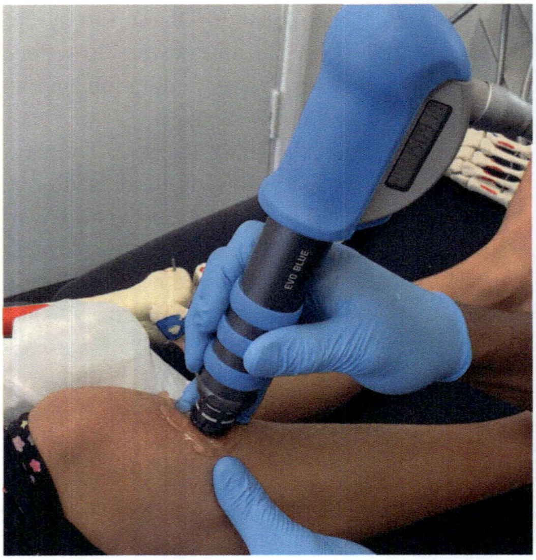

Fig. 20.3 A therapist using extracorporeal shockwave therapy

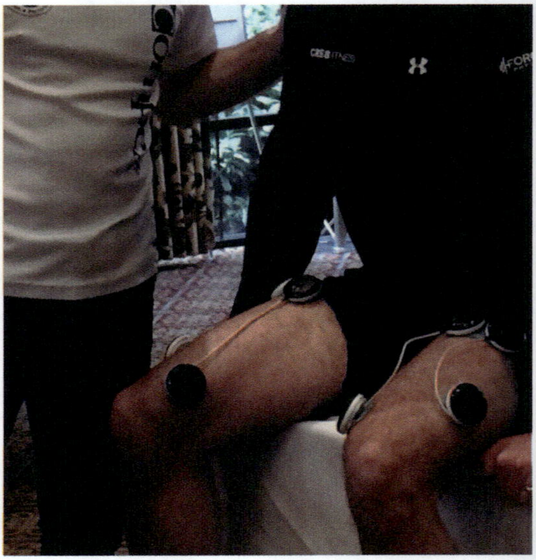

Fig. 20.4 Electrical stimulation used to superimpose muscle contractions on normal movement patterns for motor relearning

- Magnetic and pulsed magnetic (PEMF) therapies
- Extracorporeal shockwave therapy (ESWT) (Fig. 20.3)
- Vibration

Electrical stimulation
- Transcutaneous electrical nerve stimulation (TENS)
- Neuromuscular electrical stimulation (Fig. 20.4)
- Functional electrical stimulation
- Alternating currents including interferential therapy (IFT) and Russian stimulation
- Electrical stimulation for wounds

Considerations for practitioners include selection of electrophysical modality and device, target tissue, energy parameters (device settings), dose and regimen. Failure to dose within the therapeutic window may result in harmful (toxic) or subtherapeutic effects. An electrophysical therapy may produce different physiological actions by altering the parameters and dose of the energy. If used appropriately, electrophysical therapies are considered less harmful than medical interventions such as drugs, especially to provide symptomatic relief. Watson and Nussbaum [53] and the International Society for Electrophysical Agents in Physical Therapy [54] provides comprehensive evidence-based coverage of the main electrophysical therapies used in clinical practice.

Professional bodies recommend only a minority of electrophysical therapies as adjuncts to core treatment for specific medical conditions (e.g. TENS for osteoarthritis and rheumatoid arthritis). Hypothetically, PSWT, US, LLLT and electrical currents may increase the rate of division of malignant tissue and for other rapidly dividing tissue (embryo/foetus), so they are not applied over areas of malignancy in treatable cancer. Generally, a history of malignancy is not a contraindication. Pregnancy is not a contraindication provided that energy is not applied close to embryonic/foetal tissue. Nevertheless, if used appropriately, electrophysical therapies are considered less harmful than medical interventions such as drugs, especially to provide symptomatic relief. A comprehensive resource of contraindications and precautions have been reported previously [55].

The principles of treatment are to assess contraindications and appropriateness of treatment modality for the patient and to seek informed consent. Further considerations include:

- Technique: The sequence of actions necessary to administer the treatment
- Dose: Quantity of treatment to be administered at one time
- Dosage: Size (amount/concentration), frequency and number of doses of the treatment (per treatment session)
- Regimen: A program of treatment sessions, e.g. the course of treatment, how many sessions and how often

20.7.2 Extracorporeal Shockwave Therapy

Extracorporeal shockwave therapy (ESWT) is a therapy that has recently gained popularity in tendon-related disorders in clinical practice and within the literature.

ESWT is an electrotherapeutic modality that uses acoustic wave formations to deposit high energy into musculoskeletal tissues (see Fig. 20.5) [56]. It is a form of mechanotransduction which aids the regeneration and reparative process of bones, tendons and connective tissues [57]. ESWT is an increasingly popular therapeutic modality in the management of tendinopathy and connective tissues disorders [58]. From a physiological stand point, the acoustic waves utilise both positive and negative phases of the shockwave. The positive phase creates direct mechanical forces, and the negative phase generates high speed cavitation's [41]. The proposed mechanisms of action include the promotion of neovascularisation at the tendon bone interface, the simulation of tenocytes, increased leukocyte infiltration and the proliferation of growth factor and protein synthesis [56]. This may stimulate collagen synthesis and tissue remodelling [57].

Electrophysical therapies may be beneficial when used as an adjunct to core treatment or within a package of care. For example, electrical stimulation therapies (see Fig. 20.4) can be used to prevent muscle atrophy after injury [42] or to superimpose muscle contractions on normal movement patterns for motor relearning [37]. These modalities likely yield the most benefit during the early stages of rehabilitation when musculoskeletal tissues may be load compromised due to pain and/or effusion and/or immobilisation [38, 39]. However, as the athlete progresses through the stages of rehabilitation and physical preparation, these modalities become less important, and exercise-based loading regimes take on greater significance.

Glossary of terms: electrical qualities

Current	Flow of positive electric charge. The strength of current flow in any medium is related to voltage differences in that medium, as well as the electrical properties of the medium, and is measured in amperes
Voltage	Measure of the difference in electric potential between two points in space, a material, or an electric circuit, expressed in volts
Amp	Unit of electric current. One ampere corresponds to a certain number of electrons passing a fixed point each second
Resistance	Opposition offered by a body to the passage through it of a steady electric current
Impedance	Measure of the total opposition to current flow in an alternating current circuit, made up of two components, ohmic resistance and reactance
Reactance	Opposition to the flow of alternating current caused by the inductance and capacitance in a circuit rather than by resistance
Inductance	Property in an electrical circuit where a change in the electric current through that circuit induces an electromotive force (EMF) that opposes the change in current
Capacitance	Measure of the ability of a configuration of materials to store electric charge. In a capacitor, capacitance depends on the size of the plates, the type of insulator and the amount of space between the plates
Wavelength	Distance between one peak or crest of a wave of light, heat or other energy and the next corresponding peak or crest

Frequency	Number of crests of a wave that move past a given point in a given unit of time. The most common unit of frequency is the hertz (Hz), corresponding to one crest per second. The frequency of a wave can be calculated by dividing the speed of the wave by the wavelength. Thus, in the electromagnetic spectrum, the wavelengths decrease as the frequencies increase and vice versa. More simply the number of occurrences of an event per unit time
Intensity	Magnitude, as of energy or a force per unit of area, volume, time
Pulse frequency	Pulses per second
Pulse duration/ width	Length of time for each pulse
Pulse shape (waveform)	Refers to the build-up and reduction in intensity during a single pulse

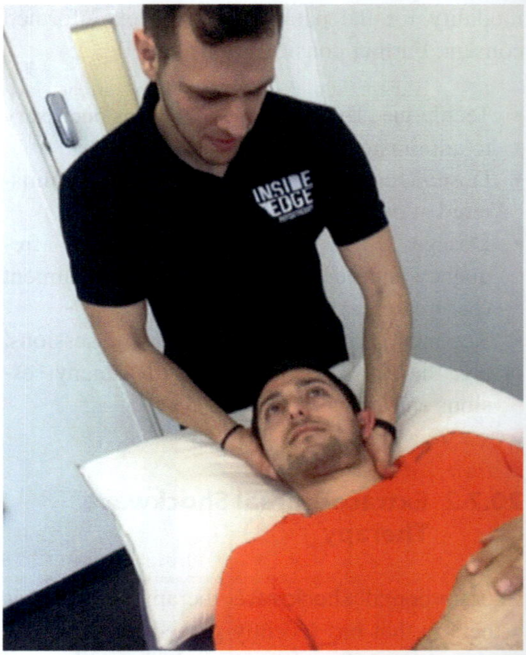

Fig. 20.5 A therapist using spinal manipulation as a treatment modality

20.7.3 Manual Therapies

Manual therapies are approaches and techniques that involve the skilled use of the hands to rub, knead, mobilise and manipulate musculoskeletal and neural tissue in and around joints [33]. Manual therapies are primarily used to treat conditions affecting musculoskeletal structures. They do this by increasing tissue extensibility and decreasing muscle tone and tension which improves range of motion at joints and movements that are restrictive. Manual therapies are also used to reduce swelling, inflammation and pain [59]. There are basic physiological principles underpinning manual therapy and diversity in philosophies, concepts, approaches and techniques according to specialism.

There is a paucity of robust research evidence to judge whether there are differences in benefit and harm between manual therapy specialisms or between different approaches and techniques within specialisms. Consequently, practitioners often base clinical decisions on their previous experience, resulting in inconsistencies in approaches and techniques selected for patients with similar clinical presentations.

20.7.4 Joint Mobilisation and Manipulation

Joint mobilisation and manipulation are manual therapy techniques used in isolation or in combination to restore normal function at spinal and peripheral joints.

Joint mobilisations are active and/or passive physiological movements and/or low velocity passive accessory movements of varying amplitude within or at the end of range of motion and are used primarily to restore movement by inducing sympathetic nervous system excitation [60]. Joint manipulations (see Fig. 20.5) are high velocity, low amplitude, thrust techniques, within or at the end of range of motion and are used primarily to restore movement by inducing cavitation and the separation of articular surfaces at the joint [33].

Treatment effects are broad, may be immediate and/or long-term and include reduction in pain, lengthening of periarticular tissue, increase range of motion and reduced joint stiffness.

Contraindications and precautions should be checked prior to treatment and include fractures, dislocations, cauda equina, rheumatological conditions, osteoporosis and vertebral artery insufficiency (cervical mobilisations). Joint mobilisation is considered a safer adjunct to core treatment. Potential risks are generally considered to be low and include increased pain, proximal and/or distal to the treatment site [61]. A variety of mechanisms of action have been offered to explain the therapeutic outcomes for joint mobilisations. In 2020, a systematic review and meta-analysis of 17 studies provided evidence that joint mobilisation elicited sympathetic-excitatory effects [62], such effects may reduce pain and stiffness.

20.7.5 Evidence for Clinical Efficacy

There is a paucity of robust clinical evidence to establish optimal technique for joint mobilisations and manipulations. Thus, in clinical practice, techniques vary considerably according to the background and previous clinical experience of the practitioner and poses challenges when deciding appropriate techniques to use in randomised controlled clinical trials, and when synthesising findings from previous research. In 2013, a systematic review of 22 studies found moderate evidence to support the use of joint mobilisations for lateral epicondylalgia and insufficient evidence to support use in conditions of the wrist and hand [63]. In 2019, a meta-analysis of seven studies provided strong evidence that joint mobilisations improve pain and/or functional grip in people with lateral epicondyle tendinopathy [64].

20.7.6 Summary

Joint mobilisations and manipulations are often used in conjunction with other treatments; it is therefore difficult to ascertain the effect of mobilisations per se. Nevertheless, evidence supports consideration of their use as an adjunct therapy to reduce pain and increase range of motion in musculoskeletal conditions.

20.8 Specific Conditions

20.8.1 Lower Limb

20.8.1.1 Acute Hamstring Injury

This is often associated with a heel hook mechanism of injury and is characterised by an overstrain of the contractile fibres of the hamstring muscle group [65]. This can inculde the bicep femoris, semimembranosus and semi tendinosis muscles [66]. The mechanism of injury is often an acute presentation with the climber complaining of a sudden onset of pain and weakness during the heel hook technique [67]. This hamstring injury is unique in that the mechanism of injury is a concentric muscle action [68] which contrasts the eccentric mechanism of injury often seen in high speed sprinting and team sports [69].

20.8.1.2 Initial Management

Initial management involves the standard protection, rest, ice, compression and elevation protocol (PRICE) (see Table 20.1) [70]. The injured muscle group can be taped or splinted into a shorten position to avoid excessive tensional forces during the first 1–5 days dependent upon the grade of injury [71]. The injury severity is generally graded from grade 1–3. Grade 1 is associated with minimal

Table 20.1 PRICE protocol [70]

Protection	Protection of the injured area from further insult avoids healing delay
Rest	Relative rest of the injured area is encouraged to allow healing. This includes avoidance of activities that are stressful to the injured area. Complete rest is often to be avoided as controlled pain-free mobilization of injured tissues can aid recovery
Ice	Cooling reduces local tissue metabolism and limits the extent of injury. This may reduce secondary cell necrosis. Cooling of tissues may provide pain modulation via the active pain gate mechanism
Compression	Compression of the injured area helps to reduce swelling and oedema
Elevation	Elevation of the injured area reduces blood flow to the injured area and reduces swelling

disruption of the contractile fibres and connective tissues, grade 2 is associated with more significant disruption of the contractile and connective tissues, and grade 3 is associated with catastrophic failure of the tissues leading to complete rupture [14].

20.8.1.3 Definitive Managment

The cornerstone of optimal management for an acute hamstring injury is a progressive tissue strengthening programme [68, 72]. The programme should aim to address the force-generating capacity of the muscle with a bias towards concentric muscle actions. In the initial phase of management, strength training of the injured tissues can begin with isometric exercises but should rapidly progress to concentric exercise as pain, irritability and tissue healing allow [69]. In the early stages, when tissues are load compromised, adjuncts such as blood flow restriction can be used to facilitate high intensity muscular adaptations with modest training loads [49]. In the latter stages of the rehabilitation programme, maximal strength in both isometric and concentric muscle actions take priority (Fig. 20.6). The programming of high velocity strength training will be specific to the performance requirements of the climber. For example, there may be a greater performance requirement for high velocity strength training for a boulder and speed climber than a lead climber [70, 71]. At the return to climbing stage, the athlete should have no pain or hesitancy during the heel hook manoeuvrer in various climbing-specific positions. Additionally, there should be no more than a 5% deficit or less in concentric muscle strength between the affected and contralateral side [29, 72].

Fig. 20.6 Single leg hamstring muscle strengthening with core muscle integration

20.8.2 Ankle Sprains

Injury to the ligaments of the ankle and foot is a common cause of chronic disability and recurrent instability in athletes [73]. Sprains of the lateral ankle ligamentous complex involving the anterior talofibular ligament, +/− the lateral and posterior calcaneofibular ligaments are often injured when the ankle is inverted and planter flexed at the time of injury [74].

This injury often involves a high-energy mechanism of injury in all but a few cases. The mechanism of injury is normally associated with a fall from height when bouldering [75] or an uncontrolled fall during sport or traditional climbing [76]. The injury mechanics classically involve inversion for the midtarsal and subtalar joints and planter flexion of the talocrural joint.

The ankle complex is unique in its anatomical configuration in that the shape of its bones are irregular conforming to very precise anatomical relationships. This provides a friction-free environment for weight bearing locomotion [77]. The ligamentous structures act as somatosensory force transducers to control precise joint motion. Injury to the ligaments affects the mechanical properties of the collagenous tissues and also the mechanoreceptors responsible for conveying information, influencing joint position sense and kinaesthesia to the central nervous system [78]. Poor initial management and rehabilitation of ankle sprains can lead to pain, disability and functional instability [79]. Chronic functional ankle instability is associated with lower satisfaction, lower rates of return to sports participation and increased referral rates for surgical opinion [80]. Therefore, we believe that appropriate initial management of ankle ligamentous injuries and subsequent rehabilitation is critical for the functional restoration of the climbing athlete.

20.8.2.1 Initial Management

The initial management of an ankle sprain should follow the principles of PRICE to prevent further insult to the injured tissues and to create optimal conditions for the restoration of tissue homeostasis. Appropriate ankle bracing or taping to minimise mechanical directional stress on the injured ligamentous structures has been shown to promote early mobilisation [81].

20.8.2.2 Definitive Management

The management approach should include a comprehensive rehabilitation programme which addresses all deficits in range of motion, muscle strength and proprioceptive control [82]. Appropriate range of motion into both dorsal flexion and planter flexion is important for shock absorption during landings and for propulsion during climbing. Failure to restore functional dorsal flexion range of motion may inhibit the transfer of momentum [79] during activities such as rockovers and prevent optimal mechanics during activities such as Egyptians and frog positions. Ankle planter flexion is part of the coordinated system of lower quadrant triple extension for dynamic propulsive activities such as dynos [83]. Failure to restore this movement range of motion frequently leads to reduced ability to overcome the effects of gravity such as is the case when executing dynamic movements.

The restoration of balanced strength development is also important from a force absorption and force generation perspective [12]. Pain inhibition and disuse is a frequent cause of impaired muscle performance in the injured climber. Often this impairment is exacerbated by inappropriate muscle conditioning and too rapid a return to sports participation [84]. Therefore, we recommend that muscle strength of the planter flexor, dorsal flexor, inversion and eversion muscle groups be within 5% of the contralateral side before return to unrestricted climbing. Ankle sensory motor retraining is also important after injury to ligamentous structures (see Fig. 20.7). This is most important for optimal foot posture during landing to prevent reinjury [85] and for effective and precise footwork when climbing. Therefore, consideration should be given to progressive training of landing technique on stable and then unstable surfaces [80]. This progression should take account of standard rehabilitation progressions involving landing from low heights to progressively higher heights [78]. This will increase the ground reaction forces and load tolerance through the ankle complex in a progressive manner. Consideration should also be given to skill-related training to develop sensory motor control of precise footwork during simulated conditions and while climbing. All rehabilitation systems should be process driven with therapeutic exercise progressing from low load to high load, simple movements to complex movements and small ranges of motion to large ranges of motion [10].

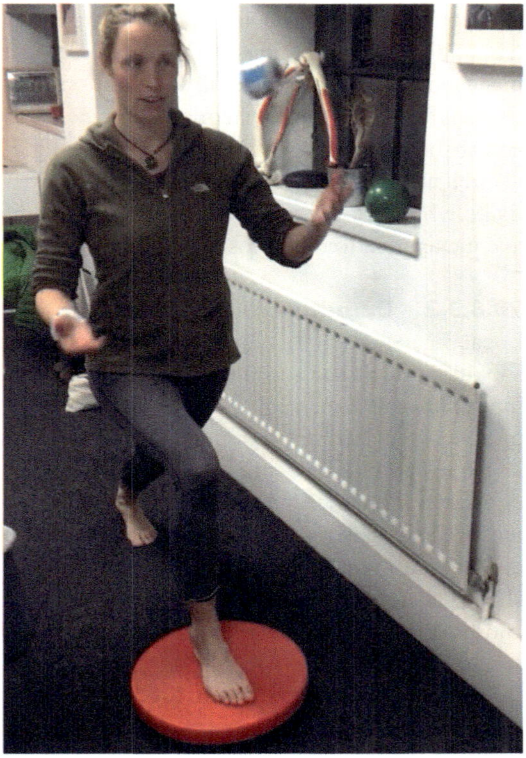

Fig. 20.7 Climber participating in ankle sensory motor retraining

20.8.3 Upper Limb

20.8.3.1 The Painful Wrist

The wrist is a common area of pain and dysfunction in climbing populations [76]. Wrist pain in climbers is often situated around the radiocarpal joint and typified by a chronic aetiology of impaired functional capacity during mechanical load [86]. In chronic wrist pain, it is often difficult to identify a specific anatomical target to focus rehabilitation interventions [87]. Therefore, local management of the chronically painful wrist is unlikely to change the systemic load transferred and absorbed by the complex during climbing. It is often more pragmatic for the clinician to investigate why the wrist complex has failed to manage the applied loads imposed by climbing [88].

20.8.3.2 Definitive Management

Local management of pain-sensitive structures should be instituted to reduce the effect of pain inhibition and localised inflammation if that is present. Standard procedures utilising PRICE and nonsteroidal anti-inflammatories may have a place in the initial management of the climber with persistent wrist pain [88]. In addition, the climbers training and competition volumes should be modified to unload the pain-sensitive and load-intolerant structures [26].

20.8.3.3 Local Assessment

A local assessment of the wrist should include the following:

1. Palpation of the radioulnar joint lines and associated palmer and volar ligaments
2. Palpation of the carpal bones and associated ligaments
3. Assessment of muscle performance of the following:
   - Flexor carpi radialis
   - Flexor carpi ulnaris
   - Extensor carpi radialis longus and brevis
   - Extensor carpi ulnaris
   - Abductor pollicis longus
   - Abductor digiti minimi
4. Distal and proximal radioulnar joint accessory movements
5. Joint position sense in low and high load situations

20.8.3.4 Global Assessment

A comprehensive assessment of the climber's technique on the wall should be examined to establish its relationship to movement efficiency. This can be done by the therapist depended upon their expertise or in conjunction with a climbing coach. The practitioner is looking for evidence of movement inefficiency that might be contributing to excessive load through the hand and wrist complex [33], refer to Chap. 21.

The neuromuscular assessment should examine the strength of the shoulder girdle in particular the rotator cuff muscles and axial scapula adductors and retractors. A general assessment of static shoulder girdle and trunk posture should be undertaken to establish muscular balance between the anterior and posterior upper body musculature [57]. We have noted an anecdotal relationship between the poor muscular development of the shoulder girdle abductors muscles when compared to the shoulder adductors and internal rotators in climbers with chronic wrist and elbow pain.

The rotational profile of the proximal and distal radioulnar joints should also be assessed for total range of motion into supination and pronation. Loss of range of motion and excessive soft tissue tightness can lead to impaired function of the wrist complex during climbing. The loss of pronation range of motion can theoretically lead to increased shear forces at the wrist complex. This is because the radiocarpal joint attempts to compensate for the reduced rotatory range of motion at the proximal and distal radio ulna joints [89]. The radiocarpal joint has two degrees of freedom: flexion and extension and ulna and radial deviation. Rotation occurs functionally at the superior and distal radioulnar joints [90]. Therefore, assessment of soft tissue compliance at bicep brachialis and supinator muscles is indicated in addition to the accessory movement of the proximal and distal radioulnar joints.

20.8.3.5 Rehabilitation

The rehabilitation programme should be prescriptive based upon the impairments found after a comprehensive evaluation of the climber's wrist, hand and upper quadrant. Rehabilitation should then aim to restore optimal function where this can be extrapolated to load intolerance of the wrist complex (Fig. 20.8). There is often an underlying cause for tissue incompetence and persistent symptoms, for example, poor joint position sense or reduced proximal radioulnar range of motion.

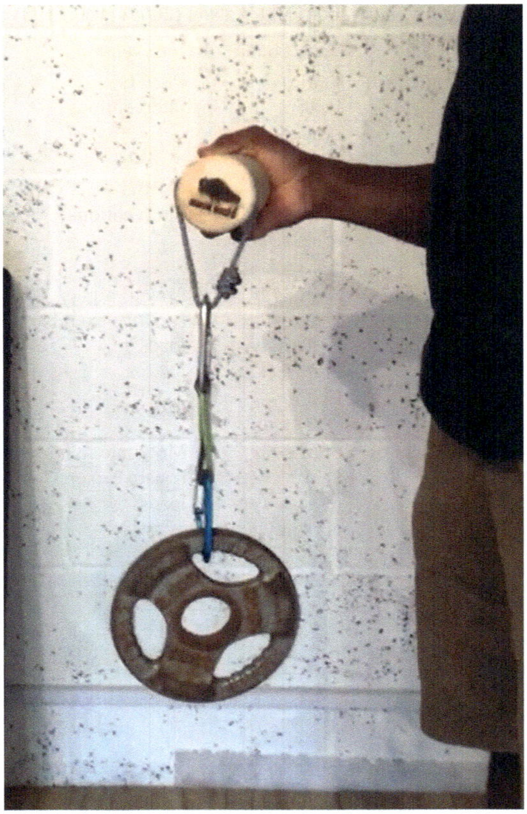

Fig. 20.8 The restoration of mechanical wrist function using resistive loading

Significant impairments must be addressed based upon the principles of rehabilitation and physical preparation already discussed to ensure optimal management and return to climbing.

20.8.4 Closed Traumatic Flexor Tendon Pulley Rupture

Injuries to the annular flexor pulley ligaments are one the most common injuries to afflict climbing populations [91]. This type of injury is particularly debilitating because of the significant impairment in climbing performance associated with rupture [3]. The pulley ligaments are connective tissue structures composed primarily of type 1 collagen [92]. The pulley ligament provides a critical biomechanical role ensuring optimal force transfer from the strong flexor tendons to the finger and climbing interface [93]. Anatomically, the flexor pulley ligaments are extracapsular thickenings of the joint capsule and are supported by the cruciate ligaments laterally [83]. There are five annular pulleys which alongside the cruciate ligaments and fibrous capsular thickening act to constrain the movement trajectory of the long flexor tendons [94].

20.8.4.1 Initial Management

The initial management regardless of the severity of the injury will follow the standard principle of PRICE [70]. The management of pain and the prevention of further soft tissue insult is the primary objective in the first 24–48 h after the injury. The suppression of inflammation at the onset of injury is less important and only becomes a priority if it continues beyond an acceptable time frame [95]. The protection of injured structures with rest and appropriate splinting [96] or taping [97] is helpful to facilitate the approximation of the bone tendon interface [96]. This theoretically assists with scar tissue formation and helps to prevent excessive bowstringing of the tendon which may delay healing. The number of pulley ligaments injured and the degree of disability will determine whether surgical or conservative management is chosen [98].

20.8.4.2 Definitive Management

The management process for rehabilitation of the partial or ruptured pulley ligament follows the principle of progressive overload [95]. In the initial stages, mechanical load can be administered with electrotherapeutic modalities such as ultrasound and laser therapies [34]. However, the cornerstone of load management is active resistance training [16, 41, 99]. This should be applied judiciously in the early stages of tissue healing (see Table 20.2) with resistance training to isolate load to the digit of the injured pulley. Loading progressions can start with active flexion at the proximal interphalangeal joint (IPJ), middle IPJ and then distal IPJ. Resistance training bands are a pragmatic method to titrate mechanical load to the recovering pulley ligament via the injured digits. Light and then eventually moderate to heavy resistance bands can be progressively added as tissue healing progresses. In the early stages of rehabilitation climbing should not be permitted

Table 20.2 Stages of soft tissue healing (Adapted from Jimenez and Jimenez (2004) and Watson (2008) [100, 101])

Phase	Description	Time course	Interventions
Bleeding phase	This phase occurs after the initial injury and is characterised by an increase in vascularity at the site of injury	Start 0–1 h Peak re-activity 4–6 h	PRICE interventions
Inflammatory phase	The inflammatory phase is a necessary component of soft tissue healing. It is composed of both a vascular and cellular cascade which have direct effects on the tissue repair sequence	Start 0–4 h Peak re-activity 1–3 days	PRICE interventions
Proliferative phase	This phase begins with the formation of granulation tissue which eventually matures to become scar tissue. The catalyst for this phase is via cytokine mediators from the inflammatory phase	Start 24 h Peak re-activity 2–3 weeks End 4–6 months	Low to higher resistive exercise loads
Remodelling phase	The remodelling phase is characterised by the refinement of the collagen from immature to more robust tissue. The deposition of collagen fibrils with appropriate loading contributes to orientational alignment along the lines of stress	Start 24 h Peak re-activity not known End 12 months	Moderate to high resistive exercise loads

[102]. The maintenance of soft tissue mobility to promote an extensible scar and fibroblastic reorganisation along the lines of mechanical stress can be achieve with mobility interventions such as tendon gliding exercises [103].

In accordance with Wolff's law of mechanical loading and adaptation, resistance loads should be increased for continued adaptation [16]. Loading stimulates and increase the morphological and material properties of the connective tissue [16], muscle [104] and tendon [105]. Although this has not been shown directly in vivo, there is substantial evidence from animal studies [106], in vivo tendon studies [107] and cross-sectional studies [108] in competitive climbers which suggest that mechanical load induces morphological changes in connective tissues such as ligaments.

Therefore, the aim of rehabilitation and physical preparation is the restoration of tissue strength to preinjury levels. As the stage of tissue healing progresses from the proliferative to the remodelling stages, the degree of loading can increase to include training modalities which apply greater concentrated loads to the digits with fingerboards utilising holds with progressively smaller contact areas [108, 109] (Fig. 20.9). This will thereby

Fig. 20.9 Training modality with concentrated mechanical loading for the fingers

encourage further tissue adaptation. Resistance bands can eventually be replaced with traditional weighted pulley systems that stress the injured digit in insolation and collectively.

Participation in climbing-specific activities should be reserved for the intermediate stages of the climber's physical preparation programme. It is often tempting to return to climbing early with an open grip hand position because the pulley tendons (A2 and A4) are loaded to a significantly lesser degree than the full crimp position [110]. This is often the anecdotal rational for early return to climbing using an open grip. However, this strategy is problematic.

The A2 and A4 pulleys are at most risk of injury because of their larger internal moment arms compared to the A3, A1 and A5 pulleys ligaments [111]. Therefore, although under less load during open hand grips, the A2 and A4 are still vulnerable in the early stages of rehabilitation. It is also practically difficult to quantify tendon and pulley mechanical loading during climbing because of the various configurations of hold types [105]. We therefore suggest it is safer to restrict climbing until the injured digit can be loaded with no symptomatic reaction during four finger isolated finger strengthening in a half crimp position. This reduces the risk of over training and reinjury and ensures a degree of tissue strength [88, 94]. Return to climbing too early increases the risk of overzealous risk taking which could cause accidents such as foot slips and excessive application of loads when negotiating difficult climbing problems [102].

Progressive resistance training principles should be followed [32]. We recommend isolated digit strengthening proceed climbing specific activities. The injured digit should be loaded in insolation starting with isometric muscle actions, progressing to concentric actions and then finally to eccentric actions [12]. Each muscle actions must be accompanied with symptom-free participation of the affected digit before progression to the next muscle action [10]. This rational is borne out by the high mechanical forces attributed to eccentric loads in the contemporary literature [112, 113] which may predispose the ligamentous tissues to excessive load if used prematurely.

The resumption of climbing should begin when the affected digit is able to tolerate concentric forces within 5% of the contralateral digit. The range of motion along the course of the flexor tendons in both flexion and extension should be full range and asymptomatic. On palpation of the soft tissues overlying the pulley ligaments this region should not reproduce the patient's original symptoms. This return to sport criteria is in accordance with other published literature for other regions of the body [29, 114, 115].

The resumption of climbing is a phased approach with a progressive increase in volume load over several weeks and months depending upon the grade of injury. In the initial phases, the duration of climbing, the grade and number of hold types is limited. For example, in the first 1–2 weeks of climbing practice, the climber may only climb 2× per week restricted to only an open grip, accumulating up to a maximum of 100–150 isolated hand grips in each session At week 5–6, the frequency of climbing may remain the same, but the climber is allowed to use half crimp or open crimp holds in addition to open grip holds accumulating between 150 and 200 isolated hand grips during a session. As the rehabilitation and physical preparation progresses, there will be a progressive increase in the volume of climbing each week, the use of full crimp holds, duration of each climbing session and inclusion of steeper over hanging terrain. The cardinal rule which must be enforced during this phase is that climbing at each stage must be symptom free.

20.9 Shoulder

20.9.1 Chronic Subacromial Pain Syndromes

The shoulder complex is often a common source of pain and dysfunction in climbers. Traditionally, pain within the shoulder complex has been managed from an anatomical perspective. This aims to establish which of many pain-sensitive structures is the source of the patient's symptoms. While this may be possible with a traumatic injury, this is often not possible in cases of persistent shoulder pain [116]. It is difficult to differentiate which structures are at fault with physical examination tests because of their low diagnostic accuracy [117]. Most physical examination tests for the shoulder complex are pain provocation tests. This renders it difficult to obtain a structurally specific diagnosis [118]. Other evidence supportive of this position is diagnostic imaging studies. Research from imaging interventions such as magnetic resonance imaging [119] and ultrasound sound studies [120] often report incidental findings in asymptomatic populations.

In patients with chronic persistent shoulder pain, it is highly probable that several structures will be symptomatic acting in combination as a source because of the close anatomical relations of pain-sensitive structures such as bursa, tendons, capsule and joint articular surfaces [121]. For example, the supraspinatus tendon, bicipital tendon and hypertrophied subacromial bursa can all be symptomatic during common climbing positions which require shoulder elevation. In respect to this difficulty, treatment approaches based upon specific tissue pathology and structural diagnosis may be therefore problematic [122].

Rehabilitation of the climber with persistent shoulder pain should be based upon management of the underlying physical impairments which have contributed to tissue sensitivity and load intolerance [123]. It is therefore important that the clinician establish the dominate features that are contributing to the climber's symptom provocation [124]. A comprehensive upper quadrant assessment should be undertaken including [125]:

1. Functional demonstration of the activities or climbing movements producing symptoms
2. Contractile function of the rotator cuff musculature and axioscapular muscles (middle trapezius and rhomboids)
3. Soft tissue muscle length of the shoulder adductors and internal rotators (pectoral major, teres major and latissimus dorsi)
4. Glenohumeral joint accessory movement assessment
5. Static upper quadrant postural assessment for muscle balance

An assessment of the trunk musculature, lower quadrant, the neck and its associated nerve plexus may be indicated if there is evidence from the patient's history and/or evidence from the patient's functional climbing demonstration that these factors may be a contributing to shoulder dysfunction. The assessment of the symptomatic regions should be compared to the contralateral side where applicable and used to decide the management approach based upon the impairments present. This approach to the management of shoulder pain is distinctly different from a traditional sports medicine approach of structural tissue failure [126]. It does not attempt to isolate a singular pathological tissue in which to direct medical management. Instead, this approach utilises a holistic approach identifying regions and tissues which are functioning poorly and potentially contributing to ongoing tissue irritation and mechanical sensitivity during high load situations [125]. Rehabilitation interventions are then directed at the areas of failed load tolerance using progressive exercise, adjuncts and electrotherapeutic modalities to restore optimal tissue function and capacity.

20.10 Traumatic Anterior Glenohumeral Dislocation

In first-time traumatic anterior glenohumeral dislocations (FTAGD) in young adults, therapeutic management is often conservative physiotherapy and rehabilitation [127]. Surgery controversial for FTAGD however there are combinations of patient - specific factors such as the presence of menaingful bone loss, significant apprehension and athletes athletes involved in contact sports [128].

Rehabilitation of the FTAGD is predicated upon the stages of soft tissue healing and remodelling (see Table 20.2). The initial stages of rehabilitation should involve appropriate conservative measures to protect the injured tissues from further insult [70]. This will encourage unhindered fibroblastic and proliferation activity. Modalities and adjuncts such as splints and braces to maintain optimal soft tissue length should be emphasised. Bracing the glenohumeral joint in a shortened position to minimise tensile forces through the injured tissues should be standardised 1–5 days post injury depending upon its severity and degree of associated soft tissue pathology [129]. After a period of initial immobilisation which should be short to prevent complications such as contracture and joint stiffness, a reactivation process should be enacted to allow progressive loading of the injured tissues.

Mechanical loading of the shoulder complex is a cornerstone of rehabilitation at this stage: to reverse arthrogenic muscle inhibition and associated muscle atrophy [130]. It is well accepted that an effusion within or in close proximity to a joint can lead to muscle inhibition [131]. For example, atrophy of the vastus medialis oblique muscle (VMO) in the knee is initiated by minimal effusion. This muscle has similar physiological [132] and morphological [133] characteristics to that of the rotator cuff group in the shoulder. Therefore, muscle activation strategies to preferentially target intramuscular coordination should be instituted as early as possible [133]. Muscle stimulation has been shown to preferentially activate high threshold motor units in atrophied muscles [43].

Isometric muscle training is a safe and effective method to stimulate motor unit recruitment without excessive mechanical load. This is important in the initial 1–2 weeks after the injury because the ligamentous structures are still load compromised [128]. In the middle stage of rehabilitation, higher mechanical loads to stimulate development of connective and muscular tissues is of paramount importance. The mechanical loads should be progressed from isometric to concentric muscle training. Initially, this should be in mid-ranges of motion and positions to lessen the mechanical load on healing structures [134]. This can gradually progress to end of range of motion strengthening in accordance with the climber's functional restrictions, location, direction of apprehension and the mechanism of injury. Low force to higher loading forces is the approach of choice to facilitate appropriate force production and increases in the tissue's material and morphological properties for joint stability [135]. Eccentric loading at a low level can be started with appropriate regard for the stage of soft tissue healing.

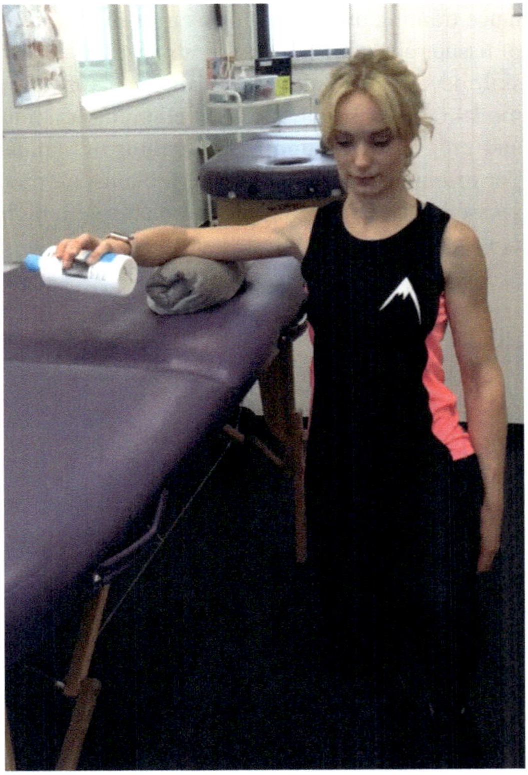

Fig. 20.10 Sensory motor retraining of the glenohumeral joint

Joint stability is dependent upon effective sensory motor control [134]. Injury to the capsule and ligamentous structures affects the afferent feedback, regulation of joint stability and upper quadrant posture and movement [128]. Proprioception offers protection against potentially injurious forces subjected to the shoulder during climbing-related activities [136]. Therefore, sensory motor retraining should be interwoven throughout the rehabilitation after FTAGD (Fig. 20.10). Therapeutic exercise is the primary adjunct. The exercise programme will utilise training exercises that challenge joint position sense in various positions and ranges of motion which are relevant to the climber's return to sport [135]. Late stage rehabilitation will focus on higher level physical conditioning which is specific to climbing. This will include off the wall resistance training modalities to ensure a high degree of tissue robustness [130]. Climbing practice and skill development similar to other rehabilitation modalities starts with low load climbing training and progresses to high load climbing activities. For example, simple climbing problems are progressed to complex climbing problems with high ranges of motion and force demands.

20.11 Elbow

20.11.1 Tendinopathy

The elbow is a common site of pathology in climbers [137]. The largest contributor to this sequelae is tendinopathic pathology [91]. Tendinopathy occurs along a continuum from an acute exacerbation because of abusive mechanical overload due to unaccustomed activity to chronic disrepair and irreversible degenerative changes [138]. The tendons in this region most affected by tendinopathic changes are the common extensor tendons and the common flexor tendons of the elbow [81]. The pathophysiology of tendinopathy is worthy of some consideration because an understanding of the causation will impact the subsequent management. Tendinopathy is associated with pain and disordered collagen within the tendon which lowers its ability to manage mechanical load [139]. It is important to note that tendinopathy represents a non-inflammatory condition with degenerative changes within the body of the tendon and suboptimal collagen synthesis [140]. In this regard, tendinopathy is characterised by a failure of the tendon to remodel after unfavourable mechanical loading. There are several non-mechanical risk factors that can contribute to an impaired tissue response, but this is beyond the scope of this chapter. Tendons are designed to withstand mechanical forces, and appropriate loading has been shown to produce favourable adaptations [141].

The management principles for tendinopathy in the elbow region can be applied to both the common extensor tendons and the common flexor tendons [142]. The point on the continuum where the climber resides determines the management approach.

In the acute or reactive stage, the tendon has been subject to acute overload or microtrauma, and inflammatory cytokines may be present because of the need for matrix synthesis and degradation [143]. There is a concomitant increase in proteoglycans which attract water and promote changes in the tendon matrix. This has been shown to cause a thickening of the tendon which may or may not serve to reduce further stress and allow some degree of adaptation [144]. However, this thickening is a different response from what would be expected from normal progressive loading. Normal progressive loading results in a stiffer tendon with less thickening [145]. The management of climbers in this stage is one of immediate load reduction. Palliative measures such as the PRICE protocol in the initial stages and taping techniques to offload the pain-sensitive structures are indicated [122, 146]. The reactive tendon then requires a period of slow and progressive mechanical loading over 1–3 weeks. Care must be taken to identify the causative factors that led up to the onset of symptoms, and the climber should be counselled on prevention strategies.

The disrepair stage is a progression of the reactive stage in that the tendon has attempted to heal, but the degree of extracellular matrix degradation exceeds adaptation [147]. This stage is characterised by an increase in fibroblasts and chondrocytes resulting in enhanced proteoglycan production. This leads to a reduction in loading capacity because of disorganisation of the matrix and displacement of the collagen fibres. There is also evidence of neovascularization and neuronal ingrowth [148].

Clinical and empirical evidence appears to suggest that resistance training modalities that promote appropriate mechanical loading consistently perform well [17, 149]. However, no loading scheme (e.g. eccentric, isometric or concentric) has been consistently shown to be superior to another loading scheme. However, there is some evidence that eccentric loading is efficacious when compared to other non-exercise modalities such as ice and therapeutic ultrasound

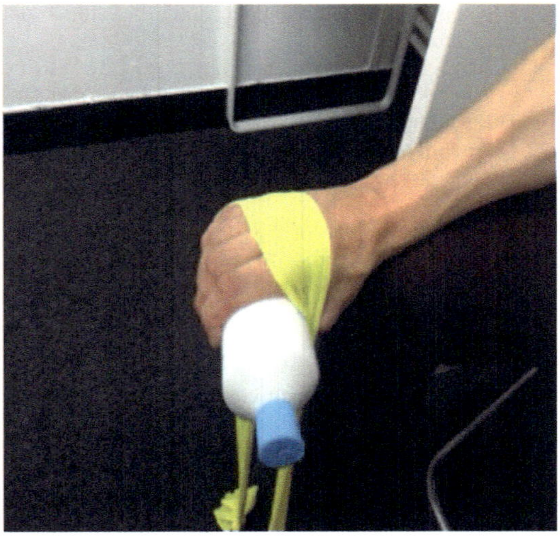

Fig. 20.11 Resistive training of the forearm extensors using a functional grip position

[150]. There is limited evidence that eccentric loading is superior to any other loading scheme [46]. Therefore, we recommend a combined concentric—eccentric training approach with due regard for symptom provocation. The tendon should be loaded with isolated wrist flexion or extension exercises (dependent upon the site of injury) which cause muscular fatigue at 8–10 repetitions (Fig. 20.11). Using this repetition maximum theorem equates to approximately 75–82% of the muscle's maximum capacity before failure [151]. This degree of loading is appropriate for improving contractile function, reducing force deficits and improving the material and morphological properties of tendon [152].

The training programme should be progressive and adapted to the needs of the climber's functional goals [153]. For example, a bouldering specialist will have greater need for strength, power and shock loading energy storage training [22] during their rehabilitation than an alpine climber. The alpine climber will need tendons that are conditioned to withstand long periods of lower demand loading because of the sport's high endurance demands [8]. Extracorporeal shock wave therapy is an evidence-based modality that has shown efficacy in the management of tendon related disorders [41, 56]. In the climber with a protracted history of persistent pain and reduced function, this modality can be beneficial as an adjunct to exercise-based training interventions [58].

Acknowledgments We would like to thank our models, Bison Grips, Compex UK, Swiss Dolorclast and the Depot Climbing Centre Birmingham UK for their assistance with this chapter.

References

1. Schöffl V, Popp D, Küpper T, Schöffl I. Injury trends in rock climbers: evaluation of a case series of 911 injuries between 2009 and 2012. Wilderness Environ Med. 2015;26(1):62–7.
2. Morrison S, Ward P, duManoir GR. Energy system development and load management through the rehabilitation and return to play process. Int J Sports Phys Ther. 2017;12(4):697–710. https://www.ncbi.nlm.nih.gov/pmc/articles/PMC5534159/
3. Jones G, Schöffl V, Johnson MI. Incidence, diagnosis, and management of injury in sport climbing and bouldering: a critical review. Curr Sports Med Rep. 2018;17(11):396–401.
4. Brooks G, Almquist J. Rehabilitation of musculoskeletal injuries in young athletes. Adolesc Med State Art Rev. 2015;26(1):100–15.
5. Spencer S, Wolf A, Rushton A. Spinal-exercise prescription in sport: classifying physical training and rehabilitation by intention and outcome. J Athl Train. 2016;51(8):613–28.
6. Wilk KE, Arrigo CA, Hooks TR, Andrews JR. Rehabilitation of the overhead throwing athlete: there is more to it than just external rotation/internal rotation strengthening. PM R. 2016;8(3 Suppl):S78–90.
7. Wilk KE, Meister K, Andrews JR. Current concepts in the rehabilitation of the overhead throwing athlete. Am J Sports Med. 2002;30(1):136–51.
8. Fanchini M, Violette F, Impellizzeri FM, Maffiuletti NA. Differences in climbing-specific strength between boulder and lead rock climbers. J Strength Cond Res. 2013;27(2):310–4.
9. Kraemer W, Denegar C, Flanagan S. Recovery from injury in sport: considerations in the transition from medical care to performance care. Sports Health. 2009;1(5):392–5.
10. Reiman MP, Lorenz DS. Integration of strength and conditioning principles into a rehabilitation program. Int J Sports Phys Ther. 2011;6(3):241–53.
11. Barratt A. Evidence based medicine and shared decision making: the challenge of getting both evidence and preferences into health care. Patient Educ Couns. 2008;73(3):407–12.

12. Brumitt J, Cuddeford T. Current concepts of muscle and tendon adaptation to strength and conditioning. Int J Sports Phys Ther. 2015;10(6):748–59.
13. Nielsen JB, Willerslev-Olsen M, Christiansen L, Lundbye-Jensen J, Lorentzen J. Science-based neurorehabilitation: recommendations for neurorehabilitation from basic science. J Mot Behav. 2015;47(1):7–17.
14. Valle X, Alentorn-Geli E, Tol JL, Hamilton B, Garrett WE, Pruna R, et al. Muscle injuries in sports: a new evidence-informed and expert consensus-based classification with clinical application. Sports Med. 2017;47(7):1241–53.
15. Dines JS, Bedi A, Williams PN, Dodson CC, Ellenbecker TS, Altchek DW, et al. Tennis injuries: epidemiology, pathophysiology, and treatment. J Am Acad Orthop Surg. 2015;23(3):181–9.
16. Frank CB, Hart DA, Shrive NG. Molecular biology and biomechanics of normal and healing ligaments—a review. Osteoarthr Cartil. 1999;7(1):130–40.
17. Kjaer M, Langberg H, Heinemeier K, Bayer ML, Hansen M, Holm L, et al. From mechanical loading to collagen synthesis, structural changes and function in human tendon. Scand J Med Sci Sports. 2009;19(4):500–10.
18. Behm DG, Bambury A, Cahill F, Power K. Effect of acute static stretching on force, balance, reaction time, and movement time. Med Sci Sports Exerc. 2004;36(8):1397–402. http://www.ncbi.nlm.nih.gov/pubmed/15292749.
19. Di Cagno A, Baldari C, Battaglia C, Gallotta MC, Videira M, Piazza M, et al. Preexercise static stretching effect on leaping performance in elite rhythmic gymnasts. J Strength Cond Res. 2010;24(8):1995–2000. http://www.ncbi.nlm.nih.gov/pubmed/20634743.
20. Richardson E, Lewis JS, Gibson J, Morgan C, Halaki M, Ginn K, et al. Role of the kinetic chain in shoulder rehabilitation: does incorporating the trunk and lower limb into shoulder exercise regimes influence shoulder muscle recruitment patterns? Systematic review of electromyography studies. BMJ Open Sport Exerc Med. 2020;6(1):e000683.
21. Grant S, Hynes V, Whittaker A, Aitchison T. Anthropometric, strength, endurance and flexibility characteristics of elite and recreational climbers. J Sports Sci. 1996;14(4):301–9.
22. Macdonald JH, Callender N. Athletic profile of highly accomplished boulderers. Wilderness Environ Med. 2011;22(2):140–3.
23. Ovid. Core training: evidence translating to better performance and injury prevention. Strength Cond J. 2010;32(3):33–46. http://ovidsp.tx.ovid.com/sp-3.8.1a/ovidweb.cgi?&S=IKLJFPBHDBDDKLLBNCOKDAIBKEKKAA00&Link+Set=S.sh.41%7c19%7csl_10.
24. Scott R, Yang HS, James CR, Sawyer SF, Sizer PS. Volitional preemptive abdominal contraction and upper extremity muscle latencies during D1 flexion and scaption shoulder exercises. J Athl Train. 2018;53(12):1181–9.
25. Chu SK, Jayabalan P, Kibler WB, Press J. The kinetic chain revisited: new concepts on throwing mechanics and injury. PM R. 2016;8(3 Suppl):S69–77.
26. Fournier M. Principles of rehabilitation and return to sports following injury. Clin Podiatr Med Surg. 2015;32(2):261–8.
27. American College of Sports Medicine. American College of Sports Medicine position stand. Progression models in resistance training for healthy adults. Med Sci Sports Exerc. 2009;41(3):687–708.
28. Cormie P, McGuigan MR, Newton RU. Developing maximal neuromuscular power: part 2—training considerations for improving maximal power production. Sports Med. 2011;41(2):125–46.
29. Mendiguchia J, Brughelli M. A return-to-sport algorithm for acute hamstring injuries. Phys Ther Sport. 2011;12(1):2–14.
30. Hoover DL, VanWye WR, Judge LW. Periodization and physical therapy: bridging the gap between training and rehabilitation. Phys Ther Sport. 2016;18:1–20.
31. Wilk V, Abrams D. Sports medicine meets musculoskeletal medicine. Aust Fam Physician. 2007;36(6):394–6.
32. Dhillon H, Dhillon S, Dhillon MS. Current concepts in sports injury rehabilitation. Indian J Orthop. 2017;51(5):529–36.
33. Lu T-W, Chang C-F. Biomechanics of human movement and its clinical applications. Kaohsiung J Med Sci. 2012;28(2 Suppl):S13–25.
34. Huang C, Holfeld J, Schaden W, Orgill D, Ogawa R. Mechanotherapy: revisiting physical therapy and recruiting mechanobiology for a new era in medicine. Trends Mol Med. 2013;19(9):555–64.
35. Ng JL, Kersh ME, Kilbreath S, Knothe TM. Establishing the basis for mechanobiology-based physical therapy protocols to potentiate cellular healing and tissue regeneration. Front Physiol. 2017;8:303.
36. Dunn SL, Olmedo ML. Mechanotransduction: relevance to physical therapist practice-understanding our ability to affect genetic expression through mechanical forces. Phys Ther. 2016;96(5):712–21.
37. Ingber DE. Mechanobiology and diseases of mechanotransduction. Ann Med. 2003;35(8):564–77.
38. Wang N, Naruse K, Stamenović D, Fredberg JJ, Mijailovich SM, Tolić-Nørrelykke IM, et al. Mechanical behavior in living cells consistent with the tensegrity model. Proc Natl Acad Sci U S A. 2001;98(14):7765–70.
39. Schmidt CE, Horwitz AF, Lauffenburger DA, Sheetz MP. Integrin-cytoskeletal interactions in migrating fibroblasts are dynamic, asymmetric, and regulated. J Cell Biol. 1993;123(4):977–91.
40. Ingber DE. Integrins, tensegrity, and mechanotransduction. Gravit Space Biol Bull. 1997;10(2):49–55.

41. Duymaz T, Sindel D. Comparison of radial extracorporeal shock wave therapy and traditional physiotherapy in rotator cuff calcific tendinitis treatment. Arch Rheumatol. 2019;34(3):281–7.
42. Bickel CS, Gregory CM, Dean JC. Motor unit recruitment during neuromuscular electrical stimulation: a critical appraisal. Eur J Appl Physiol. 2011;111(10):2399–407.
43. Hainaut K, Duchateau J. Neuromuscular electrical stimulation and voluntary exercise. Sports Med. 1992;14(2):100–13.
44. Buckthorpe M, La Rosa G, Villa FD. Restoring knee extensor strength after anterior cruciate ligament reconstruction: a clinical commentary. Int J Sports Phys Ther. 2019;14(1):159–72.
45. Bialosky JE, Beneciuk JM, Bishop MD, Coronado RA, Penza CW, Simon CB, et al. Unraveling the mechanisms of manual therapy: modeling an approach. J Orthop Sports Phys Ther. 2018;48(1):8–18.
46. Bohm S, Mersmann F, Arampatzis A. Human tendon adaptation in response to mechanical loading: a systematic review and meta-analysis of exercise intervention studies on healthy adults. Sports Med Open. 2015;1(1):7.
47. Scott BR, Loenneke JP, Slattery KM, Dascombe BJ. Blood flow restricted exercise for athletes: a review of available evidence. J Sci Med Sport. 2016;19(5):360–7.
48. Grønfeldt BM, Lindberg Nielsen J, Mieritz RM, Lund H, Aagaard P. Effect of blood-flow restricted vs heavy-load strength training on muscle strength: systematic review and meta-analysis. Scand J Med Sci Sports. 2020;30(5):837–48.
49. Yow BG, Tennent DJ, Dowd TC, Loenneke JP, Owens JG. Blood flow restriction training after Achilles tendon rupture. J Foot Ankle Surg. 2018;57(3):635–8.
50. Hughes L, Paton B, Rosenblatt B, Gissane C, Patterson SD. Blood flow restriction training in clinical musculoskeletal rehabilitation: a systematic review and meta-analysis. Br J Sports Med. 2017;51(13):1003–11.
51. Vanwye WR, Weatherholt AM, Mikesky AE. Blood flow restriction training: implementation into clinical practice. Int J Exerc Sci. 2017;10(5):649–54.
52. Dankel SJ, Jessee MB, Abe T, Loenneke JP. The effects of blood flow restriction on upper-body musculature located distal and proximal to applied pressure. Sports Med. 2016;46(1):23–33.
53. Electrophysical Agents—13th Edition [Internet]. [cited 2021 Jan 17]. https://www.elsevier.com/books/electrophysical-agents/watson/978-0-7020-5151-7.
54. International Society for Electrophysical Agents in Physical Therapy (ISEAPT) [Internet]. World Physiotherapy. [cited 2021 Jan 17]. https://world.physio/subgroups/electrophysical.
55. ELECTROPHYSICAL AGENTS—Contraindications and precautions: an evidence-based approach to clinical decision making in physical therapy. Physiother Can. 2010;62(5):1–80. https://www.ncbi.nlm.nih gov/pmc/articles/PMC3031347/.
56. Saggini R, Di Stefano A, Saggini A, Bellomo RG. Clinical application of shock wave therapy in musculoskeletal disorders: Part I. J Biol Regul Homeost Agents. 2015;29(3):533–45.
57. Ioppolo F, Rompe JD, Furia JP, Cacchio A. Clinical application of shock wave therapy (SWT) in musculoskeletal disorders. Eur J Phys Rehabil Med. 2014;50(2):217–30.
58. Sems A, Dimeff R, Iannotti JP. Extracorporeal shock wave therapy in the treatment of chronic tendinopathies. J Am Acad Orthop Surg. 2006;14(4):195–204.
59. Bialosky JE, Bishop MD, Price DD, Robinson ME, George SZ. The mechanisms of manual therapy in the treatment of musculoskeletal pain: a comprehensive model. Man Ther. 2009;14(5):531–8.
60. Geri T, Viceconti A, Minacci M, Testa M, Rossettini G. Manual therapy: exploiting the role of human touch. Musculoskelet Sci Pract. 2019;44:102044.
61. Task Force on the Low Back Pain Clinical Practice Guidelines. American Osteopathic Association Guidelines for Osteopathic Manipulative Treatment (OMT) for patients with low back pain. J Am Osteopath Assoc. 2016;116(8):536–49.
62. Navarro-Santana MJ, Sanchez-Infante J, Gómez-Chiguano GF, Cleland JA, López-de-Uralde-Villanueva I, Fernández-de-Las-Peñas C, et al. Effects of trigger point dry needling on lateral epicondylalgia of musculoskeletal origin: a systematic review and meta-analysis. Clin Rehabil. 2020;34(11):1327–40.
63. Heiser R, O'Brien VH, Schwartz DA. The use of joint mobilization to improve clinical outcomes in hand therapy: a systematic review of the literature. J Hand Ther. 2013;26(4):297–311; quiz 311
64. Lucado AM, Dale RB, Vincent J, Day JM. Do joint mobilizations assist in the recovery of lateral elbow tendinopathy? A systematic review and meta-analysis. J Hand Ther. 2019;32(2):262–276.e1.
65. Schöffl V, Lutter C, Popp D. The 'Heel Hook'—a climbing-specific technique to injure the leg. Wilderness Environ Med. 2016;27(2):294–301.
66. Stępień K, Śmigielski R, Mouton C, Ciszek B, Engelhardt M, Seil R. Anatomy of proximal attachment, course, and innervation of hamstring muscles: a pictorial essay. Knee Surg Sports Traumatol Arthrosc. 2019;27(3):673–84.
67. Thompson RN, Hanratty B, Corry IS. 'Heel hook' rock-climbing maneuver: a specific pattern of knee injury. Clin J Sport Med. 2011;21(4):365–8.
68. Ehiogu UD, Stephens G, Jones G, Schöffl V. Acute hamstring muscle tears in climbers-current rehabilitation concepts. Wilderness Environ Med. 2020;31(4):441–53.
69. Worrell TW. Factors associated with hamstring injuries. An approach to treatment and preventative measures. Sports Med. 1994;17(5):338–45.

70. Bleakley CM, Glasgow P, MacAuley DC. PRICE needs updating, should we call the POLICE? Br J Sports Med. 2012;46(4):220–1.
71. Ahmad CS, Redler LH, Ciccotti MG, Maffulli N, Longo UG, Bradley J. Evaluation and management of hamstring injuries. Am J Sports Med. 2013;41(12):2933–47.
72. al HJ et. Criteria for progressing rehabilitation and determining return-to-play clearance following hamstring strain injury: a systematic review. PubMed - NCBI [Internet]. [cited 2018 Dec 21]. https://www.ncbi.nlm.nih.gov/pubmed/28035586.
73. Vuurberg G, Hoorntje A, Wink LM. Diagnosis, treatment and prevention of ankle sprains: update of an evidence-based clinical guideline. Vol. 52, British journal of sports medicine. Br J Sports Med. et al., 2018;52(15):956. https://pubmed.ncbi.nlm.nih.gov/29514819/.
74. Scillia AJ, Pierce TP, Issa K, et al. Low ankle sprains: a current review of diagnosis and treatment. Surg Technol Int. 2017;30:411–4. https://pubmed.ncbi.nlm.nih.gov/28537353/.
75. Lum ZC, Park L. Rock climbing injuries and time to return to sport in the recreational climber. J Orthop. 2019;16(4):361–3.
76. Grønhaug G. Self-reported chronic injuries in climbing: who gets injured when? BMJ Open Sport Exerc Med. 2018;4(1):e000406.
77. Chan KW, Ding BC, Mroczek KJ. Acute and chronic lateral ankle instability in the athlete. Bull NYU Hosp Jt Dis. 2011;69(1):17–26. https://pubmed.ncbi.nlm.nih.gov/21332435/.
78. Weber JM, Maleski RM. Conservative treatment of acute lateral ankle sprains. Clin Podiatr Med Surg. 2002;19(2):309–18. https://pubmed.ncbi.nlm.nih.gov/12123147/.
79. Medina McKeon JM, Hoch MC. The ankle-joint complex: a kinesiologic approach to lateral ankle sprains. J Athl Train. 2019;54(6):589–602. https://pubmed.ncbi.nlm.nih.gov/31184957/.
80. de Vries JS, Krips R, Sierevelt IN, Blankevoort L, van Dijk CN. Interventions for treating chronic ankle instability. Cochrane Database Syst Rev. 2011;2011(8):CD004124.
81. Agres AN, Chrysanthou M, Raffalt PC. The effect of ankle bracing on kinematics in simulated sprain and drop landings: a double-blind, placebo-controlled study. Am J Sports Med. 2019;47(6):1480–7. https://pubmed.ncbi.nlm.nih.gov/31042441/.
82. McKeon PO, Donovan L. A perceptual framework for conservative treatment and rehabilitation of ankle sprains: an evidence-based paradigm shift. J Athl Train. 2019;54(6):628–38. https://pubmed.ncbi.nlm.nih.gov/31135210/.
83. Cushion EJ, Goodwin JE, Cleather DJ. Relative intensity influences the degree of correspondence of jump squats and push jerks to countermovement jumps. J Strength Cond Res. 2016;30(5):1255–64.
84. D'Hooghe P, Cruz F, Alkhelaifi K. Return to play after a lateral ligament ankle sprain. Curr Rev Musculoskelet Med. 2020;13(3):281–8.
85. de Vasconcelos GS, Cini A, Sbruzzi G, Lima CS. Effects of proprioceptive training on the incidence of ankle sprain in athletes: systematic review and meta-analysis. Clin Rehabil. 2018;32(12):1581–90. https://pubmed.ncbi.nlm.nih.gov/29996668/.
86. Heuck A, Hochholzer T, Keinath C. [MRT of the hand and wrist of sport climbers. Imaging of injuries and consequences of stress overload]. Radiologe. 1992;32(5):248–254.
87. Hincapie OL, Elkins JS, Vasquez-Welsh L. Proprioception retraining for a patient with chronic wrist pain secondary to ligament injury with no structural instability. J Hand Ther. 2016;29(2):183–90.
88. Karagiannopoulos C, Michlovitz S. Rehabilitation strategies for wrist sensorimotor control impairment: from theory to practice. J Hand Ther. 2016;29(2):154–65.
89. Kleinman WB. Stability of the distal radioulna joint: biomechanics, pathophysiology, physical diagnosis, and restoration of function what we have learned in 25 years. J Hand Surg Am. 2007;32(7):1086–106.
90. Soubeyrand M, Assabah B, Bégin M, Laemmel E, Dos Santos A, Crézé M. Pronation and supination of the hand: anatomy and biomechanics. Hand Surg Rehabil. 2017;36(1):2–11.
91. Schöffl V, Morrison A, Schöffl I, Küpper T. The epidemiology of injury in mountaineering, rock and ice climbing. Med Sport Sci. 2012;58:17–43.
92. Crowley TP. The flexor tendon pulley system and rock climbing. J Hand Microsurg. 2012;4(1):25–9.
93. King EA, Lien JR. Flexor tendon pulley injuries in rock climbers. Hand Clin. 2017;33(1):141–8. https://linkinghub.elsevier.com/retrieve/pii/S0749071216300749.
94. Zafonte B, Rendulic D, Szabo RM. Flexor pulley system: anatomy, injury, and management. J Hand Surg Am. 2014;39(12):2525–32; quiz 2533
95. Paoloni JA, Milne C, Orchard J, Hamilton B. Non-steroidal anti-inflammatory drugs in sports medicine: guidelines for practical but sensible use. Br J Sports Med. 2009;43(11):863–5.
96. Schneeberger M, Schweizer A. Pulley ruptures in rock climbers: outcome of conservative treatment with the pulley-protection splint-a series of 47 cases. Wilderness Environ Med. 2016;27(2):211–8.
97. Schöffl I, Einwag F, Strecker W, Hennig F, Schöffl V. Impact of taping after finger flexor tendon pulley ruptures in rock climbers. J Appl Biomech. 2007;23(1):52–62.
98. Schöffl V, Hochholzer T, Winkelmann H-P, Strecker W. [Therapy of injuries of the pulley system in sport climbers]. Handchir Mikrochir Plast Chir. 2004;36(4):231–236.

99. Gomez MA, Woo SL, Amiel D, Harwood F, Kitabayashi L, Matyas JR. The effects of increased tension on healing medical collateral ligaments. Am J Sports Med. 1991;19(4):347–54.
100. Electrotherapy - 12th Edition [Internet]. [cited 2021 Jan 4]. https://www.elsevier.com/books/electrotherapy/watson/978-0-443-10179-3.
101. Jimenez PA, Jimenez SE. Tissue and cellular approaches to wound repair. Am J Surg. 2004;187(5A):56S–64S.
102. Schöffl VR, Schöffl I. Injuries to the finger flexor pulley system in rock climbers: current concepts. J Hand Surg Am. 2006;31(4):647–54.
103. Buckwalter JA. Effects of early motion on healing of musculoskeletal tissues. Hand Clin. 1996;12(1):13–24.
104. Franchi MV, Reeves ND, Narici MV. Skeletal muscle remodeling in response to eccentric vs. concentric loading: morphological, molecular, and metabolic adaptations. Front Physiol. 2017;8:447.
105. Kjaer M. Role of extracellular matrix in adaptation of tendon and skeletal muscle to mechanical loading. Physiol Rev. 2004;84(2):649–98.
106. Woo SL, Inoue M, McGurk-Burleson E, Gomez MA. Treatment of the medial collateral ligament injury. II: structure and function of canine knees in response to differing treatment regimens. Am J Sports Med. 1987;15(1):22–9.
107. Fleming BC, Beynnon BD. In vivo measurement of ligament/tendon strains and forces: a review. Ann Biomed Eng. 2004;32(3):318–28.
108. Schreiber T, Allenspach P, Seifert B, Schweizer A. Connective tissue adaptations in the fingers of performance sport climbers. Eur J Sport Sci. 2015;15(8):696–702.
109. López-Rivera E, González-Badillo JJ. Comparison of the effects of three hangboard strength and endurance training programs on grip endurance in sport climbers. J Hum Kinet. 2019;66:183–95.
110. Schweizer A. Biomechanical properties of the crimp grip position in rock climbers. J Biomech. 2001;34(2):217–23. http://www.sciencedirect.com/science/article/pii/S0021929000001846.
111. Schweizer A. Biomechanical properties of the crimp grip position in rock climbers. J Biomech. 2001;34(2):217–23.
112. Vogt M, Hoppeler HH. Eccentric exercise: mechanisms and effects when used as training regime or training adjunct. J Appl Physiol. 2014;116(11):1446–54.
113. LaStayo P, Marcus R, Dibble L, Frajacomo F, Lindstedt S. Eccentric exercise in rehabilitation: safety, feasibility, and application. J Appl Physiol. 2014;116(11):1426–34.
114. Wright AR, Richardson AB, Kikuchi CK, Goldberg DB, Marumoto JM, Kan DM. Effectiveness of accelerated recovery performance for post-ACL reconstruction rehabilitation. Hawaii J Health Soc Welf. 2019;78(11 Suppl 2):41–6.
115. Cools AM, Borms D, Castelein B, Vanderstukken F, Johansson FR. Evidence-based rehabilitation of athletes with glenohumeral instability. Knee Surg Sports Traumatol Arthrosc. 2016;24(2):382–9.
116. Consigliere P, Haddo O, Levy O, Sforza G. Subacromial impingement syndrome: management challenges. Orthop Res Rev. 2018;10:83–91.
117. Gismervik SØ, Drogset JO, Granviken F, Rø M, Leivseth G. Physical examination tests of the shoulder: a systematic review and meta-analysis of diagnostic test performance. BMC Musculoskelet Disord. 2017;18(1):41.
118. Physical tests for shoulder impingements and local lesions of bursa, tendon or labrum that may accompany impingement. PubMed [Internet]. [cited 2020 Sep 25]. https://pubmed.ncbi.nlm.nih.gov/23633343/.
119. Tran G, Cowling P, Smith T, Bury J, Lucas A, Barr A, et al. What imaging-detected pathologies are associated with shoulder symptoms and their persistence? A systematic literature review. Arthritis Care Res (Hoboken). 2018;70(8):1169–84.
120. Daghir AA, Sookur PA, Shah S, Watson M. Dynamic ultrasound of the subacromial-subdeltoid bursa in patients with shoulder impingement: a comparison with normal volunteers. Skelet Radiol. 2012;41(9):1047–53.
121. Lewis J. Rotator cuff related shoulder pain: assessment, management and uncertainties. Man Ther. 2016;23:57–68.
122. Lewis JS. Rotator cuff tendinopathy: a model for the continuum of pathology and related management. Br J Sports Med. 2010;44(13):918–23.
123. Perez-Palomares S, Oliván-Blázquez B, Arnal-Burró AM, Mayoral-Del Moral O, Gaspar-Calvo E, de-la-Torre-Beldarraín ML, et al. Contributions of myofascial pain in diagnosis and treatment of shoulder pain. A randomized control trial. BMC Musculoskelet Disord. 2009;10:92.
124. Seitz AL, McClure PW, Finucane S, Boardman ND, Michener LA. Mechanisms of rotator cuff tendinopathy: intrinsic, extrinsic, or both? Clin Biomech (Bristol, Avon). 2011;26(1):1–12.
125. Spargoli G. Treatment of rotator cuff tendinopathy as a contractile dysfunction. A clinical commentary. Int J Sports Phys Ther. 2019;14(1) 148–58.
126. Lewis J, McCreesh K, Roy J-S, Ginn K. Rotator cuff tendinopathy: navigating the diagnosis-management conundrum. J Orthop Sports Phys Ther. 2015;45(11):923–37.
127. Avila Lafuente JL, Moros Marco S, García Pequerul JM. Controversies in the management of the first time shoulder dislocation. Open Orthop J. 2017;11:1001–10.
128. Savoie FH, O'Brien MJ. Anterior instability in the throwing shoulder. Sports Med Arthrosc Rev. 2014;22(2):117–9.
129. Dreinhöfer KE, Schüler S, Schäfer M, Ohly T. Rehabilitation concepts and return to sport

after interventions on the shoulder. Orthopade. 2014;43(3):256–64.
130. Ma R, Brimmo OA, Li X, Colbert L. Current concepts in rehabilitation for traumatic anterior shoulder instability. Curr Rev Musculoskelet Med. 2017;10(4):499–506.
131. Palmieri-Smith RM, Villwock M, Downie B, Hecht G, Zernicke R. Pain and effusion and quadriceps activation and strength. J Athl Train. 2013;48(2):186–91.
132. Escamilla RF, Yamashiro K, Paulos L, Andrews JR. Shoulder muscle activity and function in common shoulder rehabilitation exercises. Sports Med. 2009;39(8):663–85.
133. McNair PJ, Marshall RN, Maguire K, Brown C. Knee joint effusion and proprioception. Arch Phys Med Rehabil. 1995;76(6):566–8.
134. Jaggi A, Alexander S. Rehabilitation for shoulder instability—current approaches. Open Orthop J. 2017;11:957–71.
135. Bleichert S, Renaud G, MacDermid J, Watson L, Faber K, Lenssen R, et al. Rehabilitation of symptomatic atraumatic degenerative rotator cuff tears: a clinical commentary on assessment and management. J Hand Ther. 2017;30(2):125–35.
136. Salles JI, Velasques B, Cossich V, Nicoliche E, Ribeiro P, Amaral MV, et al. Strength training and shoulder proprioception. J Athl Train. 2015;50(3):277–80.
137. Holtzhausen LM, Noakes TD. Elbow, forearm, wrist, and hand injuries among sport rock climbers. Clin J Sport Med. 1996;6(3):196–203.
138. Stasinopoulos D, Stasinopoulou K, Johnson MI. An exercise programme for the management of lateral elbow tendinopathy. Br J Sports Med. 2005;39(12):944–7.
139. Vicenzino B. Tendinopathy: evidence-informed physical therapy clinical reasoning. J Orthop Sports Phys Ther. 2015;45(11):816–8.
140. Rathleff MS, Thorborg K. 'Load me up, Scotty': mechanotherapy for plantar fasciopathy (formerly known as plantar fasciitis). Br J Sports Med. 2015;49(10):638–9.
141. Reeves ND. Adaptation of the tendon to mechanical usage. J Musculoskelet Neuronal Interact. 2006;6(2):174–80.
142. Hotchkiss RN. Epicondylitis—lateral and medial. A problem-oriented approach. Hand Clin. 2000;16(3):505–8.
143. Thomopoulos S, Parks WC, Rifkin DB, Derwin KA. Mechanisms of tendon injury and repair. J Orthop Res. 2015;33(6):832–9.
144. Andarawis-Puri N, Flatow EL, Soslowsky LJ. Tendon basic science: development, repair, regeneration, and healing. J Orthop Res. 2015;33(6):780–4.
145. Docking SI, Cook J. How do tendons adapt? Going beyond tissue responses to understand positive adaptation and pathology development: a narrative review. J Musculoskelet Neuronal Interact. 2019;19(3):300–10.
146. Coombes BK, Bisset L, Vicenzino B. Management of lateral elbow tendinopathy: one size does not fit all. J Orthop Sports Phys Ther. 2015;45(11):938–49.
147. Cook JL, Rio E, Purdam CR, Docking SI. Revisiting the continuum model of tendon pathology: what is its merit in clinical practice and research? Br J Sports Med. 2016;50(19):1187–91.
148. Plinsinga ML, Brink MS, Vicenzino B, van Wilgen CP. Evidence of nervous system sensitization in commonly presenting and persistent painful tendinopathies: a systematic review. J Orthop Sports Phys Ther. 2015;45(11):864–75.
149. Fouré A, Nordez A, McNair P, Cornu C. Effects of plyometric training on both active and passive parts of the plantarflexors series elastic component stiffness of muscle–tendon complex. Eur J Appl Physiol. 2011;111(3):539–48. http://link.springer.com/article/10.1007/s00421-010-1667-4.
150. Cook JL, Purdam CR. Is tendon pathology a continuum? A pathology model to explain the clinical presentation of load-induced tendinopathy. Br J Sports Med. 2009;43(6):409–16.
151. Beyer RP, Kongsgaard M, Kjaer BHP, Ohlenschlaeger T, Kjaer MM, Magnusson PSP. Heavy slow resistance versus eccentric training as treatment for Achilles tendinopathy: a randomized controlled trial. J Sports Med. 2015;43(7):1704–11.
152. Franchi MV, Atherton PJ, Reeves ND, Flück M, Williams J, Mitchell WK, et al. Architectural, functional and molecular responses to concentric and eccentric loading in human skeletal muscle. Acta Physiol (Oxf). 2014;210(3):642–54.
153. Stasinopoulos D, Stasinopoulos I. Comparison of effects of eccentric training, eccentric-concentric training, and eccentric-concentric training combined with isometric contraction in the treatment of lateral elbow tendinopathy. J Hand Ther. 2017;30(1):13–9.

Injury Prevention

21

G. Jones, Uzo Dimma Ehiogu, and M. I. Johnson

Isabelle Schöffl in „Feuerball", Loch, Frankenjura, Germany, Photo Enrico Haase

21.1 Introduction

The International Olympic Committee consensus statement recommends the accurate monitoring of training load to reduce injury risk in athletes. Although a high level of physical preparedness is likely protective against injury occurrence, failure to manage athletic workloads effectively have been found to be predictive of injury [1]. The calculation of workload is reliant on the accurate recording of exposure. In the published climbing literature to date, methods used by authorship teams to record exposure and operational measures of performance are inconsistent. At present, there is no published consensus statement on design characteristics for use in epidemiological cohort studies in climbing.

In terms of injury prevention, this raises several issues. Firstly, differences in epidemiological design characteristics used by researchers impede accurate comparisons between climbing-related studies to be made. Therefore, our understanding of injury burden and associated risk factors is limited. Secondly, what are the variables that may be utilised to estimate athletic workload in climbers? Finally, given that exercise prescription is a key tenant of injury prevention, what additional

G. Jones (✉) · M. I. Johnson
School of Clinical and Applied Sciences, Leeds Becket University, Leeds, UK
e-mail: g.j.jones@leedsbeckett.ac.uk; m.johnson@leedsbeckett.ac.uk

U. D. Ehiogu
School of Clinical and Applied Sciences, Leeds Becket University, Leeds, UK

Research and Training Department, Birmingham Royal Orthopaedic Hospital, Birmingham, UK

Birmingham Medical School, College of Medical and Dental Sciences, University of Birmingham, Birmingham, UK

V. Schöffl et al. (eds.), *Climbing Medicine*, https://doi.org/10.1007/978-3-030-72184-8_21

physical preparation strategies need to be considered for the climbing athlete? In addressing these issues, effective injury prevention strategies may be developed.

The Aims of This Chapter Are to:

- Critically evaluate epidemiological design characteristics used in climbing research.
- Report methods and considerations for calculating athletic workload in climbing.
- Provide an overview of contemporary strategies used to develop soft tissue robustness and fatigue management.
- Report the role of skill development in the management of injury risk.

21.1.1 Evaluating Injury Terminology

The definition of injury used in climbing-related studies varies considerably and are not always clearly stated. The criteria used to define injury commonly include bodily damage, pain, disability, medical consultation, medical intervention, hospital admission, and withdrawal. For example, Neuhof et al. [2] only reported those injuries that required professional medical treatment intervention, whilst Bowie et al. [3] only included participants who had sustained an injury that required hospital treatment. Both injury definitions require participants to have received medical treatment but may exclude a participant based on the location of that treatment. In contrast, Van Middelkoop et al. [4] stated injury to be 'any damage as a result of climbing that caused pain and or disability irrespective of whether a treatment or medical intervention was administered'. Importantly, this definition acknowledges the reporting of pain, defined by the International Association for the Study of Pain (IASP) as 'an unpleasant sensory and emotional experience associated with actual or potential tissue damage, or described in terms of such damage' [5]. Pain is rarely stated in injury definitions used in climbing studies, yet it is not uncommon for climbers to continue in their given activity, or modify their activity, whilst perceiving some level of pain or discomfort.

Injury definitions need to account for both time-loss and non-time loss injuries. Chronic overuse injuries often have an insidious onset which is manifested subclinically before being raised in the consciousness of the individual. Even then, initial symptoms do not usually result in the termination of activity as they are generally mild in severity. Importantly, individuals may not consider themselves to be in an injurious state or in a sufficiently injurious state to withdraw from activity regardless of the mechanism. Failure to capture such data may result in the under-reporting of injury occurrence and confounds comparison of estimates between studies.

Recurrent injuries in sport are common, and it is widely accepted that subsequent injury is strongly associated with previous injury occurrence. Authorship teams need to correctly categorise first injury, re-injury and multiple re-injury of the same or different type and identify the underlying mechanism. A study by Jones et al. [6] was the first attempt to analyse previous injury as a risk factor for re-injury in climbers. Individuals were categorised as sustaining a re-injury, if they reported an injury at the same anatomical site, precipitated by the same cause, on at least two occasions within the 12-month reporting period. Individuals were categorised as sustaining a multiple re-injury if they reported an injury at the same anatomical site, precipitated by the same cause, on at least three occasions or greater within the 12-month reporting period.

Injuries in climbing may be broadly categorised as 'acute' or 'traumatic' and 'chronic' or 'overuse'. The classification systems used need to clearly differentiate between acute impact injuries, acute non-impact injuries and chronic overuse injuries. Critical reviews by Jones et al. [7, 8] categorised climbing-related injuries as: acute impact injury caused by the climber falling onto a climbing surface and/or ground, or an object such as a rock falling on to the climber, and acute non-impact injury resulting from acute trauma to the body and chronic overuse injury of the body from repetitive climbing. The Union Internationale des Associations d'Alpinisme (UIAA) recommend the use of the MedCom Score [8] to classify injury. However, the MedCom Score does not account clearly for chronic overuse injuries and therefore may fail to capture this type of data. Inconsistency and ambiguity of classification between studies limit our ability to correctly identify and attribute an injury mechanism. The percentage of climbers sustaining at least one injury due to acute impact is difficult to establish because many studies combine acute impact and acute non-impact injury data. For example, Backe et al. [9] reported a high percentage of lower limb injuries inclusive of contusions and lacerations and categorised these as traumatic. This does not accurately account for the aetiology of injury, which is likely a result of impact with the climbing surface and/or ground and/or climbing equipment, for example, the rope. Failure to inform the reader of the exact category of injury makes interpretation of the findings of limited use as individuals may sustain injury multiple times and at multiple body sites.

21.1.2 Evaluating Study Design and Data Collection Procedures

Studies determining incidence and prevalence of injury in climbing utilise prospective and retrospective cross-sectional survey methods. Although prospective cohort studies are considered the gold standard of observational research and should provide accurate and reliable data of injury and exposure, they are difficult to conduct in large sample climbing populations. Conversely, retrospective studies are particularly prone to measurement error. Jones et al. [7] reviewed four retrospective cohort studies that estimated incidence rate based on data captured by postal survey [10], memory recall [2], medical records and interview [3] and national park registrant data on accident occurrence [12]. Relying on the memory of participants to recall the number of injuries that they have experienced over a period of time is particularly error prone, especially when the recall period extends in excess of 12 months. However, retrospective studies that use survey methods do allow relative ease of access to large sample populations. Jones et al. [12] surveyed a large sample of climbers: Injury and performance behaviour questions were framed to remind participants of the recall period, i.e. 'how many times in the last 12 months'. In contrast, participants in a study by Neuhof et al. [2] were asked to recall injuries that occurred over a 5-year period. All study designs can introduce measurement error due to imprecise recording, interpretation, calculation and recollection of climbing exposure and injury occurrence.

The UIAA Medical Commission recommends that incidence rate of injury in climbing be expressed as injuries per 1000 h of exposure to control for variation in exposure, especially between different types of climbing activity [8]. However, reporting injuries per 1000 h of exposure is an imprecise measure because it may not account for non-climbing activities such as preparation, rest periods between attempts and belaying a fellow climber. The Medical Commission further recommends studies that do not measure the hours of exposure record 4 h for sport climbing outdoors and traditional climbing and 2 h for any indoor climbing activity per day. Outdoor bouldering is not accounted for in this recommendation; moreover, calculating climbing exposure using such methods is likely to introduce significant error into estimates. Performing secondary analysis of data to produce estimates of exposure per 1000 h is also likely to produce errors. Schöffl et al. [13] performed a secondary analysis of data reported by Limb [10], Bowie et al. [3] and Schussman et al. [11] to estimate the respective incidence rate of injury per 1000 h of climbing exposure. The survey methods used by each study were significantly different and raise a legitimate concern in regard of conducting analysis to generate such data on heterogeneous studies.

The incidence rate in climbing may additionally be expressed as the number of new injuries in a specified time period (e.g. per year), number of new injuries per number of visits to a climbing venue (e.g. injuries per one million visits) or number of new injuries per number of athletic exposures (injuries/100 participants). The calculation total is inclusive of first injury, multiple injuries and re-injuries. In their systematic review

Woolings et al. [14] reported incidence as a function of athlete exposures. For example, incidence of injury over the entire career was reported as 300 (95% CI 250, 357) injuries/100 participants [15] and incidence in regard of a type of climbing behaviour as 103 (95% CI 71,146) and 127 (95% CI 85,184) injuries/100 participants/year for outdoor and indoor bouldering respective [16]. Such measures are perceived as a useful method for estimating resource utilisation for healthcare providers and clinicians [17]. However, the problem with reporting incidence of injury in a specified time period, in relation to visits to a climbing venue or as a function of athlete exposures, is that the duration of individual exposures will vary and again introduce error into estimates. Therefore, such methods may not reliably indicate the extent of the injury problem.

In a study by Jones et al., individual climbing exposure was captured using estimates of the frequency of ascent [12]. Climbers were also asked to provide additional information regarding consistency of their performance standard. This better reflects actual participant exposure in climbing and importantly controls for performance standard as a potential confounder in the calculation of risk. To date this published work is the only study that has considered both frequency of ascent and the operational standard of ascent to calculate risk of injury across a wide range of climbing behaviours. As such, this method may be used to predict risk based on an individual climber's profile of climbing behaviours and athletic load.

Precise information about the environment, climbing behaviour and practice of climbing populations is needed. Correctly categorising exposure allows a direct link to be made between the specific situation and injury occurrence. Study reports are often not explicit in how other climbing behaviours of participants are accounted or controlled for. Moreover, even when precise information about the cause of the injury is captured, it is possible that other climbing activity may have contributed to the injury. For example, a traditional climber who was sampled in an outdoor climbing setting may be undertaking indoor bouldering far more frequently, and this may be the precipitating factor for injury occurrence. Inconsistency in categorising the type of climbing activity between studies can cause under- or over-reporting of a particular type of injury to a climbing behaviour. It is imperative that injury data is captured with the exact climbing behaviour and practice that caused the injury.

21.1.3 Evaluating Data Processing

A variety of different grading systems exist worldwide to report the operational standard of climbing performance. Although the use of number-based scales is commonly reported by authorship teams, inconsistencies in the conversion of identical operational standards of performance exist. As a consequence, the International Rock Climbing Research Association (IRCRA) produced a positional statement in regard of comparative grading scales for future use in climbing research [3, 18]. The authors developed a reporting scale to standardise the conversion of climbing performance, regardless of behaviour, into a numerical value suitable for data analysis. The authors acknowledged a limitation of the proposed scale is the use of the British technical grade for traditional climbing only. Traditional climbing in Britain is graded using a combined system that assigns both an adjectival and technical grade, for example, Very Severe 4c. The adjectival grade provides information about the level of difficulty, overall seriousness and potential risks to the climber. The corresponding technical grade provides information about the hardest technical movement required to complete the climb. The comparative grading scale proposed by IRCRA [20] contains considerable overlap between the British technical grade and the recommended reporting value, for example, British technical grade 6a may be recorded as 13, 14, 15, 16 or 17. Therefore, the use of the IRCRA scale in its current format may introduce significant measurement error when applied to sample populations of British traditional climbers. An amendment to the IRCRA comparative grading scale was presented to the 4th International Rock

Table 21.1 Amendment to IRCRA comparative grading scale [19]

IRCRA Reporting scale	British adjectival and technical grade	French sport
1	M	1
2	D	2
3	VD	2+
4	S	3-
5	HS/VS 4a	3
6	VS 4b	3+
7	VS 4c	4
8	VS 5a/HVS 4c	4+
9	HVS 5a	5
10	HVS 5b/E1 5a	5+
11	E1 5b	6a
12	E1 5c/E2 5b	6a+
13	E2 5c	6b
14	E3 5c	6b+
15	E3 6a	6c
16	E4 6a	6c+
17	E4 6b	7a
18	E5 6b	7a+
19	E6 6b	7b
20	E6 6c	7b+
21	E7 6c	7c
22	E7 7a	7c+
23	E8 6c	8a
24	E8 7a	8a+
25	E9 6c	8b
26	E9 7a	8b+
27	E9 7b/E10 7a	8c
28	E11 7a	8c+
29	E11 7b	9a
30	E11 7c	9a+
31	E12 7b	9b
32	E12 7c	9b+
33	E13 8a	9c

Climbing Research Congress in 2018 [18] to address this issue (see Table 21.1).

21.1.4 Summary

Inconsistency in the use of injury terminology, data collection procedures, calculation of exposure and operational measures of performance by researchers exist. Such inconsistencies likely contribute to the large variance in the incidence and prevalence of injury reported. Continued reporting of heterogeneous results in population samples limits meaningful comparison of studies to be made and an understanding of injury burden and risk factors to be known. Standardising the criteria used to attribute injury and climbing activity, coupled with more accurate methods of calculating exposure, will overcome such limitations.

21.2 Monitoring Athletic Load

21.2.1 Introduction

An International Olympic Committee consensus statement defined load as, 'the sport and non-sport burden (single or multiple physiological, psychological or mechanical stressors) as a stimulus that is applied to a human biological system (including subcellular elements, a single cell, tissues, one or multiple organ systems, or the individual)' [21]. The key variables that are required to accurately calculate load include the type, duration, frequency and intensity of activity. Load can be further quantified as external load, i.e. the objective work undertaken (training and competition), and internal load, i.e. an individual's physiological and perceptual response [20] The differentiation of load type is important as an identical external load stimulus can elicit a range of stressors. Furthermore, an individual's response to the same external stimulus may differ, at different time points [20].

Challenges in accurately calculating load and interpreting the evidence of associations between load, injury and illness exist. Primarily, athletes undertake high training loads to prepare for the demands of competition. A high level of physical preparedness likely mitigates some injury risk, but failure to manage load effectively could be detrimental to the athlete's health.

21.2.2 Session Rating of Perceived Exertion (Session-RPE)

Session-RPE is a simple method of calculating athletic load by multiplying the session intensity (normally measured using a modified Category

Ratio 10 scale) by the duration of the individual session measure in minutes [22]. The subsequent calculation is considered a quantity of total load and measured in 'arbitrary' units. It is recommended that session intensity recording using a modified Category Ratio 10 scales (CR 10) be undertaken 30 min after cessation of activity and familiarisation of the athlete with the scale necessary prior to use [22]. A review of session-RPE by Haddad et al. [23] reported the validity, reliability and internal consistency of session-RPE across a wide range of sports and physical activities. RPE is appropriate to use as a measure of internal load [20] and valid to use with men and women of different ages, including children and adolescents [23]. Session-RPE may be used as a standalone method of calculating load but can be combined with other factors to create a sport-specific measure [24].

Derivative characteristics of session-RPE are 'monotony' and 'strain' and are suggested to relate the onset of overtraining [23]. Monotony is a measure of training load fluctuations, and strain is a measure of how hard an athlete is working for a fixed time period, usually a week. Monotony and strain are characteristics of training variability derived from session-RPE and are suggested to relate to the onset of overtraining [23]. The research underpinning these measures used illness as a proxy marker of overtraining syndrome: currently insufficient evidence and data to quantify the risk of illness in response load fluctuations exist, and further studies are required [22].

Measuring the intensity of a climbing session using time likely provides an imprecise statistic that fails to capture significant performance data. In order to create a sport-specific sessional measure of external load, we suggest the following factors to be considered and appropriate weightings applied: sum of ascents; grade of ascents; climbing behaviour; nature of ascent, e.g. redpoint; and completions to non-completion ratio. The IRCRA comparative grading scale could be used to provide a single value that accounts for both operational performance grade and climbing behaviour. Internal load could be captured using the CR10 scale and multiplied with external load to calculate total load.

21.2.3 Acute Chronic Workload Ratios

High levels of physical preparedness and musculoskeletal adaptation likely protect against injury, but it is vital athletes achieve this in a controlled and systematic manner. The acute/chronic workload ratio (ACWR) is proposed as an effective method of monitoring training and competition load by means of modelling the relationship between changes in load and injury risk in athletes [23]. The ACWR is calculated by dividing an athlete's current training load (acute), usually gathered over the last 7 days by the typical training load the athlete has completed (chronic), usually gathered over the last 4 weeks. The typical training load may be calculated using a rolling average method or an exponentially weighted moving average. A ratio of greater than 1.5 is suggested to indicate an increased risk of injury; a ratio of 0.8–1.3 is suggested optimal and indicates a reduced risk of injury [23].

The IOC consensus group reviewed data in relation to relative load, rapid changes in load and injury risk in athletes and reported that team-sport athletes reacted significantly better when imposed load variations were controlled and relatively small in magnitude [24]. Furthermore, the consensus group reported ACWR to be applicable for use with individual sports participants [24]. A review of 22 studies supported the use of ACWR as part of a range of measures to monitor training load in athletes but concluded that further research across a range of sports is needed [25]. Legitimate conceptual and methodological concerns in regard of ACWR to predict injury risk have been raised. Impellizzeri et al. [26] and the Australian Institute of Sport now advise it should not be used as an indicator of injury risk.

21.2.4 Summary

Athletes need to be physically prepared to fully meet the demands imposed upon them.

Despite a paucity of empirical research evidence in climbing populations and methodological concerns of the efficacy of ACWR to predict injury risk, we recommend coaches and medical teams use monitoring protocols to better plan for the training and competitive requirements of their climbing athletes.

21.3 Contemporary Strategies Used to Develop Soft Tissue Robustness and Fatigue Management

21.3.1 Physical Preparation and Athletic Development

Injury is an inherent consequence of professional and amateur sport. The cause of injury is often multifactorial and rarely the preserve of one independent factor [23]. Therefore, the prevention of injury in sport per se is governed by logical principles that provide the athlete, healthcare professional and coaching team with direction. In the sport of climbing, these concepts are no different. Climbing, due to its heavy burden on the musculoskeletal system, clearly lends itself to an injury profile which is bias towards injury of the upper quadrant, e.g. the shoulder elbow, wrist and hand [27], but must also consider the lower limb as the spectrum of injury changes with better injury surveillance [25] The principles of acute and chronic load management, protective equipment, skill development and physical conditioning act as the cornerstone of injury prevention strategies similar to other sports [26].

Injury to musculoskeletal tissues causes a cascade of events which ultimately lead to tissue trauma. The insult to musculoskeletal tissues is often the cause of acute or chronic manifestations of mechanical load to biological tissues [27]. The biological load imposed upon musculoskeletal tissues may lead to adaptation, maintenance or maladaptation [28]. Adaptation occurs when tissue is subjected to load and modifies to the imposed demands through positive changes in its material, morphological and/or physiological properties [29]. In maintenance the load applied to tissues is within its biological capacity such that no appreciable changes occur. In the maladaptive state, the tissues are subjected to mechanical and/or physiological loads that exceed the capacity of the tissue to tolerate the imposed demand [32]. This may cause a chronic injury state from excessive cumulative microtrauma, leading to insidious tissue disruption before being raised into the consciousness of the athlete. Contrastingly, injury can be the result of acute traumatic insult in which the force profile causes rapid catastrophic failure of tissues. The underpinning theories which encapsulate these ideas of biological adaptability to mechanical and physiological stress are numerous [31, 34, 35]. However, all purport to optimally elevate tissue capacity to higher levels of performance and ultimately protection of the athlete.

Pragmatically, healthcare professionals and members of the multidisciplinary team from an injury prevention perspective should familiarise themselves with the following strategies:

- Strategies to increase the ultimate load profile of tissues to mechanical stress
- Strategies to increase the metabolic capacity of the athlete to encourage fatigue resistance
- Strategies to decrease the acute and chronic stress imposed upon tissues
- Strategies to manage fatigue without affecting performance

21.3.2 Strategies to Increase the Ultimate Load Profile of Tissues to Mechanical Stress

Although, it may be self - evident that reducing the mechanical stress imposed upon musculoskeletal tissues in the sport of climbing can be injury protective. It is important to realise that mechanical load is a potent stimulus for positive adaptations of biological tissues [35]. In order for the athlete to develop physical expertise in bio-motor capacities (such as strength, rate of force development, endurance, work capacity and power) which provide a performance advantage, there must be physiological stress [36]. The adaptive capabilities of the athlete's musculoskeletal tissues are influenced by the process of mechanotransduction.

Mechanotransduction is the biological processes in which tissues respond to mechanical load at cellular level [37]. The cells of mechanically responsive tissues sense mechanical stimuli at the extracellular tissue level. This causes a cascade of events in most but not all musculoskeletal tissues. In responsive tissues, this leads to the deposition of collagen in architectural arrangements aligned to stress adaptation [38]. The stress adaptation is governed by the application of appropriate mechanical stress which does not exceed the capabilities of the target tissues. This leads to positive adaptations in the tissues tolerance to manage stress and impose loads associated with physical activity [39]. Therefore, with appropriate physical training and athletic preparation, it is possible to alter the material and morphological properties of biological tissues. In the literature, this has been shown to be beneficial in both muscular [38], bone [39], tendon [42] and connective tissue models [41] both in vivo and in vitro and in cross-sectional studies in upper limb-based sports [42] and [45] climbing.

From a pragmatic perspective, the ability of a tissue to generate force provides an obvious performance advantage in a sport such as climbing [46]. Morphological changes in muscle, tendon, bone and connective tissue are also associated with improved muscular force capacity [39]. The nature of adaptations sought should be determined by the specific needs of the athlete and the injury profile of the sport [47]. For example, increasing the stress tolerance of the wrist and forearm flexor muscles with resistance training may reduce injury risk in boulderers that are required to generate high levels of force at high velocities (Fig. 21.1). However, compare, for example, the alpine climber in which their sports-specific conditioning needs lend itself to lower force-velocity requirements.

Furthermore, it is important to realise that the technical demands of climbing are such that a significant reduction in chronic training load may not be advantageous for the climber's yearly performance progression. A chronic reduction in training load would cause insufficient adaptive stress and also reduce the opportunity for technical improvement. This technical improvement or

Fig. 21.1 Bouldering athlete displaying high force output during a bouldering problem

skill development is important in body weight-dominated sports [48]. The opportunity to refine movement skills and tactical awareness can be lost if too great a reduction in training load is used as a strategy for injury prevention.

21.3.3 Strategies to Decrease the Acute and Chronic Stress Imposed upon Tissues

Climbing is a sport based upon the skilful application of force to optimise athletic performance [48]. Climbing requires the awareness and application of temporal and spatial relationships between the centre of gravity and the base of support. The centre of gravity is a point of equilibrium in all directions and a focal point for the earths gravitational pull on the body [49]. The climber must be cognisant of their line of gravity and the orientation of the body to this line. This understanding of the centre of mass in relation to the base of support allows the climber to minimise the effect of gravity and make progress during climbing [50]. These basic biomechanical concepts affect balance and movement efficiency of the climber. Balance and efficiency are governed by skill development [51]. A skill is an action or task directed towards achieving a specific goal [52]. In climbing, a motor skill requires voluntary movement of the athlete's body segments to achieve a specific task. Motor skills in climbing require both gross motor skills involving large muscle groups and fine motor skills involving small muscle groups. Climbing is a continuous motor skill which is distinguished by its arbitrary beginning and end points [53].

Skill development utilises the application of physics to ensure effective force generation, force transfer and force absorption [54]. Skilful movement is a strategy that aims to reduce the stress placed upon the musculoskeletal system. This is achieved by optimising energy-efficient movement patterns [55]. In essence the objective is to ensure that the climber does not place excessive force through structures that are either not suited to the role or lack the capacity to adapt to load because of their biology. This requires an in-depth understanding of both functional anatomy and clinical biomechanics to make a reasoned hypothesis about the potential effects of movement inefficiency as a basis for pathomechanics [56]. In this regard, it is important to understand that strategies to decrease acute and chronic stress on musculoskeletal tissues are often interrelated. For example, the climber's body position during movement is important because it influences the centre of gravity in relation to the base of support and hence the climber's degree of balance on the wall (Fig. 21.2). These issues ultimately affect the metabolic cost of climbing and likely the performance outcome.

The mechanical output of an activity is supported by the body's metabolism and the energetics of exercise [57]. Therefore, the energy cost of a given mechanical output to ascend a climbing route is dependent upon the climber's efficiency of movement. Inefficiency has the potential to increase the metabolic cost of mechanical work leading to premature fatigue [58]. Fundamentally, poor body positioning and technique regardless of its cause can affect the interplay between biomechanics, injury and performance [59]. Biomechanical moment arms both at a whole body, body segment and local joint muscle region are important in human movement. Moment arms influence the magnitudes of force which must be overcome and generated by the climber during all activities. The musculoskeletal system generally works at a mechanical disadvantage when compared to the external environment. This often means the muscular system is required to generate significant forces to overcome external resistances because of this disadvantageous arrangement [60]. This is why movement efficiency is theoretically of critical importance in delaying the onset of fatigue. Fatigue has been shown consistently to cause a reduction in muscle force [61], joint stability [60], impaired decision making [63] and reduced proprioception [64]. Fatigue of the muscular system may place greater demands on noncontractile neuromuscular components such as bone, connective tissues and articular cartilage. This may manifest as a challenge to soft tissue integrity and the maintenance of optimal anatomical relationships within and between joints. This may potentially contribute to increased injury risk.

Fig. 21.2 Lead climber displaying technical skill whilst ascending a competition route

21.3.4 Strategies to Increase the Metabolic Capacity of the Athlete to Encourage Fatigue Resistance

Epidemiological evidence consistently reports a high occurrence of injuries to the upper limbs [27, 65]. Mechanical load imposes significant stressors on key structures such as muscle, tendon, peri-articular connective tissues and bone [64]. Contemporary models of training in climbing reinforce modalities which stimulate tissue adaptations using mechanical load [48]. There is an abundance of research and contemporary thought recommending the use of physical preparation modalities which condition the finger and forearm musculature [67]. Activities such as finger board training, use of campus boards, system boards and climbing-based activities are common [48]. However, while this has significant sports specificity, such modalities impose high mechanical loads on musculoskeletal tissues [69–71]. Repetitive high mechanical load will likely result in injury unless planned appropriately and utilised judiciously.

Broadly speaking climbing with the exception of speed climbing is by definition near maximal intermittent exercise interspersed with periods of submaximal exercise [71]. This suggests participants need to have a well-developed capacity to support both aerobic and anaerobic metabolism. Therefore, alternative modalities that stimulate positive training adaptations should be considered as an injury prevention strategy for the climber. The ability to endure mechanical work through effective training of the metabolic system is arguably an important parameter differentiating optimal and suboptimal performance [72]. The climber, whose body is conditioned to offset fatigue yet maintain optimal force output over the duration of a climbing route, will often determine sucess [46]. The ability to resist fatigue, regardless of the event duration, is associated with an effective metabolic system [71].

The physiological attributes associated with climbing has been extensively reported elsewhere [68, 75]. It is fundamentally important that the basic energetic requirements of climbing are adequately understood by healthcare professionals and support teams when developing training programmes. An appropriate training programme is one which prepares the climber for the demands of their specific discipline [75]. The metabolic demands of climbing are varied by the rate and duration of energy utilisation undertaken by the respective disciplines. For example, lead climbing and speed climbing imposed very different physiological demands upon the climber's metabolism to sustain mechanical work. The liberation of chemical energy for mechanical work occurs by the resynthesise of adenosine triphosphate (ATP) within the muscle cell [50]. The biochemical process by which this energy source is liberated is dependent upon the rate and duration of mechanical work.

The energetics of exercise broadly falls into three different categories which ensure optimal energy production for sports performance. Alactic, anaerobic glycolytic and oxidative phosphorylation energetics are the primary systems [77]. These systems while discrete in their configuration are interrelated and active to varying degrees during all activities. However, during specific activities, there is often a strong predominance of one system over another. For example, speed climbing because of its short duration (<10 s) and mechanical force output at high velocities utilises predominantly alactic energy systems of ATP and stored phosphagens [78]. Contrastingly, sport climbing routes utilise predominantly slow glycolytic and oxidative energy systems [79]. This contrast in energetics is attributed to the longer durations of physical activity associated with this type of climbing. There is an inverse relationship between the duration of a physical activity and the rate of energy production permissible and hence the system utilised for the resynthesise of ATP [80].

Traditionally, training to improve the metabolic capacity of the climber has focused upon modalities which target the sports-specific qualities of climbing. This has traditionally included climbing-based activities and off the wall training, for example, finger boards and campus boards [51, 80] (Figs. 21.3, 21.4, and 21.5). This type of training arguably provides sports specific-

Fig. 21.3 Climber using a fingerboard to develop finger and forearm muscle capacity

Fig. 21.4 Climber using a campus board to develop high velocity climbing-specific mechanical loading

ity because of the similarity in energetics, movement patterns, neuromuscular force profiles and specific muscle groups used. However, while this is sports specific, it imposes a high mechanical load on musculoskeletal tissues. Contrastingly, other modalities whilst not climbing specific may be relevant in regard to the underpinning physical qualities that are important for climbing performance [73, 80]. These modalities stimulate central and peripheral adaptations in the metabolic pathways that support climbing performance without the mechanical load [70]. The aerobic oxidative system is the primary system responsible for exercise at submaximal work rates. The efficiency of this system is of particular importance for achieving peak exercise performance in most sports [81]. The oxidative system has been suggested to contribute up to 50% of the energy requirement for force production after 75 seconds of maximal exhaustive exercise [81]. Oxygen uptake ($V0_{2max}$) ranging from 54 to 55 mL kg^{-1} min has been shown in climbers during treadmill running [82] which is consistent with that seen in team sports and gymnastics [73]. However, peak VO_2 of 43.8 mL kg^{-1} min has been reported during treadmill climbing [83]. This lower oxygen uptake is possibly attributed to the smaller muscle mass associated with climbing when compared to running and cycling. Therefore, this data, may suggest central factors such as cardiac output and oxidative capacity are not limiting factors affecting climbing performance. However, cardiac output is the delivery mechanism for oxygen and nutrition to exercising muscles [82]. This system facilitates the resynthesis of ATP between bouts of high-intensity exercise and sustains high submaximal work rates [79]. At the site of muscular tissue, the skeletal muscle cell needs energy to perform mechanical work [78].

Fig. 21.5 Climber using a circuit board with additional system mass (weighted backpack) to mechanically overload the skill of climbing

High-intensity exercise close to VO_2 peak using various modalities has been shown to improve left ventricular contractile force and increased cardiac filling pressures [83]. Metabolic efficiency in skeletal muscle has also been shown to cause up regulation of oxidative and glycolytic enzymes [86]. Increase capillarisation within the muscle has been reported in addition to increased mitochondria protein transcription [85]. These central and peripheral adaptations collectively lead to a greater reliance on oxidative metabolism at any given workload [82]. This for the climber is suggestive of a reduced reliance on anaerobic pathways because the point of transition to glycolytic metabolism is delayed. This adaptation is performance enhancing due to the delay in fatigue. However, of greater relevance is that these adaptations may be achieved using training modalities which minimise mechanical stress through musculoskeletal structures. High intensity interval training (HIIT) has been shown to be a viable method for developing both aerobic and anaerobic performance in various upper and lower limb-dominant sports [88, 89]. HIIT can be used for climbing-related training [77]. However, from an injury reduction perspective, we would recommend a boarder remit in which it is used as an adjunct to stimulate both central and peripheral adaptations to enhance energetics in non-climbing activities. This can involve programming long and short HIIT interval training to target cardiopulmonary and oxidative muscle fibres, the glycolytic and alactic phosphate systems [73, 85]. This might include intervals sessions based upon predetermined work to rest ratios using, for example, medicines ball throws, battle rope conditioning and power bag drags for upper body conditioning or cycling and/or running bases activities to upregulate central adaptations [90] (Fig. 21.6).

Fig. 21.6 An athlete using battle ropes to develop high intensity metabolic fitness

21.3.5 Strategies to Management Fatigue Without Affecting Performance

A critical component underpinning injury prevention in a physical preparation model for the recreational climber or climbing athlete is the systematic management of training-related fatigue. Stress reduction in the musculoskeletal system can be achieved by the management of fatigue [91]. The practical application of fatigue management over a training and competition year is the planned variation in training load on a daily, weekly and monthly basis to reduce the monotony of training [92]. Over a training and competition year, monotony can influence the climber's risk of injury and illness. However, fatigue is an important part of the training and adaptation process and therefore must be managed appropriately.

Periodization is a system and philosophy of training management whose aim is to apply the manipulation of intensity and volume during the course of a training or competition cycle [34]. The cycle may be one month, one year or a quadrennial cycle such as used for Olympic sports planning. Planning the training loads and sequencing periodic recovery to allow biological adaptation is a critical step in the process [91]. Performance in any sport is often dependent upon on the interplay between the mechanical and metabolic work capacity the athlete can tolerate during physical activity. Periodization from a performance perspective has been shown to produce superior results for developing athletic performance [92]. This approach is driven by long-term planning to maximise the probability of physical preparedness of the climber. This minimises the risk of suboptimal underload or overload of bio-motor capacities deleterious to optimal performance and tissue robustness [93].

Periodization of mechanical and metabolic load is defined by training phases. In general, there are two major phases in the training system: the preparatory phase and the competitive phase. The preparatory phase can be further subdivided based upon the needs of the climber into the general physical training phase and the sports-specific training phase. The subphases allow the detailed manipulation of training loads and volumes based upon the intended physiological adaptations sought for optimal performance at a time point in the future [37, 90, 94]. There are various periodization approaches within the literature based on the needs of the athlete and philosophies of athletic development. However, we would recommend a linear approach to programming because this appears to offer a straightforward system for managing fatigue and load in a climbing athletes programme.

The methodology is operationalised by the structured variation in bio-motor capacities over the training year [94]. The principle of phase potentiation is an important construct of this method. Phased potentiation is the sequencing and ordering of physical qualities into training blocks. This ordering of bio-motor capacities is designed to ensure that physical qualities trained prior support the next phase of metabolic and mechanical loading of the climber. The manipulation of volume, load, work to rest ratios and progression of bio-motor capacities (e.g. muscle strength/endurance >strength >power) potentates the adaptive process [95]. The strategy is foundational in nature setting the framework for subsequent phases of training. This is designed to minimise inappropriate fatigue while optimising athletic performance [96]. From this position, the material and morphological (mechanical) drivers for developing tissues robustness are achieved to support higher work demands for subsequent phases of the climber's physical preparation plan. This type of systematic planning can also be applied to the metabolic development of the climber. In a similar manner to mechanical loading, the phases of development also are structured in a way that lay the foundations for subsequent sports-specific metabolic conditioning [99]. Central cardio respiratory adaptations may be used early in a training year to develop a foundation for sports-specific conditioning and recovery in later months [90]. The need to develop intra-muscular adaptations in sports-specific muscles for high levels of performance may be less relevant at this stage of the training year [84]. However, as the training year progresses, the emphasis will change to target the energetics of sport-specific musculature [100].

In the competitive phase, training modalities should mimic the kinetics and kinematic profile of climbing and the climber's end performance goals [101]. The training modalities will be highly sports specific and involve the refinement of climbing under realistic loads. The power and force output profiles should closely relate to either speed, sport, boulder, multi-pitch or other climbing speciality with identical work to recovery ratios. At this juncture of the training cycle, actual climbing will be the primary method of metabolic and mechanical development. This, for example, may involve competition simulation at or slightly higher intensities than normal for competitive athletes. Correspondingly, it may involve long indoor routes laden with gear to simulate the demands of alpine climbing for the alpine climber. The specifics of this phase will be determined by the needs of the climber, type of climbing and performance level sought [95, 103, 104]. The underpinning theoretical rational which supports a periodisation-based approach to training is the fitness fatigue model [34] and the general adaptation syndrome [31]. These theories elucidate how organisms adapt to training stress with positive or negative physiological adaptations.

21.3.6 Summary

Athletic development and physical preparation strategies are a cornerstone of climbing performance and injury prevention. The ability of musculoskeletal tissues to adapt its material, morphological and physiological properties to the imposed demands provides a performance advantage. However, this load must not exceed the physical capacity of the tissues. The principles of skill development, fatigue management, metabolic development and effective tissue loading underpin injury prevention management.

Acknowledgements We would like to thank the International Federation of Sport Climbing and Mr. Eddie Fowke for permission to use the photos in this chapter.

References

1. Soligard T, Schwellnus M, Alonso J-M, Bahr R, Clarsen B, Dijkstra HP, et al. How much is too much? (Part 1) International Olympic Committee consensus statement on load in sport and risk of injury. Br J Sports Med. 2016;50(17):1030–41. https://bjsm.bmj.com/content/50/17/1030.
2. Neuhof A, Hennig FF, Schöffl I, Schöffl V. Injury risk evaluation in sport climbing. Int J Sports Med. 2011;32(10):794–800. https://pubmed.ncbi.nlm.nih.gov/21913158/.
3. Bowie WS, Hunt TK, Allen HA. Rock-climbing injuries in Yosemite National Park. West J Med. 1988;149(2):172–7.
4. van Middelkoop M, Bruens ML, Coert JH, Selles RW, Verhagen E, Bierma-Zeinstra SMA, et al. Incidence and risk factors for upper extremity climbing injuries in indoor climbers. Int J Sports Med. 2015;36(10):837–42.
5. Treede R-D. The International Association for the Study of Pain definition of pain: as valid in 2018 as in 1979, but in need of regularly updated footnotes. Pain Rep 2018;3(2). https://www.ncbi.nlm.nih.gov/pmc/articles/PMC5902252/.
6. Jones G, Llewellyn D, Johnson MI. Previous injury as a risk factor for reinjury in rock climbing: a secondary analysis of data from a retrospective cross-sectional cohort survey of active rock climbers. BMJ Open Sport Exerc Med 2015;1(1):bmjsem-2015-000031. https://bmjopensem.bmj.com/lookup/doi/10.1136/bmjsem-2015-000031.
7. Jones G, Johnson MI. A critical review of the incidence and risk factors for finger injuries in rock climbing. Curr Sports Med Rep. 2016;15(6):400–9.
8. Schöffl V, Morrison A, Hefti U, Ullrich S, Küpper T. The UIAA Medical Commission injury classification for mountaineering and climbing sports. Wilderness Environ Med. 2011;22(1):46–51.
9. Backe S, Ericson L, Janson S, Timpka T. Rock climbing injury rates and associated risk factors in a general climbing population. Scand J Med Sci Sports. 2009;19(6):850–6.
10. Limb D. Injuries on British climbing walls. Br J Sports Med. 1995;29(3):168–70. https://www.ncbi.nlm.nih.gov/pmc/articles/PMC1332307/.
11. Schussman LC, Lutz LJ, Shaw RR, Bohnn CR. The epidemiology of mountaineering and rock climbing accidents. J Wilderness Med. 1990;1(4):235–48. http://www.sciencedirect.com/science/article/pii/S0953985990713381.
12. Jones G, Asghar A, Llewellyn DJ. The epidemiology of rock-climbing injuries. Br J Sports Med. 2008;42(9):773–8.
13. Schöffl V, Morrison A, Schwarz U, Schöffl I, Küpper T. Evaluation of injury and fatality risk in rock and ice climbing. Sports Med. 2010;40(8):657–79.
14. Woollings KY, McKay CD, Emery CA. Risk factors for injury in sport climbing and bouldering: a systematic review of the literature. Br J Sports Med. 2015;49(17):1094–9.
15. Rohrbough JT, Mudge MK, Schilling RC. Overuse injuries in the elite rock climber. Med Sci Sports Exerc. 2000;32(8):1369–72.
16. Josephsen G, Shinneman S, Tamayo-Sarver J, Josephsen K, Boulware D, Hunt M, et al. Injuries in bouldering: a prospective study. Wilderness Environ Med. 2007;18(4):271–80.
17. Knowles SB, Marshall SW, Guskiewicz KM. Issues in estimating risks and rates in sports injury research. J Athl Train. 2006;41(2):207–15.

18. Nick D, Giles D, Schöffl V, Fuss F, Watts P, Wolf P, et al. Comparative grading scales, statistical analyses, climber descriptors and ability grouping: International Rock Climbing Research Association position statement. Sports Technology. 2016;1–7
19. Draper N, Giles D, Schöffl V, Konstantin Fuss F, Watts P, Wolf P et al. Comparative grading scales, statistical analyses, climber descriptors and ability grouping: International Rock Climbing Research Association position statement. Sports Technology. 2015;8(3-4):88–94.
20. PRIME PubMed | Monitoring athlete training loads: consensus statement. [cited 2021 Jan 22]. https://www.unboundmedicine.com/medline/citation/28463642/Monitoring_Athlete_Training_Loads:_Consensus_Statement.
21. Haddad M, Stylianides G, Djaoui L, Dellal A, Chamari K. Session-RPE method for training load monitoring: validity, ecological usefulness, and influencing factors. Front Neurosci. 2017;11:612. https://www.ncbi.nlm.nih.gov/pmc/articles/PMC5673663/.
22. Schwellnus M, Soligard T, Alonso J-M, et al. How much is too much? (Part 2) International Olympic Committee consensus statement on load in sport and risk of illness. Br J Sports Med. 2016;50(17):1043–52. https://bjsm.bmj.com/content/50/17/1043.
23. Gabbett TJ. The training-injury prevention paradox: should athletes be training smarter and harder? Br J Sports Med. 2016;50(5):273–80.
24. Soligard T, Schwellnus M, Alonso J-M, et al. How much is too much? (Part 1) International Olympic Committee consensus statement on load in sport and risk of injury. Br J Sports Med. 2016;50(17):1030–41. https://bjsm.bmj.com/content/50/17/1030.
25. Griffin A, Kenny IC, Comyns TM, Lyons M. The association between the acute: chronic workload ratio and injury and its application in team sports: a systematic review. Sports Med. 2020;50(3):561–80.
26. Schöffl V, Morrison A, Schöffl I, Küpper T. The Epidemiology of Injury in Mountaineering, Rock and Ice Climbing. Epidemiol Injury Adv Extreme Sports. 2012;58:17–43. https://www.karger.com/Article/FullText/338575.
27. Lutter C, Tischer T, Hotfiel T, Frank L, Enz A, Simon M, et al. Current trends in sport climbing injuries after the inclusion into the Olympic program. Analysis of 633 injuries within the years. 2017/18:10.
28. Ehiogu UD, Stephens G, Jones G, Schöffl V. Acute hamstring muscle tears in climbers-current rehabilitation concepts. Wilderness Environ Med. 2020;31(4):441–53.
29. Emery CA, Pasanen K. Current trends in sport injury prevention. Best Pract Res Clin Rheumatol. 2019;33(1):3–15.
30. Frost HM. Wolff's Law and bone's structural adaptations to mechanical usage: an overview for clinicians. Angle Orthod. 1994;64(3):175–88.
31. Maestroni L, Read P, Bishop C, Turner A. Strength and power training in rehabilitation: underpinning principles and practical strategies to return athletes to high performance. Sports Med. 2019;50
32. Sharma P, Maffulli N. Biology of tendon injury: healing, modeling and remodeling. J Musculoskelet Neuronal Interact. 2006;6(2):181–90.
33. Selye H. Stress and the general adaptation syndrome. Br Med J. 1950;1(4667):1383–92.
34. Busso T, Candau R, Lacour JR. Fatigue and fitness modelled from the effects of training on performance. Eur J Appl Physiol Occup Physiol. 1994;69(1):50–4.
35. American College of Sports Medicine. American College of Sports Medicine position stand. Progression models in resistance training for healthy adults. Med Sci Sports Exerc. 2009;41(3):687–708.
36. Mujika I, Halson S, Burke LM, Balagué G, Farrow D. An integrated, multifactorial approach to periodization for optimal performance in individual and team sports. Int J Sports Physiol Perform. 2018;13(5):538–61.
37. Dunn SL, Olmedo ML. Mechanotransduction: relevance to physical therapist practice-understanding our ability to affect genetic expression through mechanical forces. Phys Ther. 2016;96(5):712–21.
38. Ingber DE. Integrins, tensegrity, and mechanotransduction. Gravit Space Biol Bull. 1997;10(2):49–55.
39. Kjaer M, Jørgensen NR, Heinemeier K, Magnusson SP. Exercise and regulation of bone and collagen tissue biology. Prog Mol Biol Transl Sci. 2015;135:259–91.
40. Franchi MV, Reeves ND, Narici MV. Skeletal muscle remodeling in response to eccentric vs. concentric loading: morphological, molecular, and metabolic adaptations. Front Physiol. 2017;8:447.
41. Brown GN, Sattler RL, Guo XE. Experimental studies of bone mechanoadaptation: bridging in vitro and in vivo studies with multiscale systems. Interface Focus. 2016;6(1):20150071.
42. Bohm S, Mersmann F, Arampatzis A. Human tendon adaptation in response to mechanical loading: a systematic review and meta-analysis of exercise intervention studies on healthy adults. Sports Med Open. 2015;1(1):7.
43. Maniotis AJ, Chen CS, Ingber DE. Demonstration of mechanical connections between integrins, cytoskeletal filaments, and nucleoplasm that stabilize nuclear structure. Proc Natl Acad Sci U S A. 1997;94(3):849–54.
44. Calbet JA, Moysi JS, Dorado C, Rodríguez LP. Bone mineral content and density in professional tennis players. Calcif Tissue Int. 1998;62(6):491–6.
45. Schreiber T, Allenspach P, Seifert B, Schweizer A. Connective tissue adaptations in the fingers of performance sport climbers. Eur J Sport Sci. 2015;15(8):696–702.
46. Grant S, Hasler T, Davies C, Aitchison TC, Wilson J, Whittaker A. A comparison of the anthropometric, strength, endurance and flexibility characteristics

of female elite and recreational climbers and non-climbers. J Sports Sci. 2001;19(7):499–505.
47. Stien N, Saeterbakken AH, Hermans E, Vereide VA, Olsen E, Andersen V. Comparison of climbing-specific strength and endurance between lead and boulder climbers. PLoS One. 2019;14(9):e0222529.
48. Orth D, Kerr G, Davids K, Seifert L. Analysis of relations between spatiotemporal movement regulation and performance of discrete actions reveals functionality in skilled climbing. Front Psychol. 2017;8:1744.
49. Vigotsky AD, Zelik KE, Lake J, Hinrichs RN. Mechanical misconceptions: have we lost the 'mechanics' in 'sports biomechanics'? J Biomech. 2019;93:1–5.
50. Mabe J, Butler SL. Analysis of contemporary anaerobic sport specific training techniques for rock climbing. The Sport J. 2020; https://thesportjournal.org/article/analysis-of-contemporary-anaerobic-sport-specific-training-techniques-for-rock-climbing/
51. Orth D, Davids K, Seifert L. Coordination in climbing: effect of skill, practice and constraints manipulation. Sports Med. 2016;46(2):255–68.
52. Handford C, Davids K, Bennett S, Button C. Skill acquisition in sport: some applications of an evolving practice ecology. J Sports Sci. 1997;15(6):621–40.
53. Seifert L, Wattebled L, L'hermette M, Bideault G, Herault R, Davids K. Skill transfer, affordances and dexterity in different climbing environments. Hum Mov Sci. 2013;32(6):1339–52.
54. Cowley JC, Gates DH. Proximal and distal muscle fatigue differentially affect movement coordination. PLoS One 2017;12(2). https://www.ncbi.nlm.nih.gov/pmc/articles/PMC5325574/.
55. Chu SK, Jayabalan P, Kibler WB, Press J. The kinetic chain revisited: new concepts on throwing mechanics and injury. PM R. 2016;8(3 Suppl):S69–77.
56. Kibler WB, Wilkes T, Sciascia A. Mechanics and pathomechanics in the overhead athlete. Clin Sports Med. 2013;32(4):637–51.
57. Artioli GG, Bertuzzi RC, Roschel H, Mendes SH, Lancha AH, Franchini E. Determining the contribution of the energy systems during exercise. J Vis Exp 2012;(61). https://www.ncbi.nlm.nih.gov/pmc/articles/PMC3415169/.
58. Noakes TD. Physiological models to understand exercise fatigue and the adaptations that predict or enhance athletic performance. Scandinavian J Med Sci Sports. 2000;10(3):123–45.
59. Lu T-W, Chang C-F. Biomechanics of human movement and its clinical applications. Kaohsiung J Med Sci. 2012;28(2 Suppl):S13–25.
60. Keogh J, Lake J, Swinton P. Practical applications of biomechanical principles in resistance training: moments and moment arms. J Fitness Res. 2013;2:39–48.
61. Wan J, Qin Z, Wang P, Sun Y, Liu X. Muscle fatigue: general understanding and treatment. Exp Mol Med. 2017;49(10):e384. https://www.nature.com/articles/emm2017194.
62. Abboud J, Lardon A, Boivin F, Dugas C, Descarreaux M. Effects of muscle fatigue, creep, and musculoskeletal pain on neuromuscular responses to unexpected perturbation of the trunk: a systematic review. Front Hum Neurosci 2017;10. https://www.ncbi.nlm.nih.gov/pmc/articles/PMC5209383/.
63. Weippert M, Rickler M, Kluck S, Behrens K, Bastian M, Mau-Moeller A, et al. It's harder to push, when I have to push hard—physical exertion and fatigue changes reasoning and decision-making on hypothetical moral dilemmas in males Front Behav Neurosci 2018;12. https://www.ncbi.nlm.nih.gov/pmc/articles/PMC6276357/.
64. Lee H-M, Liau J-J, Cheng C-K, Tan C-M, Shih J-T. Evaluation of shoulder proprioception following muscle fatigue. Clin Biomech (Bristol, Avon). 2003;18(9):843–7.
65. Grønhaug G. Self-reported chronic injuries in climbing: who gets injured when? BMJ Open Sport Exerc Med. 2018;4(1):e000406.
66. Holtzhausen LM, Noakes TD. Elbow, forearm, wrist, and hand injuries among sport rock climbers. Clin J Sport Med. 1996;6(3):196–203.
67. Medernach JPJ, Kleinöder H, Lötzerich HHH. Fingerboard in competitive bouldering: training effects on grip strength and endurance. J Strength Cond Res. 2015;29(8):2286–95.
68. Hahn F, Erschbaumer M, Allenspach P, et al. Physiological bone responses in the fingers after more than 10 years of high-level sport climbing: analysis of cortical parameters. Wilderness Environ Med. 2012;23(1):31–6. https://www.ncbi.nlm.nih.gov/pubmed/22441086.
69. Vigouroux L, Quaine F, Labarre-Vila A, Moutet F. Estimation of finger muscle tendon tensions and pulley forces during specific sport-climbing grip techniques. J Biomech. 2006;39(14):2583–92.
70. Schöffl VR, Hochholzer T, Imhoff AB, Schöffl I. Radiographic adaptations to the stress of high-level rock climbing in junior athletes: a 5-year longitudinal study of the German Junior National Team and a group of recreational climbers. Am J Sports Med. 2007;35(1):86–92.
71. de Moraes Bertuzzi RC, Franchini E, Kokubun E, Kiss MAPDM. Energy system contributions in indoor rock climbing. Eur J Appl Physiol. 2007;101(3):293–300. https://doi.org/10.1007/s00421-007-0501-0.
72. Buchheit M, Laursen PB. High-intensity interval training, solutions to the programming puzzle. Part II: anaerobic energy, neuromuscular load and practical applications. Sports Med. 2013;43(10):927–54.
73. Watts PB. Physiology of difficult rock climbing. Eur J Appl Physiol. 2004;91(4):361–72.
74. Giles LV, Rhodes EC, Taunton JE. The physiology of rock climbing. Sports Med. 2006;36(6):529–45.
75. Turner A, Comfort P, McMahon J, Bishop C, Chavda S, Read P, et al. Developing powerful athletes part 2: practical applications. Strength Cond J. 2020;43:23–31.

76. Bonora M, Patergnani S, Rimessi A, De Marchi E, Suski JM, Bononi A, et al. ATP synthesis and storage. Purinergic Signal. 2012;8(3):343–57.
77. Neufer PD. The bioenergetics of exercise. Cold Spring Harb Perspect Med 2018;8(5).
78. Brooks GA. Bioenergetics of exercising humans. Compr Physiol. 2012;2(1):537–62.
79. Phillips KC, Sassaman JM, Smoliga JM. Optimizing rock climbing performance through sport-specific strength and conditioning. Strength Cond J. 2012;34(3):1–18. https://journals.lww.com/nsca-scj/Fulltext/2012/06000/Optimizing_Rock_Climbing_Performance_Through.1.aspx.
80. Morrison S, Ward P, duManoir GR. Energy system development and load management through the rehabilitation and return to play process. Int J Sports Phys Ther. 2017;12(4):697–710. https://www.ncbi.nlm.nih.gov/pmc/articles/PMC5534159/.
81. Gastin P. Energy system interaction and relative contribution during maximal exercise. Sports Med (Auckland, NZ). 2001;31:725–41.
82. Billat V, Palleja P, Charlaix T, Rizzardo P, Janel N. Energy specificity of rock climbing and aerobic capacity in competitive sport rock climbers. J Sports Med Phys Fitness. 1995;35:20–4.
83. Booth J, Marino F, Hill C, Gwinn T. Energy cost of sport climbing in elite performers. Br J Sports Med. 1999;33:14–8.
84. Buchheit M, Laursen PB. High-intensity interval training, solutions to the programming puzzle: Part I: cardiopulmonary emphasis. Sports Med. 2013;43(5):313–38.
85. Lepretre P-M, Koralsztein J-P, Billat VL. Effect of exercise intensity on relationship between VO2max and cardiac output. Med Sci Sports Exerc. 2004;36(8):1357–63.
86. Harmer A, Mckenna M, Sutton J, Snow R, Ruell P, Booth J, et al. Skeletal muscle metabolic and ionic adaptations during intense exercise following sprint training in humans. J Appl Phys (Bethesda, MD: 1985). 2000;89:1793–803.
87. Hoier B, Hellsten Y. Exercise-induced capillary growth in human skeletal muscle and the dynamics of VEGF. Microcirculation. 2014;21(4):301–14.
88. Fernandez-Fernandez J, Zimek R, Wiewelhove T, Ferrauti A. High-intensity interval training vs. repeated-sprint training in tennis. J Strength Cond Res. 2012;26(1):53–62.
89. Buchheit M, Mendez-Villanueva A, Quod M, Quesnel T, Ahmaidi S. Improving acceleration and repeated sprint ability in well-trained adolescent handball players: speed versus sprint interval training. Int J Sports Physiol Perform. 2010;5(2):152–64.
90. Science and application of high intensity interval training. Human Kinetics. 2021. https://www.human-kinetics.co.uk/9781492552123/science-and-application-of-high-intensity-interval-training.
91. Bolling C, Mellette J, Pasman HR, van Mechelen W, Verhagen E. From the safety net to the injury prevention web: applying systems thinking to unravel injury prevention challenges and opportunities in Cirque du Soleil. BMJ Open Sport Exerc Med. 2019;5(1):e000492.
92. Mølmen KS, Øfsteng SJ, Rønnestad BR. Block periodization of endurance training—a systematic review and meta-analysis. Open Access J Sports Med. 2019;10:145–60.
93. Hoover DL, VanWye WR, Judge LW. Periodization and physical therapy: bridging the gap between training and rehabilitation. Phys Ther Sport. 2016;18:1–20.
94. Issurin VB. Benefits and limitations of block periodized training approaches to athletes' preparation: a review. Sports Med. 2016;46(3):329–38.
95. Issurin VB. New horizons for the methodology and physiology of training periodization. Sports Med. 2010;40(3):189–206.
96. Hartmann H, Wirth K, Keiner M, Mickel C, Sander A, Szilvas E. Short-term periodization models: effects on strength and speed-strength performance. Sports Med. 2015;45(10):1373–86.
97. Turner A. The science and practice of periodization: a brief review. Strength Cond J. 2011;33(1):34–46. http://ovidsp.ovid.com/ovidweb.cgi?T=JS&CSC=Y&NEWS=N&PAGE=fulltext&D=ovftl&AN=00126548-201102000-00006.
98. Gamble P. Periodization of training for team sports athletes. Strength Cond J. 2006;28(5):56–66.
99. Smith DDJ. A framework for understanding the training process leading to elite performance. Sports Med. 2003;33(15):1103–26. https://doi.org/10.2165/00007256-200333150-00003.
100. Young WB. Transfer of strength and power training to sports performance. Int J Sports Physiol Perform. 2006;1(2):74–83.
101. Brearley S, Bishop C. Transfer of training: how specific should we be? Strength Cond J. 2019;41(3):97–109. insights.ovid.com.
102. Issurin V. Block periodization versus traditional training theory: a review. J Sports Med Phys Fitness. 2008;48(1):65–75.
103. Kelly VG, Coutts AJ. Planning and monitoring training loads during the competition phase in team sports. Strength Cond J. 2007;29(4):32–7.
104. Issurin_New horizons for the methodology and physiology of training periodization.pdf.

Taping

22

Volker Schöffl and Christoph Lutter

Isabelle Schöffl, Fontainebleau, Photo Enrico Haase

V. Schöffl (✉)
Department of Orthopedic and Trauma Surgery, Center of Sportsmedicine, Klinikum Bamberg, Bamberg, Germany

School of Clinical and Applied Sciences, Leeds Becket University, Leeds, UK

Section of Wilderness Medicine, Department of Emergency Medicine, University of Colorado School of Medicine, Denver, CO, USA

Department of Orthopedic and Trauma Surgery, Friedrich-Alexander Universität Erlangen-Nürnberg, Erlangen, Germany

C. Lutter
Department of Orthopedic Surgery, University Hospital Rostock, Rostock, Germany

V. Schöffl et al. (eds.), *Climbing Medicine*, https://doi.org/10.1007/978-3-030-72184-8_22

22.1 Basics

Taping the hand and fingers for crack climbing has long been a commonplace practice to prevent skin injuries. Professional boxers, football players, and basketball players often use tape to prevent injuries and to allow an early return to competition after an injury. Taping can help an athlete to work at his sport with some protection while not completely immobilizing the injured area as with a cast. Using tape correctly requires a proper diagnosis of the injury and the correct application of the tape. Several procedures used as treatment may also be used to prevent injury or to support recovery [1]. In the following chapter, we will focus on therapeutic, static taping. Dynamic Kinesio® Taping (k-taping), which has a completely different ratio and approach, is used less frequently in rock climbing and will therefore not be included. Protective taping of the skin for crack climbing or for some finger pockets will help to reduce pain and skin damage; nevertheless, general prophylactic taping is not recommended. Surprisingly, a study by Woollings et al. [1] even highlighted it to be a risk factor for finger injuries! Nevertheless, in the rehabilitation after finger injuries, it is an essential tool when returning to climbing. So far, little scientific work has been performed concerning taping and climbing [2–9]; thus, some of the information in this chapter may have only level 4 evidence. Nevertheless, we try to present this topic with as much scientific evidence as possible.

Taping is used in climbing for various reasons:

1. A preventative (protective) taping against skin abrasions and injuries is performed in crack climbing as well as for certain finger pockets. In certain, repetitive work on one hard move in a climb, a preventive joint capsule tape may be used [10]. A general prophylactic finger taping is not recommended and was even found to increase injury incidence [1].
2. A therapeutic taping represents one part of therapy for (finger) overstrain injuries or during return to (climbing) sports after an acute (finger) injury. Also, some acute finger injuries (e.g., collateral ligament injuries) can receive therapeutic tapings for partial immobilization.

Applying static tapings in climbing should follow certain guidelines as insufficient taping can be ineffective or even harmful.

General Taping Guidelines [10]:

- Tape should never be applied to the skin directly off the roll. To tape correctly, a strip of tape should be torn off at the desired length and then carefully applied to the injured area. Taping an injury straight from the roll usually makes the application too tight and causes skin breaks.
- The width of the tape must relate to the size of the finger. It is crucial to apply the correct tension while applying the tape. This being said, it is also necessary to create soft tissue "windows." Thus, the subcutaneous tissue can release in these areas without being excessively compressed underneath the tape. A finger that is fully covered underneath one or several layers of tape is not advisable and harmful.
- As a general principle, taping should not be circular around a limb, as to not compromise blood flow. The fingers are an exception because, otherwise, the tape would not be stable enough and would loosen while climbing. Nevertheless, the tape should not restrict the blood flow or damage the finger nerves.
- During the application of the tape, the finger must be in a functional position, depending on what effect the tape should accomplish. For example, if a complete extension of the PIP joint should be prohibited, as in the case of a joint capsule injury, the tape must be applied in a 30° flexion of the PIP joint.

- The respective body region should not be shaved directly before taping, as this may cause micro skin abrasions which can cause infections.
- The skin should be dry and clean before taping.

22.2 Pulley Taping (Pulley Injury, Tenosynovitis) (H-taping)

Indication: Pulley injury, Tenosynovitis

After closed pulley injuries in climbers, a specific technique of finger taping is recommended to decrease the increased tendon bone distance (bowstring). Originally, circular taping around the base of the phalanx was recommended [7, 11]. Schweizer et al. [8] determined a more distal and oblique tape position to be more effective. Further biomechanical studies by Roloff et al. [12] lead to the invention of an even more effective technique, the H-tape [6]. This effectiveness of this technique was confirmed by a cadaver study in which human fingers were loaded with various tapings in various grip positions [6]. The new taping method decreased the tendon bone distance in the injured finger significantly by 16% and was superior to other taping methods. The strength development was significantly better with the new tape for the crimp grip position (+13%), while there was no significant improvement for the hanging position [6]. The study could not observe any increase in strength in the healthy finger after having applied the tape, which undermines the observations of Schweizer et al. [8] and Warme and Brooks [7] that taping can never approximate or improve upon the effectiveness of an intact pulley system [6].

This taping technique is also effective in the therapy of tenosynovitis of the finger flexor tendons as it decreases the deflection of the tendons at the distal rim of the A2 and the proximal rim of the A4 pulley, thus decreasing friction [13, 14].

In a pulley injury, the flexor tendons experience an increased distance to the bone while flexing the finger. This is called the "bowstring phenomenon," and it increases with the grade of finger flexion. The forces acting on the pulley are strongest in a flexed position. The force being applied on the finger flexor tendon pulley system is a function of the tension being developed in the flexor tendons and the angle between the tendon and the pulley [12]. In order to decrease the force on the pulley (potentially causing the rupture), the goal must be to diminish this angle since the tendon tension cannot be influenced [12]. While the A2 and A4 pulleys are rather stiff, the A3 pulley is weak, thin, and very flexible as it is attached on the palmar plate, which increases its flexibility during motion, as shown by an MRI study by Bayer et al. [15]. In an A2 or A4 injury, the flexor tendons separate from the bone (bowstring) as the A3 stretches and do not influence the path taken by the tendon.

A biomechanically effective tape thus needs to counteract the bowstring of the tendon, which is most effective at the point of the most pronounced distance between the tendon and bone. The maximum distance is situated at the point of maximum joint flexion of the PIP which coincides with the A3 pulley and on the outside, the crease in the skin over the PIP joint on the palmar aspect of the finger. Additionally, at the level of the pulleys, there is a substantial amount of subcutaneous tissue, whereas at the level of the flexor crease at the PIP joint, this is minimal. In consequence, tape applied over the crease of the skin at the PIP joint can deliver maximum support to the flexor tendons in comparison to other anatomical landmarks. After the initial study and invention of the H-tape, multiple variations have been discussed online. However, only the original technique has been investigated scientifically so far [10].

For the application of a correct H-tape, a 1.5 cm-wide and 8–10 cm-long strip of tape is cut (size varies according to finger size). The strip is then split longitudinally from both ends in such a way that a central "bridge" of 1 cm width remains (branches ~0.75 cm wide, Fig. 22.1) [13]. The central "bridge" of the tape is pressed onto the flexor fold of the PIP joint. The two proximal branches are then applied rather tightly onto the base phalanx. With the PIP joint flexed at about 45–60°, the two distal branches are then tightened over the middle phalanx (Fig. 22.2). The tape should be rather tight; thus, it should only be applied directly before climbing to minimize

Fig. 22.1 The tape is split from both ends, and the branches should not be too wide. Normally, we use a 1.5 cm-wide and 8–10 cm-long strip of tape; thus, the branches are about 0.75 cm wide each (Photo: Michael Simon)

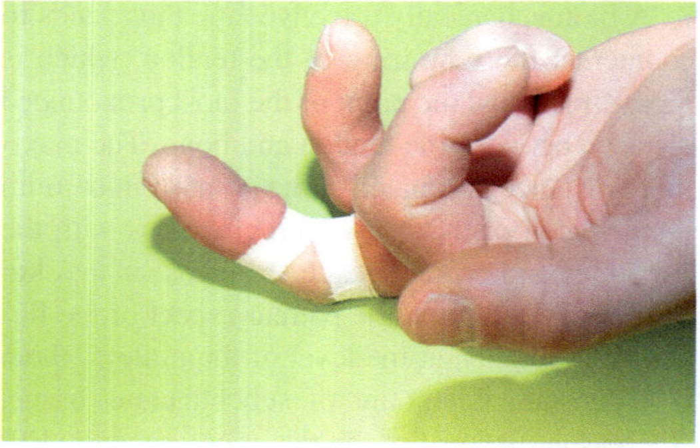

Fig. 22.2 Applying the tape: the PIP joint is flexed, and first the distal branches and then the proximal branches are fixed (Photo: Michael Simon)

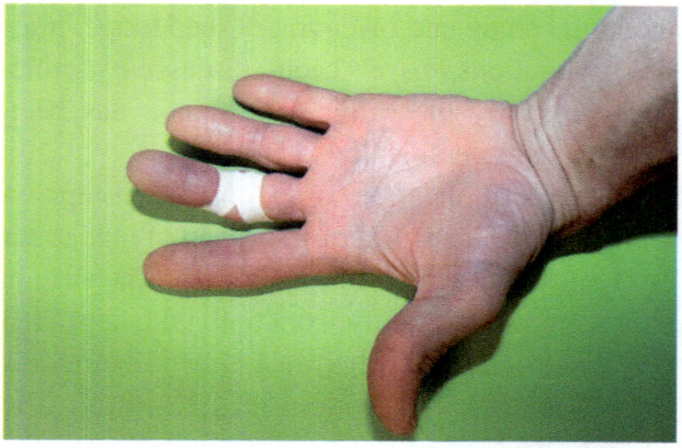

Fig. 22.3 The bridge of the tape is directly upon the flexor fold of the PIP joint (Photo: Michael Simon)

potential blood flow restriction (Fig. 22.3). It is crucial to know that the tape will loosen during climbing. To enhance the tape, a second layer (a thin figure-eight-shaped tape) can be applied on top of the H-tape to protect the base-layer [10].

1. As the tape is exactly positioned over the finger middle joints folding crease, it accomplishes various effects. The finger's subcutaneous fat and tissue is minimal at this point; the tape's pressure can directly work on the force transmission of the flexor tendons.
2. The tape supports the A3 pulley. The A3 pulley is thinner and weaker, thus more flexible than the A2 and A4 pulleys. Additionally, it is not attached directly to the bone but to the palmar plate, which increases its flexibility during motion.
3. The tape indirectly decreases the forces acting on the injured A2 or A4 pulley [13] (Chap. 6.3.1. Figs. 11 and 12).

The effect of the H-tape also works similarly for tenosynovitis. During the crimping grip position, the flexor tendons are bent at the distal end of the A2 pulley and at the proximal end of the A4 pulley [16, 17]. This leads to an increased flexion of the tendons and thus increased friction on the rim of the pulley [13, 17]. Repetition of this position can lead to a chronic inflammatory reaction of the tendon sheath [18]. As the H-tape decreases the angle of flexor tendon deflection, friction will be reduced, and the tenosynovitis can recover [14, 18]. In an ultrasound study, Schöffl et al. [6] showed that only the H-tape can significantly decrease the tendon bone distance and increase finger strength taping after finger pulley rupture. No benefit was found for any tape in an intact pulley system. After a pulley rupture, taping during climbing should be performed for a few months (see Chap. 6 "Finger Injuries"). This helps to reduce the risk that symptoms will become chronic or an onset of tenosynovitis based on finger pulley injury. While the H-tape is found to work effectively in tenosynovitis [13, 14], circular tape at the base of the proximal phalanx can also be effective in cases of minor tenosynovitis [10]. Alternatively, a more distal taping method as shown by Schweizer et al. [8] can be used for both pathologies.

22.3 Finger Middle Joint (PIP Joint)

Indication: Middle joint capsule injury, capsulitis, strain of the flexor tendons, and lumbricals

The proximal interphalangeal joint can be best supported by a "figure eight" (or 8-shaped) tape. It is important to apply it in a functional finger position, meaning a 45° flexion of the PIP joint, so the tape reduces the overall extension, while the full finger flexion is still attainable. Fist closure, rope and belay device management, and other hand functionalities are guaranteed [10]. The finger middle joint's capsule undergoes the most stress in passive overextension. The same applies to the flexor tendons, which are sensitive to overextension. Thus, the aim of the tape is to avoid finger hyperextension, which is why it needs to be applied in a flexed finger position. If the tape were to be applied in an extended finger position, it would be ineffective.

Starting at the base of the finger, the first wrap of tape should go around the base phalanx, as close to the palm as possible. While crossing the middle joint, it is important to keep 30–45° of flexion in this joint. The tape then encircles the middle phalanx and comes back to the base phalanx, crossing palmar sided at the PIP joint. At this point, the 30–45° flexion is highly important. The tape is finished by encircling the base phalanx. Similar to the H-tape, it is important that the tape branches are not too wide, enabling stress to be applied to certain areas, while other areas work as stress release for (sub-)cutaneous tissue. Therefore, a soft tissue window should be kept free of tape on the palmar side of the proximal joint. As previously stated, full coverage of the finger with tape is insufficient and harmful [10] (Figs.22.4, 22.5, 22.6, 22.7, and 22.8).

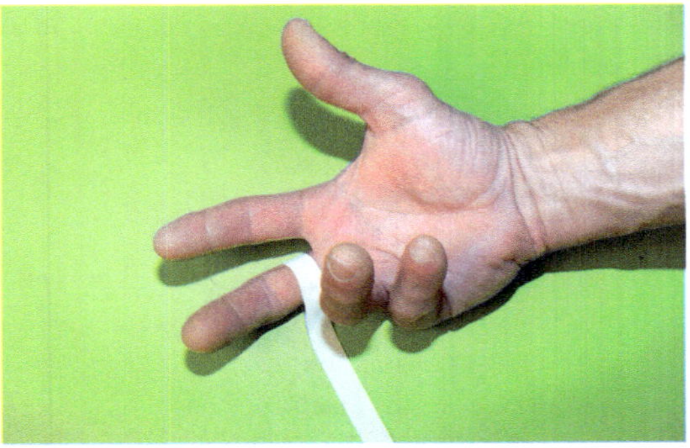

Fig. 22.4 Taping of the finger middle joint: starting at the base of the finger, the first wrap of tape should go around the phalanx as close to the palm as possible (Photo: Michael Simon)

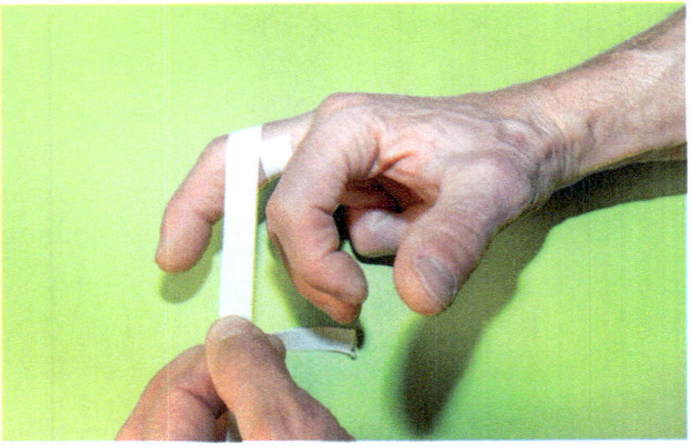

Fig. 22.5 While crossing the middle joint, it is important to keep 30–45° of flexion in this joint (Photo: Michael Simon)

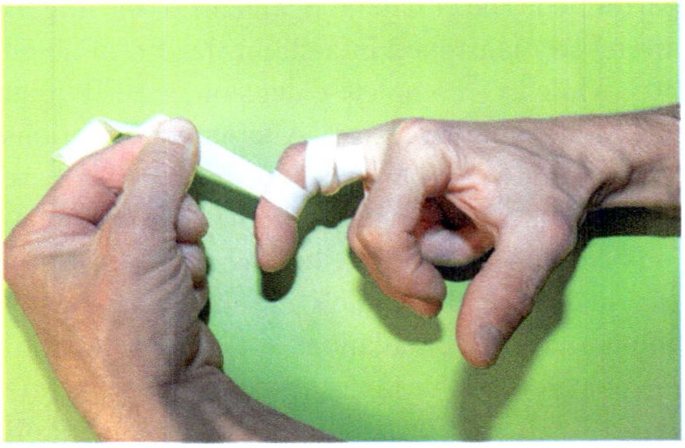

Fig. 22.6 The tape circles around the middle phalanx (Photo: Michael Simon)

22.4 Finger Middle Joint Collateral Ligament Injury

Indication: Collateral Ligament Injury PIP

A collateral ligament injury to the PIP joint can also be effectively supported by a modified figure eight tape. In the initial phase of the injury, a splint immobilization is advised followed by a buddy-taping [18]. In the next phase of recovery, a modified figure eight tape is used. Therefore, the crossing of the eight is not applied to the palmar side at the PIP joint but onto the injured lateral side. It is highly important that the finger is

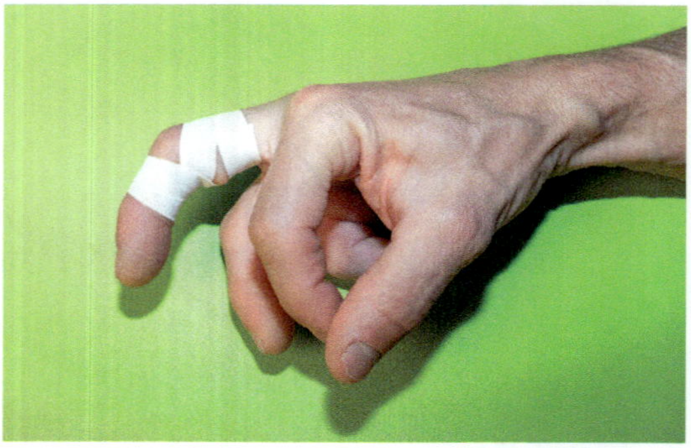

Fig. 22.7 It then crosses again palmar sided the PIP joint (Photo: Michael Simon)

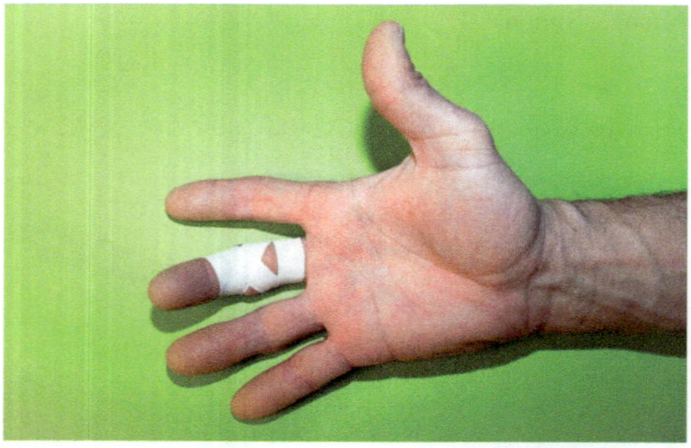

Fig. 22.8 A soft tissue release window is important (Photo: Michael Simon)

flexed at 30–45°, while the lateral crossings are performed. The collateral ligaments of the fingers are mostly tensioned in a slight flexion of about 30°; while in a complete extension, the PIP joint has some bony lockage. Alternatively, various single straps can be taped out by doing a base circulation at the finger's middle and base phalanx combined with longitudinal straps based on top of both sides which are finally fixed by another layer of circulations at the base and middle phalanx. As this tape consists of multiple layers, it tends to protrude more and also tears off more easily. We therefore recommend the lateral-sided figure eight [10] (Figs. 22.9, 22.10, 22.11, and 22.12).

22.5 Finger End Joint (DIP Joint)

Indication: Joint capsular injury, capsulitis, ganglion cyst

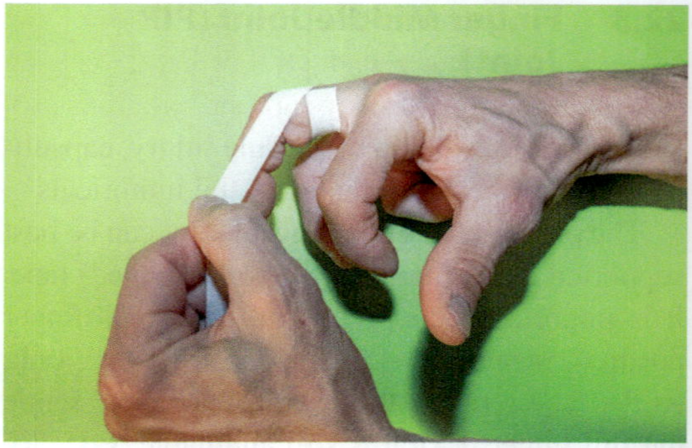

Fig. 22.9 In a collateral ligament injury, the tape starts as with a classic figure of eight tape close to the fingers base but crosses the PIP joint in 30–45° flexion at the respective side of the injured ligament (Photo: Michael Simon)

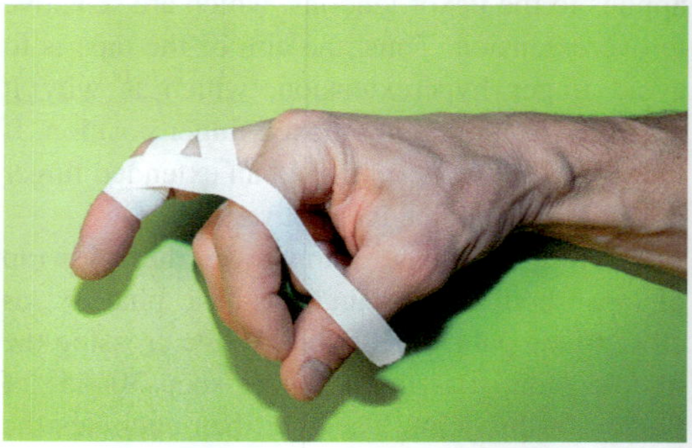

Fig. 22.10 After circling around the middle phalanx, the tape is lead back to the base phalanx, crossing the PIP joint sideways, which is in a 30–45° flexion (Photo: Michael Simon)

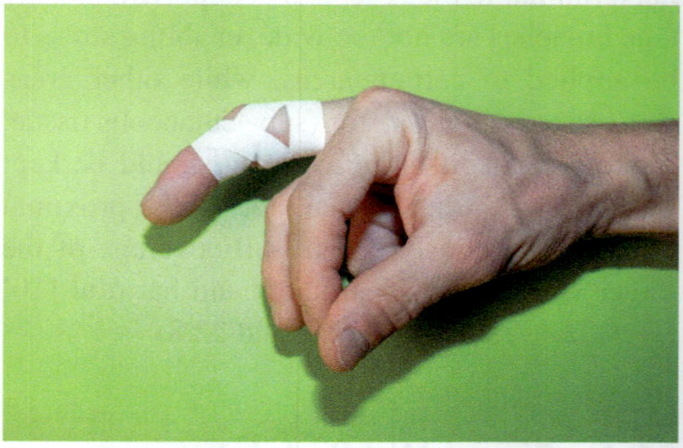

Fig. 22.11 The tapes' crossing is sideways at the PIP joint (Photo: Michael Simon)

As the distal interphalangeal joint (DIP) is tiny, a figure eight taping does not work there. Additionally, it would cover too much of the fingertip and decrease friction while climbing. Thus, for the finger end joint, we recommend plain circular tape directly around the joint [10] (Figs. 22.13 and 22.14).

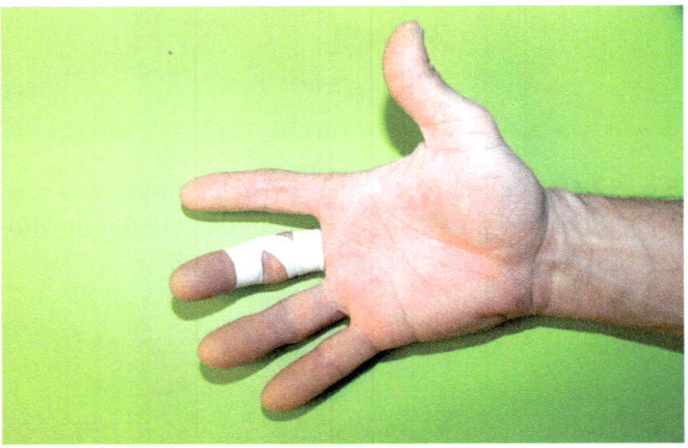

Fig. 22.12 The tapes' crossing is sideways at the PIP joint (Photo: Michael Simon)

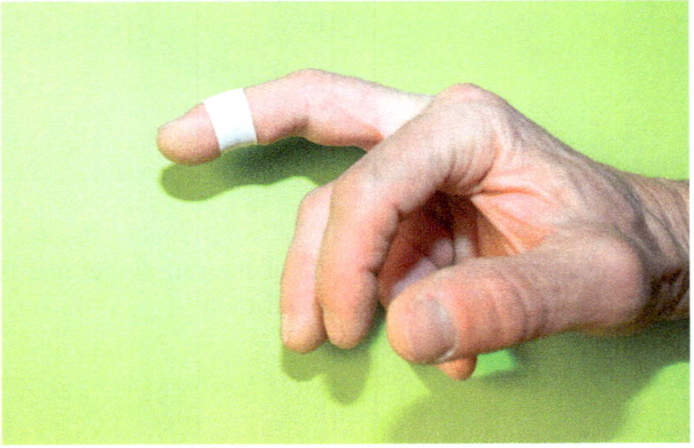

Fig. 22.13 To stabilize the DIP joint, it is directly encircled by a thin tape layer (Photo: Michael Simon)

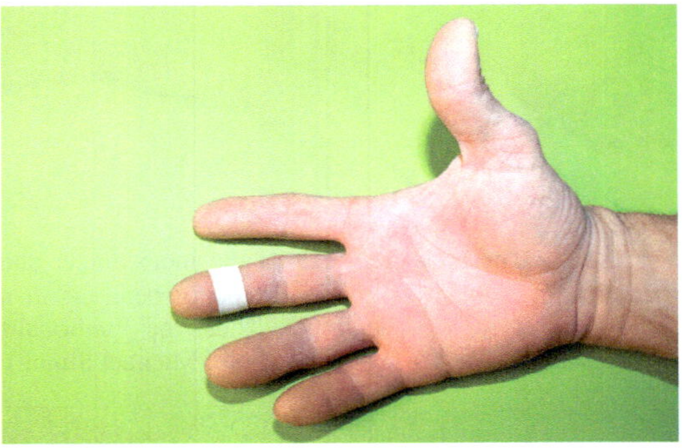

Fig. 22.14 To stabilize the DIP joint, it is directly encircled by a thin tape layer (Photo: Michael Simon)

22.6 Finger Base Joint (MCP Joint)

Indication: Joint capsular injury, capsulitis, ligament injuries, injuries to the connexus intertendineus of the extensor tendons

Taping of the finger base joint (MCP joint) is a bit trickier but can be effective. For this taping technique, the skin must be dry and clean, as the tape has free ends. During its application, the MCP joint is flexed at about 20°. The branches of the tape should cross each other, one distal and the other proximal to the knuckle (head of the metacarpal). The ends of the tape are only pressed onto the dorsal aspect of the hand, without any additional reinforcement (tape sticks better to tape than to skin). We found that these free ends still hold onto the skin well and do not come off too easily [10] (Figs. 22.15 and 22.16).

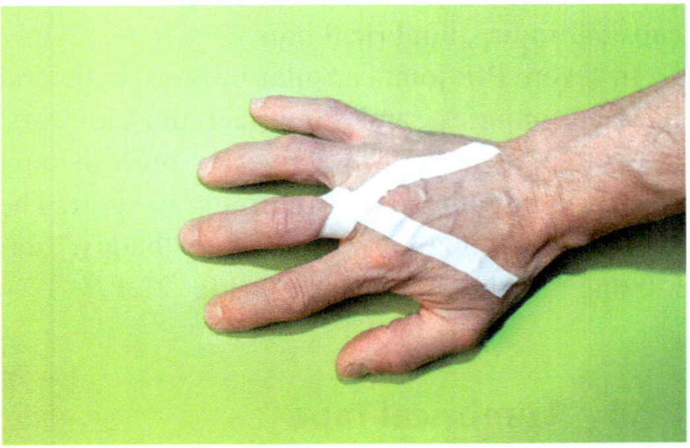

Fig. 22.15 Tape stabilization of the MCP joint needs to be performed in 30° flexion with the tape branches crossing each another. These crossings are distal and proximal of the "knuckle" (Photo: Michael Simon)

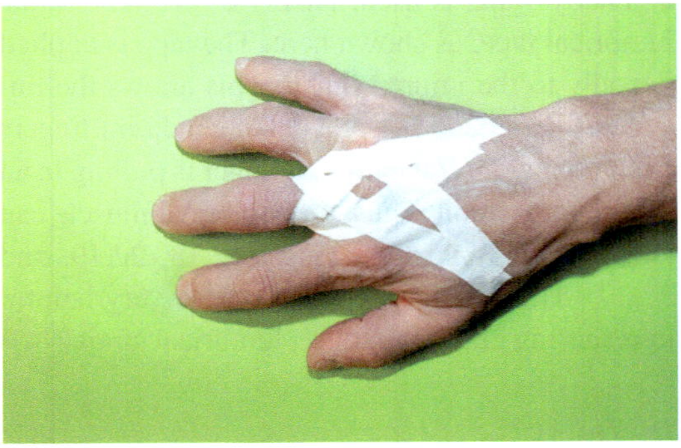

Fig. 22.16 Tape stabilization of the MCP joint needs to be performed in 30° flexion with the tape branches crossing each another. These crossings are distal and proximal of the "knuckle" (Photo: Michael Simon)

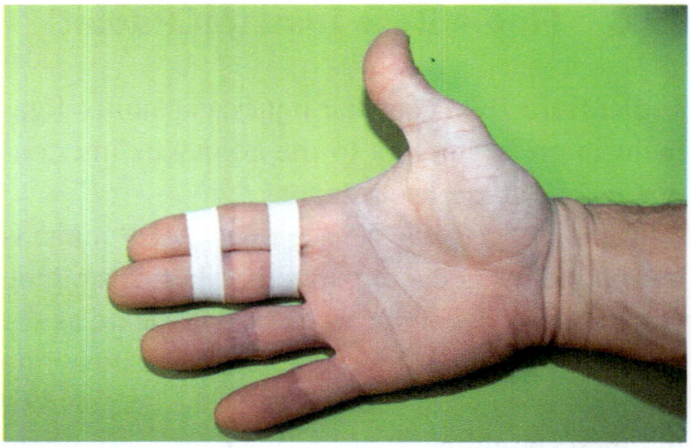

Fig. 22.17 One tape fixes the base phalanx, and the other fixes the middle phalanx to the adjunct finger. If the lengths of the phalanges are very different, these strips must be attached obliquely (Photo: Michael Simon)

22.7 Buddy-Taping

Indication: Collateral ligament injury PIP, joint capsular injury, lumbrical injury

In severe PIP joint capsular injuries, collateral ligament injuries, and lumbrical muscle tears, buddy-taping is very effective for both therapy and during the process of returning to sport. The healthy finger thereby works as the "buddy," stabilizing the injured finger [10] (Fig. 22.17).

22.8 Lumbrical Tape

Indication: Lumbrical muscle strain or tear

Lumbrical muscle injuries can be very effectively treated with taping [19]. In the acute phase, a "buddy"-tape is used; later, we recommend the lumbrical tape, as shown here. The tape is applied dorsally to the injured finger. This allows the finger to freely bend but prevents the adjoint fingers from flexing far if the injured finger is pulling in an extended position. Thus, the lumbrical muscle can be relieved from stress (Figs. 22.18, 22.19, and 22.20). This taping technique is also very helpful in restarting climbing after a lumbrical strain.

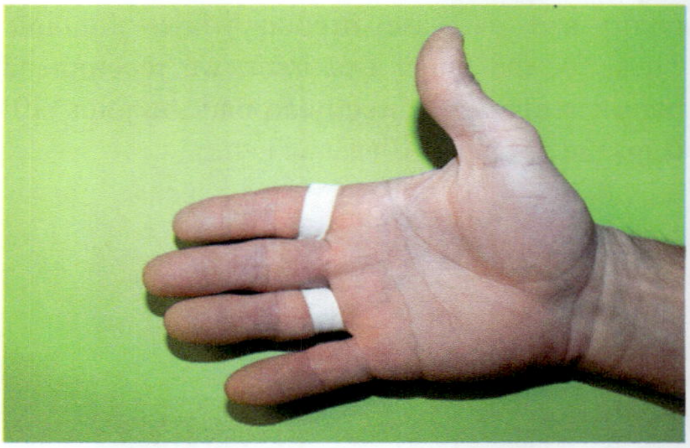

Fig. 22.18 The tape starts at the index finger and runs dorsally behind the (injured) middle finger to the ring finger (Photo: Michael Simon)

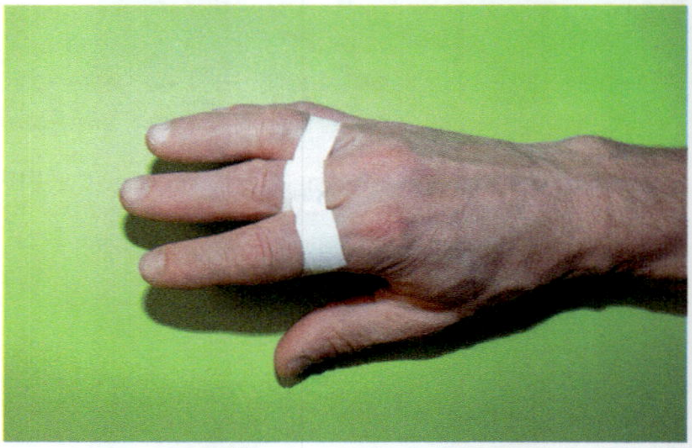

Fig. 22.19 After encircling the ring finger, the tape comes back to the extensor side of the ring finger and is attached to itself (Photo: Michael Simon)

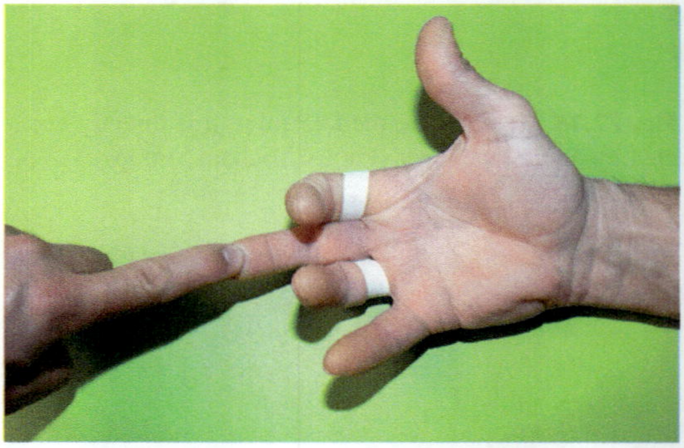

Fig. 22.20 The tape then ends at the index finger and prevents the adjunct fingers from being bent far in a mono-pull of the middle finger. Thus, the lumbricalis muscles are relieved from stress (Photo: Michael Simon)

22.9 Wrist

Indication: Tenosynovitis, discus lesion, synovitis in the ulnocarpal recesses

With the increasing number of people who have become interested in climbing, climbing-related injuries of the wrist and the forearm are currently seen more frequently than they have been in the past. As more and more athletes perform the sport on a level which had previously only been reached by a few top athletes, injuries such as hamate fractures or other pathologies of the wrist have become more common [20–22]. Circular wrist taping is a technique commonly used by athletes in different sports to both strengthen and stabilize the wrist for a high-compression load (e.g., weightlifting [23, 24], gymnastics [25, 26]) or to stabilize the wrist for tensile stress as present in climbing when the athlete is pulling on his/her hand [10, 20]. However, in their 1997 study, Rettig et al. proved that circular wrist tape, with or without additional taping of the fingers, does not increase wrist strength [9]. Contrary to the perceptions of the athletes, Takahashi et al. even showed that tight-circular wrist tape can even slightly decrease grip strength when applied too tightly. However, no climbing-specific data on grip strength variations under wrist tape has been published so far.

Climbers suffering from a feeling of instability in the wrist or nonspecific wrist pain are often diagnosed with injuries of the ligaments and capsules or even with bone marrow edema of the carpal bones or carpal fractures [10, 20]. The pain and discomfort mainly arise in radial/ulnar abducted positions of the hand or while grasping an undercling (position of the hand in maximum supination) [10, 20].

During climbing or bouldering, the majority of thenar and hypothenar muscular strength is transferred to the carpal bones and the distal part of the radius by the transverse carpal ligament, causing frequent reactions of the lunate and the distal radius [22, 28]. If maximum strength is applied in a slightly ulnar abducted and dorsally flexed position of the hand, the carpal bones (especially the lunate) are in a relatively unstable position, as biomechanical analyses have shown [4, 28]. Lutter et al. [20] published a study which showed that high stress to the flexor tendons can even cause fractures of the carpal bones. It is comprehensible that radioulnar joint instability or anatomic predispositions such as incongruent wrists (pos./neg. ulnar variance) can exacerbate the development of problems in this region [4, 22].

22.9.1 Taping Technique

Unlike finger tapes, for which thin strips of tapes are used [6], wider tapes are recommended for circular wrist tapes [4]. As climbers need both stability within the wrist and a high level of flexibility, the tape strips are recommended to be approximately 2.5 cm wide. This width allows for an increase of pressure within the carpal while maintaining the flexibility of the wrist. Due to the fact that this taping technique increases the carpal and thus adds stability within the wrist and not the distal part of the forearm, it is recommended to place the tape distally to the styloid process of the ulna (skin fold of the wrist) [4] (Fig. 22.21). This supports the carpal architecture, including the transverse carpal ligament. In contrast to Takahashi et al. and Rettig et al., who investigated athletes directly after taping the wrist, we recommend applying the circular wrist tape rather firmly (two to three layers) due to the fact that even non-elastic tape loosens a little bit within the first minutes of climbing [4, 9, 10]. Additionally, in mantle and stemming moves, the climber applies high pressure to the distal radioulnar joint, which allows some grade of divergent movement. If the tape is too tight, this would be painful. It must be considered that not all wrist conditions benefit from circular taping, which

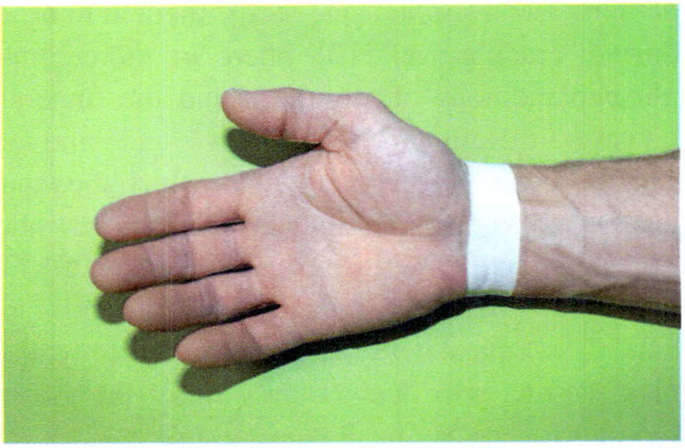

Fig. 22.21 The wrist is encircled one and a half to two times with a 2 cm-wide tape, as far distally as possible (Photo: Michael Simon)

may be explained by the intra-articular increase of pressure due to the tape [4]. If the climber's pain worsens with the tape, it should not be continued [4].

22.10 Base Joint of the Thumb

Indication: Joint capsule and collateral ligament injuries, hypermobility

Strains or tears of the ulnar collateral ligament of the thumb, commonly suffered in falls while skiing, can cause a severe pain in the respective area. If a serious injury or complications (e.g., Stener lesion) are excluded, taping the thumb can help reduce pain and offer stabilization and protection. The tape starts with two base layers around the wrist and the proximal phalanx of the thumb. The following layers are then applied like roof tile, with one layer stacking over the previous one. The crossing point of the tapes should be at the point of the injury; thus, the tape can be varied accordingly. This technique creates support for the thumb similar to a soft splint (Figs. 22.22, 22.23, and 22.24) [10].

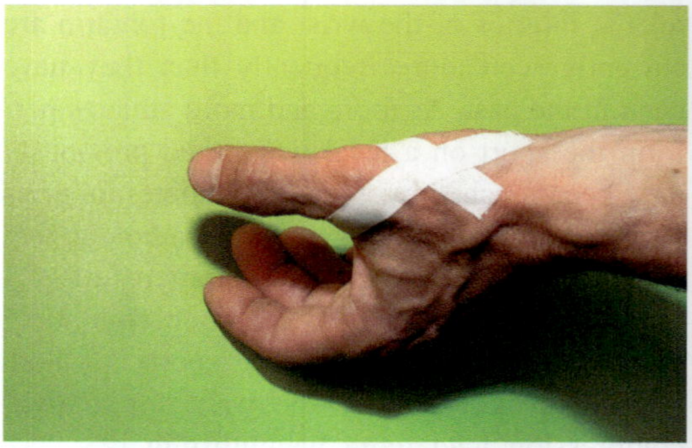

Fig. 22.22 The tape branches will be set so they cross each another at the dorsum, starting distally and continuing stepwise proximally, partially overlapping (Photo: Michael Simon)

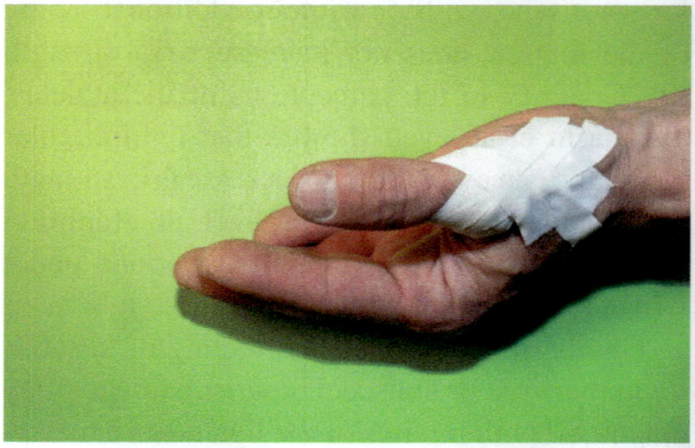

Fig. 22.23 This is continued proximally (Photo: Michael Simon)

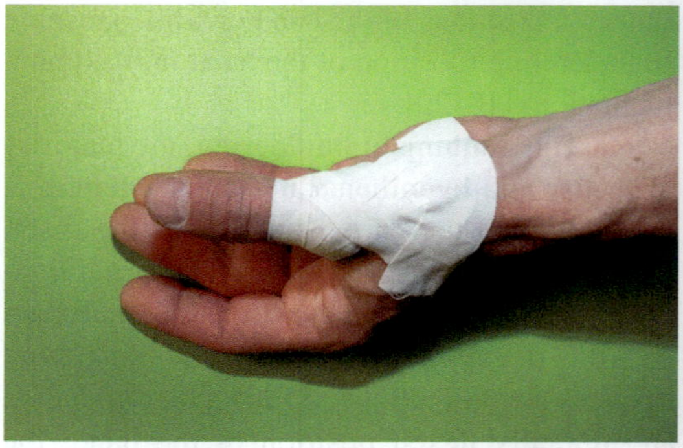

Fig. 22.24 Finally, an anchoring strip is placed on top horizontally (Photo: Michael Simon)

22.11 Hand Taping in Crack Climbing

Indication: Prophylactic in crack climbing

There are a number of products that help protect the hands when crack climbing. The traditional form of protection is athletic tape; we demonstrate a simple "one-strap" method to produce a "crack glove" [10]. There are more complicated methods of taping around the fingers; however, this method is effective and simple. This is one of the few methods in which the tape is attached to the hand directly off from the roll. The tape is pulled out from the roll in 20–30 cm increments before being wrapped around the hand. The tape should not be pulled while wrapping, as this will inadvertently make the glove too tight.

To start, the end of the tape is placed on the back of the wrist with the fingers spread as wide as possible. The tape glove is started with one to two circular wraps around the wrist and then continued diagonally across the back of the hand. At the base of the finger, the tape makes a slight turn, the hand is turned over, and the tape is placed at the very end of the palm. The tape is then brought

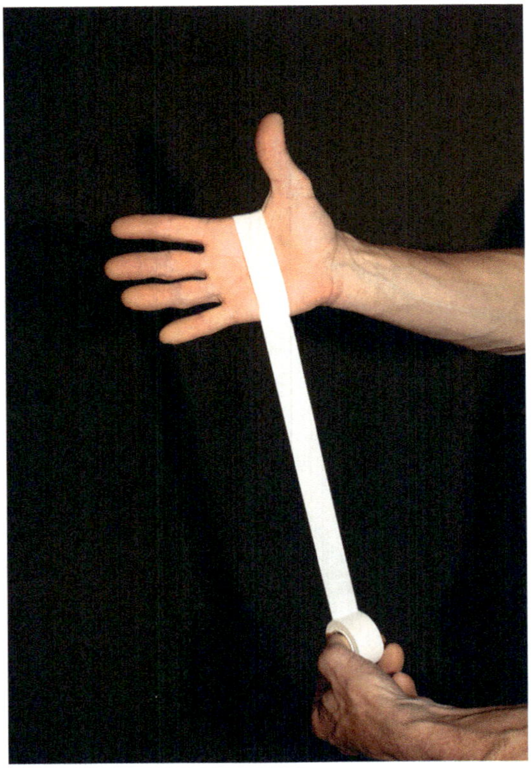

Fig. 22.25 The tape starts with one to two circulations around the wrist and then runs dorsally diagonally over the back of the hand. Then, it wraps around the palm twice and comes back diagonally over the back of the hand to the wrist, where it is wrapped around again (Photo: Michael Simon)

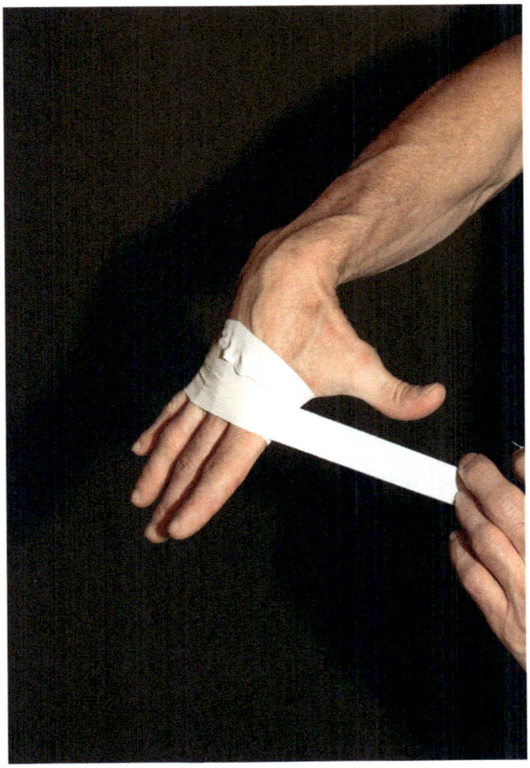

Fig. 22.26 The tape starts with one to two circulations around the wrist and then runs dorsally diagonally over the back of the hand. Then, it wraps around the palm twice and comes back diagonally over the back of the hand to the wrist, where it is wrapped around again (Photo: Michael Simon)

back to the dorsal aspect of the hand, placed just over the MCP joints and again across the same area of the palm. Rounding the blade of the hand, the tape is pulled toward the wrist. Another wrap is performed around the wrist, finishing the glove. This method covers part of the palm, but that area usually separates from the tape while climbing, allowing the tape to roll into a wide strip that is independent of the palm (Figs. 22.25, 22.26, 22.27, 22.28, 22.29, and 22.30) [10].

22.12 Elbow Taping

Indication: Epicondylitis, hypermobility of the ulnar nerve, brachialis tendinosis

Taping the elbow is used as therapy and prophylaxis for epicondylitis (medial or lateral), brachialis tendinosis, and an irritated, hypermobile ulnar nerve. As many larger taping techniques and external braces reduce venous blood backflow and increase the feeling of "pumped" forearms, we use a singular tape directly over the epicondyles at the elbow's flexor fold [10]. This technique is applied by wrapping a 2–2.5 cm-wide strip of tape around the elbow. It is important to flex the elbow at 90° and tense the biceps when closing the tape. With elbow flexion and biceps tension, the diameter of the elbow increases; tape applied onto a relaxed muscle in extension would cause an excessively tight tape that is prone to rupture. Correctly applied tape should be rather loose and lay in folds on the skin. When the elbow is flexed and the bicep muscle is tensed, pressure is applied to the epicondyles. For this taping technique, the effective

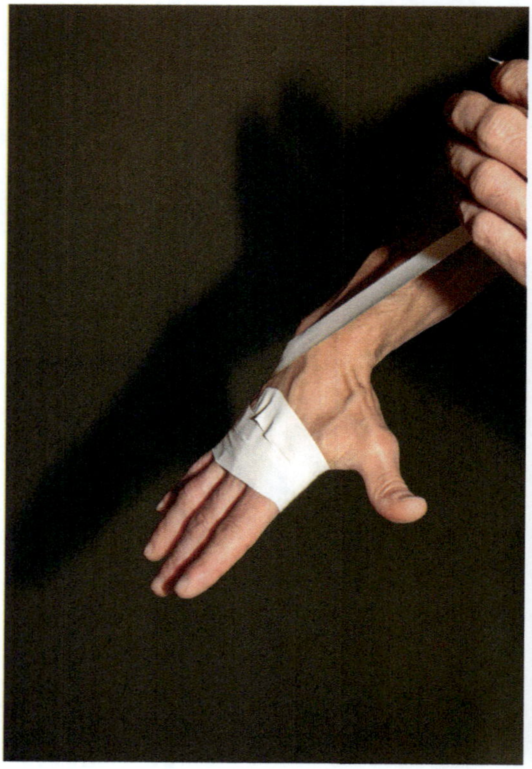

Fig. 22.27 The tape starts with one to two circulations around the wrist and then runs dorsally diagonally over the back of the hand. Then, it wraps around the palm twice and comes back diagonally over the back of the hand to the wrist, where it is wrapped around again (Photo: Michael Simon)

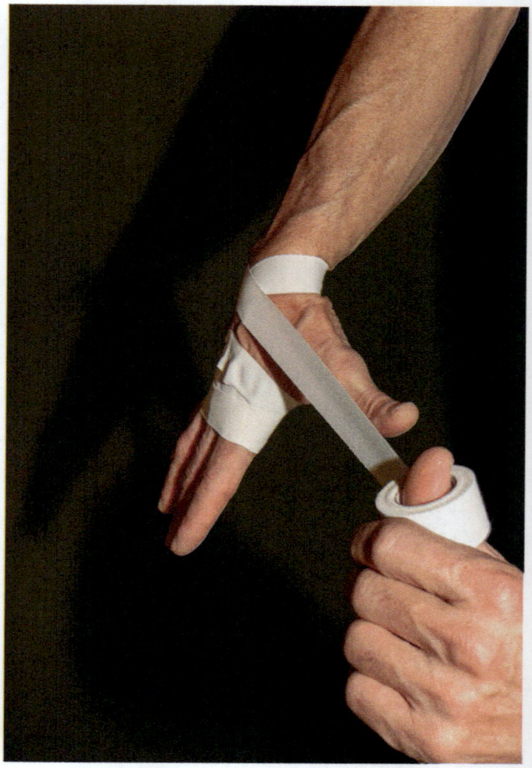

Fig. 22.28 The tape starts with one to two circulations around the wrist and then runs dorsally diagonally over the back of the hand. Then, it wraps around the palm twice and comes back diagonally over the back of the hand to the wrist, where it is wrapped around again (Photo: Michael Simon)

mechanism is more through stimulation of the proprioceptors than a pure mechanical effect. Overall, this taping technique is more accommodating during climbing than external braces, as these decrease blood flow and thus increase "pumped" forearms (Figs. 22.31, 22.32, 22.33, and 22.34) [10].

22.13 Taping of the Ankle

Indication: Sprains, chronic instability

Ankle injuries (strains and sprains) are frequent, especially in bouldering, and supportive taping can be used as therapy, depending on the extent of the injury, as well as during the process of returning to sport. There are many variations of ankle-joint tapings presented in literature. We demonstrate an easy and reproducible technique.

First, a base layer (anchor strips) of two semicircular straps is applied at the forefoot and the lower leg. Then, the actual holding strips, which are "U"-shaped (stirrup), are applied to the front and to the lower leg. There should be three strips of each, applied alternatingly. The most important strip is then applied, fixing the foot in a pronation through elevation of the lateral foot edge. While this being applied, it is crucial that the foot is kept in a 90° flexion and pronation. Finally, the tape can be covered with optional semicircular sheeting. We would definitely apply the sheeting to severe injuries (following the acute phase), but

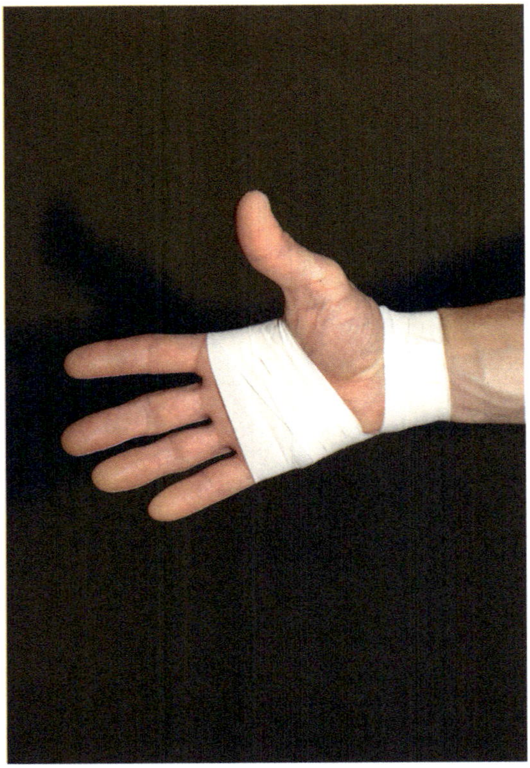

Fig. 22.29 The tape starts with one to two circulations around the wrist and then runs dorsally diagonally over the back of the hand. Then, it wraps around the palm twice and comes back diagonally over the back of the hand to the wrist, where it is wrapped around again (Photo: Michael Simon)

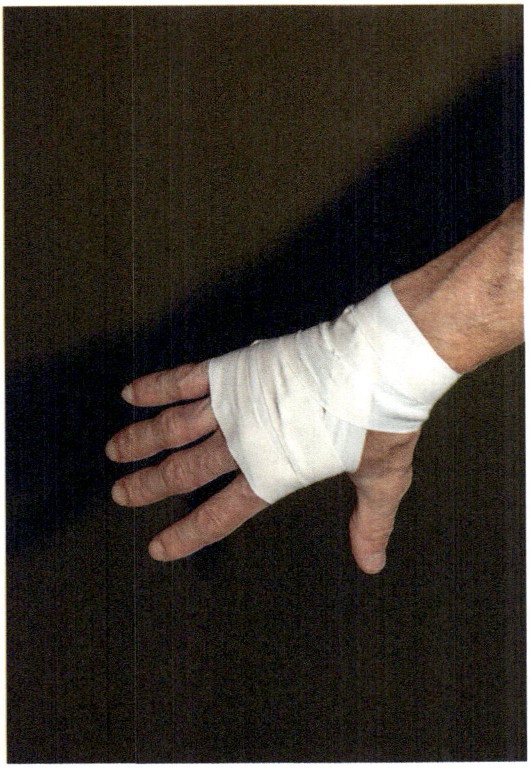

Fig. 22.30 The tape starts with one to two circulations around the wrist and then runs dorsally diagonally over the back of the hand. Then, it wraps around the palm twice and comes back diagonally over the back of the hand to the wrist, where it is wrapped around again (Photo: Michael Simon)

if the tape is supposed to support the foot in a climbing shoe, it may be too thick. It must be mentioned that this taping technique is circular and cannot be used directly after an acute trauma, as this may cause compartment syndrome. After completion of this taping technique, the distal tape is cut from distal to proximal for about 1–1.5 cm in the first interdigital space. In a normal standing position, the metatarsals yield transversally to the weight, and the actual diameter of the forefoot increases; thus, the tape would be too tight (Figs. 22.35, 22.36, 22.37, 22.38, 22.39, 22.40, 22.41, 22.42, 22.43, 22.44, 22.45, 22.46, and 22.47).

22.14 Conclusion

Taping techniques, especially for the hand and fingers, are important therapeutic tools for climbers, especially during rehabilitation and the process of returning to. Specific biomechanical tapes can relieve stress on the finger flexor tendon pulleys, the tendons, and the joint capsule. Correct application of these tapings is essential, and soft tissue "windows" are important in order to ensure that a certain pressure can be applied to various areas, while other areas release the (sub-)cutaneous tissue. Fully covering a finger with tape is ineffective and harmful.

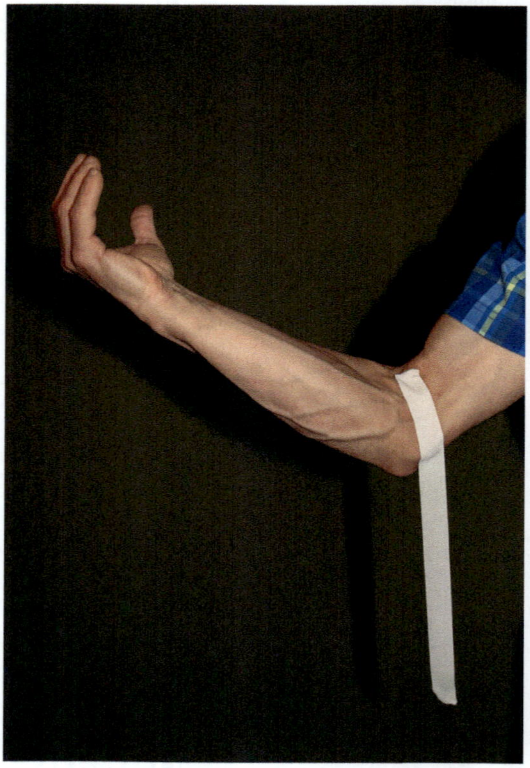

Fig. 22.31 The tape runs directly over the ulnar epicondyle, goes around the back of the arm, and comes to the lateral epicondyle. It is then closed at the front. For closure, the biceps must be tensed and the tape must be applied very firmly. In a relaxed arm, the tape should have folds which these unfold as the tape is under tension during muscle activation and flexion. Thus, it works with direct pressure onto the epicondyles and muscle insertions while guaranteeing free blood flow when the arm is relaxed (Photo: Michael Simon)

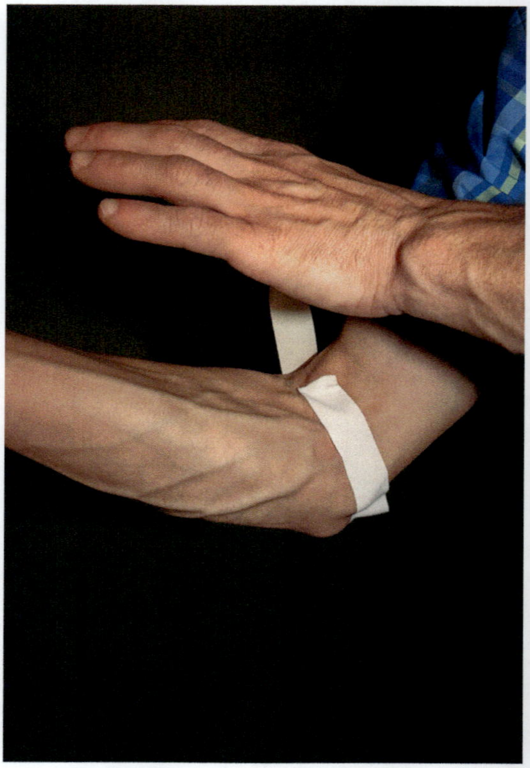

Fig. 22.32 The tape runs directly over the ulnar epicondyle, goes around the back of the arm, and comes to the lateral epicondyle. It is then closed at the front. For closure, the biceps must be tensed and the tape must be applied very firmly. In a relaxed arm, the tape should have folds which these unfold as the tape is under tension during muscle activation and flexion. Thus, it works with direct pressure onto the epicondyles and muscle insertions while guaranteeing free blood flow when the arm is relaxed (Photo: Michael Simon)

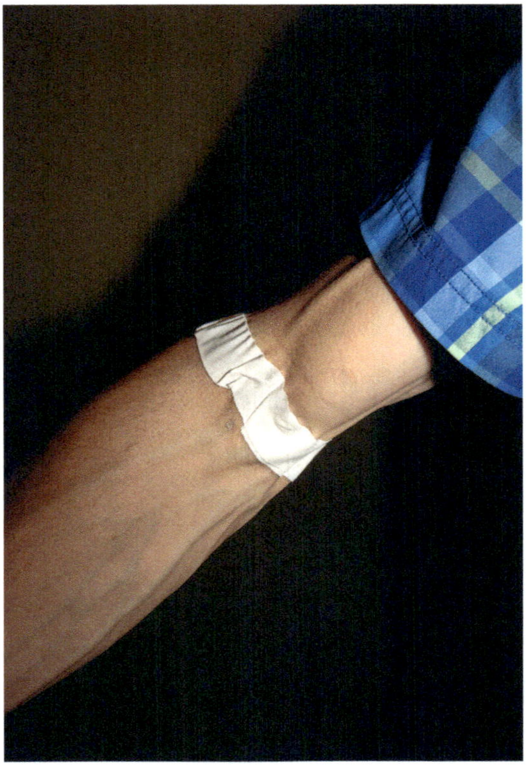

Fig. 22.33 The tape runs directly over the ulnar epicondyle, goes around the back of the arm, and comes to the lateral epicondyle. It is then closed at the front. For closure, the biceps must be tensed and the tape must be applied very firmly. In a relaxed arm, the tape should have folds which these unfold as the tape is under tension during muscle activation and flexion. Thus, it works with direct pressure onto the epicondyles and muscle insertions while guaranteeing free blood flow when the arm is relaxed (Photo: Michael Simon)

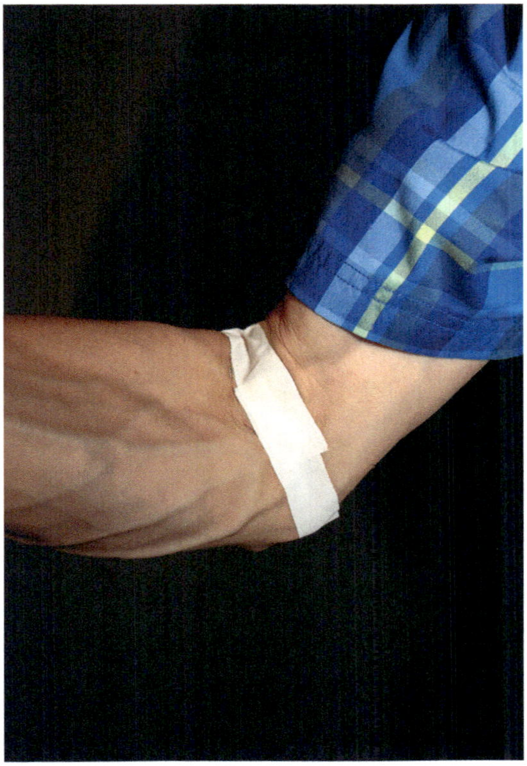

Fig. 22.34 The tape runs directly over the ulnar epicondyle, goes around the back of the arm, and comes to the lateral epicondyle. It is then closed at the front. For closure, the biceps must be tensed and the tape must be applied very firmly. In a relaxed arm, the tape should have folds which these unfold as the tape is under tension during muscle activation and flexion. Thus, it works with direct pressure onto the epicondyles and muscle insertions while guaranteeing free blood flow when the arm is relaxed (Photo: Michael Simon)

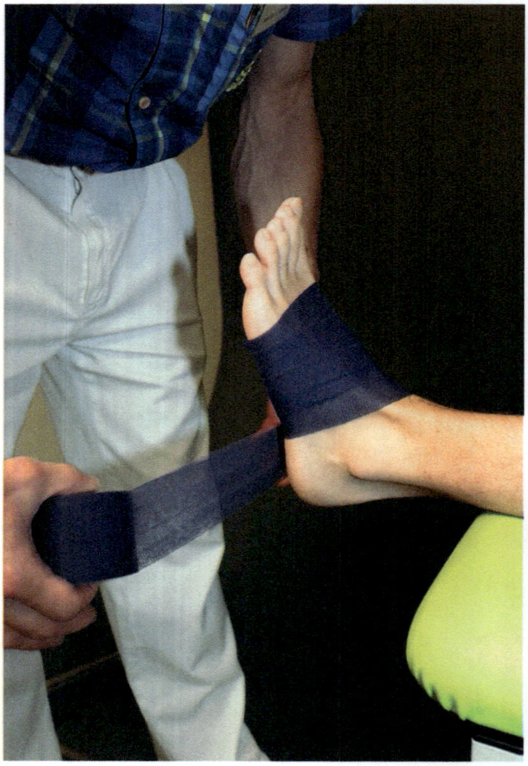

Fig. 22.35 A base layer with an adhesive bandage is applied to avoid skin alterations (Photo: Michael Simon)

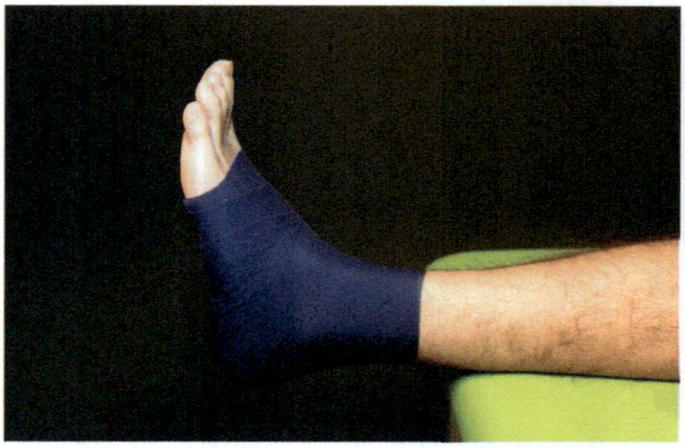

Fig. 22.36 A base layer with an adhesive bandage is applied to avoid skin alterations (Photo: Michael Simon)

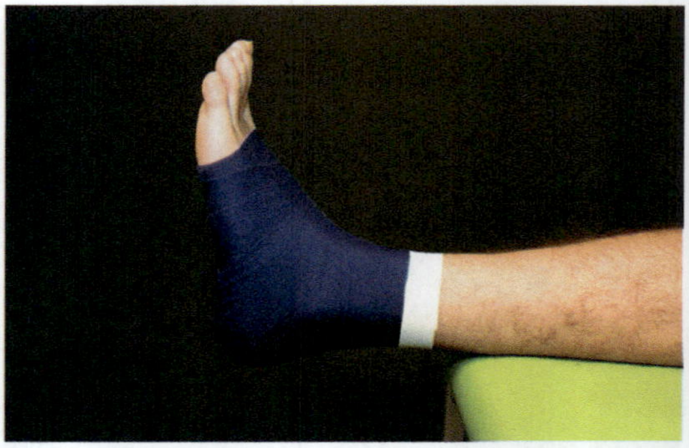

Fig. 22.37 Proximal and distal tape anchors are fixed in two half strips (Photo: Michael Simon)

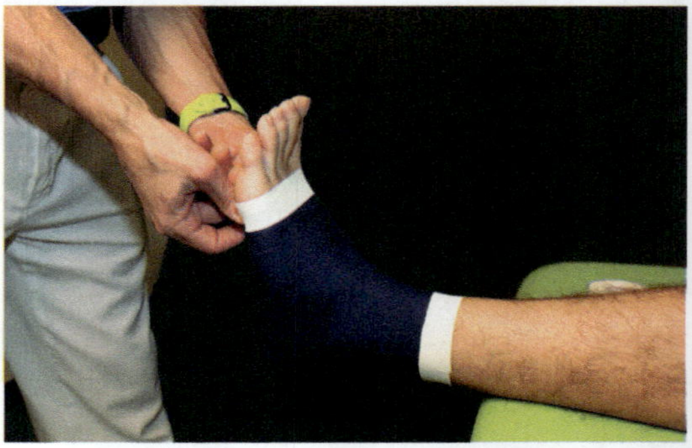

Fig. 22.38 Proximal and distal tape anchors are fixed in two half strips (Photo: Michael Simon)

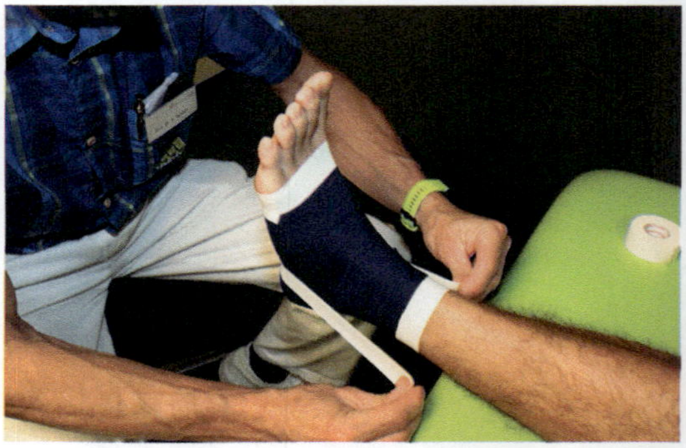

Fig. 22.39 The first retaining strap is applied in a "U"-shape (stirrup). It starts at the plantar aspect and is then attached at the lower leg. In the next step, the rest of the tape is molded onto the base layer with the palms (Photo: Michael Simon)

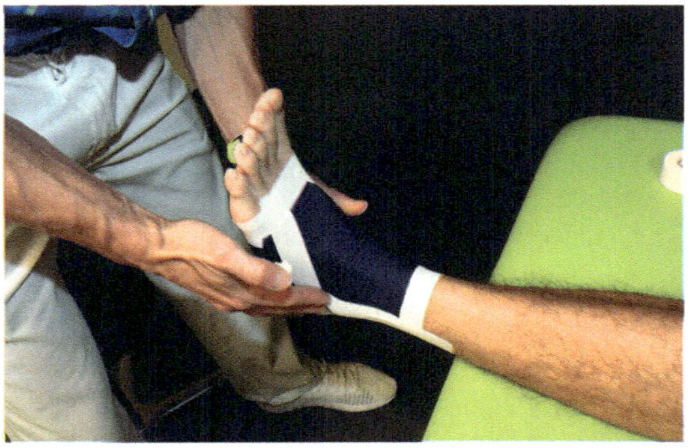

Fig. 22.40 The second retaining strap comes dorsally to secure the talus. Three of these two respective strips are applied, alternating and partially overlapping each another (Photo: Michael Simon)

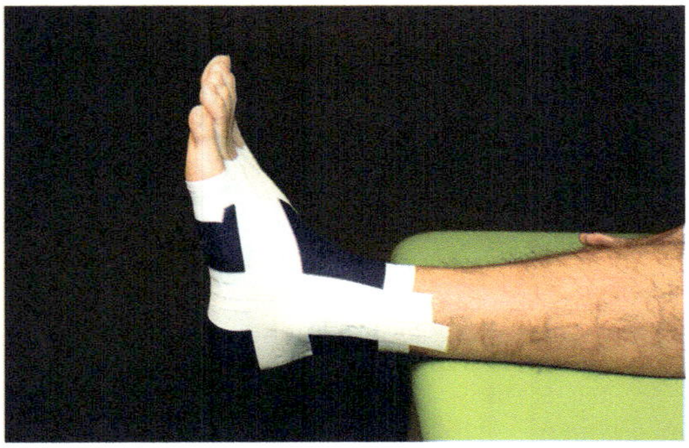

Fig. 22.41 It is crucial that the foot is bent at 90° and is in pronation during the taping (Photo: Michael Simon)

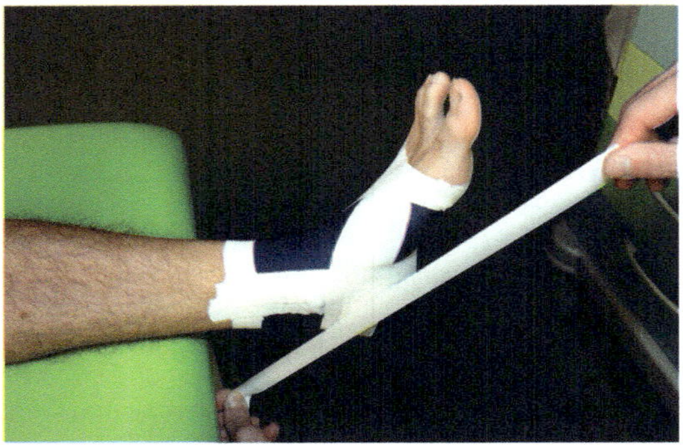

Fig. 22.42 The next strap is the most important for stability and runs obliquely underneath the medial malleolus (Photo: Michael Simon)

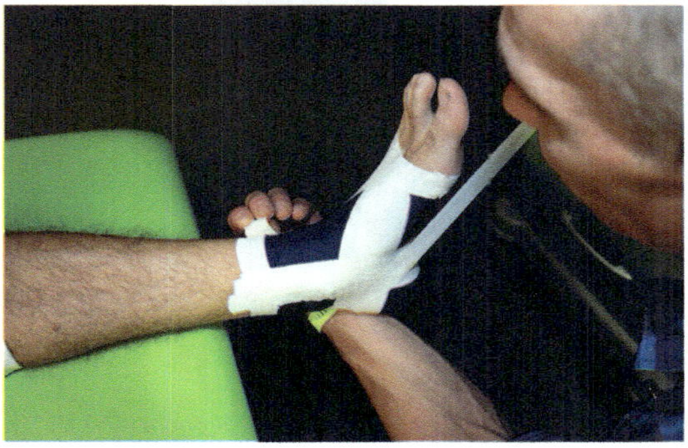

Fig. 22.43 The tape starts medially proximally (Photo: Michael Simon)

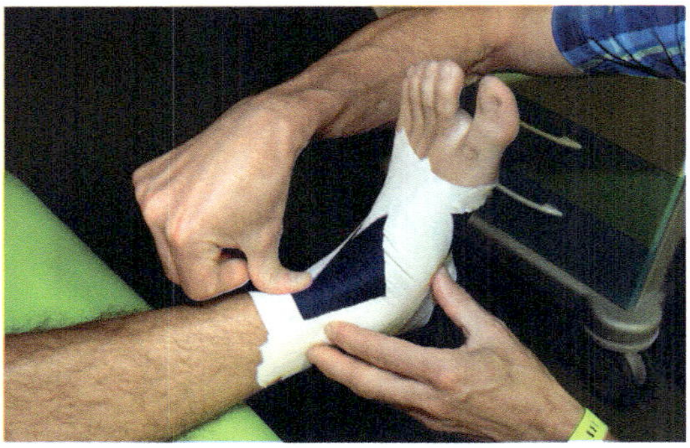

Fig. 22.44 The tape runs around the lateral side of the foot to work as a pronation strap and is continued to the cranial aspect (Photo: Michael Simon)

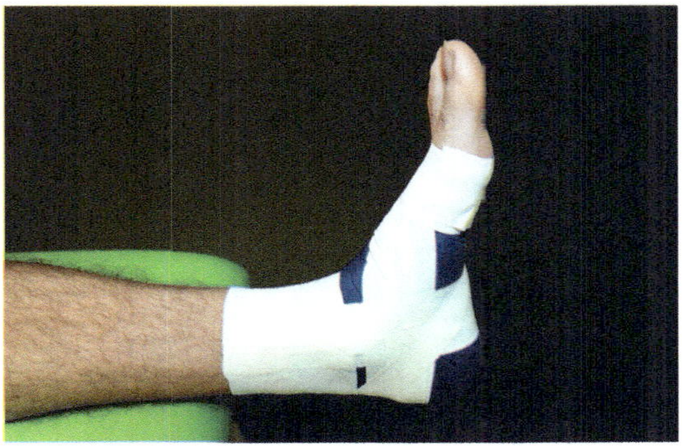

Fig. 22.45 Optional sheeting is applied with half wraps (Photo: Michael Simon)

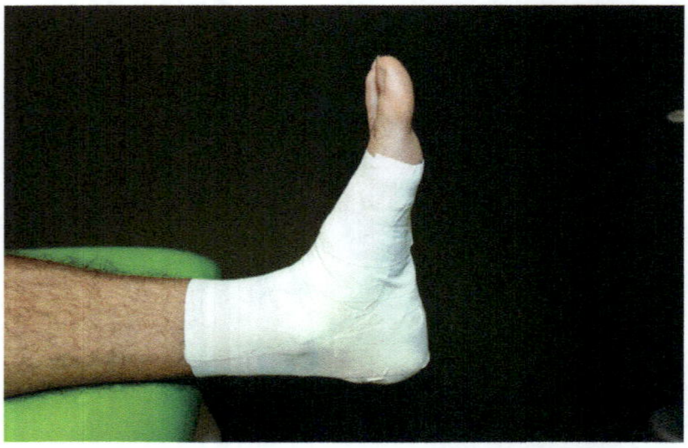

Fig. 22.46 Optional sheeting is applied with half wraps (Photo: Michael Simon)

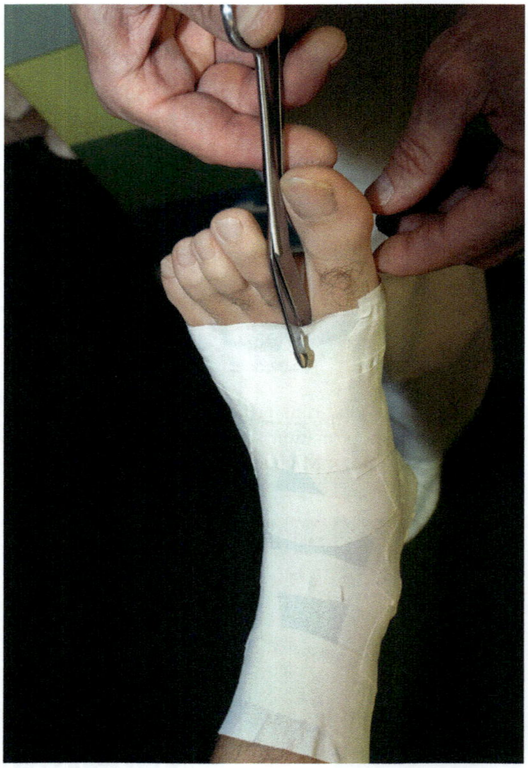

Fig. 22.47 Finally, the tape is cut between the first and the second phalanges to ensure a spread within the forefoot (Photo: Michael Simon)

References

1. Woollings KY, McKay CD, Kang J, Meeuwisse WH, Emery CA. Incidence, mechanism and risk factors for injury in youth rock climbers. Br J Sports Med. 2015;49(1):44–50.
2. Limmer M, Buck S, de Marees M, Roth R. Acute effects of kinesio taping on muscular strength and endurance parameters of the finger flexors in sport climbing: a randomised, controlled crossover trial. Eur J Sport Sci. 2019:1–10.
3. Dykes B, Johnson J, San Juan JG. Effects of finger taping on forearm muscle activation in rock climbers. J Electromyogr Kinesiol. 2019;45:11–7.
4. Lutter C, Schöffl V. Circular wrist tape—how to correctly use this technique in climbing and bouldering. Sports Orthop Traumatol. 2017;33:410–3.
5. Takahashi K, Demura S-I. Effects of different taping pressures of wrist taping using rigid tape on grip strength exertion. Am J Med Sci Med. 2014;2(6):122–5.
6. Schöffl I, Einwag F, Strecker W, Hennig F, Schöffl V. Impact of taping after finger flexor tendon pulley ruptures in rock climbers. J Appl Biomech. 2007;23(1):52–62.
7. Warme WJ, Brooks D. The effect of circumferential taping on flexor tendon pulley failure in rock climbers. Am J Sports Med. 2000;28(5):674–8.
8. Schweizer A. Biomechanical effectiveness of taping the A2 pulley in rock climbers. J Hand Surg (Br). 2000;25(1):102–7.
9. Rettig AC, Stube KS, Shelbourne KD. Effects of finger and wrist taping on grip strength. Am J Sports Med. 1997;25(1):96–8.
10. Schöffl V, Hochholzer T, Lightner S. One move too many. Boulder, CO: Sharp End Publishing; 2016.
11. Schöffl V, Hochholzer T, Elnwag F. Tapeverbände. In: Schöffl V, Hochholzer T, editors. Sportklettern—Aktuelle sportmedizinische Aspekte. Ebenhausen: Lochner-Verlag; 2004. p. 106–14.
12. Roloff I, Schöffl VR, Vigouroux L, Quaine F. Biomechanical model for the determination of the forces acting on the finger pulley system. J Biomech. 2006;39(5):915–23.
13. Schöffl I, Einwag F, Strecker W, Hennig F, Schöffl V. Impact of taping after finger flexor tendon pulley ruptures in rock climbers. J Apl Biomech. 2007;23:52–62.
14. Schöffl V, Strohm P, Lutter C. Efficacy of corticosteroid injection in rock climber's tenosynovitis. Hand Surg Rehabil. 2019;38(5):317–22.

15. Bayer T, Adler W, Schweizer A, Schöffl I, Uder M, Janka R. Evaluation of finger A3 pulley rupture in the crimp grip position-a magnetic resonance imaging cadaver study. Skelet Radiol. 2015;44(9):1279–85.
16. Schöffl I, Oppelt K, Jungert J, Schweizer A, Neuhuber W, Schöffl V. The influence of the crimp and slope grip position on the finger pulley system. J Biomech. 2009;42(13):2183–7.
17. An KN, Berglund L, Uchiyama S, Coert JH. Measurement of friction between pulley and flexor tendon. Biomed Sci Instrum. 1993;29:1–7.
18. Schöffl VR, Schöffl I. Finger pain in rock climbers: reaching the right differential diagnosis and therapy. J Sports Med Phys Fitness. 2007;47(1):70–8.
19. Lutter C, Schweizer A, Schöffl V, Römer F, Bayer T. Lumbrical muscle tear: clinical presentation, imaging findings and outcome. J Hand Surg Eur. 2018;43(7):767–75.
20. Lutter C, Schweizer A, Hochholzer T, Bayer T, Schöffl V. Pulling harder than the hamate tolerates: evaluation of hamate injuries in rock climbing and bouldering. Wilderness Environ Med. 2016;27(4):492–9.
21. Schöffl V, Popp D, Küpper T, Schöffl I. Injury trends in rock climbers: evaluation of a case series of 911 injuries between 2009 and 2012. Wilderness Environ Med. 2015;26(1):62–7.
22. Schöffl V, Lutter C. Acute and chronic instabilities of the hand and wrist. Sport Sport. 2017;33(1):8–15.
23. Keogh JW, Winwood PW. The epidemiology of injuries across the weight-training sports. Sports Med. 2017;47(3):479–501.
24. Siewe J, Rudat J, Rollinghoff M, Schlegel UJ, Eysel P, Michael JW. Injuries and overuse syndromes in powerlifting. Int J Sports Med. 2011;32(9):703–11.
25. Kox LS, Kuijer PP, Kerkhoffs GM, Maas M, Frings-Dresen MH. Prevalence, incidence and risk factors for overuse injuries of the wrist in young athletes: a systematic review. Br J Sports Med. 2015;49(18):1189–96.
26. Wolf MR, Avery D, Wolf JM. Upper extremity injuries in gymnasts. Hand Clin. 2017;33(1):187–97.
28. Hochholzer TBT, Straub G, Schöffl V. Overuse bone marrow edema of the hands in sport climbers. Sport Orthop Traumatol. 2013;29:219–24.

Part VIII

Future Perspectives

Future Aspects: Climbing in the Olympics

23

Christoph Lutter and Volker Schöffl

Volker Schöffl, „Carybdis“, Frankenjura, Germany, Photo Enrico Haase

C. Lutter (✉)
Department of Orthopedic Surgery, University Hospital Rostock, Rostock, Germany

V. Schöffl
Department of Orthopedic and Trauma Surgery, Center of Sportsmedicine, Klinikum Bamberg, Bamberg, Germany

School of Clinical and Applied Sciences, Leeds Becket University, Leeds, UK

Section of Wilderness Medicine, Department of Emergency Medicine, University of Colorado School of Medicine, Denver, CO, USA

Department of Orthopedic and Trauma Surgery, Friedrich-Alexander Universität Erlangen-Nürnberg, Erlangen, Germany

V. Schöffl et al. (eds.), *Climbing Medicine*, https://doi.org/10.1007/978-3-030-72184-8_23

23.1 Competitive Sport Climbing

Climbing is currently experiencing a "boom" that has rapidly transformed it from a niche sport to a widely practiced popular sport [1]. The worldwide "boom" of the sub-discipline of bouldering and increasingly exciting competition formats in competitive sports has certainly played a decisive role in this development (Figs. 23.1 and 23.2) [2]. The great public interest is emphasized by the World Cup livestreams of all climbing disciplines, the participation in major sporting events such as the "World Games," and the inclusion in the Olympic program. The IFSC (International Federation of Sport Climbing) World Cup circuit as well as major national and international competitions (world championships, Olympic qualifying competitions, etc.) has recently been attracting an increasingly large audience (Fig. 23.3). One of the main reasons for this is a new competition format adapted to the interests of worldwide (TV) viewers, in which two or more athletes no longer climb simultaneously but start one after the other. Optimized camera settings and specialized commentators provide additional excitement. Spectator records at the competition sites are regularly broken, and TV/livestream ratings continue to increase.

Fig. 23.1 Competitive athlete in a boulder final 2020. (M. Kleesattel, Germany; Photo: Lutter)

Fig. 23.2 Jernej Kruder (left) and Adam Ondra (right) during the IFSC Boulder World Cup in Munich 2019. (Photo: Lutter)

Fig. 23.3 A large crowd of visitors during the qualification day of the IFSC Boulder World Cup in Munich 2019. (Photo: Lutter)

After the initial euphoria, the inclusion in the Olympic program caused a certain amount of uncertainty in the climbing community, especially among the athletes [1]. Due to the strict specifications of the organizing committee, a "climbing triathlon" was initially chosen as the competition format for Tokyo 2020, combining the previously separate competition formats of bouldering, lead and speed climbing (so-called Combined). For the 2024 Games in Paris, a separation of the sub-disciplines is planned. Based on the changed demands on the athletes, well-thought-out support of new talents is indispensable in order to avoid known climbing injuries and to prevent new injury patterns and overload reactions associated with the competition formats [3].

In addition, a trend that has been apparent for some time is likely to continue in the coming years; the average age of the top athletes is continuously decreasing, and both on real rock and in competition formats, most records are set by young or even very young athletes [1]. The inclusion in the Olympic program could reinforce this trend. Therefore, special attention must be paid to typical climbing injuries and overuse injuries in young athletes [4]. Injuries in the growth plates of the fingers and eating disorders (anorexia athletica) are the two most frequent and important entities among young athletes.

The authors therefore recommend the following steps and measures to ensure that the sport is safe and free of injuries in the future:

1. Structured, evidence-based, sport-specific medical supervision for all ambitious athletes
2. Supervision and adjustment of training and competition sites according to official recommendations and standards
3. Further development and consistent use of low-injury climbing holds and adapted route construction
4. Age-specific training and competition adjustment, especially for "Combined" athletes [1]

If these measures are considered and consistently implemented, it should also be possible to ensure largely injury-free training and competition conditions for competitive athletes in the future of climbing sports.

23.2 Anti-Doping

For athletes, trainers, and fans of the sport, the rapid rise of climbing as a competitive sport is an exciting development. However, the new Olympic status as well as the increasing professionalization comes with added responsibility. With the growing pressure to perform in competitive sports, the doping problem will inevitably become more important. In modern sport climbing, performance-enhancing substances can be used for both physical and mental performance enhancement. Competition climbing federations have adapted anti-doping rules or adopted regulations that had been standardized in other sports many years ago. So far, there is no corresponding scientific basis or evidence regarding the doping problem in climbing sport [2, 5]. This is in contrast to other new Olympic sports such as baseball or karate and reveals an urgent need to catch up. The state of knowledge about performance-enhancing substances in modern climbing is therefore still manageable and is largely based on anecdotal evidence and assumptions. Since a high strength-to-body weight ratio is crucial in climbing, low body weight has always been one of the decisive performance-limiting factors. Various substances such as appetite suppressants, diuretics, and other supplements have been used by climbers in the past. To prevent anorexic disorders in athletes, body weight or body mass index limits have been introduced in competition. However, severe sanctions have not yet been implemented. Instead, recommendations and warnings are issued to the relevant federations in the event of abnormalities. There is still little evidence of the use of stimulants to prevent fatigue, optimize training, or improve performance in competition. However, prohibited substances used in other sports also allow more intense and frequent training sessions in climbing, a shortened recovery period, and improved performance in competition.

Special importance is attached to anxiolytic and disinhibitory substances such as THC, which have positive psychological effects on the athletic performance of climbers and are unfortunately widespread in parts of the climbing community. The use of so-called low psychoactive cannabinoids (CBD oils, etc.) must also be viewed critically, even though these substances are not currently considered doping substances by the World Anti-Doping Organization (WADA) [2].

A further problem within the climbing sport is the partially insufficiently developed sports-medical supervision of (top) athletes. While internationally competing athletes from larger federations receive appropriate support from their national federations, athletes from smaller nations without sufficient funds sometimes do not receive sufficient sports-medical support at all (especially anti-doping).

The following measures are therefore recommended:

1. Increased awareness of sports ethics in climbing
2. Increased anti-doping education of fans, athletes, trainers, and officials
3. Sports-medical supervision for all competitive athletes
4. Cooperation between national sports organizations in anti-doping efforts
5. Increased monitoring and anti-doping controls (training and competition)

23.3 Newbie Syndrome

Due to the enormous popularity of bouldering, there has been a significant shift in the number of climbing sport injuries in the last 5 years [6]. Among these injuries, more severe injuries have been reported, with the emphasis on indoor bouldering. In contrast to older surveys of injury severity in climbing and bouldering, a slightly higher number of more severe climbing injuries are currently being observed. When analyzing this current trend, the change of the bouldering sport and the development of modern bouldering gyms must be especially considered. With the emergence of bouldering competitions, the design of bouldering routes has changed. The attractiveness of the sport for athletes and spectators is improved by adapting the route construction, demanding more and more spectacular and acrobatic movements. Modern boulder problems are often a combination of classical climbing and bouldering and acrobatic elements as seen in parkour. This requires different skills than in the past, both from competitive athletes and from recreational athletes who often do similarly spectacular boulder problems. Most of the movements are dynamic and are often performed in a three-dimensional, overhanging wall structure. Special foot techniques such as so-called heel hooks are increasing frequently, which explains the increase of knee and hip injuries as a logical consequence of the changes in the sport. While the climbing or bouldering style and thus the athletic requirements change, another development can be observed parallel to this: a change in the athlete collective. Bouldering gyms have become popular meeting points and destinations for various events; now, company events and meetings, children's birthday parties, etc. are sometimes held in boulder gyms. Visiting a bouldering gym has also become an important leisure activity for many people. This leads to a new injury mechanism, which we recently called "newbie" syndrome [7]. Among the large number of absolute beginners, there are some physically unfit and coordinatively limited climbers who are interested in climbing. In such cases, even small falls can cause serious injuries, such as vertebral or luxation fractures (Fig. 23.4). Balance, body awareness, and control are also less developed in absolute beginners than in athletes who have been practicing the sport for a long time and who had started climbing before acrobatic elements became common. Therefore, the current standards of modern boulder gyms must be adapted for injury prevention. Structured training (correct falling and rolling, etc.) and an introduction to the sport and its hazards is of utmost importance for beginner athletes.

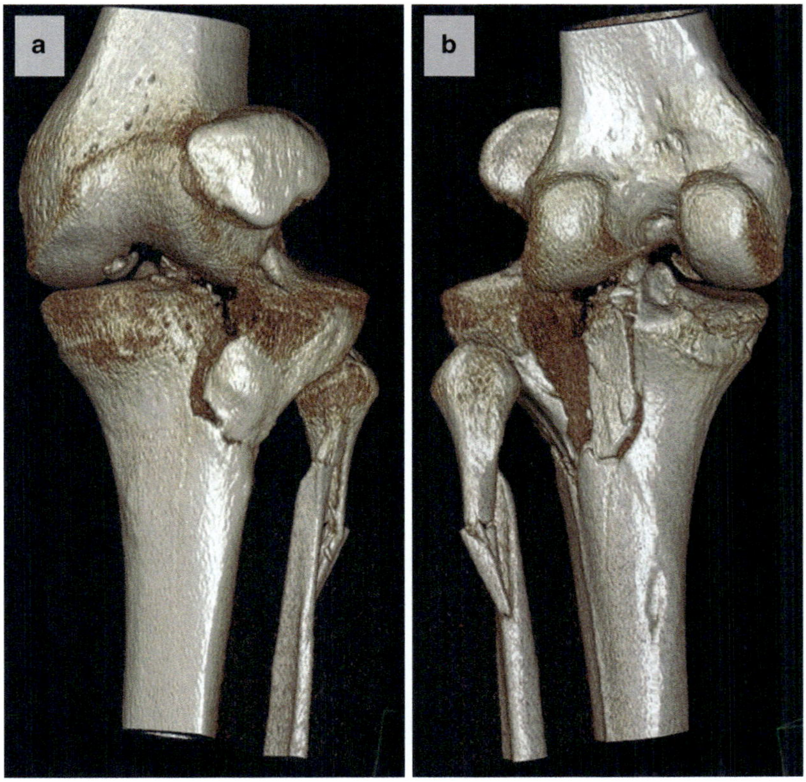

Fig. 23.4 Tibial plateau fracture (AO 41-B3, Schatzker II) with luxation of the patella and fracture of the proximal fibula (right knee joint) in a 36-year-old athlete after a fall during indoor bouldering despite sufficient floor mats. (**a**) Ventral view of the knee joint, (**b**) posterior view of the knee

23.3.1 Conclusion

Regarding the large stage upon which climbing sports can and will present itself in the future, high demands are placed on sports physicians. Not only must injury patterns and overuse damage be quickly recognized and treated, but the sports physicians also play an important role in the implementation of preventive measures and anti-doping work. Coaches, parents, and athletes rely on appropriate advice; only then can this inspiring sport continue to be valued and practiced by everyone as a spectacular, easily accessible, and health-promoting sport.

References

1. Lutter C, El-Sheikh Y, Schöffl I, Schöffl V. Sport climbing: medical considerations for this new Olympic discipline. Br J Sports Med. 2017;51(1):2–3.
2. Lutter C, Tischer T, El-Sheikh Y, Schöffl V. Doping in sport climbing: status quo in a new olympic discipline. Curr Sports Med Rep. 2019;18(10):351–2.
3. Schöffl I, Lutter C, Schöffl V. What to examine in youth climbing athletes: yearly examination of the german climbing team and its consequences. Sportverletz Sportschaden. 2019;33(4):212–5.
4. Schöffl I, Schöffl V, Dötsch J, Dörr HG, Jüngert J. Correlations between high level sport-climbing and the development of adolescents. Pediatr Exerc Sci. 2011;23(4):477–86.
5. Schöffl V, Hochholzer T, Lightner S. One move too many. Ebenhausen: Lochner; 2016.
6. Lutter C, Tischer T, Hotfiel T, Frank L, Enz A, Simon M, et al. Current trends in sport climbing injuries after the inclusion into the Olympic program. Analysis of 633 injuries within the years 2017/18. Muscles, Ligaments Tendons J. 2020;10(2):201–10.
7. Schoffl V, Lutter C. The "newbie" syndrome. Wilderness Environ Med. 2017;28(4):377–80.

GPSR Compliance

The European Union's (EU) General Product Safety Regulation (GPSR) is a set of rules that requires consumer products to be safe and our obligations to ensure this.

If you have any concerns about our products, you can contact us on ProductSafety@springernature.com

In case Publisher is established outside the EU, the EU authorized representative is:

Springer Nature Customer Service Center GmbH
Europaplatz 3
69115 Heidelberg, Germany

Batch number: 10731385

Printed by Printforce, the Netherlands